REMOTE SENSING AND IMAGE INTERPRETATION

Fifth Edition

REMOTE SENSING AND IMAGE INTERPRETATION

Fifth Edition

Thomas M. Lillesand
University of Wisconsin—Madison

Ralph W. Kiefer
University of Wisconsin—Madison

Jonathan W. Chipman
University of Wisconsin—Madison

WILEY

ACQUISITIONS EDITOR   Ryan Flahive
ASSISTANT EDITOR   Denise Powell
MARKETING MANAGER   Clay Stone
SENIOR PRODUCTION EDITOR   Norine M. Pigliucci
SENIOR DESIGNER   Dawn Stanley
PRODUCTION MANAGEMENT SERVICES   UG / GGS Information Services, Inc.
COVER CREDIT   NASA/Goddard Space Flight Center, Scientific Visualization Studio

This book was set in 10/12 New Aster by UG / GGS Information Services, Inc. and printed and bound by
R. R. Donnelley-Crawfordsville. The cover was printed by The Lehigh Press.

This book is printed on acid-free paper. ∞

The paper in this book was manufactured by a mill whose forest management programs include sustained
yield harvesting of its timberlands. Sustained yield harvesting principles ensure that the numbers of trees
cut each year does not exceed the amount of new growth.

ISBN: 0-471-15227-7
WIE ISBN: 0-471-45152-5

Printed in the United States of America

10 9 8 7 6 5 4 3 2 1

To the peaceful application of remote sensing in order to maximize the scientific, social, and commercial benefits of this technology for all humankind.

PREFACE

This book is designed to be primarily used in two ways: as a textbook in introductory courses in remote sensing and image interpretation, and as a reference for the burgeoning number of practitioners who use geospatial information and analysis in their work. Remote sensing and its kindred technologies, such as geographic information systems (GIS) and the Global Positioning System (GPS), are having a pervasive impact on the conduct of science, government, and business alike. Because of the wide range of academic and professional settings in which this book might he used, we have made this discussion "discipline neutral." That is, rather than writing a book heavily oriented toward a single field such as business, ecology, engineering, forestry, geography, geology, urban and regional planning, or water resource management, we approach the subject in such a manner that students and practitioners in any discipline should gain a clear understanding of remote sensing systems and their virtually unlimited applications. In short, anyone involved in geospatial data acquisition and analysis should find this book to be a valuable text and reference.

Work on the first edition of this book began more than 26 years ago. To say much has changed in this time period is an understatement. In this edition we still provide the basics of analog image analysis, but place greater emphasis on digitally based systems and analysis techniques. Hence, the subject matter contained herein ranges from the principles of airphoto interpretation and photogrammetry, to a description of next-generation satellite systems and recent developments in digital image processing. We also underscore the close interaction among the related areas of remote sensing, GIS, GPS, digital image processing, and environmental modeling.

There is enough material in this book for it to be used in many different ways—in many different course settings. These include courses in remote sensing, image interpretation, photogrammetry, and digital image processing. Some courses may omit certain chapters and use the book in a one-semester or one-quarter course; the book may also be used in a two-course sequence. Others may use this discussion in a series of modular courses, or in a short-course/workshop format. We have designed the book with these different potential uses in mind.

Those familiar with the previous editions of this book will recognize that the fifth edition has a new co-author, Dr. Jonathan W. Chipman. Though this addition of authorship brings "new blood" to this work, the book follows the same general organizational structure as its predecessors. However, we have made many additions and changes that both update and improve the coverage of the previous editions. New or expanded coverage is provided on such topics as the GPS Wide-Area Augmentation System (WAAS), digital cameras, new satellite systems, image interpretation for natural disaster assessment, hyperspectral sensing, lidar, change detection, and biophysical modeling via remote sensing data. We also include a new appendix on radiation measurement concepts, terminology, and units, which should prove to be very useful to those interested in the more quantitative aspects of remote sensing. Similarly, we have added a new appendix dealing with sample coordinate transformation and resampling procedures. Additional appendix material includes sources of remote sensing data and information, remote sensing periodicals, online glossaries, and online tutorials. This edition also has many new line drawings and photographic illustrations, and a greatly expanded number of color plates.

We express our sincere thanks to the many individuals who have contributed to this edition.

Preliminary reviews of the outline for this edition were provided by: Professor Kelley A. Crews-Meyer, University of Texas—Austin; Professor Jonathan Haskett, University of Maryland; Professor Michael Keables, Denver University; Professor Andrew Klein, Texas A & M University; and Professor Steven Steinberg, Humbolt State University.

The following individuals provided reviews of all or major portions of the original draft manuscript for this edition: Professor Marvin Bauer, University of Minnesota; Professor Russell Congalton, University of New Hampshire; Professor Lloyd Queen, University of Montana; and Professor Lynn Usery, University of Georgia.

Portions of the fifth edition manuscript were also reviewed by the following (some of whom also supplied illustrations): Mr. Michael Renslow, Spencer B. Gross, Inc., and Ms. DeAnne Stevens, Alaska Division of Geological and Geophysical Surveys.

Special illustration materials and suggestions were provided by: Mr. Cody Benkleman, Positive Ststems, Inc.; Mr. Donald Garofalo, EPA Environmental Photographic Interpretation Center; Mr. Gerald Kinn and Mr. Michael Burnett, Emerge; Mr. Bryan Leavitt, Dr. Al Peters, and Dr. Donald Runquist, University of Nebraska—Lincoln; Dr. John Lyon, EPA; Mr. Roger Pacey, Leica Geosystems; Mr. Mark Radin and Mr. Robert Shone, Thales-Optem, Inc.; Ms. Denise Rothman and Mr. Peter Stack, FLIR Systems, Inc.; Mr. Alistar Stuart, Z/I Imaging Corp.; Mr. C. Wicks, USGS; and Professor Howard A. Zebker, Stanford University.

We also thank the many faculty, academic staff, and graduate and undergraduate students at the University of Wisconsin—Madison who made valuable contributions to this edition. Specifically, we acknowledge the assistance of Professor Frank Scarpace; Nancy Podger, Jeffrey Schmaltz, and the numerous other participants in the NASA Affiliated Research Center (ARC) and Regional Earth Science Applications Center (RESAC) programs as well as the NSF North Temperate Lakes Long Term Ecological Research (LTER) program. Much of what is presented in this new edition is research brought directly into the classroom from these programs.

Marcia Verhage played a key role in the physical preparation of this manuscript.

We are also grateful to the various individuals, instrument manufacturers, government agencies, and commercial firms who provided background materials and many of the illustrations used in this book.

Special recognition is due our families for their patient understanding and encouragement while this edition was in preparation.

Finally, we want to encourage you, the reader, to use the knowledge of remote sensing that you might gain from this book to literally make the world a better place. Remote sensing technology has proven to provide numerous scientific, commercial, and social benefits. Among these is not only the efficiency it brings to the day-to-day decision-making process in an ever-increasing range of applications, but also the potential this field holds for improving the stewardship of earth's resources and the global environment. This book is intended to provide a technical foundation for you to aid in making this tremendous potential a reality.

Thomas M. Lillesand
Ralph W. Kiefer
Jonathan W. Chipman

CONTENTS

xi

**3
Basic Principles of
Photogrammetry   126**

**4
Introduction to Visual Image
Interpretation   193**

**5
Multispectral, Thermal, and
Hyperspectral Sensing   330**

6
Earth Resource Satellites Operating in the Optical Spectrum   397

7
Digital Image Processing   491

REMOTE SENSING AND IMAGE INTERPRETATION

Fifth Edition

1

CONCEPTS AND FOUNDATIONS OF REMOTE SENSING

1.1 INTRODUCTION

Remote sensing is the science and art of obtaining information about an object, area, or phenomenon through the analysis of data acquired by a device that is not in contact with the object, area, or phenomenon under investigation. As you read these words, you are employing remote sensing. Your eyes are acting as sensors that respond to the light reflected from this page. The "data" your eyes acquire are impulses corresponding to the amount of light reflected from the dark and light areas on the page. These data are analyzed, or interpreted, in your mental computer to enable you to explain the dark areas on the page as a collection of letters forming words. Beyond this, you recognize that the words form sentences and you interpret the information that the sentences convey.

In many respects, remote sensing can be thought of as a reading process. Using various sensors, we remotely collect *data* that may be analyzed to obtain *information* about the objects, areas, or phenomena being investigated. The remotely collected data can be of many forms, including variations in force distributions, acoustic wave distributions, or electromagnetic energy distributions. For example, a gravity meter acquires data on variations in the distribution of the force of gravity. Sonar, like a bat's navigation system, obtains data on variations in acoustic wave distributions. Our eyes acquire data on variations in electromagnetic energy distributions.

This book is about *electromagnetic* energy sensors that are currently being operated from airborne and spaceborne platforms to assist in inventorying, mapping, and monitoring earth resources. These sensors acquire data on the way various earth surface features emit and reflect electromagnetic energy, and these data are analyzed to provide information about the resources under investigation.

Figure 1.1 schematically illustrates the generalized processes and elements involved in electromagnetic remote sensing of earth resources. The two basic processes involved are *data acquisition* and *data analysis*. The elements of the data acquisition process are energy sources (*a*), propagation of energy through the atmosphere (*b*), energy interactions with earth surface features (*c*), retransmission of energy through the atmosphere (*d*), airborne and/or spaceborne sensors (*e*), resulting in the generation of sensor data in pictorial and/or digital form (*f*). In short, we use sensors to record variations in the way earth surface features reflect and emit electromagnetic energy. The data analysis process (*g*) involves examining the data using various viewing and interpretation devices to analyze pictorial data and/or a computer to analyze digital sensor data. Reference data about the resources being studied (such as soil maps, crop statistics, or field-check data) are used when and where available to assist in the data analysis. With the aid of the reference data, the analyst extracts information about the type, extent, location, and condition of the various resources over which the sensor data were collected. This information is then compiled (*h*), generally in the form of hardcopy maps and tables or as computer files that can be merged with other "layers" of information in a *geographic information system* (*GIS*). Finally, the information is presented to users (*i*) who apply it to their decision-making process.

In the remainder of this chapter, we discuss the basic principles underlying the remote sensing process. We begin with the fundamentals of electromagnetic energy and then consider how the energy interacts with the atmosphere and with earth surface features. We also treat the role that reference data play in the data analysis procedure and describe how the spatial location of reference data observed in the field is often determined using *Global Positioning System* (*GPS*) methods. These basics will permit us to conceptualize an "ideal" remote sensing system. With that as a framework, we consider the limitations encountered in "real" remote sensing systems. We also discuss briefly the rudiments of GIS technology. At the end of this chapter, the reader should have a grasp of the general concepts and foundations of remote sensing and an appreciation for the close relationship among remote sensing, GPS methods, and GIS operations.

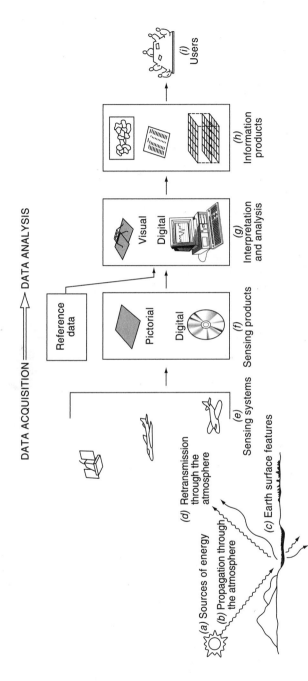

Figure 1.1 Electromagnetic remote sensing of earth resources.

1.2   ENERGY SOURCES AND RADIATION PRINCIPLES

Visible light is only one of many forms of electromagnetic energy. Radio waves, heat, ultraviolet rays, and X-rays are other familiar forms. All this energy is inherently similar and radiates in accordance with basic wave theory. As shown in Figure 1.2, this theory describes electromagnetic energy as traveling in a harmonic, sinusoidal fashion at the "velocity of light," c. The distance from one wave peak to the next is the *wavelength* λ, and the number of peaks passing a fixed point in space per unit time is the wave *frequency* ν.

From basic physics, waves obey the general equation

$$c = \nu\lambda \tag{1.1}$$

Since c is essentially a constant (3×10^8 m/sec), frequency ν and wavelength λ for any given wave are related inversely, and either term can be used to characterize a wave. In remote sensing, it is most common to categorize electromagnetic waves by their wavelength location within the *electromagnetic spectrum* (Figure 1.3). The most prevalent unit used to measure wavelength along the spectrum is the *micrometer* (μm). A micrometer equals 1×10^{-6} m. (Tables of units used frequently in this book are included inside the back cover.)

Although names (such as "ultraviolet" and "microwave") are generally assigned to regions of the electromagnetic spectrum for convenience, there is no clear-cut dividing line between one nominal spectral region and the next. Divisions of the spectrum have grown from the various methods for sensing each type of radiation more so than from inherent differences in the energy characteristics of various wavelengths. Also, it should be noted that the portions of

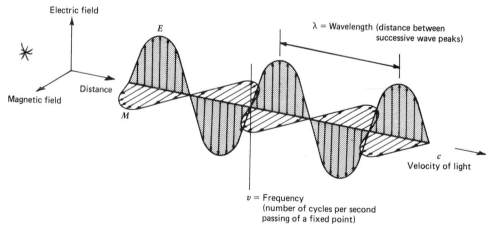

Figure 1.2   Electromagnetic wave. Components include a sinusoidal electric wave (E) and a similar magnetic wave (M) at right angles, both being perpendicular to the direction of propagation.

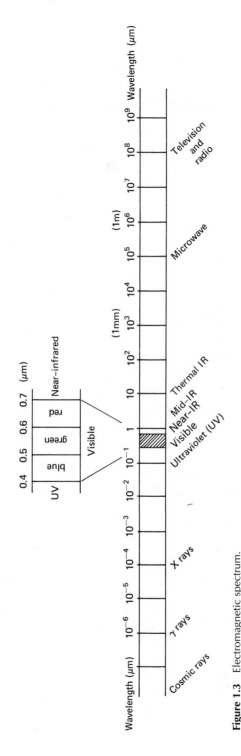

Figure 1.3 Electromagnetic spectrum.

5

the electromagnetic spectrum used in remote sensing lie along a continuum characterized by magnitude changes of many powers of 10. Hence, the use of logarithmic plots to depict the electromagnetic spectrum is quite common. The "visible" portion of such a plot is an extremely small one, since the spectral sensitivity of the human eye extends only from about 0.4 μm to approximately 0.7 μm. The color "blue" is ascribed to the approximate range of 0.4 to 0.5 μm, "green" to 0.5 to 0.6 μm, and "red" to 0.6 to 0.7 μm. *Ultraviolet* (*UV*) energy adjoins the blue end of the visible portion of the spectrum. Adjoining the red end of the visible region are three different categories of *infrared* (*IR*) waves: *near IR* (from 0.7 to 1.3 μm), *mid IR* (from 1.3 to 3 μm), and *thermal IR* (beyond 3 to 14 μm). At much longer wavelengths (1 mm to 1 m) is the *microwave* portion of the spectrum.

Most common sensing systems operate in one or several of the visible, IR, or microwave portions of the spectrum. *Within the IR portion of the spectrum, it should be noted that only thermal IR energy is directly related to the sensation of heat; near- and mid-IR energy are not.*

Although many characteristics of electromagnetic radiation are most easily described by wave theory, another theory offers useful insights into how electromagnetic energy interacts with matter. This theory—the particle theory—suggests that electromagnetic radiation is composed of many discrete units called *photons* or *quanta*. The energy of a quantum is given as

$$Q = h\nu \tag{1.2}$$

where

    Q = energy of a quantum, joules (J)
    h = Planck's constant, 6.626×10^{-34} J sec
    ν = frequency

We can relate the wave and quantum models of electromagnetic radiation behavior by solving Eq. 1.1 for ν and substituting into Eq. 1.2 to obtain

$$Q = \frac{hc}{\lambda} \tag{1.3}$$

Thus, we see that the energy of a quantum is inversely proportional to its wavelength. *The longer the wavelength involved, the lower its energy content.* This has important implications in remote sensing from the standpoint that naturally emitted long wavelength radiation, such as microwave emission from terrain features, is more difficult to sense than radiation of shorter wavelengths, such as emitted thermal IR energy. The low energy content of long wavelength radiation means that, in general, systems operating at long wavelengths must "view" large areas of the earth at any given time in order to obtain a detectable energy signal.

The sun is the most obvious source of electromagnetic radiation for remote sensing. However, *all* matter at temperatures above absolute zero (0 K, or −273°C) continuously emits electromagnetic radiation. Thus, terrestrial objects are also sources of radiation, though it is of considerably different magnitude and spectral composition than that of the sun. How much energy any object radiates is, among other things, a function of the surface temperature of the object. This property is expressed by the *Stefan–Boltzmann law*, which states that

$$M = \sigma T^4 \qquad (1.4)$$

where
  M = total radiant exitance from the surface of a material, watts (W) m^{-2}
  σ = *Stefan–Boltzmann constant*, 5.6697×10^{-8} W m^{-2} K^{-4}
  T = absolute temperature (K) of the emitting material

The particular units and the value of the constant are not critical for the student to remember, yet it is important to note that the total energy emitted from an object varies as T^4 and therefore increases very rapidly with increases in temperature. Also, it should be noted that this law is expressed for an energy source that behaves as a *blackbody*. A blackbody is a hypothetical, ideal radiator that totally absorbs and reemits all energy incident upon it. Actual objects only approach this ideal. We further explore the implications of this fact in Chapter 5; suffice it to say for now that the energy emitted from an object is primarily a function of its temperature, as given by Eq. 1.4.

Just as the total energy emitted by an object varies with temperature, the spectral distribution of the emitted energy also varies. Figure 1.4 shows energy distribution curves for blackbodies at temperatures ranging from 200 to 6000 K. The units on the ordinate scale (W m^{-2} μm^{-1}) express the radiant power coming from a blackbody per 1-μm spectral interval. Hence, the *area* under these curves equals the total radiant exitance, M, and the curves illustrate graphically what the Stefan–Boltzmann law expresses mathematically: the higher the temperature of the radiator, the greater the total amount of radiation it emits. The curves also show that there is a shift toward shorter wavelengths in the peak of a blackbody radiation distribution as temperature increases. The *dominant wavelength*, or wavelength at which a blackbody radiation curve reaches a maximum, is related to its temperature by *Wien's displacement law*,

$$\lambda_m = \frac{A}{T} \qquad (1.5)$$

where
  λ_m = wavelength of maximum spectral radiant exitance, μm
  A = 2898 μm K
  T = temperature, K

Thus, for a blackbody, the wavelength at which the maximum spectral radiant exitance occurs varies inversely with the blackbody's absolute temperature. We observe this phenomenon when a metal body such as a piece of iron

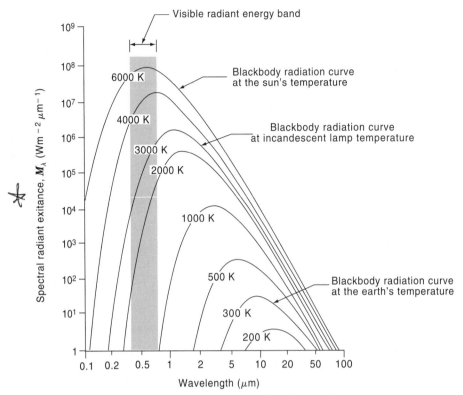

Figure 1.4 Spectral distribution of energy radiated from blackbodies of various temperatures. (Note that spectral radiant exitance M_λ is the energy emitted per unit wavelength interval. Total radiant exitance M is given by the area under the spectral radiant exitance curves.)

is heated. As the object becomes progressively hotter, it begins to glow and its color changes successively to shorter wavelengths—from dull red, to orange, to yellow, and eventually to white.

The sun emits radiation in the same manner as a blackbody radiator whose temperature is about 6000 K (Figure 1.4). Many incandescent lamps emit radiation typified by a 3000 K blackbody radiation curve. Consequently, incandescent lamps have a relatively low output of blue energy, and they do not have the same spectral constituency as sunlight.

The earth's ambient temperature (i.e., the temperature of surface materials such as soil, water, and vegetation) is about 300 K (27°C). From Wien's displacement law, this means the maximum spectral radiant exitance from earth features occurs at a wavelength of about 9.7 μm. Because this radiation correlates with terrestrial heat, it is termed "thermal infrared" energy. This energy can neither be seen nor photographed, but it can be sensed with such thermal devices as radiometers and scanners (described in Chapter 5). By comparison, the sun has a much higher energy peak that occurs at about 0.5 μm, as indicated in Figure 1.4. Our eyes—and photographic film—are sensitive to energy of this magnitude and wavelength. Thus, when the sun is present, we can observe earth features by virtue of *reflected* solar energy. Once again, the longer wavelength energy *emitted* by ambient earth features can be observed only with a nonphotographic sensing system. The general dividing line between reflected and emitted IR wavelengths is approximately 3 μm. Below this wavelength, reflected energy predominates; above it, emitted energy prevails.

Certain sensors, such as radar systems, supply their own source of energy to illuminate features of interest. These systems are termed "active" systems, in contrast to "passive" systems that sense naturally available energy. A very common example of an active system is a camera utilizing a flash. The same camera used in sunlight becomes a passive sensor.

1.3   ENERGY INTERACTIONS IN THE ATMOSPHERE

Irrespective of its source, all radiation detected by remote sensors passes through some distance, or *path length*, of atmosphere. The path length involved can vary widely. For example, space photography results from sunlight that passes through the full thickness of the earth's atmosphere twice on its journey from source to sensor. On the other hand, an airborne thermal sensor detects energy emitted directly from objects on the earth, so a single, relatively short atmospheric path length is involved. The net effect of the atmosphere varies with these differences in path length and also varies with the magnitude of the energy signal being sensed, the atmospheric conditions present, and the wavelengths involved.

Because of the varied nature of atmospheric effects, we treat this subject on a sensor-by-sensor basis in other chapters. Here, we merely wish to introduce the notion that the atmosphere can have a profound effect on, among other things, the intensity and spectral composition of radiation available to any sensing system. These effects are caused principally through the mechanisms of atmospheric *scattering* and *absorption*.

Scattering

Atmospheric scattering is the unpredictable diffusion of radiation by particles in the atmosphere. *Rayleigh scatter* is common when radiation interacts with atmospheric molecules and other tiny particles that are much smaller in diameter than the wavelength of the interacting radiation. The effect of Rayleigh scatter is inversely proportional to the fourth power of wavelength. Hence, there is a much stronger tendency for short wavelengths to be scattered by this mechanism than long wavelengths.

A "blue" sky is a manifestation of Rayleigh scatter. In the absence of scatter, the sky would appear black. But, as sunlight interacts with the earth's atmosphere, it scatters the shorter (blue) wavelengths more dominantly than the other visible wavelengths. Consequently, we see a blue sky. At sunrise and sunset, however, the sun's rays travel through a longer atmospheric path length than during midday. With the longer path, the scatter (and absorption) of short wavelengths is so complete that we see only the less scattered, longer wavelengths of orange and red.

Rayleigh scatter is one of the primary causes of "haze" in imagery. Visually, haze diminishes the "crispness," or "contrast," of an image. In color photography, it results in a bluish-gray cast to an image, particularly when taken from high altitude. As we see in Chapter 2, haze can often be eliminated, or at least minimized, in photography by introducing, in front of the camera lens, a filter that does not transmit short wavelengths.

Another type of scatter is *Mie scatter*, which exists when atmospheric particle diameters essentially equal the wavelengths of the energy being sensed. Water vapor and dust are major causes of Mie scatter. This type of scatter tends to influence longer wavelengths compared to Rayleigh scatter. Although Rayleigh scatter tends to dominate under most atmospheric conditions, Mie scatter is significant in slightly overcast ones.

A more bothersome phenomenon is *nonselective scatter*, which comes about when the diameters of the particles causing scatter are much larger than the wavelengths of the energy being sensed. Water droplets, for example, cause such scatter. They commonly have a diameter in the range 5 to 100 μm and scatter all visible and near- to mid-IR wavelengths about equally. Consequently, this scattering is "nonselective" with respect to wavelength. In the visible wavelengths, equal quantities of blue, green, and red light are scattered; hence fog and clouds appear white.

Absorption

In contrast to scatter, atmospheric absorption results in the effective loss of energy to atmospheric constituents. This normally involves absorption of energy at a given wavelength. The most efficient absorbers of solar radiation in this regard are water vapor, carbon dioxide, and ozone. Because these gases tend to absorb electromagnetic energy in specific wavelength bands, they strongly influence "where we look" spectrally with any given remote sensing system. The wavelength ranges in which the atmosphere is particularly transmissive of energy are referred to as *atmospheric windows*.

Figure 1.5 shows the interrelationship between energy sources and atmospheric absorption characteristics. Figure 1.5*a* shows the spectral distribu-

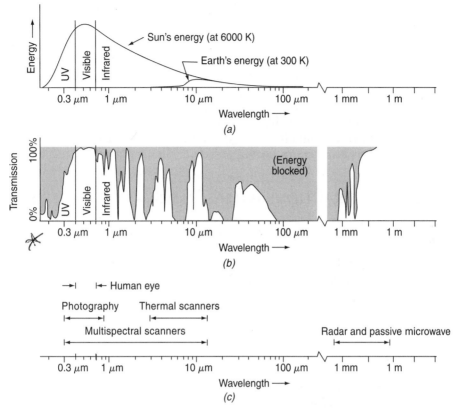

Figure 1.5 Spectral characteristics of (*a*) energy sources, (*b*) atmospheric transmittance, and (*c*) common remote sensing systems. (Note that wavelength scale is logarithmic.)

tion of the energy emitted by the sun and by earth features. These two curves represent the most common sources of energy used in remote sensing. In Figure 1.5*b*, spectral regions in which the atmosphere blocks energy are shaded. Remote sensing data acquisition is limited to the nonblocked spectral regions, the atmospheric windows. Note in Figure 1.5*c* that the spectral sensitivity range of the eye (the "visible" range) coincides with both an atmospheric window and the peak level of energy from the sun. Emitted "heat" energy from the earth, shown by the small curve in (*a*), is sensed through the windows at 3 to 5 μm and 8 to 14 μm using such devices as *thermal scanners*. *Multispectral scanners* sense simultaneously through multiple, narrow wavelength ranges that can be located at various points in the visible through the thermal spectral region. *Radar* and *passive microwave systems* operate through a window in the region 1 mm to 1 m.

The important point to note from Figure 1.5 is the interaction and the interdependence between the primary sources of electromagnetic energy, the atmospheric windows through which source energy may be transmitted to and from earth sur-face features, and the spectral sensitivity of the sensors available to detect and record the energy. One cannot select the sensor to be used in any given remote sensing task arbitrarily; one must instead consider (1) the spectral sensitivity of the sensors available, (2) the presence or absence of atmospheric windows in the spectral range(s) in which one wishes to sense, and (3) the source, magnitude, and spectral composition of the energy available in these ranges. Ultimately, however, the choice of spectral range of the sensor must be based on the manner in which the energy interacts with the features under investigation. It is to this last, very important, element that we now turn our attention.

1.4 ENERGY INTERACTIONS WITH EARTH SURFACE FEATURES

When electromagnetic energy is incident on any given earth surface feature, three fundamental energy interactions with the feature are possible. This is illustrated in Figure 1.6 for an element of the volume of a water body. Various fractions of the energy incident on the element are *reflected, absorbed,* and/or *transmitted.* Applying the principle of conservation of energy, we can state the interrelationship among these three energy interactions as

$$E_I(\lambda) = E_R(\lambda) + E_A(\lambda) + E_T(\lambda) \tag{1.6}$$

where
E_I = incident energy
E_R = reflected energy
E_A = absorbed energy
E_T = transmitted energy

with all energy components being a function of wavelength λ.

Equation 1.6 is an energy balance equation expressing the interrelationship among the mechanisms of reflection, absorption, and transmission. Two

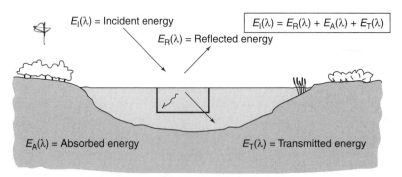

Figure 1.6 Basic interactions between electromagnetic energy and an earth surface feature.

points concerning this relationship should be noted. First, the proportions of energy reflected, absorbed, and transmitted will vary for different earth features, depending on their material type and condition. These differences permit us to distinguish different features on an image. Second, the wavelength dependency means that, even within a given feature type, the proportion of reflected, absorbed, and transmitted energy will vary at different wavelengths. Thus, two features may be indistinguishable in one spectral range and be very different in another wavelength band. Within the visible portion of the spectrum, these spectral variations result in the visual effect called *color*. For example, we call objects "blue" when they reflect more highly in the blue portion of the spectrum, "green" when they reflect more highly in the green spectral region, and so on. Thus, the eye utilizes spectral variations in the magnitude of reflected energy to discriminate between various objects.

Because many remote sensing systems operate in the wavelength regions in which reflected energy predominates, the reflectance properties of earth features are very important. Hence, it is often useful to think of the energy balance relationship expressed by Eq. 1.6 in the form

$$E_R(\lambda) = E_I(\lambda) - [E_A(\lambda) + E_T(\lambda)] \tag{1.7}$$

That is, the reflected energy is equal to the energy incident on a given feature reduced by the energy that is either absorbed or transmitted by that feature.

The geometric manner in which an object reflects energy is also an important consideration. This factor is primarily a function of the surface roughness of the object. *Specular* reflectors are flat surfaces that manifest mirrorlike reflections, where the angle of reflection equals the angle of incidence. *Diffuse* (or *Lambertian*) reflectors are rough surfaces that reflect uniformly in all directions. Most earth surfaces are neither perfectly specular nor diffuse reflectors. Their characteristics are somewhat between the two extremes.

Figure 1.7 illustrates the geometric character of specular, near-specular, near-diffuse, and diffuse reflectors. The category that characterizes any given

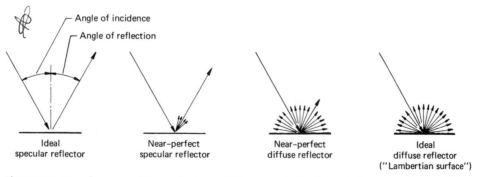

Figure 1.7   Specular versus diffuse reflectance. (We are most often interested in measuring the diffuse reflectance of objects.)

surface is dictated by the surface's roughness *in comparison to the wavelength of the energy incident upon it*. For example, in the relatively long wavelength radio range, a sandy beach can appear smooth to incident energy, whereas in the visible portion of the spectrum, it appears rough. In short, when the wavelength of incident energy is much smaller than the surface height variations or the particle sizes that make up a surface, the reflection from the surface is diffuse.

Diffuse reflections contain spectral information on the "color" of the reflecting surface, whereas specular reflections do not. *Hence, in remote sensing, we are most often interested in measuring the diffuse reflectance properties of terrain features.*

The reflectance characteristics of earth surface features may be quantified by measuring the portion of incident energy that is reflected. This is measured as a function of wavelength and is called *spectral reflectance*, ρ_λ. It is mathematically defined as

$$\rho_\lambda = \frac{E_R(\lambda)}{E_I(\lambda)}$$

$$= \frac{\text{energy of wavelength } \lambda \text{ reflected from the object}}{\text{energy of wavelength } \lambda \text{ incident upon the object}} \times 100 \qquad (1.8)$$

where ρ_λ is expressed as a percentage.

A graph of the spectral reflectance of an object as a function of wavelength is termed a *spectral reflectance curve*. The configuration of spectral reflectance curves gives us insight into the spectral characteristics of an object and has a strong influence on the choice of wavelength region(s) in which remote sensing data are acquired for a particular application. This is illustrated in Figure 1.8, which shows highly generalized spectral reflectance curves for deciduous versus coniferous trees. Note that the curve for each of these object types is plotted as a "ribbon" (or "envelope") of values, not as a single line. This is because spectral reflectances vary somewhat within a given material class. That is, the spectral reflectance of one deciduous tree species and another will never be identical. Nor will the spectral reflectance of trees of the same species be exactly equal. We elaborate upon the variability of spectral reflectance curves later in this section.

In Figure 1.8, assume that you are given the task of selecting an airborne sensor system to assist in preparing a map of a forested area differentiating deciduous versus coniferous trees. One choice of sensor might be the human eye. However, there is a potential problem with this choice. The spectral reflectance curves for each tree type overlap in most of the visible portion of the spectrum and are very close where they do not overlap. Hence, the eye might see both tree types as being essentially the same shade of "green" and might confuse the identity of the deciduous and coniferous trees. Certainly one could improve things somewhat by using spatial clues to each tree type's

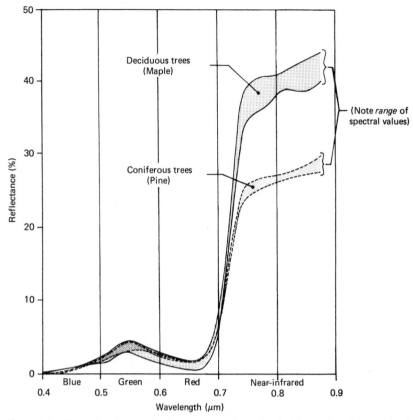

Figure 1.8 Generalized spectral reflectance envelopes for deciduous (broad-leaved) and coniferous (needle-bearing) trees. (Each tree type has a range of spectral reflectance values at any wavelength.) (Adapted from Kalensky and Wilson, 1975.)

identity, such as size, shape, site, and so forth. However, this is often difficult to do from the air, particularly when tree types are intermixed. How might we discriminate the two types on the basis of their spectral characteristics alone? We could do this by using a sensor system that records near-IR energy. A camera loaded with black and white IR film is just such a system. On black and white IR photographs, deciduous trees (having higher IR reflectance than conifers) generally appear much lighter in tone than do conifers. This is illustrated in Figure 1.9, which shows stands of coniferous trees surrounded by deciduous trees. In Figure 1.9*a* (visible spectrum), it is virtually impossible to distinguish between tree types, even though the conifers have a distinctive conical shape whereas the deciduous trees have rounded crowns. In Figure 1.9*b* (near IR), the coniferous trees have a distinctly darker tone. On such an image, the task of delineating deciduous versus coniferous trees becomes almost trivial. In fact, if we were to electronically scan this type of image and

Figure 1.9 Low altitude oblique aerial photographs illustrating deciduous versus coniferous trees. (a) Panchromatic photograph recording reflected sunlight over the wavelength band 0.4 to 0.7 μm. (b) Black and white infrared photograph recording reflected sunlight over 0.7 to 0.9 μm wavelength band.

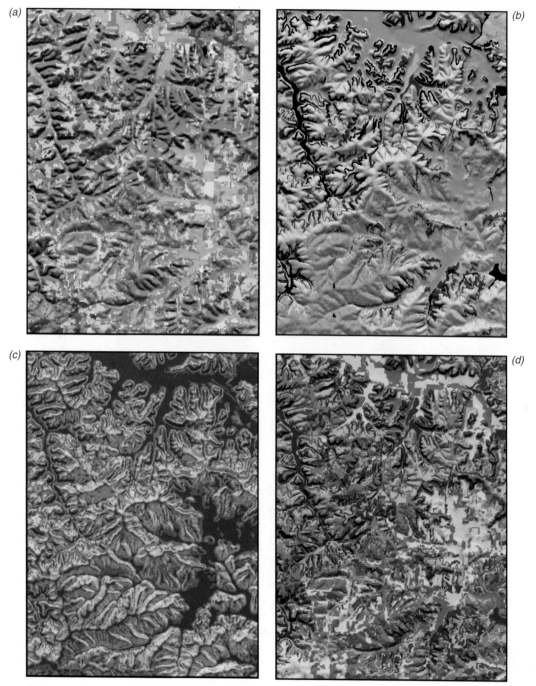

Plate 1 Integration of remote sensing data in a geographic information system, Dane County, WI: (*a*) land cover; (*b*) soil erodibility; (*c*) slope; (*d*) soil erosion potential. (Relief shading added in order to enhance interpretability of figure.) Scale 1 : 150,000. (For major discussion, see Section 1.11.)

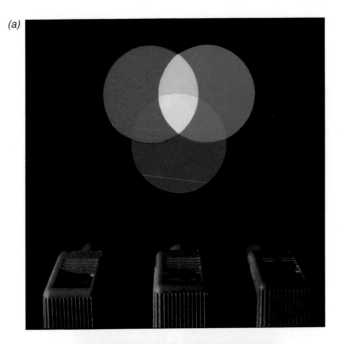

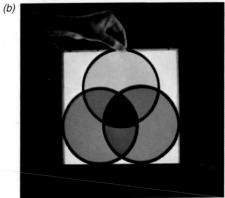

Plate 2 Color-mixing processes. (*a*) Color *additive* process—operative when *lights* of different colors are superimposed. (*b*) Color *subtractive* process—operative when *dyes* of different colors are superimposed. (Courtesy Eastman Kodak Company.) (For major discussion, see Section 2.7.)

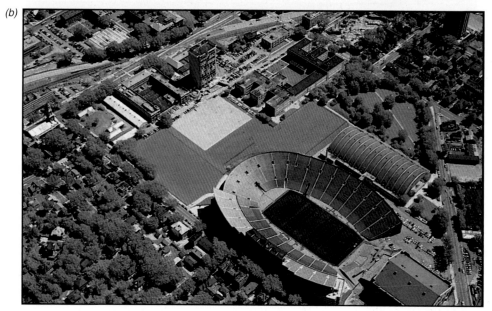

Plate 3 Oblique normal color (*a*) and color IR (*b*) aerial photographs showing a portion of the University of Wisconsin-Madison campus. The football field has artificial turf with low near-IR reflectance. (For major discussion, see Section 2.9.)

(a)

(b)

Plate 4 Oblique normal color (*a*) and color IR (*b*) aerial photographs showing flowing lava on the face of Kilauea Volcano, HI. The orange tones on the color IR photograph represent IR energy *emitted* from the flowing lava. The pink tones represent sunlight *reflected* from the living vegetation. (For major discussion, see Section 2.8.)

(a)

(b)

Plate 5 Color (*a*) and color IR (*b*) digital camera images, Nashua, NH, mid-June, scale 1:2000. Images were acquired on the same day, but at different times: (*a*) was exposed under sunlit conditions, and (*b*) was exposed under overcast conditions. (Courtesy of Emerge.) (For major discussion, see Section 2.12.)

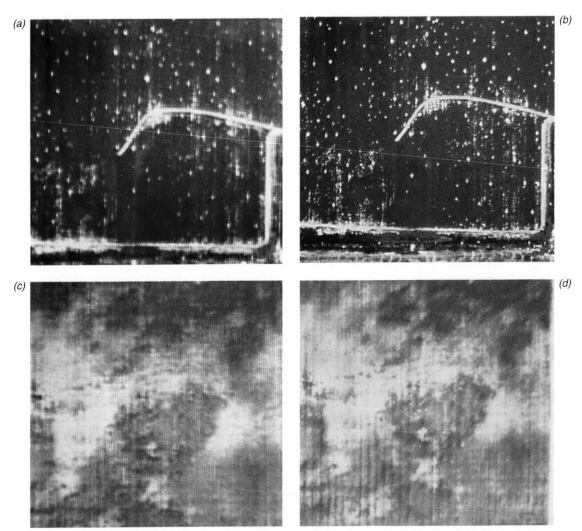

Plate 6 Color IR video composite images (*a*) and (*c*) compared with color IR photographs (*b*) and (*d*) of a cotton field infested with harvester ants (*a* and *b*) and a cotton field with saline soils (*c* and *d*). (Copyright © 1992, American Society for Photogrammetry and Remote Sensing. Reproduced with permission.) (For major discussion, see Section 2.13.)

(a)

(b)

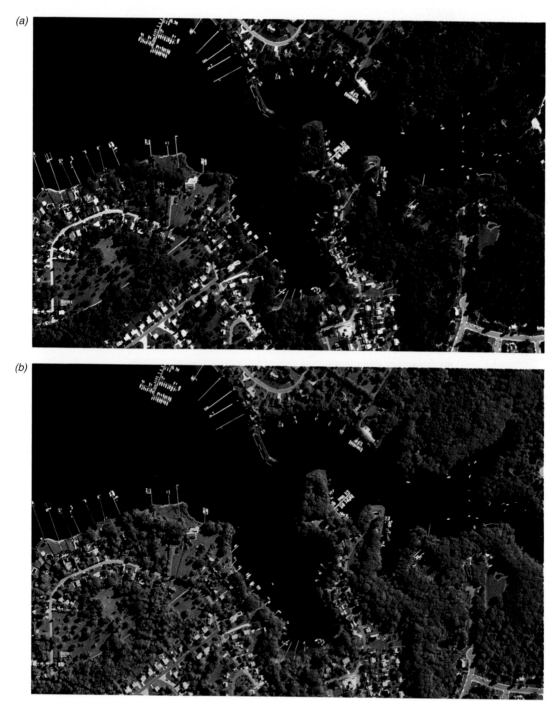

Plate 7 Color composites produced in an all-digital environment from images acquired by a multiband digital camera system *from the images shown in Figure 2.40.* Chesapeake Bay, MD, early-October. (*a*) Simulated normal color photograph. (*b*) Simulated color IR photograph. Scale 1:9000. (Courtesy Positive Systems, Inc.) (For major discussion, see Section 2.14.)

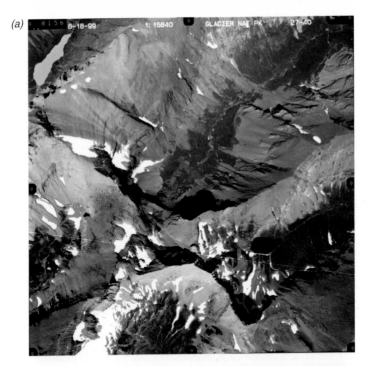

(a)

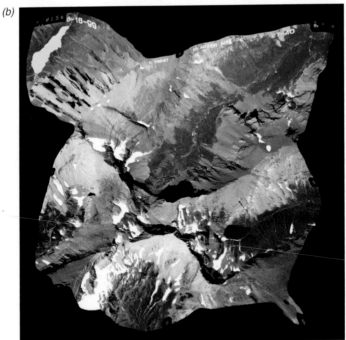

(b)

Plate 8 Uncorrected color photograph (*a*) and digital orthophoto (*b*) showing an area of high relief in Glacier National Park. Scale of (*b*) 1:64,000. (Courtesy Dr. Frank L. Scarpace and the USGS Upper Midwest Environmental Sciences Center.) (For major discussion, see Section 3.9.)

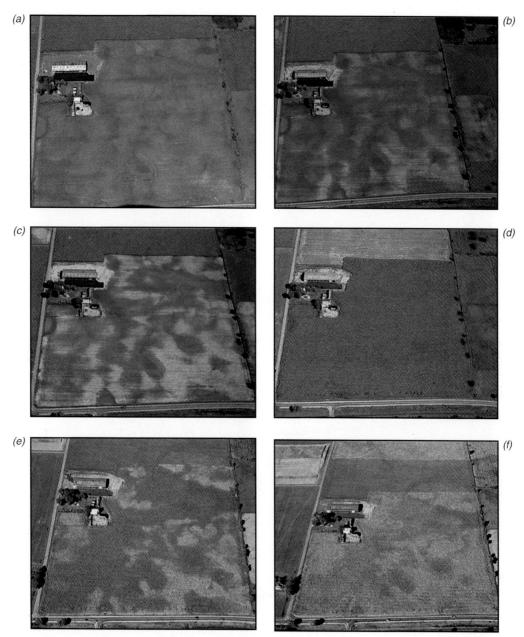

Plate 9 Oblique color IR aerial photographs illustrating the effects of date of photography: (*a*) June 30; (*b*) July 1; (*c*) July 2; (*d*) August 11; (*e*) September 17; (*f*) October 8. Dane County, WI. Approximate horizontal scale at photo center is 1:7600. (For major discussion, see Section 4.5.)

(a)

(b)

Plate 10 Landsat images illustrating population growth in the Las Vegas metropolitan area, Nevada. Scale 1:400,000. (*a*) August 1972. (*b*) August 2000. (Courtesy U.S. Environmental Protection Agency, Environmental Sciences Division—Las Vegas.) (For major discussion, see Section 4.10.)

(a)

(b)

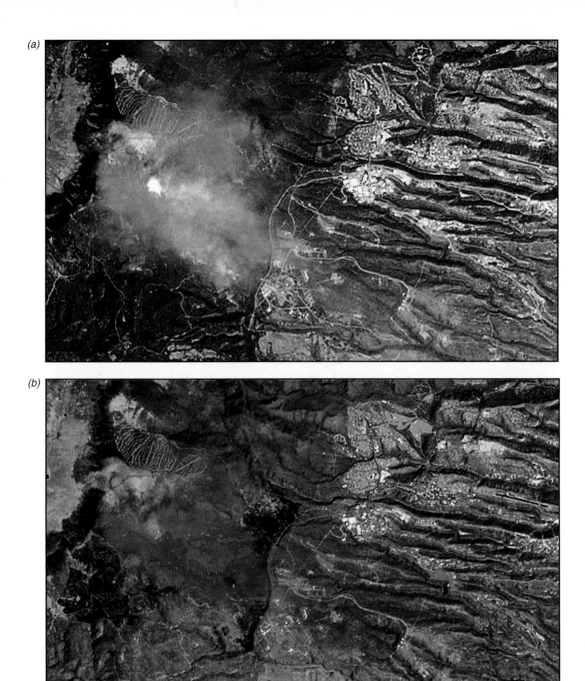

Plate 11 Landsat-7 images showing a wildfire burning near the city of Los Alamos, NM, May 2000. Scale 1:110,000. (*a*) "Normal color" composite of bands 1, 2, and 3. (*b*) "False color" composite of bands 2, 4, and 7. In (*b*), note that the hottest burning areas appear bright red-orange in color, visible because of the strong emitted energy from the fires in the 2.08 to 2.35-μm wavelength sensitivity of band 7. (NASA image.) (For major discussion of these images, see Section 4.15. For major discussion of Landsat-7, see Chapter 6.)

Plate 12   Digital camera image showing the aftermath of an F4 tornado striking Hayesville, KS, on May 3, 1999. Scale 1:2300. (Courtesy Emerge.) (For major discussion, see Section 4.15.)

Plate 13 Landsat-5 image of Mt. Etna eruption on July 21, 2002. Landsat bands 2, 5, and 7 are displayed as blue, green, and red, respectively. The bright red-orange areas are flowing lava and fissures containing molten lava. (NASA image.) (For major discussion, see Section 4.15.)

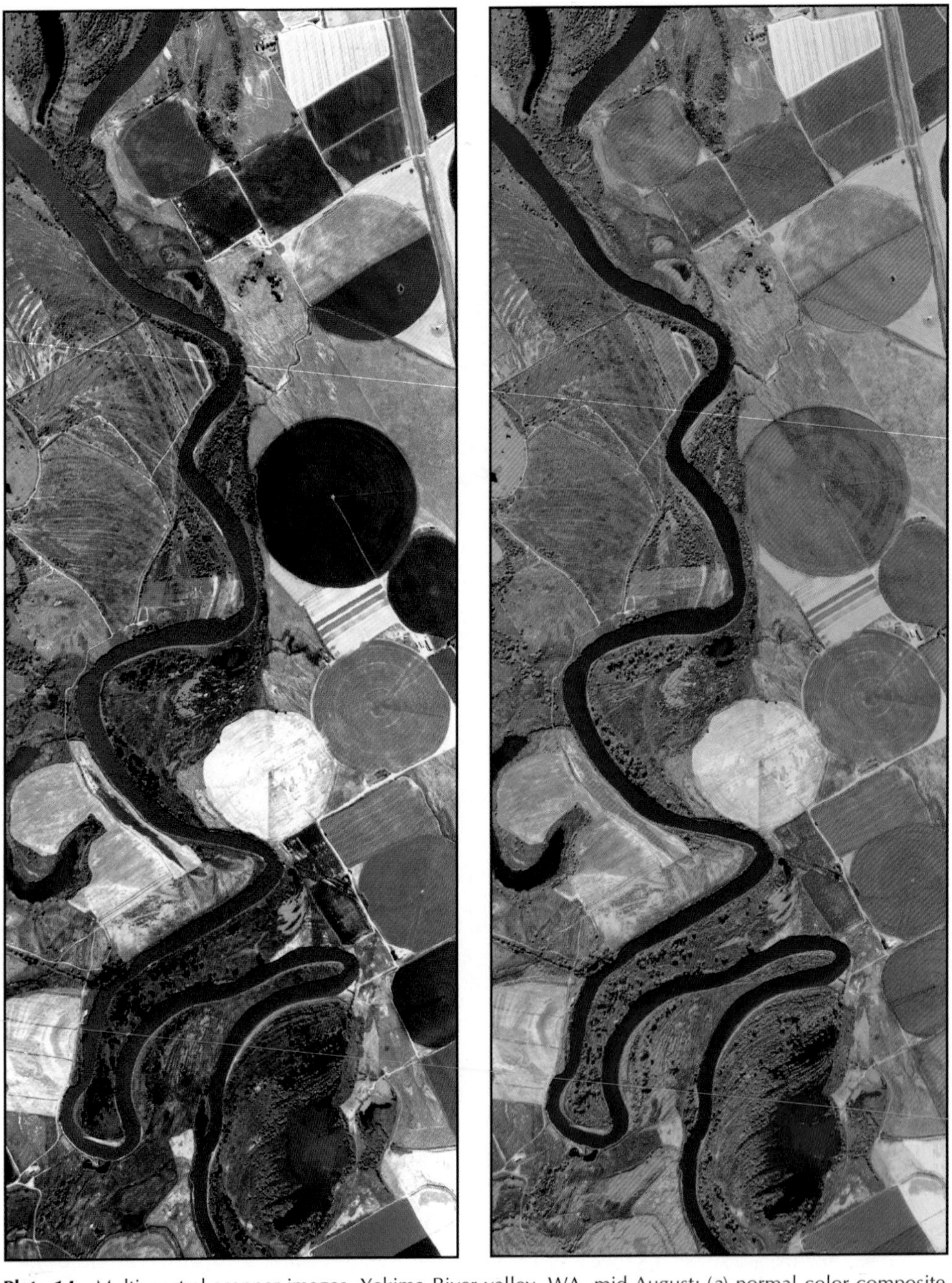

Plate 14 Multispectral scanner images, Yakima River valley, WA, mid-August: (*a*) normal color composite, bands 1, 2, and 3 (blue, green, and red) shown in Figure 5.8; (*b*) color IR composite, bands, 2, 3, and 5 (green, red, and near-IR) shown in Figure 5.8. Scale 1:34,000. (Courtesy Sensys Technologies, Inc.) (For major discussion, see Section 5.5.)

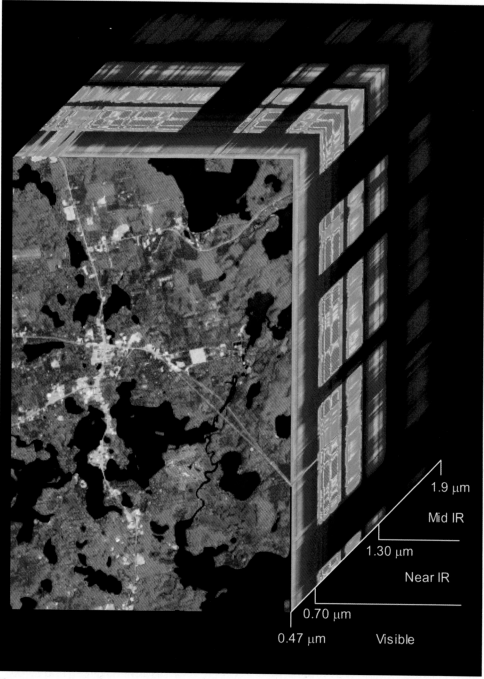

1.9 μm

Mid IR

1.30 μm

Near IR

0.70 μm

0.47 μm

Visible

Plate 15 Isometric view of a hyperspectral image cube showing the spatial and spectral characteristics of 187 bands of EO-1 Hyperion data, northern Wisconsin, August 11, 2002. (For major discussion, see Section 5.14.)

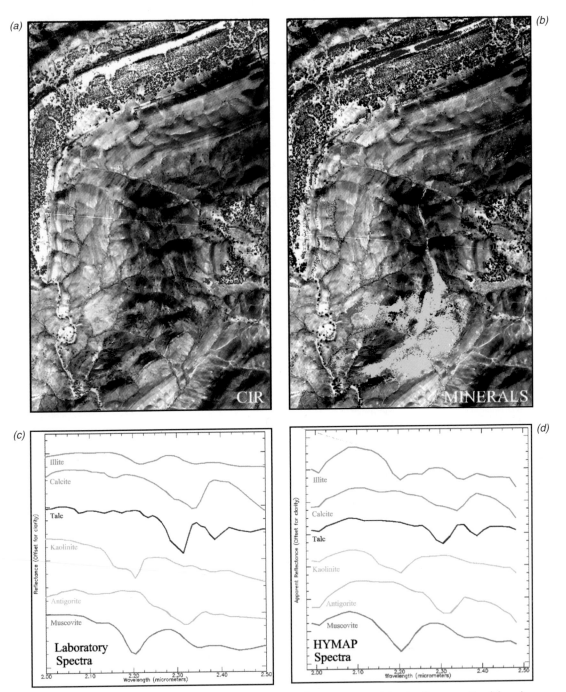

Plate 16 HyMap hyperspectral scanner data and spectral reflectance curves: (*a*) color IR composite of three hyperspectral bands; (*b*) location of six minerals displayed on a grayscale image; (*c*) laboratory spectral reflectance curves of selected minerals; (*d*) spectral reflectance curves of selected minerals as determined from hyperspectral data. Scale 1:40,000. (Courtesy American Society for Photogrammetry and Remote Sensing, Integrated Spectronics Pty Ltd., and Analytical Imaging and Geophysics.) (For major discussion, see Section 5.14.)

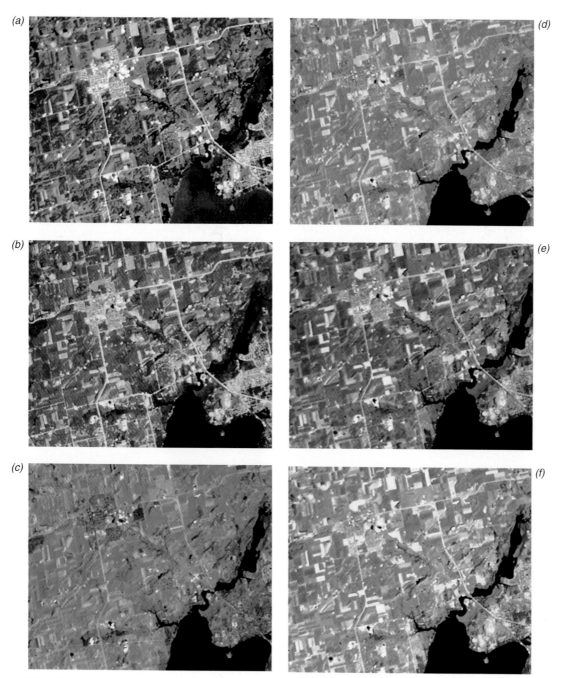

Plate 17 Color composite Landsat TM images illustrating six band-color combinations, suburban Madison, WI, late August. Scale 1:200,000. (See Section 6.8, Table 6.5, for the specific band-color combinations shown here.)

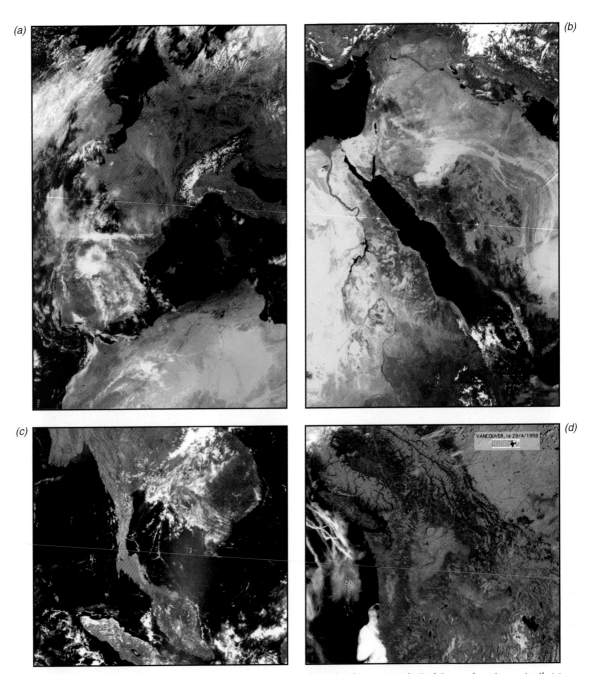

Plate 18 SPOT-4 Vegetation images: (*a*) Western Europe and North Africa, May; (*b*) Red Sea and environs, April; (*c*) southeast Asia, April; (*d*) northwest United States and western Canada, April. (*a*), (*b*), and (*c*) are composites of the blue, red, and near-IR bands, displayed as blue, green, and red, respectively. (*d*) is a composite of the red, near-IR, and mid-IR bands, displayed as blue, green, and red, respectively. (SPOT Imagery: copyright © CNES 1999. Provided by SPOT Image Corporation.) (For major discussion, see Section 6.14.)

Plate 19 Merger of IKONOS multispectral and panchromatic images, for an agricultural area in eastern Wisconsin, mid-June. (*a*) IKONOS 4-m-resolution multispectral image. (*b*) IKONOS 1-m-resolution panchromatic image. (*c*) Merged multispectral and panchromatic image, using intensity-hue-saturation transformation techniques, with 1-m effective resolution. Scale 1:10,000. (For major discussion, see Section 6.15.)

Plate 20 High resolution satellite imagery from QuickBird showing a portion of the Port of Hamburg, Germany. The image is the result of applying pan-sharpening techniques to a true-color composite of multispectral bands 1, 2, and 3 shown as blue, green, and red, respectively, and has an effective resolution of 61 cm. (Courtesy Digital-Globe.) (For major discussion, see Section 6.15.)

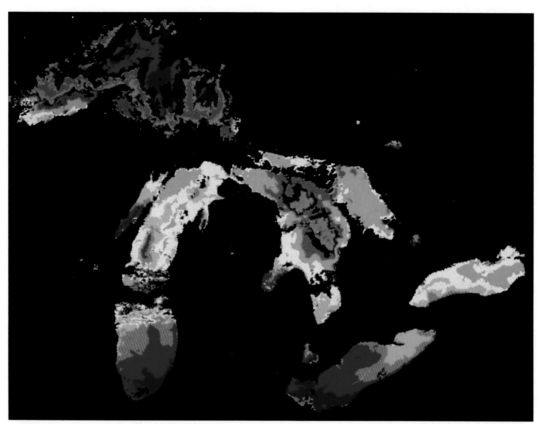

Plate 21 NOAA AVHRR band 4 thermal image of the Great Lakes, mid-July. Shorter wavelength colors (e.g., violet and blue) portray cooler temperatures and longer wavelength colors (e.g., orange and red) show warmer temperatures. Black patches and bands are cloud-covered areas (colder temperature than coolest water temperature). (NOAA data courtesy University of Wisconsin-Madison Space Science and Engineering Center, processed by the University of Wisconsin-Madison Environmental Remote Sensing Center.) (For major discussion, see Section 6.16.)

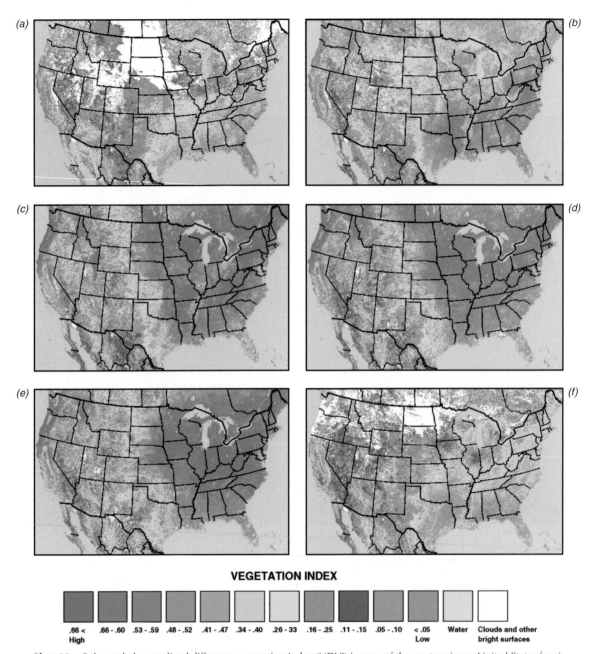

VEGETATION INDEX

| .66 <
High | .66 - .60 | .53 - .59 | .48 - .52 | .41 - .47 | .34 - .40 | .26 - 33 | .16 - .25 | .11 - .15 | .05 - .10 | < .05
Low | Water | Clouds and other
bright surfaces |

Plate 22 Color-coded normalized difference vegetation index (NDVI) images of the conterminous United States for six 2-week periods: (*a*) February 27–March 12; (*b*) April 24–May 7; (*c*) June 19–July 2; (*d*) July 17–30; (*e*) August 14–27; (*f*) November 6–19. (Courtesy USGS EROS Data Center.) (For major discussion, see Section 6.16.)

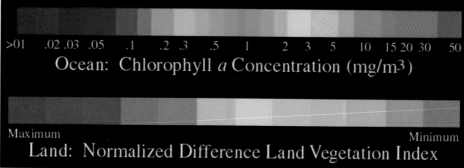

>01 .02 .03 .05   .1   .2 .3   .5   1   2   3   5   10 15 20 30   50

Ocean: Chlorophyll *a* Concentration (mg/m³)

Maximum   Minimum

Land: Normalized Difference Land Vegetation Index

Plate 23 SeaWiFS image showing ocean biogeochemistry and land vegetation. Composite image of data from September through July. Orthographic projection. (Image provided by the SeaWiFS Project, NASA/Goddard Space Flight Center, and ORBIMAGE.) (For major discussion, see Section 6.17.)

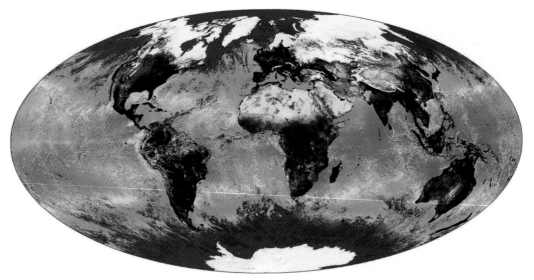

Plate 24 Global rendition of land surface reflectance and sea surface temperature, developed from composites of multiple dates of Terra MODIS imagery in March and April 2000. (Courtesy MODIS Instrument Team, NASA/GSFC.) (For major discussion, see Section 6.18.)

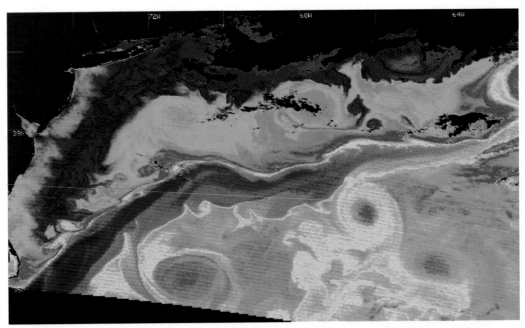

Plate 25 Sea surface temperature map of the western Atlantic, derived from Terra MODIS imagery, showing warm temperatures associated with the Gulf Stream. (Courtesy RSMAS, University of Miami.) (For major discussion, see Section 6.18.)

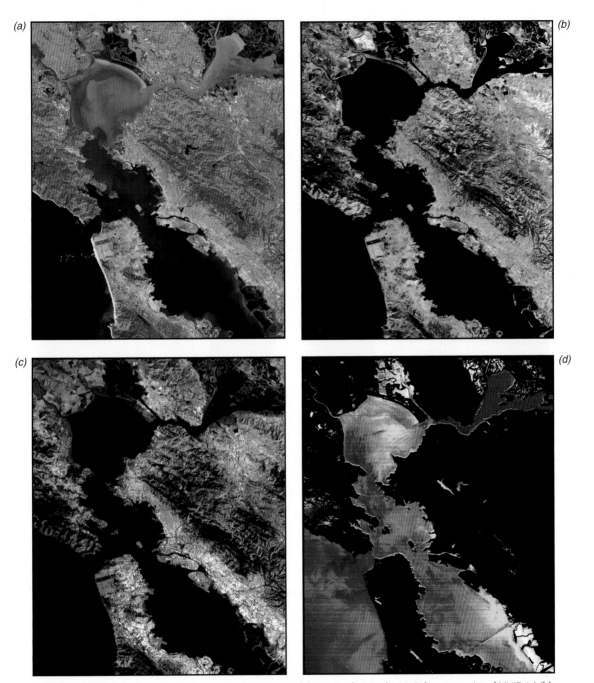

Plate 26 ASTER imagery of the San Francisco Bay area, California, early March. (*a*) Color composite of VNIR (visible and near-IR) bands. (*b*) Color composite of SWIR (mid-IR) bands. (*c*) Color composite of TIR (thermal-IR) bands. (*d*) Color-coded map of water temperature, derived from TIR bands. (Courtesy NASA.) (For major discussion, see Section 6.18.)

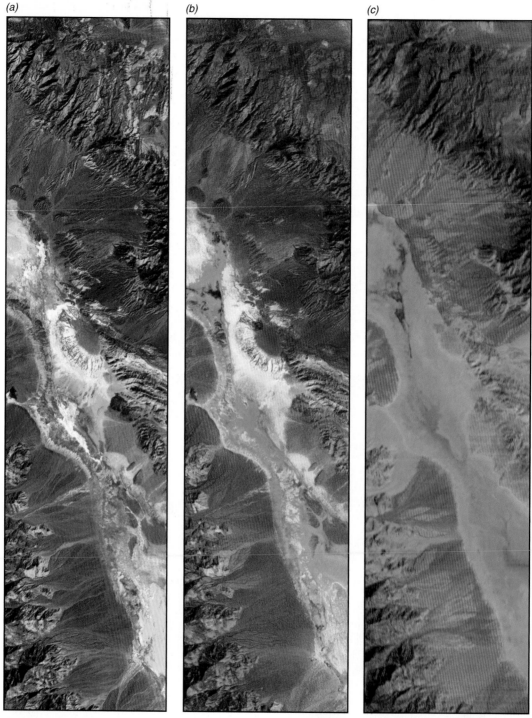

Plate 27 ASTER data, Death Valley, CA, early March. (*a*) VNIR (visible and near-IR) bands 1, 2, and 3 displayed as blue, green, and red, respectively. (*b*) SWIR (mid-IR) bands 5, 7, and 9 displayed as blue, green, and red, respectively. (*c*) TIR (thermal IR) bands 10, 12, and 13 displayed as blue, green, and red, respectively. Scale 1:350,000. (For major discussion, see Section 6.18.)

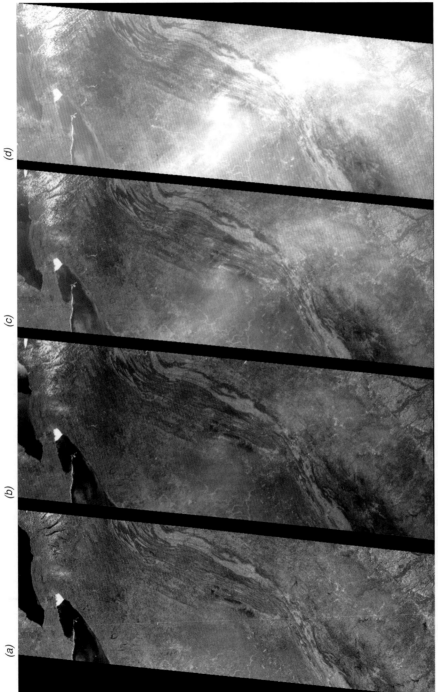

Plate 28 Multiangle images acquired by MISR over the eastern United States: (a) 0° (nadir-viewing), (b) 45.6° forward, (c) 60.0° forward, (d) 70.5° forward. (Courtesy NASA.) (For major discussion, see Section 6.18.)

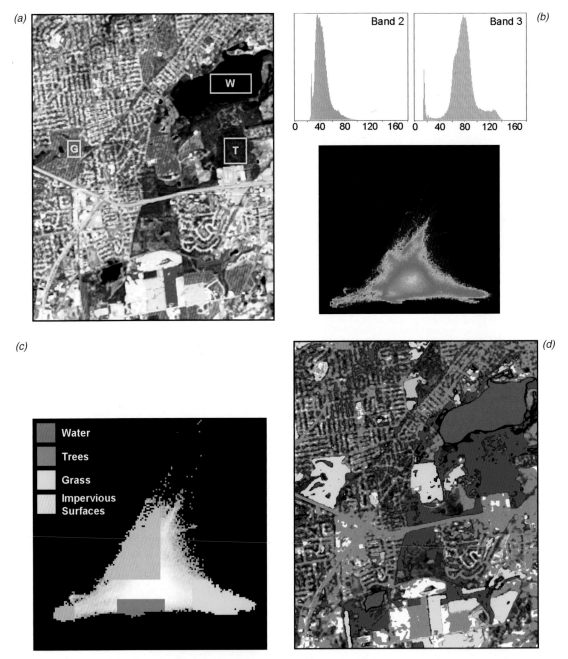

Plate 29 Sample interactive preliminary classification procedure used in the supervised training set refinement process: (*a*) original SPOT HRV color infrared composite image including selected training areas; (*b*) histograms and scatter diagram for band 2 (red) and band 3 (near-IR); (*c*) parallelepipeds associated with the initial training areas showing their locations in the band 2/band 3 scatter diagram; (*d*) partially completed classification superimposed on band 2 of the original images. (For major discussion, see Section 7.10.)

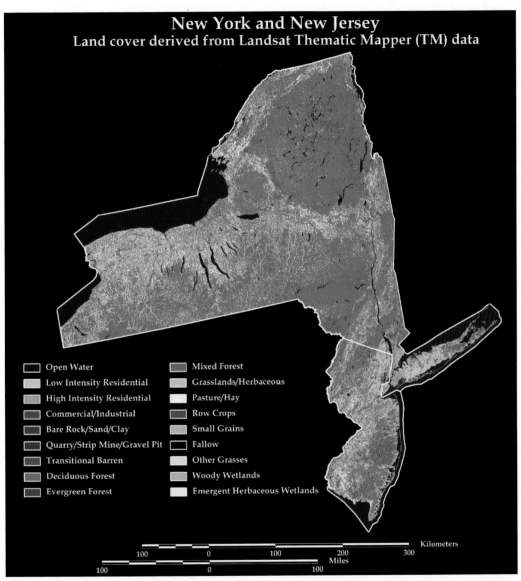

New York and New Jersey
Land cover derived from Landsat Thematic Mapper (TM) data

Open Water
Low Intensity Residential
High Intensity Residential
Commercial/Industrial
Bare Rock/Sand/Clay
Quarry/Strip Mine/Gravel Pit
Transitional Barren
Deciduous Forest
Evergreen Forest

Mixed Forest
Grasslands/Herbaceous
Pasture/Hay
Row Crops
Small Grains
Fallow
Other Grasses
Woody Wetlands
Emergent Herbaceous Wetlands

Kilometers
100   0   100   200   300
Miles
100   0   100

Plate 30 New York and New Jersey land cover derived from Landsat Thematic Mapper data. (Courtesy USGS EROS Data Center.) (For major discussion, see Section 7.14.)

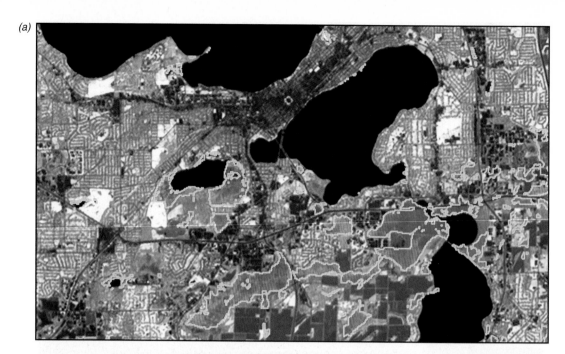

Plate 31 Multitemporal NDVI data merging. (*a*) Using multitemporal data as an aid in mapping invasive plant species, in this case reed canary grass (shown in red to pink tones) in wetlands. This color composite results from the merger of NDVI values derived from Landsat-7 ETM+ images of southern Wisconsin acquired on March 7 (blue), April 24 (green), and October 15 (red), respectively. Scale 1 : 120,000. (*b*) Monitoring algal blooms in portions of the two northernmost lakes shown. NDVI values derived from Landsat-7 ETM+ images acquired on April 24 (blue), October 31 (green), and October 15 (red), respectively. Scale 1 : 130,000. (For major discussion, see Section 7.17.)

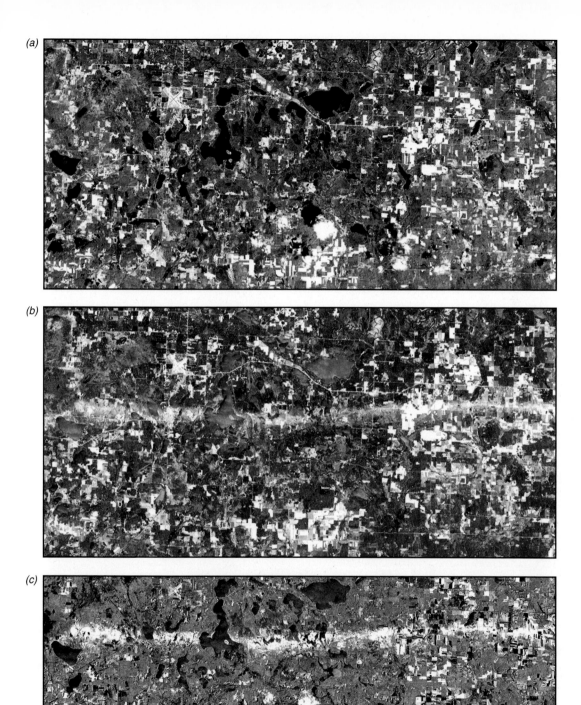

Plate 32 Use of multitemporal principal components analysis to detect change due to a tornado occurring in Burnett County in northwestern Wisconsin. (*a*) "Before" image acquired approximately one month prior to the tornado. (*b*) "After" image acquired one day after the tornado. (*c*) "Change" image depicting the damage path as portrayed in the second principal component derived from a composite of the "before" and "after" images. Scale 1:285,000. (For major discussion, see Section 7.17.)

Plate 33 IHS multisensor image merger of SPOT HRV (band 2—shown as blue; band 3—shown as green), Landsat TM (band 5—shown as red), and digital orthophoto data (shown as intensity component). The GIS overlay (shown as yellow) depicts boundaries of selected farm fields. (For major discussion, see Section 7.17.)

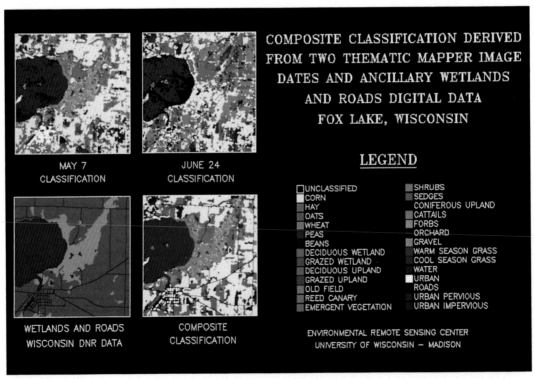

Plate 34 Use of user-defined decision rules and GIS data to improve the accuracy and detail of automated land cover classification. Upper left—May Landsat TM classification. Upper right—June Landsat TM classification. Lower left— GIS road and wetlands layer. Lower right—composite classification. (For major discussion, see Section 7.17.)

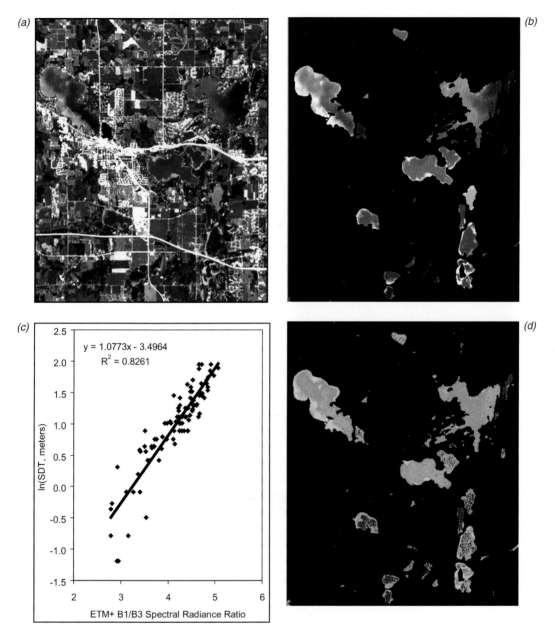

Plate 35 Biophysical modeling of lake water clarity in western Waukesha County, WI. (*a*) True-color composite of Landsat-7 ETM+ bands 1, 2, and 3 shown as blue, green, and red, respectively. (*b*) Water-only image. (*c*) Relationship between the ETM+ band 1/band 3 spectral radiance ratio and the natural logarithm of water clarity (Secchi disk transparency). (*d*) Pixel-level map of water clarity as predicted by the model based on the ETM+ imagery. Color indicates water clarity, ranging from dark blue (highest clarity) to yellow (lowest clarity). (For major discussion, see Section 7.19.)

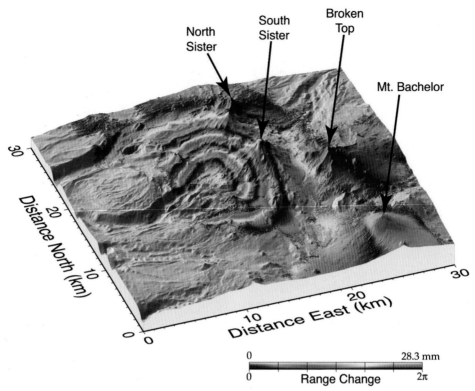

North
Sister

South
Sister

Broken
Top

Mt. Bachelor

Distance North (km)

Distance East (km)

0   28.3 mm

0   Range Change   2π

Plate 36 Radar interferogram, South Sister volcano, central Oregon Cascade Range, showing ground uplift caused by magma accumulation at depth. (Interferogram by C. Wicks, USGS.) (For major discussion, see Section 8.9.)

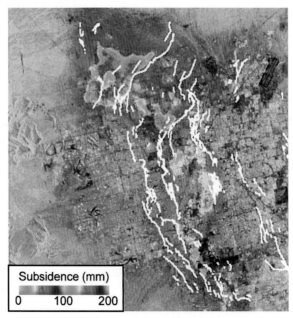

Subsidence (mm)

0   100   200

Plate 37 Radar interferogram showing subsidence in Las Vegas, NV, 1992–1997. White lines indicate location of surface faults. Scale 1:420,000. (Courtesy Stanford University Radar Interferometry Group.) (For major discussion, see Section 8.9.)

Plate 38 SIR-C color composite SAR image of a volcano-dominated landscape in central Africa, October 1994. Scale 1:360,000. (Courtesy NASA/JPL/Caltech). (For major discussion, see Section 8.12.)

(a) 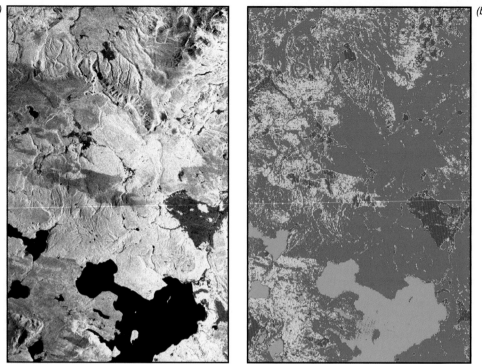 *(b)*

Plate 39 Yellowstone National Park, WY: (*a*) SIR-C image, L-band, VH-polarization; (*b*) map of estimated aboveground biomass derived from (*a*). (Courtesy NASA/JPL/Caltech.) (For major discussion, see Section 8.12.)

Plate 40 Side view of a three-dimensional model showing the position of lidar returns from a Douglas fir forest. Colors indicate elevation. (Courtesy Spencer B. Gross, Inc.) (For major discussion, see Section 8.21.)

feed the results to a computer in terms of image tone, we might "automate" our entire mapping task. Many remote sensing data analysis schemes attempt to do just that. For these schemes to be successful, the materials to be differentiated must be spectrally separable.

Experience has shown that many earth surface features of interest can be identified, mapped, and studied on the basis of their spectral characteristics. Experience has also shown that some features of interest cannot be spectrally separated. Thus, to utilize remote sensing data effectively, one must know and understand the spectral characteristics of the particular features under investigation in any given application. Likewise, one must know what factors influence these characteristics.

Spectral Reflectance of Vegetation, Soil, and Water

Figure 1.10 shows typical spectral reflectance curves for three basic types of earth features: healthy green vegetation, dry bare soil (gray-brown loam), and clear lake water. The lines in this figure represent *average* reflectance curves compiled by measuring a large sample of features. Note how distinctive the curves are for each feature. In general, the configuration of these curves is an indicator of the type and condition of the features to which they apply. Although the reflectance of individual features will vary considerably above and below the average, these curves demonstrate some fundamental points concerning spectral reflectance.

For example, spectral reflectance curves for healthy green vegetation almost always manifest the "peak-and-valley" configuration illustrated in Figure 1.10.

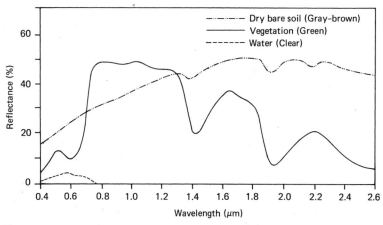

Figure 1.10 Typical spectral reflectance curves for vegetation, soil, and water. (Adapted from Swain and Davis, 1978.)

The valleys in the visible portion of the spectrum are dictated by the pigments in plant leaves. Chlorophyll, for example, strongly absorbs energy in the wavelength bands centered at about 0.45 and 0.67 μm (often called the "chlorophyll absorption bands"). Hence, our eyes perceive healthy vegetation as green in color because of the very high absorption of blue and red energy by plant leaves and the very high reflection of green energy. If a plant is subject to some form of stress that interrupts its normal growth and productivity, it may decrease or cease chlorophyll production. The result is less chlorophyll absorption in the blue and red bands. Often, the red reflectance increases to the point that we see the plant turn yellow (combination of green and red).

As we go from the visible to the near-IR portion of the spectrum at about 0.7 μm, the reflectance of healthy vegetation increases dramatically. In the range from about 0.7 to 1.3 μm, a plant leaf typically reflects 40 to 50 percent of the energy incident upon it. Most of the remaining energy is transmitted, since absorption in this spectral region is minimal (less than 5 percent). Plant reflectance in the range 0.7 to 1.3 μm results primarily from the internal structure of plant leaves. Because this structure is highly variable between plant species, reflectance measurements in this range often permit us to discriminate between species, even if they look the same in visible wavelengths. Likewise, many plant stresses alter the reflectance in this region, and sensors operating in this range are often used for vegetation stress detection. Also, multiple layers of leaves in a plant canopy provide the opportunity for multiple transmittance and reflectance. Hence, the near-IR reflectance increases with the number of layers of leaves in a canopy, with the reflection maximum achieved at about eight leaf layers (Bauer et al., 1986).

Beyond 1.3 μm, energy incident upon vegetation is essentially absorbed or reflected, with little to no transmittance of energy. Dips in reflectance occur at 1.4, 1.9, and 2.7 μm because water in the leaf absorbs strongly at these wavelengths. Accordingly, wavelengths in these spectral regions are referred to as *water absorption bands*. Reflectance peaks occur at about 1.6 and 2.2 μm, between the absorption bands. Throughout the range beyond 1.3 μm, leaf reflectance is approximately inversely related to the total water present in a leaf. This total is a function of both the moisture content and the thickness of a leaf.

The soil curve in Figure 1.10 shows considerably less peak-and-valley variation in reflectance. That is, the factors that influence soil reflectance act over less specific spectral bands. Some of the factors affecting soil reflectance are moisture content, soil texture (proportion of sand, silt, and clay), surface roughness, presence of iron oxide, and organic matter content. These factors are complex, variable, and interrelated. For example, the presence of moisture in soil will decrease its reflectance. As with vegetation, this effect is greatest in the water absorption bands at about 1.4, 1.9, and 2.7 μm (clay soils also have hydroxyl absorption bands at about 1.4 and 2.2 μm). Soil moisture content is strongly related to the soil texture: Coarse, sandy soils are usually well drained, resulting in low moisture content and relatively high reflectance; poorly drained fine-textured soils will generally have lower reflectance. In the absence of water, however, the soil itself will exhibit the reverse tendency: Coarse-textured soils will appear darker than fine-textured soils. Thus, the reflectance properties of a soil are consistent only within particular ranges of conditions. Two other factors that reduce soil reflectance are surface roughness and content of organic matter. The presence of iron oxide in a soil will also significantly decrease reflectance, at least in the visible wavelengths. In any case, it is essential that the analyst be familiar with the conditions at hand.

Considering the spectral reflectance of water, probably the most distinctive characteristic is the energy absorption at near-IR wavelengths and beyond. In short, water absorbs energy in these wavelengths whether we are talking about water features per se (such as lakes and streams) or water contained in vegetation or soil. Locating and delineating water bodies with remote sensing data are done most easily in near-IR wavelengths because of this absorption property. However, various conditions of water bodies manifest themselves primarily in visible wavelengths. The energy–matter interactions at these wavelengths are very complex and depend on a number of interrelated factors. For example, the reflectance from a water body can stem from an interaction with the water's surface (specular reflection), with material suspended in the water, or with the bottom of the depression containing the water body. Even with deep water where bottom effects are negligible, the reflectance properties of a water body are a function of not only the water per se but also the material in the water.

Clear water absorbs relatively little energy having wavelengths less than about 0.6 μm. High transmittance typifies these wavelengths with a maximum in the blue-green portion of the spectrum. However, as the turbidity of water changes (because of the presence of organic or inorganic materials), transmittance—and therefore reflectance—changes dramatically. For example, waters containing large quantities of suspended sediments resulting from soil erosion normally have much higher visible reflectance than other "clear" waters in the same geographic area. Likewise, the reflectance of water changes with the chlorophyll concentration involved. Increases in chlorophyll concentration tend to decrease water reflectance in blue wavelengths and increase it in green wavelengths. These changes have been used to monitor the presence and estimate the concentration of algae via remote sensing data. Reflectance data have also been used to determine the presence or absence of tannin dyes from bog vegetation in lowland areas and to detect a number of pollutants, such as oil and certain industrial wastes.

Many important water characteristics, such as dissolved oxygen concentration, pH, and salt concentration, cannot be observed directly through changes in water reflectance. However, such parameters sometimes correlate with observed reflectance. In short, there are many complex interrelationships between the spectral reflectance of water and particular characteristics. One must use appropriate reference data to correctly interpret reflectance measurements made over water.

Our discussion of the spectral characteristics of vegetation, soil, and water has been very general. The student interested in pursuing details on this subject, as well as factors influencing these characteristics, is encouraged to consult the various references contained in the Selected Bibliography at the end of this chapter.

Spectral Response Patterns

Having looked at the spectral reflectance characteristics of vegetation, soil, and water, we should recognize that these broad feature types are normally spectrally separable. However, the degree of separation between types is a function of "where we look" spectrally. For example, water and vegetation might reflect nearly equally in visible wavelengths, yet these features are almost always separable in near-IR wavelengths.

Because spectral responses measured by remote sensors over various features often permit an assessment of the type and/or condition of the features, these responses have often been referred to as *spectral signatures*. Spectral reflectance and spectral emittance curves (for wavelengths greater than 3.0 μm) are often referred to in this manner. The physical radiation measurements acquired over specific terrain features at various wavelengths are also referred to as the spectral signatures for those features.

Although it is true that many earth surface features manifest very distinctive spectral reflectance and/or emittance characteristics, these characteristics result in spectral "response patterns" rather than in spectral "signatures." The reason for this is that the term *signature* tends to imply a pattern that is absolute and unique. This is not the case with the spectral patterns observed in the natural world. As we have seen, spectral response patterns measured by remote sensors may be quantitative but they are not absolute. They may be distinctive but they are not necessarily unique.

Although the term *spectral signature* is used frequently in remote sensing literature, the student should keep in mind the variability of spectral signatures. This variability might cause severe problems in remote sensing data analysis if the objective is to identify various earth feature types spectrally. However, if the objective of an analysis is to identify the condition of various objects of the same type, we may have to rely on spectral response pattern variability to derive this information. This pertains to such applications as identifying stressed versus healthy vegetation within a given species. Therefore, it is extremely important to understand the nature of the ground area one is "looking at" with remote sensor data, not only to minimize unwanted spectral variability, but also to maximize this variability when the particular application requires it.

We have already looked at some characteristics of objects per se that influence their spectral response patterns. *Temporal effects* and *spatial effects* can also enter into any given analysis. Temporal effects are any factors that change the spectral characteristics of a feature over time. For example, the spectral characteristics of many species of vegetation are in a nearly continual state of change throughout a growing season. These changes often influence when we might collect sensor data for a particular application.

Spatial effects refer to factors that cause the same types of features (e.g., corn plants) at a given point in *time* to have different characteristics at different geographic *locations*. In small-area analysis the geographic locations may be meters apart and spatial effects may be negligible. When analyzing satellite data, the locations may be hundreds of kilometers apart where entirely different soils, climates, and cultivation practices might exist.

Temporal and spatial effects influence virtually all remote sensing operations. These effects normally complicate the issue of analyzing spectral reflectance properties of earth resources. Again, however, temporal and spatial effects might be the keys to gleaning the information sought in an analysis. For example, the process of *change detection* is premised on the ability to measure temporal effects. An example of this process is detecting the change in suburban development near a metropolitan area by using data obtained on two different dates.

An example of a useful spatial effect is the change in the leaf morphology of trees when they are subjected to some form of stress. For example, when a tree becomes infected with Dutch elm disease, its leaves might begin to cup and curl, changing the reflectance of the tree relative to healthy trees that surround it. So, even though a spatial effect might cause differences in the spectral reflectances of the same type of feature, this effect may be just what is important in a particular application.

Atmospheric Influences on Spectral Response Patterns

In addition to being influenced by temporal and spatial effects, spectral response patterns are influenced by the atmosphere. Regrettably, the energy recorded by a sensor is always modified to some extent by the atmosphere between the sensor and the ground. We will indicate the significance of this effect on a sensor-by-sensor basis throughout this book. For now, Figure 1.11 provides an initial frame of reference for understanding the nature of atmospheric effects. Shown in this figure is the typical situation encountered when a sensor records reflected solar energy. The atmosphere affects the "brightness," or *radiance*, recorded over any given point on the ground in two almost

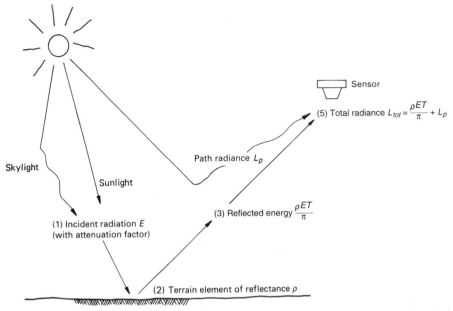

Figure 1.11   Atmospheric effects influencing the measurement of reflected solar energy. Attenuated sunlight and skylight (E) is reflected from a terrain element having reflectance ρ. The attenuated radiance reflected from the terrain element ($\rho ET/\pi$) combines with the path radiance (L_p) to form the total radiance (L_{tot}) recorded by the sensor.

contradictory ways. First, it attenuates (reduces) the energy illuminating a ground object (and being reflected from the object). Second, the atmosphere acts as a reflector itself, adding a scattered, extraneous *path radiance* to the signal detected by the sensor. By expressing these two atmospheric effects mathematically, the total radiance recorded by the sensor may be related to the reflectance of the ground object and the incoming radiation or *irradiance* using the equation

$$L_{tot} = \frac{\rho ET}{\pi} + L_p \tag{1.9}$$

where

L_{tot} = total spectral radiance measured by sensor
ρ = reflectance of object
E = irradiance on object, incoming energy
T = transmission of atmosphere
L_p = path radiance, from the atmosphere and not from the object

It should be noted that all of the above factors depend on wavelength. Also, as shown in Figure 1.11, the irradiance (E) stems from two sources: (1) directly reflected "sunlight" and (2) diffuse "skylight," which is sunlight that has been previously scattered by the atmosphere. The relative dominance of sunlight versus skylight in any given image is strongly dependent on weather conditions (e.g., sunny vs. hazy vs. cloudy). Likewise, irradiance varies with the seasonal changes in solar elevation angle (Figure 7.3) and the changing distance between the earth and sun. (Readers who might be interested in obtaining additional details about the concepts, terminology, and units used in radiation measurement may wish to consult Appendix A.)

1.5 DATA ACQUISITION AND INTERPRETATION

To this point, we have discussed the principal sources of electromagnetic energy, the propagation of this energy through the atmosphere, and the interaction of this energy with earth surface features. Combined, these factors result in energy "signals" from which we wish to extract information. We now consider the procedures by which these signals are detected, recorded, and interpreted.

The *detection* of electromagnetic energy can be performed either photographically or electronically. The process of photography uses chemical reactions on the surface of a light-sensitive film to detect energy variations within a scene. Photographic systems offer many advantages: They are relatively simple and inexpensive and provide a high degree of spatial detail and geometric integrity.

Electronic sensors generate an electrical signal that corresponds to the energy variations in the original scene. A familiar example of an electronic sensor is a video camera. Although sometimes more complex and expensive than photographic systems, electronic sensors offer the advantages of a broader spectral range of sensitivity, improved calibration potential, and the ability to electronically store and transmit data.

By developing a photograph, we obtain a *record* of its detected signals. Thus, the film acts as both the detecting and the recording medium. Electronic sensor signals are generally recorded onto some magnetic medium. Subsequently, the signals can be converted to an image on photographic film using a film recorder. In these cases, photographic film is used only as a recording medium.

In remote sensing, the term *photograph* is reserved exclusively for images that were *detected* as well as recorded on film. The more generic term *image* is used for any pictorial representation of image data. Thus, a pictorial record from a thermal scanner (an electronic sensor) would be called a "thermal image," *not* a "thermal photograph," because film would not be the original detection mechanism for the image. Because the term *image* relates to any pictorial product, all photographs are images. Not all images, however, are photographs.

A common exception to the above terminology is use of the term *digital photography*. As we describe in Section 2.12, digital cameras use electronic detectors rather than film for image detection. While this process is not "photography" in the traditional sense, "digital photography" is now the common way to refer to this technique of digital data collection.

We can see that the data interpretation aspects of remote sensing can involve analysis of pictorial (image) and/or digital data. *Visual interpretation* of pictorial image data has long been the most common form of remote sensing. Visual techniques make use of the excellent ability of the human mind to qualitatively evaluate spatial patterns in an image. The ability to make subjective judgments based on selective image elements is essential in many interpretation efforts.

Visual interpretation techniques have certain disadvantages, however, in that they may require extensive training and are labor intensive. In addition, *spectral characteristics* are not always fully evaluated in visual interpretation efforts. This is partly because of the limited ability of the eye to discern tonal values on an image and the difficulty for an interpreter to simultaneously analyze numerous spectral images. In applications where spectral patterns are highly informative, it is therefore preferable to analyze *digital*, rather than pictorial, image data.

The basic character of digital image data is illustrated in Figure 1.12. Though the image shown in (*a*) appears to be a continuous tone photograph, it is actually composed of a two-dimensional array of discrete *picture elements*, or *pixels*. The intensity of each pixel corresponds to the average brightness, or radiance, measured electronically over the ground area corresponding to each pixel. A total of 500 rows and 400 columns of pixels are shown in Figure 1.12*a*. Whereas the individual pixels are virtually impossible to discern in (*a*), they are readily observable in the enlargements shown in (*b*) and (*c*). These enlargements correspond to subareas located near the center of (*a*). A 100 row × 80 column enlargement is shown in (*b*) and a 10 row × 8 column enlargement is included in (*c*). Part (*d*) shows the individual *digital number* (*DN*) corresponding to the average radiance measured in each pixel shown in (*c*). These values are simply positive integers that result from quantizing the original electrical signal from the sensor into positive integer values using a process called *analog-to-digital* (*A-to-D*) signal conversion.

Figure 1.13 is a graphical representation of the A-to-D conversion process. The original electrical signal from the sensor is a continuous analog signal (shown by the continuous line plotted in the figure). This continuous signal is sampled at a set time interval (ΔT) and recorded numerically at each sample point ($a, b, \ldots, i, k$). The sampling rate for a particular signal is determined by the highest frequency of change in the signal. The sampling rate must be at least twice as high as the highest frequency present in the original signal in order to adequately represent the variation in the signal.

In Figure 1.13 we illustrate the incoming sensor signal in terms of an electrical voltage value ranging between 0 and 2 V. The DN output values are inte-

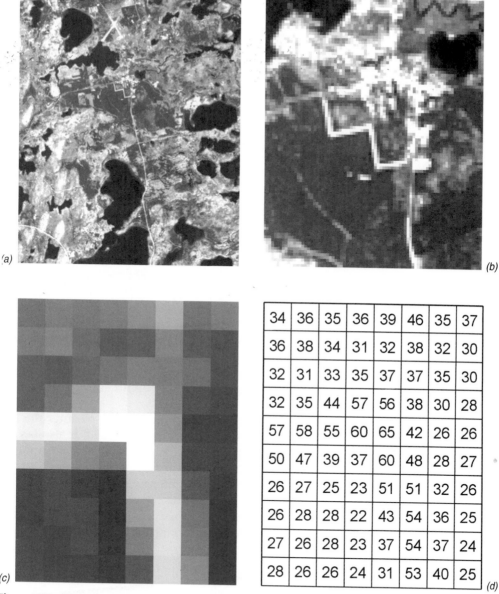

Figure 1.12 Basic character of digital image data. (*a*) Original 500 row × 400 column digital image. Scale 1 : 200,000. (*b*) Enlargement showing 100 row × 80 column area of pixels near center of (*a*). Scale 1 : 40,000. (*c*) 10 row × 8 column enlargement. Scale 1 : 4000. (*d*) Digital numbers corresponding to the radiance of each pixel shown in (*c*).

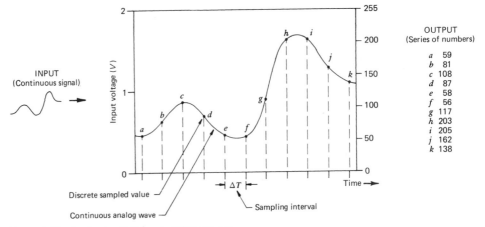

Figure 1.13 Analog-to-digital conversion process.

gers ranging from 0 to 255. Accordingly, a sampled voltage of 0.46 recorded by the sensor [at (*a*) in Figure 1.13] would be recorded as a DN of 59. (The DNs measured at the other sampling points along the signal are shown along the right side of the figure.)

Typically, the DNs constituting a digital image are recorded over such numerical ranges as 0 to 255, 0 to 511, 0 to 1023, or higher. These ranges represent the set of integers that can be recorded using 8-, 9-, and 10-bit binary computer coding scales, respectively. (That is, $2^8 = 256$, $2^9 = 512$, and $2^{10} = 1024$.) In such numerical formats, the image data can be readily analyzed with the aid of a computer.

The use of computer-assisted analysis techniques permits the spectral patterns in remote sensing data to be more fully examined. It also permits the data analysis process to be largely automated, providing cost advantages over visual interpretation techniques. However, just as humans are somewhat limited in their ability to interpret spectral patterns, computers are somewhat limited in their ability to evaluate spatial patterns. Therefore, visual and numerical techniques are complementary in nature, and consideration must be given to which approach (or combination of approaches) best fits a particular application.

1.6 REFERENCE DATA

As we have indicated in the previous discussion, rarely, if ever, is remote sensing employed without the use of some form of *reference data*. The acquisition of reference data involves collecting measurements or observations about the objects, areas, or phenomena that are being sensed remotely. These data can take on any of a number of different forms and may be derived from a number of sources. For example, the data needed for a particular analysis might be derived from a soil survey map, a water quality laboratory report, or an aerial photograph. They may also stem from a "field check" on the identity, extent, and condition of agricultural crops, land uses, tree species, or water pollution problems. Reference data may also involve field measurements of temperature and other physical and/or chemical properties of various features. The geographic positions at which such field measurements are made are often noted on a map base to facilitate their location in a corresponding remote sensing image. Usually, GPS receivers are used to determine the precise geographic position of field observations and measurements (as described in Section 1.7).

Reference data are often referred to by the term *ground truth*. This term is not meant literally, since many forms of reference data are not collected on the ground and can only approximate the truth of actual ground conditions. For example, "ground" truth may be collected in the air, in the form of detailed aerial photographs used as reference data when analyzing less detailed high altitude or satellite imagery. Similarly, the "ground" truth will actually be "water" truth if we are studying water features. In spite of these inaccuracies, ground truth is a widely used term for reference data.

Reference data might be used to serve any or all of the following purposes:

1. To aid in the analysis and interpretation of remotely sensed data.
2. To calibrate a sensor.
3. To verify information extracted from remote sensing data.

Hence, reference data must be collected in accordance with the principles of statistical sampling design appropriate to the particular application.

Reference data can be very expensive and time consuming to collect properly. They can consist of either *time-critical* and/or *time-stable* measurements. Time-critical measurements are those made in cases where ground conditions change rapidly with time, such as in the analysis of vegetation condition or water pollution events. Time-stable measurements are involved when the materials under observation do not change appreciably with time. For example, geologic applications often entail field observations that can be conducted at any time and that would not change appreciably from mission to mission.

One form of reference data collection is the ground-based measurement of the reflectance and/or emittance of surface materials to determine their spectral response patterns. This might be done in the laboratory or in the field using the principles of *spectroscopy*. Spectroscopic measurement procedures can involve the use of a variety of instruments. Often, a *spectroradiometer* is used in such measurement procedures. This device measures, as a function of wavelength, the energy coming from an object within its view. It is used primarily to prepare spectral reflectance curves for various objects.

In laboratory spectroscopy, artificial sources of energy might be used to illuminate objects under study. In the laboratory, other field parameters such as viewing geometry between object and sensor are also simulated. More often, therefore, in situ field measurements are preferred because of the many variables of the natural environment that influence remote sensor data that are difficult, if not impossible, to duplicate in the laboratory.

In the acquisition of field measurements, spectroradiometers may be operated in a number of modes, ranging from hand held to helicopter or aircraft mounted. Figure 1.14 illustrates a highly portable instrument that is well suited for hand-held operation. Through a fiber-optic input, this particular system acquires a continuous spectrum by recording data in over 1000 narrow bands simultaneously (over the range 0.35 to 2.5 μm). The unit incorporates a built-in notebook computer, which provides for flexibility in data acquisition, display, and storage. For example, as a standard feature, re-

(a)

(b)

Figure 1.14 Analytical Spectral Devices FieldSpec FR Spectroradiometer: (a) the instrument; (b) instrument shown in field operation. (Courtesy Analytical Spectral Devices, Inc.)

flectance spectra are displayed in real time, as are computed reflectance values within the wavelength bands of various satellite systems. In-field calculation of band ratios and other computed values is also possible. One such calculation might be the normalized difference vegetation index (NDVI), which relates the near-IR and visible reflectance of earth surface features (Chapter 7). Another option is matching measured spectra to a library of previously measured samples.

Figure 1.15 shows a versatile all-terrain instrument platform designed primarily for collecting spectral measurements in agricultural cropland environments. The system provides the high clearance necessary for making measurements over mature row crops and the tracked wheels allow access to difficult landscape positions. Several measurement instruments can be suspended from the system's telescopic boom. Typically, these include a spectroradiometer, a remotely operated digital camera system, and a global position system (Section 1.7). While designed primarily for data collection in agricultural fields, the long reach of the boom makes this device a useful tool for collecting spectral data over such targets as emergent vegetation found in wetlands as well as small trees and shrubs.

Figure 1.15 All-terrain instrument platform designed for collecting spectral measurements in agricultural cropland environments. (Courtesy University of Nebraska-Lincoln Center for Advanced Land Management Information Technologies.)

Using a spectroradiometer to obtain spectral reflectance measurements is normally a three-step process. First, the instrument is aimed at a *calibration panel* of known, stable reflectance. The purpose of this step is to quantify the incoming radiation, or irradiance, incident upon the measurement site. Next, the instrument is suspended over the target of interest and the radiation reflected by the object is measured. Finally, the spectral reflectance of the object is computed by ratioing the reflected energy measurement in each band of observation to the incoming radiation measured in each band. Normally, the term *reflectance factor* is used to refer to the result of such computations. A reflectance factor is defined formally as the ratio of the radiant flux actually reflected by a sample surface to that which would be reflected into the same sensor geometry by an ideal, perfectly diffuse (Lambertian) surface irradiated in exactly the same way as the sample.

Another term frequently used to describe the above type of measurement is *bidirectional reflectance factor*: one direction being associated with the sample viewing angle (usually 0° from normal) and the other direction being that of the sun's illumination (defined by the solar zenith and azimuth angles). In the bidirectional reflectance measurement procedure described above, the sample and the reflectance standard are measured sequentially. Other approaches exist in which the incident spectral irradiance and reflected spectral radiance are measured simultaneously.

What is important to emphasize here is that it is often assumed that surface features are truly Lambertian in character (e.g., reflect uniformly in all directions). In reality, many, if not most, surfaces are not Lambertian. Their reflectance properties vary with wavelength, illumination angle, and viewing angle. The *bidirectional reflectance distribution function (BRDF)* is a mathematical description of how reflectance varies for all combinations of illumination and viewing angles at a given wavelength (Schott, 1997). The BRDF for any given feature can approximate that of a Lambertian surface at some angles and be non-Lambertian at other angles. Similarly, the BRDF can vary considerably with wavelength.

Figure 1.16 shows a model of the BRDF for a soybean field with the sun at a zenith angle of 30°. This model, which is for wavelengths of 0.4 to 0.7 μm, shows percent reflectance (on the vertical axis) as a function of viewing angle (on the two horizontal axes). The center of the BRDF model represents a zenith angle of 0°, in which the hypothetical sensor is looking vertically downward at the surface of the field. For any other point on the model, the viewing zenith angle (ranging from 0° to 90°) is represented by the radial distance from the center of the model. The viewing azimuth angle (ranging from 0° to 360°) is represented by the angle between a line from the center of the plot to a given point and the solar azimuth (0°, located at the right side of the model).

Several features of the BRDF model shown in Figure 1.16 should be noted. First, the reflectance generally increases toward the right side of the model; thus, reflectance is generally higher when viewing the surface on the same side from which it is being illuminated by the sun. Second, the prominent peak

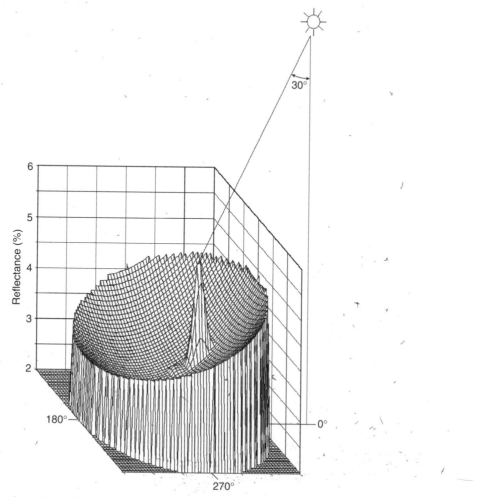

Figure 1.16 Bidirectional reflectance distribution function (over the range of 0.4 to 0.7 μm) for a soybean field illuminated with the sun at a 30° zenith angle.

just to the right of the center of the model is referred to as the "hotspot." This is an area of increased reflectance that occurs when the azimuth and zenith angles of the sensor are the same as those of the sun (in this case, an azimuth angle of 0° and a zenith angle of 30°). The existence of the hotspot is due to the fact that the sensor is then viewing only the sunlit portion of the surface, without any shadowing. Variations in bidirectional reflectance, such as the hotspot, can significantly affect the appearance of objects in remotely sensed images, causing them to appear brighter or darker solely as a result of the angular relationships among the sun, the object, and the sensor, without regard to any actual reflectance differences on the surface. A variety of mathematical models (including a provision for wavelength dependence) have been proposed to represent the BRDF (Jupp and Strahler, 1991).

1.7   THE GLOBAL POSITIONING SYSTEM

As mentioned previously, the location of field-observed reference data is usually determined using GPS methods. Originally developed for defense purposes, the U.S. Global Positioning System includes a group of 24 satellites

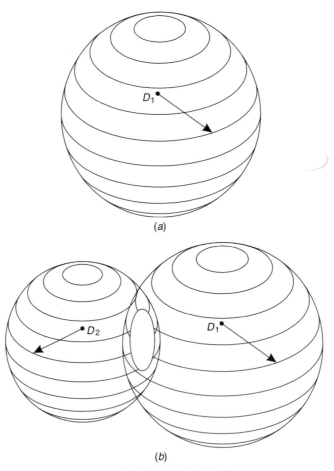

(a)

(b)

Figure 1.17   Principle of satellite ranging: (a) distance measurement from one satellite establishes position on a sphere; (b) distance measurement from two satellites establishes a circle of intersection of two spheres; (c) distance measurement from three satellites narrows position to only two possibilities.

rotating around the earth in precisely known orbits, with subgroups of four satellites operating in each of six different orbit planes. Typically, these satellites revolve around the earth approximately once every 12 hours and they operate at an altitude of approximately 20,200 km. With their positions in space precisely known at all times, these satellites can be thought of as a constellation of "artificial stars" specifically designed to aid in positioning and navigation. The satellites transmit time-encoded radio signals that are recorded by ground-based receivers. The nearly circular orbital planes of the satellites are inclined about 60° from the equator and are spaced every 60° in longitude. This means that an observer at any point on the earth's surface can receive the signal from at least four GPS satellites at any given time (day or night).

The means by which GPS signals are used to determine ground positions is called *satellite ranging*. Conceptually, the process simply involves measuring the time required for a signal transmitted by a given satellite to reach the ground receiver. Knowing that the signal travels at the speed of light (3×10^8 m/sec), the distance between the satellite and the receiver can be computed. As shown in Figure 1.17a, if the distance to only one satellite is measured (D_1), the position of the GPS receiver is only known to be somewhere on the surface of an imaginary sphere of this radius and whose center is at the known position of the satellite. If the signals from two satellites are received (Figure 1.17b), the receiver position is known to be somewhere on the circle where the two spheres of radius D_1 and D_2 intersect. Recording the signals from three satellites (Figure 1.17c) narrows the position of the receiver down

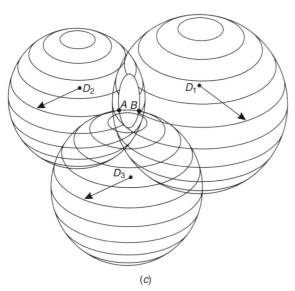

(c)

Figure 1.17   (Continued)

to only two possible points (*A* and *B*). Usually, one of these points is not reasonable (often located a great distance away, or even off the surface of the earth). Hence, the three-dimensional position of the receiver can be determined by measuring the signals from just three satellites. However, the signals from four satellites are needed to correct for *clock bias*, a lack of exact synchronization between the extremely exact atomic clocks in the satellites and the less expensive, lower accuracy clocks in GPS receivers. In practice, it is usually desirable to make repeated measurements to as many satellites as practical.

GPS measurements are potentially subject to numerous sources of error in addition to clock bias. Among these are uncertainties in the satellite orbits (known as *satellite ephemeris errors*), errors due to atmospheric conditions (signal velocity depends on time of day, season, and angular direction through the atmosphere), receiver errors (due to such influences as electrical noise and signal matching errors), and multipath errors (reflection of a portion of the transmitted signal from objects not in the straight-line path between the satellite and receiver). Such errors can be compensated for (in great part) using *differential* GPS measurement methods. In this approach, simultaneous measurements are made by a stationary base station receiver (located over a point of precisely known position) and one (or more) roving receivers moving from point to point. The positional errors measured at the base station are used to refine the position measured by the rover(s) at the same instant in time. This can be done either by bringing the data from the base and rover together in a postprocessing mode after the field observations are completed or by instantaneously broadcasting the base station corrections to the rovers. The latter approach is termed *real-time differential* GPS positioning.

The U.S. Coast Guard operates the most common differential GPS correction service. This system consists of a network of towers that receive GPS signals and transmit a correction signal using omnidirectional beacon transmitters. In order to receive the correction information, users must have a differential beacon receiver and beacon antenna in addition to their GPS. No such additional equipment is necessary with receivers that include *Wide Area Augmentation System (WAAS)* capability.

The WAAS consists of approximately 25 ground reference stations distributed across the United States that continuously monitor GPS satellite data. Two master stations, located on the East and West Coasts, collect the data from the reference stations and create a composited correction message that is location specific. This message is then broadcast through one of two *geostationary* satellites, satellites occupying a fixed position over the equator. Any WAAS-enabled GPS unit can receive these correction signals. The GPS receiver determines which correction data are appropriate at the current location.

The WAAS signal reception is ideal for open land, aircraft, and marine applications, but the position of the relay satellites over the equator makes it difficult to receive the signals when features such as trees and mountains obstruct the view of the horizon. In such situations, GPS positions can sometimes actually contain more error with WAAS correction than without. However, in unobstructed operating conditions where a strong WAAS signal is available, positions are normally accurate to within 3 m or better.

Paralleling the deployment of the WAAS system in North America are the Japanese *Multi-functional Satellite Augmentation System (MSAS)* in Asia and the *European Geostationary Navigation Overlay Service (EGNOS)* in Europe. Eventually, GPS users will have access to these and other compatible systems on a global basis.

At the time of this writing (2002), the U.S. Global Positioning System has only one operational counterpart, the Russian *GLONASS* system. However, a fully comprehensive European global satellite navigation system, *Galileo*, is scheduled for operation within 5 years. The future for these and similar systems is an extremely bright and rapidly changing one.

1.8   AN IDEAL REMOTE SENSING SYSTEM

Having introduced some basic concepts, we now have the elements necessary to conceptualize an ideal remote sensing system. In so doing, we can begin to appreciate some of the problems encountered in the design and application of the various real sensing systems examined in subsequent chapters.

The basic components of an *ideal* remote sensing system are shown in Figure 1.18. These include the following:

1. **A uniform energy source.**   This source would provide energy over all wavelengths, at a constant, known, high level of output, irrespective of time and place.

2. **A noninterfering atmosphere.**   This would be an atmosphere that would not modify the energy from the source in any manner, whether that energy were on its way to the earth's surface or coming from it. Again, ideally, this would hold irrespective of wavelength, time, place, and sensing altitude involved.

3. **A series of unique energy–matter interactions at the earth's surface.**   These interactions would generate reflected and/or emitted signals that not only are selective with respect to wavelength, but also are known, invariant, and unique to each and every earth surface feature type and subtype of interest.

4. **A supersensor.**   This would be a sensor, highly sensitive to all wavelengths, yielding spatially detailed data on the absolute brightness (or

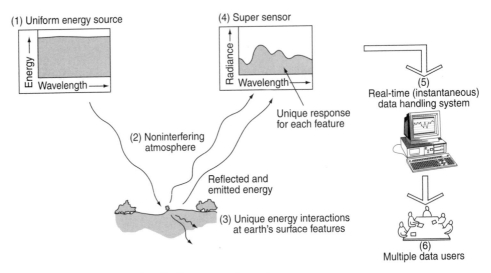

Figure 1.18 Components of an ideal remote sensing system.

radiance) from a scene as a function of wavelength, throughout the spectrum. This supersensor would be simple and reliable, require virtually no power or space, and be accurate and economical to operate.

5. **A real-time data processing and supply system.** In this system, the instant the radiance-versus-wavelength response over a terrain element was generated, it would be transmitted to the ground, geometrically and radiometrically corrected as necessary, and processed into a readily interpretable format. Each data observation would be recognized as being unique to the particular terrain element from which it came. This processing would be performed nearly instantaneously ("real time"), providing timely information. Because of the consistent nature of the energy–matter interactions, there would be no need for reference data in the analysis procedure. The derived data would provide insight into the physical–chemical–biological state of each feature of interest.

6. **Multiple data users.** These people would have knowledge of great depth, both of their respective disciplines and of remote sensing data acquisition and analysis techniques. The same set of "data" would become various forms of "information" for different users, because of their wealth of knowledge about the particular earth resources being sensed. This information would be available to them faster, at less expense, and over larger areas than information collected in any other manner. With this information, the various users would make profound, wise decisions about how best to manage the earth re-

sources under scrutiny, and these management decisions would be implemented—to everyone's delight!

Unfortunately, an ideal remote sensing system as described above does not exist. Real remote sensing systems fall far short of the ideal at virtually every point in the sequence outlined.

1.9 CHARACTERISTICS OF REAL REMOTE SENSING SYSTEMS

Let us consider some of the basic shortcomings common to all real remote sensing systems in order to better understand their general operation and utility. Regarding the elements of the ideal system we have developed, the following general shortcomings of real systems should be recognized:

1. **The energy source.**   All passive remote sensing systems rely on energy that is reflected and/or emitted from earth surface features. As already discussed, the spectral distribution of reflected sunlight and self-emitted energy is far from uniform. Solar energy levels obviously vary with respect to time and location, and different earth surface materials emit energy with varying degrees of efficiency. While we have some control over the nature of sources of energy for active systems, the sources of energy used in most real systems are generally nonuniform with respect to wavelength, and their properties vary with time and location. Consequently, we normally must calibrate for source characteristics on a mission-by-mission basis or deal with *relative* energy units sensed at any given time and location.

2. **The atmosphere.**   The atmosphere normally compounds the problems introduced by energy source variation. To some extent, the atmosphere always modifies the strength and spectral distribution of the energy received by a sensor. It restricts "where we can look" spectrally, and its effects vary with wavelength, time, and place. The importance of these effects, like source variation effects, is a function of the wavelengths involved, the sensor used, and the sensing application at hand. Elimination of, or compensation for, atmospheric effects via some form of calibration is particularly important in those applications where repetitive observations of the same geographic area are involved.

3. **The energy–matter interactions at the earth's surface.**   Remote sensing would be simple if every material reflected and/or emitted energy in a unique, known way. Although spectral response patterns (signatures) play a central role in detecting, identifying, and analyzing earth surface materials, the spectral world is full of ambiguity. Radically different material types can have great spectral similarity,

making differentiation difficult. Furthermore, the general understanding of the energy–matter interactions for earth surface features is at an elementary level for some materials and virtually nonexistent for others.

4. **The sensor.** At this point, it should come as no surprise that an ideal "supersensor" does not exist. No single sensor is sensitive to all wavelengths. All real sensors have fixed limits of *spectral sensitivity*. They also have a limit on how small an object on the earth's surface can be and still be "seen" by a sensor as being separate from its surroundings. This limit, called the *spatial resolution* of a sensor, is an indication of how well a sensor can record spatial detail.

Figure 1.19 illustrates, in the context of a digital image, the interplay between the spatial resolution of a sensor and the spatial variability present in a ground scene. In (a), a single pixel covers only a small area of the ground (on the order of the width of the rows of the crop shown). In (b), a coarser ground resolution is depicted and a single pixel integrates the radiance from both the crop rows and the soil between them. In (c), an even coarser resolution results in a pixel measuring the average radiance over portions of the two fields. Thus, depending on the spatial resolution of the sensor and the spatial structure of the ground area being sensed, digital images comprise a range of "pure" and "mixed" pixels. In general, the larger the percentage of mixed pixels, the more limited is the ability to record and extract spatial detail in an image. This is illustrated in Figure 1.20, in which the same area has been imaged over a range of different ground resolution cell sizes.

The choice of a sensor for any given task always involves trade-offs. For example, photographic systems generally have very good

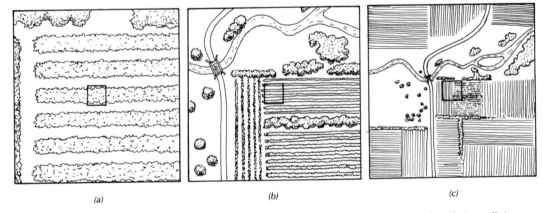

(a)   (b)   (c)

Figure 1.19 Ground resolution cell size effect: (a) small, (b) intermediate, and (c) large ground resolution cell size.

spatial resolution characteristics, but they lack the broad spectral sensitivity obtainable with nonphotographic systems which usually have poorer spatial resolution characteristics. Similarly, many nonphotographic systems (and some photographic systems) are quite complex optically, mechanically, and/or electronically. They may have restrictive power, space, and stability requirements. These requirements often dictate the type of *platform*, or vehicle, from which a sensor can be operated. Platforms can vary from stepladders to satellites. Depending on the sensor–platform combination needed in a particular application, the acquisition of remote sensing data can be a very expensive endeavor.

5. **The data processing and supply system.** The capability of current remote sensors to generate data far exceeds the capacity to handle these data. This is generally true whether we consider "manual" image interpretation procedures or digital analyses. Processing sensor data into an interpretable format can be—and often is—an effort entailing considerable thought, hardware, time, experience, and reference data. Also, many data users would like to receive their data immediately after acquisition by the sensor in order to make the timely decisions required in certain applications (e.g., agricultural crop management, disaster assessment). Regrettably, many sources of remote sensing data are unable to supply data over the exact areas and time spans that might be desired by the data user.

6. **The multiple data users.** Central to the successful application of any remote sensing system is the person (or persons) using the remote sensor data from that system. The "data" generated by remote sensing procedures become "information" only if and when someone understands their generation, knows how to interpret them, and knows how best to use them. *A thorough understanding of the problem at hand is paramount to the productive application of any remote sensing methodology. Also, no single combination of data acquisition and analysis procedures will satisfy the needs of all data users.*

Whereas the interpretation of aerial photography has been used as a practical resource management tool for nearly a century, other forms of remote sensing are relatively new, technical, and "unconventional" means of acquiring information. These newer forms of remote sensing have had relatively few satisfied users until recently. However, as new applications continue to be developed and implemented, increasing numbers of users are becoming aware of the potentials, *as well as the limitations*, of remote sensing techniques. As a result, remote sensing has become an essential tool in many aspects of science, government, and business alike.

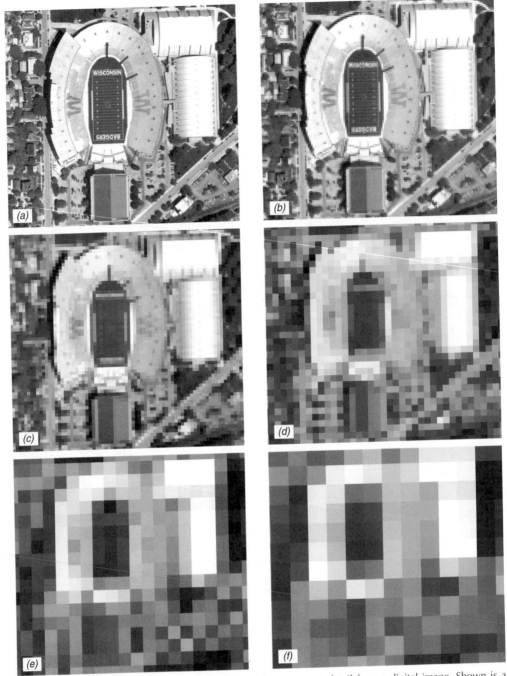

Figure 1.20 Ground resolution cell size effect on ability to extract detail from a digital image. Shown is a portion of the University of Wisconsin-Madison campus, including Camp Randall Stadium and vicinity, at a ground resolution cell size (per pixel) of: (*a*) 1 m, (*b*) 2.5 m, (*c*) 5 m, (*d*) 10 m, (*e*) 20 m, and (*f*) 30 m, and an enlarged portion of the image at (*g*) 0.5 m, (*h*) 1 m, and (*i*) 2.5 m. (Courtesy University of Wisconsin-Madison, Environmental Remote Sensing Center, and NASA Affiliated Research Center Program.)

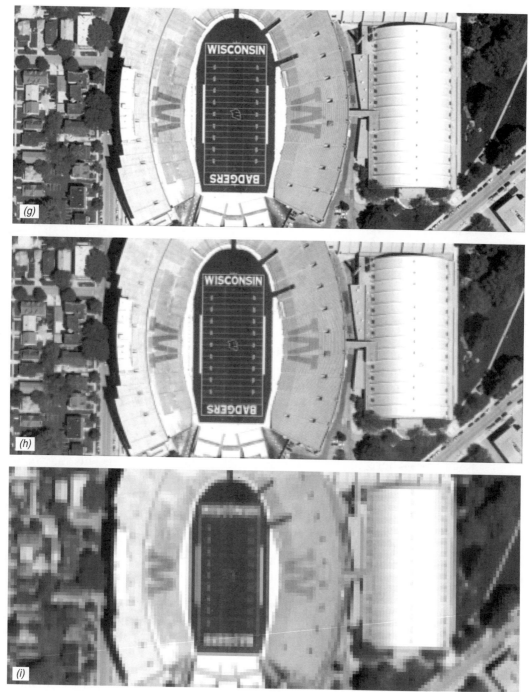

Figure 1.20   *(Continued)*

1.10 SUCCESSFUL APPLICATION OF REMOTE SENSING

The student should now begin to appreciate that successful application of remote sensing is premised on the *integration* of multiple, interrelated data sources and analysis procedures. No single combination of sensor and interpretation procedure is appropriate to all resource inventorying and environmental monitoring applications. In fact, *many inventorying and monitoring problems are not amenable to solution by means of remote sensing at all.* Among the applications appropriate, a wide variety of data acquisition and analysis approaches exist. Conceptually, however, all designs of successful remote sensing efforts involve, at a minimum, (1) clear definition of the problem at hand, (2) evaluation of the potential for addressing the problem with remote sensing techniques, (3) identification of the remote sensing data acquisition procedures appropriate to the task, (4) determination of the data interpretation procedures to be employed and the reference data needed, and (5) identification of the criteria by which the quality of information collected can be judged.

All too often, one (or more) of the above components of a remote sensing application is overlooked. The result may be disastrous. Many resource management programs exist with little or no means of evaluating the performance of remote sensing systems in terms of information quality. Many people have acquired burgeoning quantities of remote sensing data with inadequate capability to interpret them. Many occasions have occurred when remote sensing has *or* has not been used because the problem was not clearly defined. A clear articulation of the information requirements of a particular problem and the extent to which remote sensing might meet these requirements is paramount to any successful application.

The success of many applications of remote sensing is improved considerably by taking a *multiple-view* approach to data collection. This may involve *multistage* sensing wherein data about a site are collected from multiple altitudes. It may involve *multispectral* sensing whereby data are acquired simultaneously in several spectral bands. Or, it may entail *multitemporal* sensing, where data about a site are collected on more than one occasion.

In the multistage approach, satellite data may be analyzed in conjunction with high altitude data, low altitude data, and ground observations (Figure 1.21). Each successive data source might provide more detailed information over smaller geographic areas. Information extracted at any lower level of observation may then be extrapolated to higher levels of observation.

A commonplace example of the application of multistage sensing techniques is the detection, identification, and analysis of forest disease and insect problems. From space images, the image analyst could obtain an overall view of the major vegetation categories involved in a study area. Using this information, the areal extent and position of a particular species of interest could be determined and representative subareas could be studied more

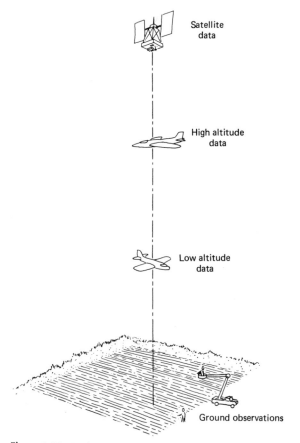

Satellite
data

High altitude
data

Low altitude
data

Ground observations

Figure 1.21 Multistage remote sensing concept.

closely at a more refined stage of imaging. Areas exhibiting stress on the second-stage imagery could be delineated. Representative samples of these areas could then be field checked to document the presence and particular cause of the stress.

After analyzing the problem in detail by ground observation, the analyst would use the remote sensor data to extrapolate assessments beyond the small study areas. By analyzing the large-area remote sensor data, the analyst can determine the severity and geographic extent of the disease problem. Thus, while the question on specifically *what* the problem is can generally be evaluated only by detailed ground observation, the equally important questions of *where, how much,* and *how severe* can often be best handled by remote sensing analysis.

In short, more information is obtained by analyzing multiple views of the terrain than by analysis of any single view. In a similar vein, multispectral imagery provides more information than data collected in any single spectral band. For example, the multispectral scanner is a sensor that acquires data from multiple spectral bands simultaneously. When the signals recorded in the multiple bands are analyzed in conjunction with each other, more information becomes available than if only a single band were used or if the multiple bands were analyzed independently. The multispectral approach forms the heart of numerous remote sensing applications involving discrimination of earth resource types and conditions.

Again, multitemporal sensing involves sensing the same area at multiple times and using changes occurring with time as discriminants of ground conditions. This approach is frequently taken to monitor land use change, such as suburban development in urban fringe areas. In fact, regional land use surveys might call for the acquisition of multisensor, multispectral, multistage, multitemporal data to be used for multiple purposes!

In any approach to applying remote sensing, not only must the right mix of data acquisition and data interpretation techniques be chosen, but the right mix of remote sensing and "conventional" techniques must also be identified. The student must recognize that remote sensing is a tool best applied in concert with others; it is not an end in itself. In this regard, remote sensing data are currently being used extensively in computer-based GISs (Section 1.11). The GIS environment permits the synthesis, analysis, and communication of virtually unlimited sources and types of biophysical and socioeconomic data—as long as they can be geographically referenced. Remote sensing can be thought of as the "eyes" of such systems providing repeated, synoptic (even global) visions of earth resources from an aerial or space vantage point.

Remote sensing affords us the capability to literally see the invisible. We can begin to see components of the environment on an "ecosystem basis," in that remote sensing data can transcend the cultural boundaries within which much of our current resource data are collected. Remote sensing also transcends disciplinary boundaries. It is so broad in its application that nobody "owns" the field. Important contributions are made to—and benefits derived from—remote sensing by both the "hard" scientist interested in basic research and the "soft" scientist interested in its operational application.

There is little question that remote sensing will continue to play an increasingly broad and important role in natural resource management. The technical capabilities of sensors, space platforms, data communication systems, GPSs, digital image processing systems, and GISs are improving on almost a daily basis. At the same time, we are witnessing an evolution of various remote sensing procedures from being purely research activities to becoming commercially available services. Most importantly, we are becoming increasingly aware of how interrelated and fragile the elements of our global resource base really are and of the role remote sensing can play in inventorying, monitoring, and managing earth resources and in modeling and helping us understand the global ecosystem.

1.11   GEOGRAPHIC INFORMATION SYSTEMS

We anticipate that the majority of individuals using this book will at some point in their educational backgrounds and/or professional careers have experience with geographic information systems. The discussion below is provided as a brief introduction to such systems primarily for those readers who might lack such background.

Geographic information systems are computer-based systems that can deal with virtually any type of information about features that can be referenced by geographical location. These systems are capable of handling both *locational data* and *attribute data* about such features. That is, not only do GISs permit the automated mapping or display of the locations of features, but also these systems provide a capability for recording and analyzing descriptive characteristics about the features. For example, a GIS might contain not only a "map" of the locations of roads but also a database of descriptors about each road. These "attributes" might include information such as road width, pavement type, speed limit, number of traffic lanes, date of construction, and so on. Table 1.1 lists other examples of attributes that might be associated with a given point, line, or area feature.

Much of the power of a GIS comes from the database management system (DBMS) that is designed to store and manipulate these attribute data. Many GISs employ a *relational database*, consisting of tables, or "relations," with attributes in columns and data records, or "*tuples*," in rows (Table 1.2). Recently there has been interest in *object-oriented databases*, in which data objects take on more complex properties of their own. While database implementations vary, there are certain desirable characteristics that will improve the utility of a database in a GIS. These characteristics include flexibility, to allow a wide range of database queries and operations; reliability, to avoid accidental loss of data; security, to limit access to authorized users; and ease of use, to insulate the end user from the details of the database implementation. Often the user can interact with the DBMS to search for information by means of "queries," simple commands or questions in a language that can be understood by the DBMS. The most common query language is known as

TABLE 1.1   Example Point, Line, and Area Features and Typical Attributes Contained in a GIS[a]

Point feature	Well (depth, chemical constituency)
Line feature	Power line (service capacity, age, insulator type)
Area feature	Soil mapping unit (soil type, texture, color, permeability)

[a]Attributes shown in parentheses.

TABLE 1.2 **Relational Database Table Format**

ID Number[a]	Street Name	Lanes	Parking	Repair Date	. . .
143897834	"Maple Ct"	2	Yes	1982/06/10	. . .
637292842	"North St"	2	Seasonal	1986/08/22	. . .
347348279	"Main St"	4	Yes	1995/05/15	. . .
234538020	"Madison Ave"	4	No	1989/04/20	. . .

[a]Each data record, or "tuple," has a unique identification, or ID, number.

SQL, or Structured Query Language, and several accepted standards for SQL have been adopted.

One of the most important benefits of a GIS is the ability to spatially interrelate multiple types of information stemming from a range of sources. This concept is illustrated in Figure 1.22, where we have assumed that a hydrologist wishes to use a GIS to study soil erosion in a watershed. As shown, the system contains data from a range of source maps (*a*) that have been geocoded on a cell-by-cell basis to form a series of land data files, or *layers* (*b*), all in geographic registration. The analyst can then manipulate and overlay the information contained in, or derived from, the various data files. In this example, we assume that assessing the potential for soil erosion throughout the watershed involves the simultaneous cell-by-cell consideration of three types of data derived from the original data files: slope, soil erodibility, and

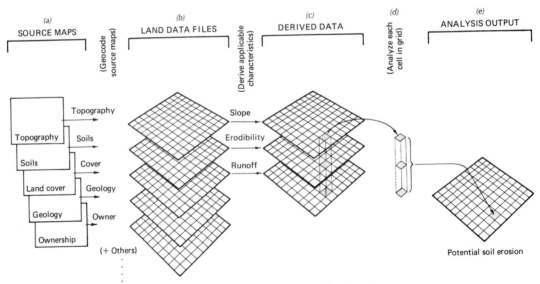

Figure 1.22 GIS analysis procedure for studying potential soil erosion.

surface runoff potential. The slope information can be computed from the elevations in the topography file. The erodibility, which is an attribute associated with each soil type, can be extracted from a relational database management system incorporated in the GIS. Similarly, the runoff potential is an attribute associated with each land cover type (land cover data can be obtained through interpretation of aerial photographs or satellite images). The analyst can use the system to interrelate these three sources of derived data (*c*) in each grid cell and use the result to locate, display, and/or record areas where combinations of site characteristics indicate high soil erosion potential (i.e., steep slopes and highly erodible soil–cover combinations).

The above example illustrates the GIS analysis function commonly referred to as *overlay analysis*. The number, form, and complexity of other data analyses possible with a GIS are virtually limitless. Such procedures can operate on the system's spatial data, the attribute data, or both. For example, *aggregation* is an operation that permits combining detailed map categories to create new, less detailed categories (e.g., combining "jack pine" and "red pine" categories into a single "pine" category). *Buffering* creates a zone of specified width around one or more features (e.g., the area within 50 m of a stream). *Network analysis* permits such determinations as finding the shortest path through a street network, determining the stream flows in a drainage basin, or finding the optimum location for a fire station. *Intervisibility* operations use elevation data to permit *viewshed mapping* of what terrain features can be "seen" from a specified location. Similarly, many GISs permit the generation of *perspective views* portraying terrain surfaces from a viewing position other than vertical.

Several constraints need to be considered when combining data from multiple sources in a GIS, as in the example illustrated in Figure 1.22. Obviously, all of the data layers must cover the same geographic area, or at least partially overlap. An equally important consideration is that the data layers must share the same geographic coordinate system. While a GIS can convert data from one coordinate system to another, it is necessary for the user to understand the characteristics of the various types of coordinate systems that may be used.

Because the shape of the earth is approximately spherical, locations on the earth's surface are often described in an angular coordinate or *geographical* system, with latitude and longitude specified in degrees (°), minutes ('), and seconds ("). This system originated in ancient Greece, and it is familiar to many people today. Unfortunately, the calculation of distances and areas in an angular coordinate system is complex. More significantly, it is impossible to accurately represent the three-dimensional surface of the earth on the two-dimensional planar surface of a map or image without introducing distortion in one or more of the following elements: shape, size, distance, and direction. Thus, for many purposes we wish to mathematically transform angular geographical coordinates into a planar, or Cartesian (*X–Y*) coordinate system. The result of this transformation process is referred to as a *map projection*.

While many types of map projections have been defined, they can be grouped into several broad categories based either on the geometric models used or on the spatial properties that are preserved or distorted by the transformation. Geometric models for map projection include cylindrical, conic, and azimuthal or planar surfaces. From a map user's perspective, the spatial properties of map projections may be more important. A *conformal* map projection preserves angular relationships, or shapes, within local areas; over large areas, angles and shapes become distorted. An *azimuthal* (or *zenithal*) projection preserves absolute directions relative to the central point of projection. An *equidistant* projection preserves equal distances, for some but not all points—scale is constant either for all distances along meridians or for all distances from one or two points. An *equal-area* (or *equivalent*) projection preserves equal areas. Since a detailed explanation of the relationships among these properties is beyond the scope of this discussion, suffice it to say that no two-dimensional map projection can accurately preserve all of these properties, but certain subsets of these characteristics can be preserved in a single projection. For example, the azimuthal equidistant projection preserves both direction and distance—but only relative to the central point of the projection; directions and distances between other points are not preserved.

In addition to the map projection associated with a given GIS data layer, it is also often necessary to consider the *datum* used for that map projection. A datum is a mathematical definition of the three-dimensional solid (generally a slightly flattened ellipsoid) used to represent the surface of the earth. The actual planet itself has an irregular shape that does not correspond perfectly to any ellipsoid. As a result, a variety of different datums have been described, some designed to fit the surface well in one particular region (such as the North American Datum of 1927, or NAD27) and others designed to best approximate the planet as a whole. Some GISs require that both a map projection and a datum be specified before performing any coordinate transformations.

Beyond the need for all data layers to share the same coordinate system, there are other more subtle constraints on the use of multiple layers in a GIS. One such constraint is that the spatial scale at which each of the original source maps was compiled must be compatible with the others. For example, in the analysis shown in Figure 1.22 it would be inappropriate to incorporate both soil data from very high resolution aerial photographs of a single township and land cover digitized from a highly generalized map of the entire nation. Another common constraint is that the compilation dates of different source maps must be reasonably close in time. For example, a GIS analysis of wildlife habitat might yield incorrect conclusions if it were based on land cover data that are many years out of date. On the other hand, since other types of spatial data are less changeable over time, the map compilation date might not be as important for a layer such as topography or bedrock geology.

Most GISs use one of two primary approaches to represent the locational component of geographic information: a *raster* (grid cell) or *vector* (polygon) format. The raster data model that was used in our soil erosion example is illustrated in Figure 1.23*b*. In this approach, the location of geographic objects or conditions is defined by the row and column position of the cells they occupy. The value stored for each cell indicates the type of object or condition that is found at that location over the entire cell. Note that the finer the grid cell size used, the more geographic specificity there is in the data file. A coarse grid requires less data storage space but will provide a less accurate geographic description of the original data. Also, when using a very coarse grid, several data types and/or attributes may occur in each cell, but the cell is still usually treated as a single homogeneous unit during analysis.

The vector data model is illustrated in Figure 1.23*c*. Using this format, feature boundaries are converted to straight-sided polygons that approximate the original regions. These polygons are encoded by determining the coordinates of their vertices, called *nodes*, which can be connected to form *arcs*. *Topological coding* includes "intelligence" in the data structure relative to the spatial relationship (connectivity and adjacency) among features. For example, topological coding keeps track of which arcs share common nodes and what polygons are to the left and right of a given arc. This information facilitates such spatial operations as overlay analysis, buffering, and network analysis.

Raster and vector data models each have their advantages and disadvantages. Raster systems tend to have simpler data structures; they afford greater computational efficiency in such operations as overlay analysis; and they represent features having high spatial variability and/or "blurred boundaries" (e.g., between pure and mixed vegetation zones) more effectively. On the other hand, raster data volumes are relatively greater; the spatial resolution of the data is limited to the size of the cells comprising the raster; and the topological relationships among spatial features are more difficult to represent. Vector data formats have the advantages of relatively lower data volumes, better spatial resolution, and the preservation of topological data relationships (making such operations as network analysis more efficient). However,

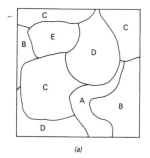
(a)

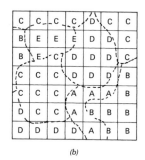
(b)

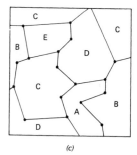
(c)

Figure 1.23 Raster versus vector data formats: (*a*) original line map, (*b*) raster format, and (*c*) vector format.

certain operations (e.g., overlay analysis) are more complex computationally in a vector data format than in a raster format.

As we discuss frequently throughout this book, digital remote sensing images are collected in a raster format. Accordingly, digital images are inherently compatible spatially with other sources of information in a raster domain. Because of this, "raw" images can be easily included directly as layers in a raster-based GIS. Likewise, such image processing procedures as automated land cover classification (Chapter 7) result in the creation of interpreted or derived data files in a raster format. These derived data are again inherently compatible with the other sources of data represented in a raster format. This concept is illustrated in Plate 1, in which we return to our earlier example of using overlay analysis to assist in soil erosion potential mapping, for an area in western Dane County, Wisconsin. Shown in (a) is an automated land cover classification that was produced by processing Landsat Thematic Mapper (TM) data of the area. (See Chapter 7 for additional information on computer-based land cover classification.) To assess the soil erosion potential in this area, the land cover data were merged with information on the intrinsic erodibility of the soil present (b) and with land surface slope information (c). These latter forms of information were already resident in a GIS covering the area. Hence, all data could be combined for analysis in a mathematical model, producing the soil erosion potential map shown in (d). To assist the viewer in interpreting the landscape patterns shown in Plate 1, the GIS was also used to visually enhance the four data sets with topographic shading based on a digital elevation model, providing a three-dimensional appearance.

For the land cover classification in Plate 1a, water is shown as dark blue, nonforested wetlands as light blue, forested wetlands as pink, corn as orange, other row crops as pale yellow, forage crops as olive, meadows and grasslands as yellow-green, deciduous forest as green, evergreen forest as dark green, low-intensity urban areas as light gray, and high-intensity urban areas as dark gray. In (b), areas of low soil erodibility are shown in dark brown, with increasing soil erodibility indicated by colors ranging from orange to tan. In (c), areas of increasing steepness of slope are shown as green, yellow, orange, and red. The soil erosion potential map (d) shows seven colors depicting seven levels of potential soil erosion. Areas having the highest erosion potential are shown in dark red. These areas tend to have row crops growing on inherently erodible soils with sloping terrain. Decreasing erosion potential is shown in a spectrum of colors from orange through yellow to green. Areas with the lowest erosion potential are indicated in dark green. These include forested regions, continuous-cover crops, and grasslands, growing on soils with low inherent erodibility, and flat terrain.

One might be tempted to conclude from the above example that remote sensing data are only useful in a GIS context if they are digital in character and the GIS is raster based. However, most GISs support conversion between raster and vector formats as well as the simultaneous *integration* of raster and vector data. Hence, a raster image can be displayed as a backdrop for a vector overlay to the image (Figure 1.24). In this way the image can be used as a pictorial mapping base for the vector overlay and the overlay can also be updated or edited based on information interpreted from the image. Image backdrops are also frequently used in the collection of GPS locational data or for vehicle navigation (wherein the position of the GPS receiver is shown on the image).

Remote sensing images certainly need not be digital in format to be of value in a GIS environment. Visual interpretation of hardcopy images is used

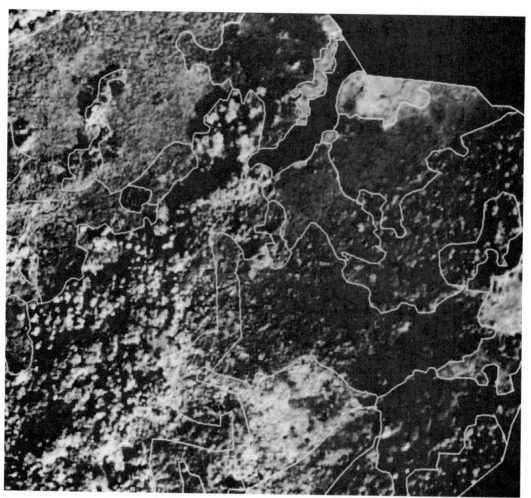

Figure 1.24 Vector overlay to a raster image. The raster image is a black and white copy of a color infrared photograph that was converted to a raster image by scanning the photograph in a microdensitometer (Section 2.5). The vector data (lines) outline wetland vegetation types.

extensively to locate specific features and conditions, which are then subsequently geocoded for inclusion in a GIS. At the same time, the information resident in a GIS can also be used to aid in a visual or digital image interpretation process. For example, GIS information on elevation, slope, and aspect might be used to aid in the classification of forest types appearing in images acquired over areas of high relief. Thus, the interaction between remote sensing and GIS techniques is two-way in nature.

Remote sensing images (and information extracted from such images), along with GPS data, have become primary data sources for modern GISs. Indeed, the boundaries between remote sensing, GIS, and GPS technology have become blurred, and these combined fields will continue to revolutionize how we inventory, monitor, and manage natural resources on a day-to-day basis. Likewise, these technologies are assisting us in modeling and understanding biophysical processes at all scales of inquiry. They are also permitting us to develop and communicate cause-and-effect "what-if" scenarios in a spatial context in ways never before possible.

The anticipated importance of remote sensing, GIS, GPS, and related information technologies in the professional careers of today's students involved in measuring, studying, and managing earth resources cannot be overstated. Those students or professionals interested in obtaining further information about GIS (or GPS) methods are encouraged to consult any one of the several excellent texts and references on these subjects in the Selected Bibliography. Here, we emphasize the principles of remote sensing and image interpretation and do not attempt to provide further background on the subjects of GIS or GPS methods. However, we do emphasize and illustrate the interrelationship between remote sensing, GIS, and GPS throughout this discussion.

1.12 ORGANIZATION OF THE BOOK

Because of the breadth of the discussion of remote sensing systems and analysis techniques presented in this book, it is important for the student to have a mental road map of the relationships among the various chapters. Chapters 2 to 4 deal primarily with photographic remote sensing. Chapter 2 describes the basic tools used in acquiring aerial photographs: films, filters, and aerial cameras. We also describe how to derive digital images from aerial photographs in this chapter. Digital imaging and videography are also treated in Chapter 2. Chapter 3 describes the photogrammetric procedures by which precise spatial measurements, maps, digital elevation models (DEMs), orthophotos, and other derived products are made from airphotos. Chapter 4 presents the basics of visual image interpretation in a broad range of applications, with an emphasis on airphoto interpretation (but also including satellite image interpretation examples).

Discussion of nonphotographic systems begins in Chapter 5, which describes the acquisition of airborne multispectral, thermal, and hyperspectral data. In Chapter 6 we describe the principal satellite systems used to collect reflected and emitted data on a global basis. These include the Landsat, SPOT, and other earth resource satellites, as well as certain meteorological satellites. Chapter 7 treats the subject of digital image processing and describes the most commonly employed procedures through which computer-assisted image interpretation is accomplished.

Chapter 8 is concerned with the collection and analysis of radar and lidar data. Both airborne and spaceborne radar systems are discussed. Included in this latter category are such systems as the ALOS, Envisat, ERS, JERS, Radarsat, and ICESat satellite systems.

In essence, this book progresses from the simplest sensing systems to the more complex. In a sense, there is also a progression from short to long wavelengths along the electromagnetic spectrum (see Figure 1.3). That is, discussion centers on photography in the UV, visible, and near-IR regions, multispectral scanning (including thermal scanning in the "heat" region), and radar sensing in the microwave region.

Throughout this book, the International System of Units (SI) is used. Tables are included on the back cover to assist the reader in converting between SI and units of other measurement systems.

Finally, three appendices are also included in the back of the book. Appendix A summarizes the various concepts, terms, and units commonly used in radiation measurement in remote sensing. Appendix B lists various sources of remote sensing data as well as remote sensing periodicals, online glossaries, and online courses and tutorials. Appendix C includes sample coordinate transformation and resampling procedures used in digital image processing.

SELECTED BIBLIOGRAPHY

Albert, D.P., W.M. Gesler, and B. Levergood, *Spatial Analysis, GIS and Remote Sensing: Applications in the Health Sciences*, Taylor & Francis, New York, 2000.

American Society for Photogrammetry and Remote Sensing (ASPRS), *Glossary of the Mapping Sciences*, ASPRS, Bethesda, MD, 1994. (Co-published by the American Congress on Surveying and Mapping, and the American Society of Civil Engineers.)

American Society for Photogrammetry and Remote Sensing, *Remote Sensing for the Earth Sciences, Manual of Remote Sensing*, 3rd ed., Vol. 3, Wiley, New York, 1999.

American Society for Photogrammetry and Remote Sensing (ASPRS), *Digital Elevation Model Technolo-*

gies and Applications: The DEM Users Manual, ASPRS, Bethesda, MD, 2001.

American Society for Photogrammetry and Remote Sensing, *Remote Sensing Core Curriculum*, on the Internet at http://www.asprs.org.

Anderson, J.E., and E.I. Robbins, "Spectral Reflectance and Detection of Iron-Oxide Precipitates Associated with Acid Mine Drainage," *Photogrammetric Engineering and Remote Sensing*, vol. 64, no. 12, 1998, pp. 1201–1208.

Atkinson, P.M., and P.J. Curran, "Choosing an Appropriate Spatial Resolution for Remote Sensing Investigations," *Photogrammetric Engineering and Remote Sensing*, vol. 63, no. 12, 1998, pp. 1345–1351.

Atkinson, P.M., and J.J. Tate, *Advances in Remote Sensing and GIS Analysis*, Wiley, New York, 2000.

Avery, T.E., and G.L. Berlin, *Fundamentals of Remote Sensing and Airphoto Interpretation*, 6th ed., Prentice Hall, Upper Saddle River, NJ, 2004.

Barrett, E.C., and L.F. Curtis, *Introduction to Environmental Remote Sensing*, Stanley Thornes Publishers, Cheltenham, Glos., UK, 1999.

Bauer, M.E., et al., "Field Spectroscopy of Agricultural Crops," *IEEE Transactions on Geoscience and Remote Sensing*, vol. GE-24, no. 1, 1986, pp. 65–75.

Bolstad, P., *GIS Fundamentals*, Eider Press, White Bear Lake, MN, 2002.

Bossler, J.D., et al., eds., *Manual of Geospatial Science and Technology*, Taylor & Francis, New York, 2001.

Bowker, D.E., et al., *Spectral Reflectances of Natural Targets for Use in Remote Sensing Studies*, NASA Ref. Publ. 1139, National Technical Information Service, Springfield, VA, 1985.

Brivio, P.A., et al., "Integration of Remote Sensing Data and GIS for Accurate Mapping of Flooded Areas," *International Journal of Remote Sensing*, vol. 23, no. 3, 2002, pp. 429–441.

Brown, S.W., et al., "Radiometric Characterization of Field Radiometers in Support of the 1997 Lunar Lake, Nevada, Experiment to Determine Surface Reflectance and Top-of-Atmosphere Radiance," *Remote Sensing of Environment*, vol. 77, no. 3, 2001, pp. 367–376.

Bukata, R.P., et al., *Optical Properties and Remote Sensing of Inland and Coastal Waters*, CRC Press, New York, 1995.

Burrough, P.A., *Principles of Geographical Information Systems for Land Resources Assessment*, 2nd ed., Oxford University Press, New York, 1998.

Campbell, G.S., and J.M. Norman, *An Introduction to Environmental Biophysics*, 2nd ed., Springer, New York, 1997.

Campbell, J.B., *Introduction to Remote Sensing*, 3rd ed., Guilford, New York, 2002.

Canadian Journal of Remote Sensing, Special Issue on Remote Sensing Calibration/Validation, vol. 23, no. 4, 1997.

Chavez, P.S., "Image-Based Atmospheric Corrections Revisited and Improved," *Photogrammetric Engineering and Remote Sensing*, vol. 62, no. 9, 1996, pp. 1025–1036.

Chen, X.W., "Using Remote Sensing and GIS to Analyse Land Cover Change and Its Impacts on Regional Sustainable Development," *International Journal of Remote Sensing*, vol. 23, no. 1, 2002, pp. 107–124.

Chrisman, N.R., *Exploring Geographic Information Systems*, 2nd ed., Wiley, New York, 2002.

Colwell, R.N., et al., "Basic Matter and Energy Relationships Involved in Remote Reconnaissance," *Photogrammetric Engineering*, vol. 29, no. 5, 1963, pp. 761–799.

Cracknell, A.P., and L.W.B. Hayes, *Introduction to Remote Sensing*, Taylor & Francis, New York, 1991.

Danson, F.M., "Teaching the Physical Principles of Vegetation Canopy Reflectance Using the SAIL Model," *Photogrammetric Engineering and Remote Sensing*, vol. 64, no. 8, 1998, pp. 809–812.

DeMers, M.N., *Fundamentals of Geographic Information Systems*, 2nd ed., Wiley, New York, 2000.

Drury, S.A., *Images of the Earth: A Guide to Remote Sensing*, 2nd ed., Oxford University Press, Oxford, UK, 1998.

Duggin, M.J., and C.J. Robinove, "Assumptions Implicit in Remote Sensing Data Acquisition and Analysis," *International Journal of Remote Sensing*, vol. 11, no. 10, 1990, pp. 1669–1694.

Elachi, C., *Introduction to the Physics and Techniques of Remote Sensing*, Wiley, New York, 1987.

Falkner, D., *Aerial Mapping: Methods and Applications*, Lewis Press, Boca Raton, FL, 1995.

Foresman, T.W., ed., *History of GIS*, Prentice-Hall, Upper Saddle River, NJ, 1998a.

Foresman, T.W., *The History of Geographic Information Systems: Perspectives from the Pioneers*, Prentice Hall, Upper Saddle River, NJ, 1998b.

Fotheringham, S., and P. Rogerson, *Spatial Analysis and GIS*, Taylor & Francis, New York, 1994.

Frohn, R.C., *Remote Sensing for Landscape Ecology*, Lewis Publishers, Boca Raton, FL, 1998.

Gabrynowicz, J.I., "Commercial High-Altitude Unpiloted Aerial Remote Sensing: Some Legal Considerations," *Photogrammetric Engineering and Remote Sensing*, vol. 62, no. 3, 1996, pp. 275–278.

Gao, J., "Integration of GPS with Remote Sensing and GIS: Reality and Prospect," *Photogrammetric Engineering and Remote Sensing*, vol. 68, no. 5, 2002, pp. 447–453.

Gibson, P., and C. Power, *Introductory Remote Sensing: Principles and Concepts*, Taylor & Francis, New York, 2000.

Goodchild, M., *Spatial Data Quality*, Taylor & Francis, New York, 2002.

Goodchild, M.F., and S. Gopal, *The Accuracy of Spatial Databases*, Taylor & Francis, New York, 1989.

Green, R.G., and J.F. Cruise, "Development of a Geographic Information System for Urban Watershed Analysis," *Photogrammetric Engineering and Remote Sensing*, vol. 62, no. 7, 1996, pp. 863–870.

Haines-Young, R., D.R. Green, and S. Cousins, *Landscape Ecology and GIS*, Taylor & Francis, Washington, DC, 1993.

Han, L., "Spectral Reflectance with Varying Suspended Sediment Concentrations in Clear and Algae-Laden Waters," *Photogrammetric Engineering and Remote Sensing*, vol. 63, no. 6, 1997, pp. 701–705.

Hall, F.G., et al., "Radiometric Rectification: Toward a Common Radiometric Response Among Multidate, Multisensor Images," *Remote Sensing of Environment*, vol. 35, no. 1, 1991.

Heric, M., C. Lucas, and C. Devine, "The Open Skies Treaty: Qualitative Utility Evaluations of Aircraft Reconnaissance and Commercial Satellite Imagery," *Photogrammetric Engineering and Remote Sensing*, vol. 62, no. 3, 1996, pp. 279–284.

Hsieh, P.F., L.C. Lee, and N.Y. Chen, "Effect of Spatial Resolution on Classification Errors of Pure and Mixed Pixels in Remote Sensing," *IEEE Transactions on Geoscience and Remote Sensing*, vol. 39, no. 12, 2001, pp. 2657–2663.

Huxhold, W.E., *An Introduction to Urban Geographic Information Systems*, Oxford University Press, New York, 1991.

Illiffe, J.C., *Datums and Map Projections: For Remote Sensing, GIS and Surveying*, CRC Press, Boca Raton, FL, 2000.

Jensen, J.R., *Remote Sensing of the Environment: An Earth Resource Perspective*, Prentice-Hall, Upper Saddle River, NJ, 2000.

Jupp, D.L.B., and A.H. Strahler, "A Hotspot Model for Leaf Canopies," *Remote Sensing of Environment*, vol. 38, 1991, pp. 193–210.

Kalensky, Z., and D.A. Wilson, "Spectral Signatures of Forest Trees," *Proceedings: Third Canadian Symposium on Remote Sensing*, 1975, pp. 155–171.

Kennedy, M., *The Global Positioning System and GIS, An Introduction*, Ann Arbor Press, Ann Arbor, MI, 1996.

Kennedy, M., *The Global Positioning System and GIS: An Introduction*, 2nd ed., Taylor & Francis, New York, 2002.

Kushwaha, S.P.S. et al., "Interfacing Remote Sensing and GIS Methods for Sustainable Rural Development," *International Journal of Remote Sensing*, vol. 17, no. 15, 1996, pp. 3055–3069.

Liu, Weidong, et al., "Relating Soil Surface Moisture to Reflectance," *Remote Sensing of Environment*, vol. 81, nos. 2–3, 2002, pp. 238–246.

Logicon, *Multispectral Users Guide*, Logicon Geodynamics, Inc., Fairfax, VA. 1997.

Lowell, K., and A. Jaton, *Spatial Accuracy Assessment Land Information Uncertainty in Natural Resources*, Taylor & Francis, New York, 2000.

Lowman, P.D., *Exploring Space, Exploring Earth: New Understanding of the Earth from Space Research*, Cambridge University Press, New York, 2002.

Lunetta, R.S., et al., "GIS-Based Evaluation of Salmon Habitat in the Pacific Northwest," *Photogrammetric Engineering and Remote Sensing*, vol. 63, no. 10, 1997, pp. 1219–1229.

Lyon, J.G., *GIS for Water Resources and Watershed Management*, Taylor & Francis, New York, 2003.

Marble, D.F., and H. Sazanami, eds., *The Role of Geographic Information Systems in Development Planning*, Taylor & Francis, Washington, DC, 1993.

Mikkola, J., and P. Pellikka, "Normalization of Bi-Directional Effects in Aerial CIR Photographs to Improve Classification Accuracy of Boreal and Subarctic Vegetation for Pollen-Landscape Calibration," *International Journal of Remote Sensing*, vol. 23, no. 21, 2002, pp. 4719–4742.

Milman, A.S., *Mathematical Principles of Remote Sensing*, Ann Arbor Press, Ann Arbor, MI, 1999.

Milman, A.S., *Mathematical Principles of Remote Sensing*, Taylor & Francis, New York, 2000.

Moran, M.S., et al., "Deployment and Calibration of Reference Reflectance Tarps for Use with Airborne Imaging Sensors," *Photogrammetric Engineering and Remote Sensing*, vol. 67, no. 3, 2001, pp. 273–286.

Mowrer, H.T., and R.G. Congalton, *Quantifying Spatial Uncertainty in Natural Resources*, Ann Arbor Press, Ann Arbor, MI, 2000.

Muehrcke, P.C., *Map Use: Reading, Analysis, and Interpretation*, 4th ed., J. P. Publications, Madison, WI, 1998.

Naesset, E., "Effects of Differential Single- and Dual-Frequency GPS and GLONASS Observations on Point Accuracy under Forest Canopies," *Photogrammetric Engineering and Remote Sensing*, vol. 67, no. 9, 2001, pp. 1021–1026.

Narayanan, R.M., and M.K. Desetty, "Effect of Spatial Resolution on Information Content Characterization in Remote Sensing Imagery Based on Classification Accuracy," *International Journal of Remote Sensing*, vol. 23, no. 3, 2002, pp. 537–553.

Narumalani, S., et al., "Aquatic Macrophyte Modeling Using GIS and Logistic Multiple Regression," *Photogrammetric Engineering and Remote Sensing*, vol. 63, no. 1, 1997, pp. 41–49.

Ni, W., C.E. Woodcock, and D.L.B. Jupp, "Variance in Bidirectional Reflectance over Discontinuous Plant Canopies," *Remote Sensing of Environment*, vol. 69, no. 1, 1999, pp. 1–15.

Photogrammetric Engineering and Remote Sensing, Special Issue on Geographic Information Systems, vol. 62, no. 11, 1996.

Photogrammetric Engineering and Remote Sensing, Special Issue on U.S./Mexico Border Region, vol. 64, no. 11, 1998.

Price, J.C., "Spectral Band Selection for Visible Near Infrared Remote Sensing: Spectral-Spatial Resolution Tradeoffs," *IEEE Transactions on Geoscience and Remote Sensing*, vol. 35, no. 5, 1997, pp. 1277–1285.

Quattrochi, D.A., and M.F. Goodchild, *Scale in Remote Sensing and GIS*, Lewis Publishers, Boca Raton, FL, 1997.

Ramsey, E.W., and J.R. Jensen, "Remote Sensing of Mangrove Wetlands: Relating Canopy Spectral to Site-Specific Data," *Photogrammetric Engineering and Remote Sensing*, vol. 62, no. 8, 1996, pp. 939–948.

Rees, W.G., *The Remote Sensing Data Book*, Cambridge University Press, New York, 1999.

Rees, W.G., *Physical Principles of Remote Sensing*, 2nd ed., Cambridge University Press, New York, 2001.

Remote Sensing of Environment, Special Issue on Physical Measurements and Signatures in Remote Sensing, vol. 41, nos. 2/3, 1992.

Remote Sensing of Environment, Special Issue on Spectral Signatures of Objects, vol. 68, no. 3, 1999.

Richter, R., "Correction of Atmospheric and Topographic Effects for High Spatial Resolution Satellite Imagery," *International Journal of Remote Sensing*, vol. 18, no. 5, 1997, pp. 1099–1111.

Robinson, A.H., et al., *Elements of Cartography*, 6th ed., Wiley, New York, 1995.

Sabins, F.F., Jr., *Remote Sensing—Principles and Interpretation*, 3rd ed., W.H. Freeman, New York, 1997.

Sanden, E.M., C.M. Britton, and J.H. Everitt, "Total Ground-Cover Estimates from Corrected Scene Brightness Measurements," *Photogrammetric Engineering and Remote Sensing*, vol. 62, no. 2, 1996, pp. 147–150.

Sandmeier, S.R., "Acquisition of Bidirectional Reflectance Factor Data with Field Goniometers," *Remote Sensing of Environment*, vol. 73, no. 3, 2000, pp. 257–269.

Schlagel, J.D., and C.M. Newton, "GIS-Based Statistical Method to Analyze Spatial Change," *Photogrammetric Engineering and Remote Sensing*, vol. 62, no. 7, 1996, pp. 839–844.

Schott, J.R., *Remote Sensing: The Image Chain Approach*, Oxford University Press, New York, 1997.

Short, N.M., *The Remote Sensing Tutorial Online Handbook*, on the Internet at http://rst.gsfc.nasa.gov. (CD-ROM, 1998, available from NASA Goddard Space Flight Center.)

Sims, D.A., and J.A. Gamon, "Relationships between Leaf Pigment Content and Spectral Reflectance Across a Wide Range of Species, Leaf Structure and Developmental Stages," *Remote Sensing of Environment*, vol. 81, nos. 2–3, 200, pp. 337–354.

Slater, P.N., *Remote Sensing: Optics and Optical Systems*, Addison-Wesley, Reading, MA, 1980.

Space Technology Applications and Research Program, *Remote Sensing Notes*, on the Internet at http://www.star.ait.ac.th.

Stanislawski, L.V., B.A. Dewitt, and R.L. Shrestha, "Estimating Positional Accuracy of Data Layers Within a GIS Through Error Propagation," *Photogrammetric Engineering and Remote Sensing*, vol. 62, no. 4, 1996, pp. 429–433.

Star, J.L., J.E. Estes, and K.C. McGwire, *Integration of GIS and Remote Sensing*, Cambridge University Press, Cambridge, 1997.

Swain, P.H., and S.M. Davis, eds., *Remote Sensing: The Quantitative Approach*, McGraw-Hill, New York, 1978.

Teillet, P.M., and G. Fedosejevs, "On the Dark Target Approach to Atmospheric Correction of Remotely

Sensed Data," *Canadian Journal of Remote Sensing*, 1995, pp. 374–387.

Usery, E.L., "A Feature-Based Geographic Information System Model," *Photogrammetric Engineering and Remote Sensing*, vol. 62, no. 7, 1996, pp. 833–838.

Varekamp, C., A.K. Skidmore, and P.A.B. Burrough, "Using Public Domain Geostatistical and GIS Software for Spatial Interpolation," *Photogrammetric Engineering and Remote Sensing*, vol. 62, no. 7, 1996, pp. 845–854.

Verbyla, D. L., *Practical GIS Analysis*, Taylor & Francis, New York, 2002.

Woodcock, C.W., and A.H. Strahler, "The Factor of Scale in Remote Sensing," *Remote Sensing of Environment*, vol. 21, no. 3, 1987, pp 311–332.

Worboys, M.F., *GIS: A Computing Perspective*, Taylor & Francis, New York, 1995.

Yang, Q., J. Snyder, and W. Tobler, *Map Projection Transformation Principles and Applications*, Taylor & Francis, New York, 2000.

Zhu, A., "Measuring Uncertainty in Class Assignment for Natural Resources Maps Under Fuzzy Logic," *Photogrammetric Engineering and Remote Sensing*, vol. 63, no. 10, 1997, pp. 1195–1202.

2 ELEMENTS OF PHOTOGRAPHIC SYSTEMS

2.1 INTRODUCTION

One of the most common, versatile, and economical forms of remote sensing is aerial photography. The basic advantages aerial photography affords over on-the-ground observation include:

1. **Improved vantage point.** Aerial photography gives a bird's-eye view of large areas, enabling us to see earth surface features in their spatial context. In short, aerial photography permits us to look at the "big picture" in which objects of interest reside. It is often difficult, if not impossible, to obtain this view of the environment through on-the-ground observation. With aerial photography, we also see the "whole picture" in that *all* observable earth surface features are recorded simultaneously. Completely different information might be extracted by different people looking at a photograph. The hydrologist might concentrate on surface water bodies, the geologist on bedrock structure, the agriculturalist on soil or crop type, and so on.

2. **Capability to stop action.** Unlike the human eye, photographs can give us a "stop action" view of dynamic conditions. For example, aerial photographs are very useful in studying dynamic phenomena such

58

as floods, moving wildlife populations, traffic, oil spills, and forest fires.

3. **Permanent recording.** Aerial photographs are virtually permanent records of existing conditions. As such, these records can be studied at leisure, under office rather than field conditions. A single image can be studied by a large number of users. Airphotos can also be conveniently compared against similar data acquired at previous times, so that changes over time can be monitored easily.

4. **Broadened spectral sensitivity.** Film can "see" and record over a wavelength range about twice as broad as that of the human eye (0.3 to 0.9 μm versus 0.4 to 0.7 μm). With photography, invisible UV and near-IR energy can be detected and subsequently recorded in the form of a visible image; hence film can see certain phenomena the eye cannot.

5. **Increased spatial resolution and geometric fidelity.** With the proper selection of camera, film, and flight parameters, we are able to record more spatial detail on a photograph than we can see with the unaided eye. This detail becomes available to us by viewing photographs under magnification. With proper ground reference data, we can also obtain accurate measurements of positions, distances, directions, areas, heights, volumes, and slopes from airphotos. In fact, most planimetric and topographic maps are currently produced using measurements extracted from airphotos.

This and the following two chapters detail and illustrate the above characteristics of aerial photographs. In this chapter, we describe the various materials and methods used to *acquire* aerial photographs (and digital images) and we deal with the process of *obtaining radiometric measurements* from airphotos. In Chapter 3 we examine various aspects of *measuring* and *mapping* with airphotos (photogrammetry). The topic of *visual image interpretation* of aerial and satellite images is treated in Chapter 4.

2.2 EARLY HISTORY OF AERIAL PHOTOGRAPHY

Photography was born in 1839 with the public disclosure of the pioneering photographic processes of Nicephore Niepce, William Henry Fox Talbot, and Louis Jacques Mande Daguerre. As early as 1840, Argo, Director of the Paris Observatory, advocated the use of photography for topographic surveying. The first known aerial photograph was taken in 1858 by a Parisian photographer named Gaspard-félix Tournachon. Known as "Nadar," he used a tethered balloon to obtain the photograph over Val de Bievre, near Paris. Balloon photography flourished after that. The earliest *existing* aerial photograph was taken from a balloon over Boston in 1860 by James Wallace Black (Figure 2.1). This photograph was immortalized by Oliver Wendell Holmes, who described it in the *Atlantic Monthly*, July 1863: "Boston, as the eagle and the wild goose see it, is a very different object from the same place as the solid citizen looks up at its eaves and chimneys" (Newhall, 1969).

As an outgrowth of their use in obtaining meteorological data, kites were used to obtain aerial photographs beginning in about 1882. The first aerial photograph taken from a kite is credited to an English meteorologist,

Figure 2.1 Balloon view of Boston photographed by James Wallace Black, October 13, 1860. This was one of the first aerial photographs taken in the United States. It was taken from a captive balloon, Professor Sam King's "Queen of the Air," at an altitude of approximately 365 m. The photograph shows a portion of the Boston business district and the masts of square-rigged ships in the adjacent harbor. (Courtesy J. Robert Quick, Wright Patterson AFB.)

E. D. Archibald. By 1890, A. Batut of Paris had published a textbook on the latest state of the art. In the early 1900s the kite photography of an American, G. R. Lawrence, brought him worldwide attention. On May 28, 1906, he photographed San Francisco shortly after the great earthquake and fire (Baker, 1989).

The airplane, which had been invented in 1903, was not used as a camera platform until 1908, when a photographer accompanied Wilbur Wright and took the first aerial motion pictures (over Le Mans, France). Obtaining aerial photographs became a much more practical matter with the airplane than it had been with kites and balloons. Photography from aircraft received heightened attention in the interest of military reconnaissance during World War I, when over one million aerial reconnaissance photographs were taken. After World War I, former military photographers founded aerial survey companies, and widespread aerial photography of the United States began. In 1934, the American Society of Photogrammetry (now the American Society for Photogrammetry and Remote Sensing) was founded as a scientific and professional organization dedicated to advancing this field.

2.3   BASIC NEGATIVE-TO-POSITIVE PHOTOGRAPHIC SEQUENCE

Many photographic procedures, particularly black and white techniques, employ a two-phase negative-to-positive sequence. In this process, the "negative" and "positive" materials are typically film and paper prints. Each of these materials consists of a light-sensitive photographic *emulsion* coated onto a *base*. The generalized cross sections of black and white film and print paper are shown in Figures 2.2a and b. In both cases, the emulsion consists of a thin layer of light-sensitive silver halide crystals, or grains, held in place by a solidified gelatin. Paper is the base material for paper prints. Various plastics are used for film bases. When exposed to light, the silver halide crystals within an emulsion undergo a photochemical reaction forming an invisible *latent image*. Upon treatment with suitable agents in the *development process*, these exposed silver salts are reduced to silver grains that appear black, forming a visible image.

The negative-to-positive sequence of black and white photography is depicted in Figure 2.3. In Figure 2.3a, the letter *F* is shown to represent a scene that is imaged through a lens system and recorded as a latent image on a film. When processed, the film crystals exposed to light are reduced to silver. The number of crystals reduced at any point on the film is proportional to the exposure at that point. Those areas on the negative that were not exposed are clear after processing because crystals in these areas are dissolved as part of the development process. Those areas of the film that were exposed become various shades of gray, depending on the amount of exposure. Hence a "negative" image of reversed tonal rendition is produced. In Figure 2.3b the negative is illuminated and reprojected through an enlarger lens so that it is focused on

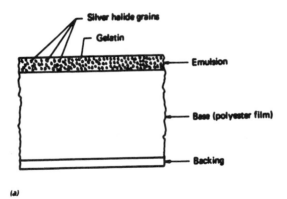

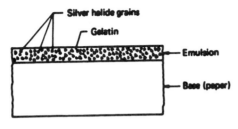

Figure 2.2 Generalized cross section of black and white photographic materials: (*a*) film and (*b*) print paper. (Adapted from Eastman Kodak Company, 1992.)

print paper, again forming a latent image. When processed, the paper print produces dark areas where light was transmitted through the negative and light areas where the illuminating light was decreased by the negative. The final result is a realistic rendering of the original scene whose size is determined by the enlarger setup. In the two-phase process of creating the final image, the negative provides an image of reversed geometry (left for right and top for bottom) and reversed brightness (light for dark and dark for light). The positive image gives a second reversal and, thus, true relative scene geometry and brightness.

Most aerial photographic paper prints are produced using the negative-to-positive sequence and a *contact printing procedure* (Figure 2.3*c*). Here, the film is exposed and processed as usual, resulting in a negative of reversed scene geometry and brightness. The negative is then placed in emulsion-to-emulsion contact with print paper. Light is passed through the negative, thereby exposing the print paper. When processed, the image on the print is a positive representation of the original ground scene at the size of the negative.

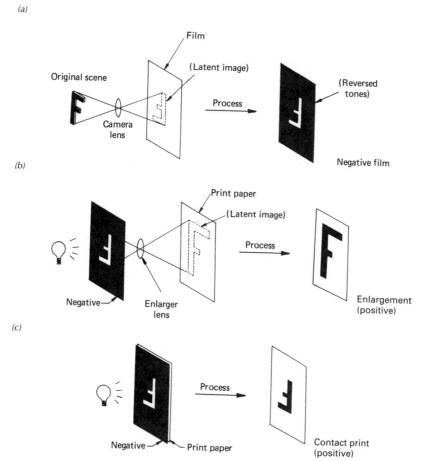

Figure 2.3 Negative-to-positive sequence of photography: (a) negative film exposure, (b) paper print enlargement, and (c) contact printing.

Positive images need not be printed on print paper. For example, transparent positives are often made on plastic-based or glass-based emulsions. These types of images are referred to as *diapositives* or *transparencies*.

2.4 FILM EXPOSURE

The exposure at any point on a photographic film depends on several factors, including the scene brightness, the diameter of the camera lens opening, the exposure time, and the camera lens focal length. In this section, we describe the interrelationships among these factors. We also describe various geometric factors influencing film exposure.

The Simple Camera

The cameras used in the early days of photography were often no more than a light-tight box with a pinhole at one end and the light-sensitive material to be exposed positioned against the opposite end (Figure 2.4*a*). The amount of exposure of the film was controlled by varying the time the pinhole was allowed to pass light. Often, exposure times were in hours because of the low sensitivity of the photographic materials available and the limited light-gathering capability of the pinhole design. In time, the pinhole camera was replaced by the simple lens camera, shown in Figure 2.4*b*. By replacing the pinhole with a lens, it became possible to enlarge the hole through which light rays from an object were collected to form an image, thereby allowing more light to reach the film in a given amount of time. In addition to the lens, an adjustable *diaphragm* and an adjustable *shutter* were introduced. The diaphragm controls the diameter of the lens opening during film exposure, and the shutter controls the duration of exposure. A more detailed illustration of these features can be seen later in Figure 2.28.

The design and function of modern adjustable cameras is conceptually identical to that of the early simple lens camera. To obtain sharp, properly exposed photographs with such systems, they must be focused and the proper exposure settings must be made. We shall describe each of these operations separately.

Focus

Three parameters are involved in focusing a camera: the focal length of the camera lens, *f*, the distance between the lens and the object to be pho-

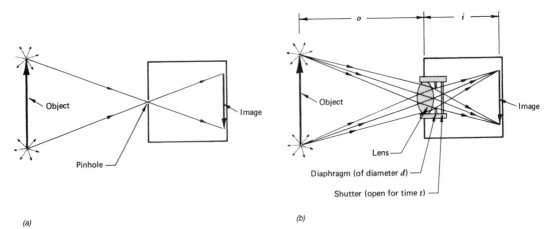

(a) *(b)*

Figure 2.4 Comparison between (*a*) pinhole and (*b*) simple lens cameras.

tographed, o, and the distance between the lens and the image plane, i. The focal length of a lens is the distance from the lens at which parallel light rays are focused to a point. (Light rays coming from an object at an infinite distance are parallel.) Object distance o and image distance i are shown in Figure 2.4b. When a camera is properly focused, the relationship among the focal length, object distance, and image distance is

$$\frac{1}{f} = \frac{1}{o} + \frac{1}{i} \qquad (2.1)$$

Since f is a constant for any given lens, as object distance o for a scene changes, image distance i must change. This is done by moving the camera lens with respect to the film plane. When focused on an object at a discrete distance, a camera can image over a range just beyond and in front of this distance with acceptable focus. This range is commonly referred to as the *depth of field*.

In aerial photography the object distances involved are effectively infinite. Hence the $1/o$ term in Eq. 2.1 goes to zero and i must equal f. Thus, most aerial cameras are manufactured with their film plane precisely located at a *fixed* distance f from their lens.

Exposure

The exposure[1] at any point in the film focal plane of a camera is determined by the irradiance at that point multiplied by the exposure time, expressed by

$$E = \frac{sd^2t}{4f^2} \qquad (2.2)$$

where

    E = film exposure, J mm^{-2}
    s = scene brightness, J mm^{-2} sec^{-1}
    d = diameter of lens opening, mm
    t = exposure time, sec
    f = lens focal length, mm

It can be seen from Eq. 2.2 that, for a given camera and scene, the exposure reaching a film can be varied by changing the camera shutter speed t and/or the diameter of the lens opening d. Various combinations of d and t will yield equivalent exposures.

---

[1]The internationally accepted symbol for exposure is H. To avoid confusion with the use of this symbol for flying height, we use E to represent "exposure" in our discussion of photographic systems. Elsewhere, E is used as the internationally accepted symbol for "irradiance."

EXAMPLE 2.1

A film in a camera with a 40-mm-focal-length lens is properly exposed with a lens opening diameter of 5 mm and an exposure time of $\frac{1}{125}$ sec (condition 1). If the lens opening is increased to 10 mm and the scene brightness does not change, what exposure time should be used to maintain proper exposure (condition 2)?

Solution
We wish to maintain the same exposure for conditions 1 and 2. Hence,

$$E_1 = \frac{s_1(d_1)^2 t_1}{4(f_1)^2} = \frac{s_2(d_2)^2 t_2}{4(f_2)^2} = E_2$$

Canceling constants, we obtain

$$(d_1)^2 t_1 = (d_2)^2 t_2$$

or

$$t_2 = \frac{(d_1)^2 t_1}{(d_2)^2} = \frac{5^2}{10^2} \cdot \frac{1}{125} = \frac{1}{500} \text{ sec}$$

The diameter of the lens opening of a camera is determined by adjusting the diaphragm to a particular *aperture setting*, or *f-stop*. This is defined by

$$F = \text{f-stop} = \frac{\text{lens focal length}}{\text{lens opening diameter}} = \frac{f}{d} \qquad (2.3)$$

As can be seen in Eq. 2.3, as the f-stop number increases, the diameter of the lens opening decreases and, accordingly, the film exposure decreases. Because the *area* of the lens opening varies as the square of the diameter, the change in exposure with f-stop is proportional to the square root of the f-stop. Shutter speeds are normally established in sequential multiples of 2 ($\frac{1}{125}$ sec, $\frac{1}{250}$ sec, $\frac{1}{500}$ sec, $\frac{1}{1000}$ sec, ...). Thus, f-stops vary as the square root of 2 (f/1.4, f/2, f/2.8, f/4, ...). Note that when the value of the f-stop is 2, it is written as f/2.

The interplay between f-stops and shutter speeds is well known to photographers. For constant exposure, an incremental change in shutter speed setting must be accompanied by an incremental change in f-stop setting. For example, the exposure obtained at $\frac{1}{500}$ sec and f/1.4 could also be obtained at $\frac{1}{250}$ sec and f/2. Short exposure times allow one to "stop action" and prevent blurring when photographing moving objects (or when the camera is moving, as in the case of aerial photography). Large lens-opening diameters (small f-stop numbers) allow more light to reach the film and are useful under low light conditions. Small lens-opening diameters (large f-stop numbers) yield greater depth of field. The f-stop corresponding to the largest lens opening diameter is called the *lens speed*. The larger the lens-opening diameter (smaller f-stop number), the "faster" the lens is.

Using f-stops, Eq. 2.2 can be simplified to

$$E = \frac{st}{4F^2} \tag{2.4}$$

where F = f-stop = f/d.

Equation 2.4 is a convenient means of summarizing the interrelationship among film exposure, scene brightness, exposure time, and f-stop. This relationship may be used in lieu of Eq. 2.2 to determine various f-stop and shutter speed settings that result in identical film exposures.

EXAMPLE 2.2

A film is properly exposed when the lens aperture setting is f/8 and the exposure time is $\frac{1}{125}$ sec (condition 1). If the lens aperture setting is changed to f/4, and the scene brightness does not change, what exposure time should be used to yield a proper film exposure (condition 2)? (Note that this is simply a restatement of the condition of Example 2.1.)

Solution

We wish to maintain the *same exposure* for conditions 1 and 2. With the scene brightness the same in each case,

$$E_1 = \frac{s_1 t_1}{4(F_1)^2} = \frac{s_2 t_2}{4(F_2)^2} = E_2$$

Canceling constants,

$$\frac{t_1}{(F_1)^2} = \frac{t_2}{(F_2)^2}$$

and

$$t_2 = \frac{t_1 (F_2)^2}{(F_1)^2} = \frac{1}{125} \cdot \frac{4^2}{8^2} = \frac{1}{500} \text{ sec}$$

Geometric Factors Influencing Film Exposure

Images are formed on film because of variations in scene brightness values over the area photographed. Ideally, in aerial photography, such variations would be related solely to variations in ground object type and/or condition. This assumption is a great oversimplification since many factors that have nothing to do with the type or condition of a ground feature can and do influence film exposure measurements. Because these factors influence exposure measurements but have nothing to do with true changes in ground cover type or condition, we term them *extraneous effects*. Extraneous effects are of two general types: geometric and atmospheric. Atmospheric effects were introduced in Section 1.3; here we discuss the major geometric effects that influence film exposure.

Probably the most important geometric effect influencing film exposure is *exposure falloff*. This extraneous effect is a variation in focal plane exposure purely associated with the distance an image point is from the image center. Because of falloff, *a ground scene of spatially uniform reflectance does not produce spatially uniform exposure in the focal plane*. Instead, for a uniform ground scene, exposure in the focal plane is at a maximum at the center of the film format and decreases with radial distance from the center.

The factors causing falloff are depicted in Figure 2.5, which shows a film being exposed to a ground area assumed to be of uniform brightness. For a beam of light coming from a point directly on the optical axis, exposure E_0 is directly proportional to the area, A, of the lens aperture and inversely proportional to the square of the focal length of the lens, f^2. However, for a beam exposing a point at an angle θ off the optical axis, exposure E_θ is reduced from E_0 for three reasons:

1. The effective light-collecting area of the lens aperture, A, decreases in proportion to $\cos \theta$ when imaging off-axis areas ($A_\theta = A \cos \theta$).

2. The distance from the camera lens to the focal plane, f_θ, increases as $1/\cos \theta$ for off-axis points, $f_\theta = f/\cos \theta$. Since exposure varies inversely as the square of this distance, there is an exposure reduction of $\cos^2 \theta$.

3. The effective size of a film area element, dA, projected perpendicular to the beam decreases in proportion to $\cos \theta$ when the element is located off-axis, $dA_\theta = dA \cos \theta$.

Combining the above effects, the overall theoretical reduction in film exposure for an off-axis point is

$$E_\theta = E_0 \cos^4 \theta \qquad (2.5)$$

where
θ = angle between the optical axis and the ray to the off-axis point
E_θ = film exposure at the off-axis point
E_0 = exposure that would have resulted if the point had been located at the optical axis

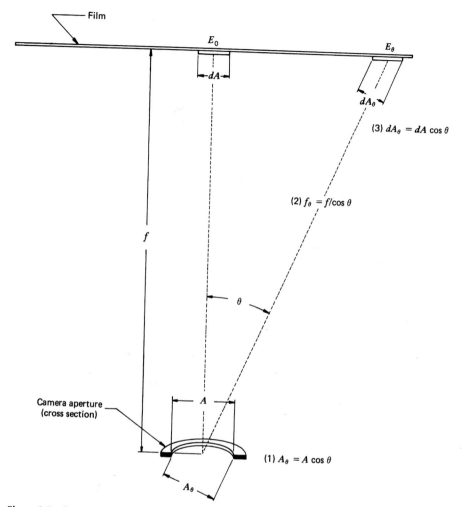

Figure 2.5 Factors causing exposure falloff.

The systematic effect expressed by the above equation is compounded by differential transmittance of the lens and by *vignetting effects* in the camera optics. Vignetting refers to internal shadowing resulting from the lens mounts and other aperture surfaces within the camera. The effect of vignetting varies from camera to camera and varies with aperture setting for any given camera.

Falloff and vignetting are normally mitigated at the time of exposure by using antivignetting filters (see Section 2.9). When such filters are not used, or when they fail to negate the exposure variations completely, it is appropriate to correct off-axis exposure values by normalizing them to the value they would possess had they been at the center of the photograph. This is done through the application of a correction model that is determined (for a given f-stop) by a radiometric calibration of the camera. This calibration essentially involves photographing a scene of uniform brightness, measuring exposure at various θ locations, and identifying the relationship that best describes the falloff. For most cameras this relationship takes on the form

$$E_\theta = E_0 \cos^n \theta \tag{2.6}$$

Because modern cameras are normally constructed in such a way that their actual falloff characteristics are much less severe than the theoretical $\cos^4$ falloff, n in the above equation is normally in the range 1.5 to 4. All exposure values measured off-axis are then corrected in accordance with the falloff characteristics of the particular camera in use.

The location of an object within a scene can also affect the resulting film exposure, as illustrated in Figures 2.6 and 2.7. Figure 2.6 illustrates the relationships that exist among *solar elevation*, *azimuth angle*, and *camera viewing angle*. Figure 2.7 illustrates geometric effects that can influence the apparent

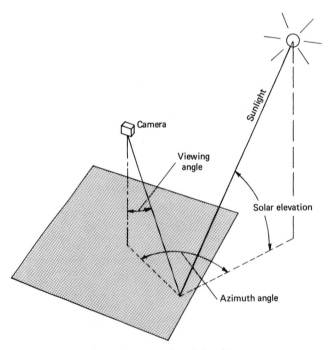

Figure 2.6 Sun–object–image angular relationship.

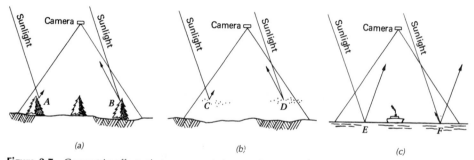

Figure 2.7 Geometric effects that cause variations in focal plane irradiance: (a) differential shading, (b) differential scattering, and (c) specular reflection.

reflectance and exposure. In (a), the effect of *differential shading* is illustrated in profile view. Because of relief displacement (Section 3.6), vertical features are imaged in an aerial photograph slightly in side view as well as in top view. Because the sides of features may be either sunlit or shaded, varied exposures can result from identical ground objects at different locations in the image. The film receives more energy from the sunlight side of the tree at B than from the shaded side of the tree at A. Differential shading is clearly a function of solar elevation and object height, with a stronger effect at low solar angles. The effect is also compounded by differences in slope and aspect (slope orientation) over terrain of varied relief. (Also see BRDF discussion, Section 1.6.)

Figure 2.7b illustrates the effect of *differential atmospheric scattering*. Backscatter from atmospheric molecules and particles adds light to that reflected from ground features. The film receives more atmospheric backscatter from area D than from area C due to geometric relationships. In some analyses, the variation in this "airlight," or path radiance, component is small and can be ignored. However, under hazy conditions, differential quantities of airlight often result in varied exposure across a photograph.

Yet another problem in many analyses is the presence of *specular* reflections in a scene. Photographs taken over water bodies often show areas of specular reflection. They represent the extreme in directional reflectance. Figure 2.7c illustrates the geometric nature of this problem. Immediately surrounding point E on the image, a considerable increase in exposure would result from specular reflection. This is illustrated in Figure 2.8, which shows areas of specular reflection from the right half of the lake shown in the image. These mirrorlike reflections normally contribute little information about the true character of the objects involved. For example, the small water bodies just below the larger lake take on a tone similar to that of some of the fields in the area. Because of the low information content of specular reflections, they are avoided in most analyses.

Figure 2.8 Aerial photograph containing areas of specular reflection from water bodies. This image is a portion of a summertime photograph taken over Green Lake, Green Lake County, WI. Scale 1 : 95,000. Cloud shadows indicate direction of sunlight at time of exposure. Reproduced from color IR original. (NASA image.)

2.5 FILM DENSITY AND CHARACTERISTIC CURVES

The *radiometric* characteristics of aerial photographs determine how a specific film—exposed and processed under specific conditions—responds to scene energy of varying intensity. Knowledge of these characteristics is often useful, and sometimes essential, to the process of photographic image analysis. This is particularly true when one attempts to establish a quantitative relationship between the tonal values on an image and some ground phenomenon. For example, one might wish to measure the darkness, or optical density, of a transparency at various image points in a corn field and correlate these measurements with a ground-observed parameter such as crop yield. If a correlation exists, the relationship could be used to predict crop yield based on photographic density measurements at other points in the scene. Such an effort can be successful only if the radiometric properties of the particular film under analysis are known. Even then, the analysis must be undertaken with due regard for such extraneous sensing effects as differing levels of illumination across a scene, atmospheric haze, and so on. If these factors can be sufficiently accounted for, considerable information can be extracted from the tonal levels expressed on a photograph. In short, image density measurements may sometimes be used in the process of determining the type, extent, and condition of ground objects. In this section, we discuss the interrelationships between film exposure and film density, and we explain how *characteristic curves* (plots of film density versus log exposure) are analyzed.

A photograph can be thought of as a visual record of the response of many small detectors to energy incident upon them. The energy detectors in a photographic record are the silver halide grains in the film emulsion, and the energy causing the response of these detectors is referred to as a film's *exposure*. During the instant that a photograph is exposed, the different reflected energy levels in the scene irradiate the film for the same length of time. A scene is visible on a processed film only because of the irradiance differences that are caused by the reflectance differences among scene elements. Thus, *film exposure at a point in a photograph is directly related to the reflectance of the object imaged at that point. Theoretically, film exposure varies linearly with object reflectance, with both being a function of wavelength.*

There are many ways of quantifying and expressing film exposure. Most photographic literature uses units of the form *meter-candle-second* (MCS) or *ergs/cm²*. The student first "exposed" to this subject might feel hopelessly lost in understanding unit equivalents in photographic radiometry. This comes about since many exposure calibrations are referenced to the sensitivity response of the human eye, through definition of a "standard observer." Such observations are termed *photometric* and result in photometric, rather than radiometric, units. To avoid unnecessary confusion over how exposure is measured and expressed in absolute terms, we will deal with *relative* exposures and not be directly concerned about specifying any absolute units.

The result of exposure at a point on a film, after development, is a silver deposit whose darkening, or light-stopping, qualities are systematically related to the amount of exposure at that point. One measure of the "darkness" or "lightness" at a given point on a film is *opacity O*. Since most quantitative remote sensing image analyses involve the use of negatives or diapositives, opacity is determined through measurement of film *transmittance T*. As shown in Figure 2.9, transmittance T is the ability of a film to pass light. At any given point p, the transmittance is

$$T_p = \frac{\text{light passing through the film at point } p}{\text{total light incident upon the film at point } p} \tag{2.7}$$

Opacity O at point p is

$$O_p = \frac{1}{T_p} \tag{2.8}$$

Although transmittance and opacity adequately describe the "darkness" of a film emulsion, it is often convenient to work with a logarithmic expression,

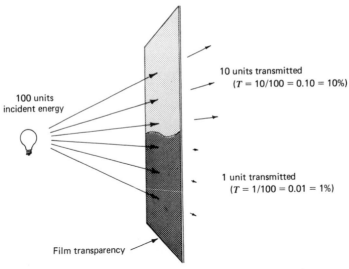

Figure 2.9   Film transmittance. To measure transmittance, a negative or positive transparency is illuminated from one side and the light transmitted through the image is measured on the other. Shown is a section of an image having a transmittance of 0.10 (or 10 percent) at one image point and 0.01 (or 1 percent) at another.

density. This is an appropriate expression, since the human eye responds to light levels nearly logarithmically. Hence, there is a nearly linear relationship between image density and its visual tone. Density D at a point p is defined as the common logarithm of film opacity at that point:

$$D_p = \log(O_p) = \log\left(\frac{1}{T_p}\right) \tag{2.9}$$

Instruments designed to measure density by shining light through film transparencies are called *transmission densitometers.* Density measurements may also be made from paper prints with a *reflectance densitometer,* but more precise measurements can be made on the original film material. When analyzing density on a transparency, the process normally involves placing the film in a beam of light that passes through it. The darker an image is, the smaller the amount of light that is allowed to pass, the lower the transmittance, the higher the opacity, and the higher the density. Some sample values of transmittance, opacity, and density are indicated in Table 2.1.

There are some basic differences between the nature of light absorptance in black and white versus color films. Densities measured on black and white film are controlled by the amount of developed silver in the image areas of measurement. In color photography, the processed image contains no silver and densities are caused by the absorption characteristics of three dye layers

TABLE 2.1 **Sample Transmittance, Opacity, and Density Values**

Percent Transmittance	T	O	D
100	1.0	1	0.00
50	0.50	2	0.30
25	0.25	4	0.60
10	0.10	10	1.00
1	0.01	100	2.00
0.1	0.001	1000	3.00

in the film: yellow, magenta, and cyan. The image analyst is normally interested in investigating the image density of each of these dye layers separately. Hence, color film densities are normally measured through each of three filters chosen to isolate the spectral regions of maximum absorption of the three film dyes.

An essential task in quantitative film analysis is to relate image density values measured on a photograph to the exposure levels that produced them. This is done to establish the cause (exposure) and effect (density) relationship that characterizes a given photograph.

Since density is a logarithmic parameter, it is convenient to also deal with exposure E in logarithmic form ($\log E$). If one plots density values as a function of the $\log E$ values that produced them, curves similar to those shown in Figure 2.10 will be obtained.

The curves shown in Figure 2.10 are for a typical black and white negative film (*a*) and a color positive film (*b*). Every film has a unique D–$\log E$ *curve*, from which many of the characteristics of the film may be determined. Because of this, these curves are known as *characteristic curves*. (Plotting D versus $\log E$ to express the nature of the photographic response was first suggested in the 1890s by Hurter and Driffield. Consequently, characteristic curves are often referred to as *H and D* as well as D–$\log E$ curves.)

Characteristic curves are different for different film types, for different manufacturing batches within a film type, and even for films of the same batch. Manufacturing, handling, storage, and processing conditions all affect the response of a film (indicated by its D–$\log E$ curve). In the case of color film, characteristic curves also differ between one emulsion layer and another.

Figure 2.11 illustrates the various film response characteristics extractable from a D–$\log E$ curve. The curve shown is typical of a black and white negative film (similar characteristics are found for each layer of a color film). There are three general divisions to the curve. First, as the exposure increases from that of point A to that of point B, the density increases from a minimum,

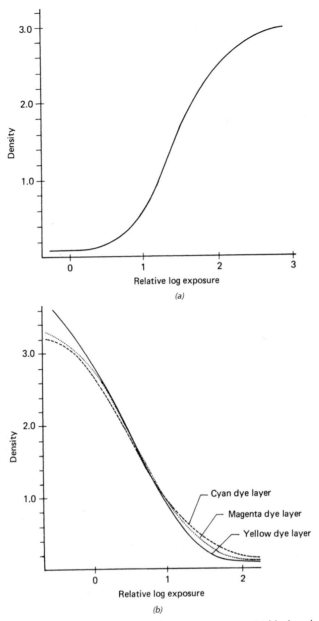

Figure 2.10 Film density versus log exposure curves: (*a*) black and white negative film; (*b*) color reversal film (positive film). (These curves are referred to as film characteristic curves, *D*–log *E* curves, or H and D curves.)

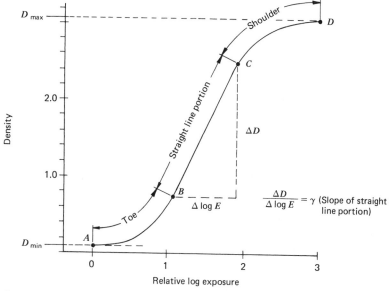

Figure 2.11   Components of a characteristic curve.

$D_{\min}$, at an increasing rate. This portion of the curve is called the *toe*. As exposure increases from point B to point C, changes in density are nearly linearly proportional to changes in log exposure. This region is called the *straight-line portion* of the curve. Finally, as log exposure increases from point C to point D, the density increases at a decreasing rate. This portion is known as the *shoulder* of the curve. The shoulder terminates at a maximum density, $D_{\max}$. Remember that this curve applies to a negative film. For a positive film (Figure 2.10*b*), the relationship is reversed. That is, density decreases with increasing exposure.

It should be noted that even in areas of a film where there is no exposure, a minimum density $D_{\min}$ results from two causes: (1) The plastic base of the film has some density D_{base} and (2) some density develops even when an unexposed emulsion is processed. This second type of density is called *fog*, or *net fog*, D_{fog}. Density $D_{\min}$ is sometimes called *gross fog* and is expressed as

$$D_{\min} = D_{\text{base}} + D_{\text{fog}} \tag{2.10}$$

The range of densities a film provides is simply the difference between $D_{\max}$ and $D_{\min}$.

Another important characteristic of D–log E curves is the slope of the linear portion of the curve. This slope is called *gamma* (γ) and is expressed as

$$\gamma = \frac{\Delta D}{\Delta \log E} \tag{2.11}$$

Gamma is an important determinant of the *contrast* of a film. While the term *contrast* has no rigid definition, in general, the higher the gamma, the higher the contrast of a film. With high contrast film, a given scene exposure range is distributed over a large density range; the reverse is true of low contrast film. For example, consider a photograph taken of a light gray and a dark gray object. On high contrast film, the two gray levels may lie at the extremes of the density scale, resulting in nearly white and nearly black images on the processed photograph. On low contrast film, both gray values would lie at nearly the same point on the density scale, showing the two objects in about the same shade of gray.

Gamma is a function of not only emulsion type but also film development conditions. For example, gamma can be varied by changing developer, development time, and/or development temperature. For any given developer, gamma is usually increased with longer development time or higher development temperature.

An important basic characteristic of a film is its *speed*, which expresses the sensitivity of the film to light. This parameter is graphically represented by the horizontal position of the characteristic curve along the $\log E$ axis. A "fast" film (more light-sensitive film) is one that will accommodate low exposure levels (i.e., it lies farther to the left on the $\log E$ axis). For a given level of scene energy, a fast film will require a shorter exposure time than will a slow film. This is advantageous in aerial photography, since it reduces image blur due to flight motion. However, high speed films are generally characterized by larger film grains, limiting the spatial resolution of images. Thus, no single film speed will be optimum in all cases.

The speed of nonaerial films is typically stated using the International Standardization Organization (ISO) system. This system is generally not used to specify the speed of aerial photographic films. Rather, for panchromatic aerial films, the American National Standard for film responsivity is the *aerial film speed* (*AFS*). By definition,

$$\text{AFS} = \frac{3}{2E_0} \tag{2.12}$$

where E_0 is the exposure (in meter-candle-seconds) at the point on the characteristic curve where the density is 0.3 above $D_{\min}$ under strictly specified processing conditions. For other processing conditions and for color and infrared-sensitive films, *effective aerial film speeds* are used to indicate sensitivity. These values are determined empirically, often by comparison with black and white films in actual flight tests.

In selecting the lens opening and shutter speed combination to give proper exposure for a given film speed, the date of photography, the latitude of the flight area, the time of day, the flight altitude, and the haze conditions must be considered.

Two other useful film characteristics can be determined from the D–$\log E$ curve. These are a film's exposure latitude and its radiometric resolution.

These characteristics are best described with reference to Figure 2.12, where the D–log E curves for two different negative films are shown.

The term *exposure latitude* expresses the range of log E values that will yield an acceptable image on a given film. For most films, good results are obtained when scenes are recorded over the linear portion of the D–log E curves and a fraction of the toe of the curve (Figure 2.12). Features recorded on the extremes of the toe or shoulder of the curve will be underexposed or overexposed. In these areas, different exposure levels will be recorded at essentially the same density, making discrimination difficult. Note in Figure 2.12 that film 2 has a much larger exposure latitude than film 1. (Also note that film 2 is a "slower" film than film 1.)

The term *exposure latitude* is also used to indicate the range of variation from the optimum camera exposure setting that can be tolerated without excessively degrading the image quality. For example, an exposure latitude of $\pm\frac{1}{2}$ stop is generally specified for certain aerial films. This means that the f-stop setting can be $\frac{1}{2}$ stop above or below the optimum setting and still produce an acceptable photograph.

As shown in Figure 2.12, *radiometric resolution* is the smallest difference in exposure that can be detected in a given film analysis. Radiometric resolution

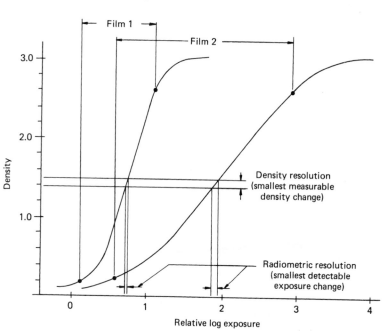

Figure 2.12 Exposure latitude and radiometric resolution of two films. (Film 2 has larger exposure latitude but poorer radiometric resolution than film 1.)

is inversely proportional to contrast, so for a given density resolvability, a higher contrast film (i.e., film 1 in Figure 2.12) is able to resolve smaller differences in exposure.

The trade-offs between contrast, exposure latitude, and radiometric resolution can now be seen. Although exceptions to the rule exist, low contrast films offer greater radiometric range (exposure latitude) at the expense of radiometric resolution. High contrast films offer a smaller exposure range but improved radiometric resolution.

Image density is measured with an instrument called a *densitometer* (or *microdensitometer* when small film areas are measured). While many varieties of densitometers exist, most have the same six basic components, shown in Figure 2.13:

1. **Light source.** Supplies energy to illuminate the image with a beam of incident radiation.

2. **Aperture assembly.** Provides for selectable spot sizes over which density measurements can be made.

3. **Filter assembly.** Allows selection of spectral bands when making density measurements on color film.

4. **Receiver.** A photoelectric device, normally a photomultiplier tube (PMT), that responds electronically to the component of the illuminating beam transmitted through the image.

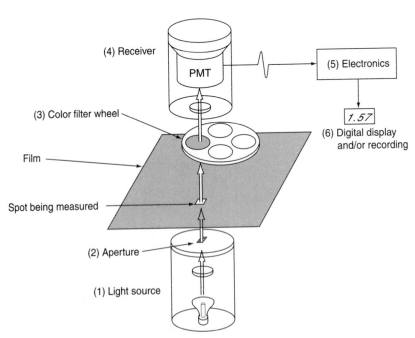

Figure 2.13 Schematic of one configuration of a densitometer.

5. **Electronics.** Amplify the output of the receiver, convert it logarithmically to a density value, and express it in a digital representation. Calibration controls enable the density readings to be referenced to a calibration standard.

6. **Digital display and/or recording.** Indicates the density value. Many densitometers have a provision for recording density values on various computer storage media.

With *spot* densitometers, different reading positions on the image are located by manually translating the image under analysis with respect to the measurement optics. These devices are convenient in applications where conventional visual interpretation is supported by taking a small number of density readings at discrete points of interest in an image. Applications calling for density measurements throughout an entire image dictate the use of *scanning densitometers*.

There are basically two types of film scanning densitometers: *rotating drum* systems and *flatbed* systems. Rotating drum scanners (Figure 2.14) accomplish the scanning task by mounting the film over a square opening in a rotating drum such that it forms a portion of the drum's circumference. The *x* coordinate scanning motion is provided by the rotation of the drum. The *y* coordinate motion comes from incremental translation of the source/receiver optics after each drum rotation. Typically, such systems employ a PMT to convert the light energy incident upon the receiver into an electronic signal.

There are several forms of flatbed film scanning systems. The most common form uses a linear array of charge-coupled devices (CCDs; Section 2.12). Optics focus the CCD array on one horizontal line of the subject at a time and step vertically to capture repeated lines.

The output from a scanning film densitometer, whether rotating drum or flatbed, is a digital image composed of pixels whose size is determined by the aperture of the source/receiving optics used during the scanning process. The output is converted to a series of digital numbers. This A-to-D conversion

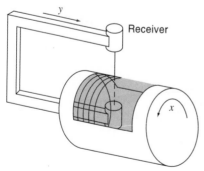

Figure 2.14 Rotating drum scanning densitometer operation.

process normally results in recording the density data on any of several forms of computer storage media.

Scanners with the highest radiometric resolution can record a film density range of 0 to 4, which means that the difference between the lightest and darkest detectable intensities is 10,000 : 1. Typically monochromatic scanning is done with a 12-bit resolution, which results in 4096 levels of gray, although 16-bit resolutions (65,536 levels) are sometimes used. Color scanning is typically 36- to 48-bit scanning (12 to 16 bits per color), resulting in many billions of colors that can be resolved.

Because scanning spot sizes as small as about 6.25 μm can be utilized with drum scanners, the process can result in immense data volumes. For example, assuming a scanning aperture and scan line spacing of 6.25 μm were used, 160 density observations per lineal millimeter in both x and y would result, yielding 25,600 observations for a 1-mm-square film area (more than one billion observations on a 230 × 230-mm film).

Figure 2.15 Z/I Imaging PhotoScan 2002 photogrammetric flatbed scanner. (Courtesy Z/I Imaging Corporation.)

Figure 2.15 illustrates a flatbed scanner that is often employed in soft-copy photogrammetric operations (Section 3.9). This particular scanner permits scanning at resolutions of 7, 14, 21, 28, 56, 112, or 224 μm over formats as large as 250 × 275 mm. Monochromatic scanning at 10-bit resolution, and color scanning at 24-bit resolution (8 bits per color), over a density range of 0.001 to 3.3, are the typical modes of scanning for this system. The geometric accuracy of this scanner is better than 2 μm in each axis of scanning.

2.6   SPECTRAL SENSITIVITY OF BLACK AND WHITE FILMS

Black and white aerial photographs are normally made with either *panchromatic* film or *infrared-sensitive* film. The generalized spectral sensitivities for each of these film types are shown in Figure 2.16. Panchromatic film has long been the "standard" film type for aerial photography. As can be seen from Figure 2.16, the spectral sensitivity of panchromatic film extends over the UV and the visible portions of the spectrum. Infrared-sensitive film is sensitive not only to UV and visible energy but also to near-IR energy.

The use of black and white IR photography to distinguish between deciduous and coniferous trees was illustrated in Figure 1.9. Many other applications of both panchromatic and black and white infrared aerial photography are described in Chapter 4. Here, we simply want the reader to become familiar with the spectral sensitivities of these materials.

It is of interest to note what determines the "boundaries" of the spectral sensitivity of black and white film materials. As indicated in Section 2.1, we

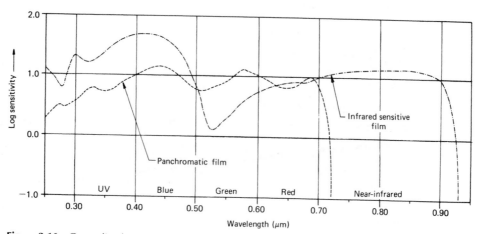

Figure 2.16   Generalized spectral sensitivities for panchromatic and black and white IR-sensitive films. (Adapted from Eastman Kodak Company, 1992.)

Figure 2.17 Comparison of panchromatic and black and white IR aerial photographs. Flooding of Bear Creek, northwest Alabama (scale 1 : 9000): (*a*) panchromatic film with a Wratten No. 12 (yellow) filter; (*b*) black and white IR film with a Wratten No. 12 filter. (Courtesy Mapping Services Branch, Tennessee Valley Authority.)

can photograph over a range of about 0.3 to 0.9 μm. The 0.9-μm limit stems from the photochemical instability of emulsion materials that are sensitive beyond this wavelength. (Certain films used for scientific experimentation are sensitive out to about 1.2 μm and form the only exception to this rule. These films are not commonly available and typically require long exposure times, making them unsuitable for aerial photography.)

Figure 2.17 shows a comparison between panchromatic and black and white IR aerial photographs. The image tones shown in this figure are very typical. Healthy green vegetation reflects much more sunlight in the near IR than in the visible part of the spectrum; therefore, it appears lighter in tone on black and white infrared photographs than on panchromatic photographs. Note, for example, that in Figure 2.17 the trees are much lighter toned in (b) than in (a). Note also that the limits of the stream water and the presence of water and wet soils in the fields can be seen more distinctly in the black and white IR photograph (b). Water and wet soils typically have a much darker tone in black and white IR photographs than in panchromatic photographs because sunlight reflection from water and wet soils in the near IR is considerably less than in the visible part of the electromagnetic spectrum.

As might be suspected from Figure 2.16, the 0.3-μm limit to photography is determined by something other than film sensitivity. In fact, virtually all photographic emulsions are sensitive in this UV portion of the spectrum. The problem with photographing at wavelengths shorter than about 0.4 μm is twofold: (1) the atmosphere absorbs or scatters much of this energy and (2) glass camera lenses absorb such energy. But photographs can be acquired in the 0.3- to 0.4-μm range if extremes of altitude and unfavorable atmospheric conditions are avoided. Furthermore, some improvement in image quality is realized if quartz camera lenses are used.

To date, the applications of aerial UV photography have been limited in number, due primarily to strong atmospheric scattering of UV energy. A notable exception is the use of UV photography in monitoring oil films on water. Minute traces of floating oil, often invisible on other types of photography, can be detected in UV photography.

2.7 COLOR FILM

Although black and white panchromatic film has long been the standard film type for aerial photography, many remote sensing applications currently involve the use of color film. The major advantage to the use of color is the fact that the human eye can discriminate many more shades of color than it can tones of gray. As we illustrate in subsequent chapters, this capability is essential in many applications of airphoto interpretation. In the remainder of this section we present the basics of how color film works. To do this, we must first consider the way in which human color vision works.

Color-Mixing Processes

Light falling on the retina of the human eye is sensed by rod and cone cells. There are about 130 million rod cells, and they are 1000 times more light sensitive than the cone cells. When light levels are low, human vision relies on the rod cells to form images. All rod cells have the same wavelength sensitivity, which peaks at about 0.55 μm. Therefore, human vision at low light levels is monochromatic. It is the cone cells that determine the colors the eye sees. There are about 7 million cone cells; some sensitive to blue energy, some to green energy, and some to red energy. The *trichromatic theory of color vision* explains that when the blue-sensitive, green-sensitive, and red-sensitive cone cells are stimulated by different amounts of light, we perceive color. When all three types of cone cells are stimulated equally, we perceive white light. Other theories of color vision have been proposed. The *opponent process of color vision* hypothesizes that color vision involves three mechanisms, each responding to a pair of so-called opposites: white–black, red–green, and blue–yellow. This theory is based on many psychophysical observations and states that colors are formed by a *hue cancellation method*. The hue cancellation method is based on the observation that when certain colors are mixed together, the resulting colors are not what would be intuitively expected. For example, when red and green are mixed together, they produce yellow, not reddish green. (For further information, see Robinson et al., 1995.)

In the remainder of this discussion, we focus on the trichromatic theory of color vision. Again, this theory is based on the concept that we perceive all colors by synthesizing various amounts of just three (blue, green, and red).

Blue, green, and red are termed *additive primaries*. Plate 2a shows the effect of projecting blue, green, and red light in partial superimposition. Where all three beams overlap, the visual effect is white because all three of the eyes' receptor systems are stimulated equally. Hence, white light can be thought of as the mixture of blue, green, and red light. Various combinations of the three additive primaries can be used to produce other colors. As illustrated, when red light and green light are mixed, yellow light is produced. Mixture of blue and red light results in the production of magenta light (bluish-red). Mixing blue and green results in cyan light (bluish-green).

Yellow, magenta, and cyan are known as the *complementary colors*, or *complements*, of blue, green, and red light. Note that the complementary color for any given primary color results from mixing the remaining two primaries.

Like the eye, color television and computer monitors operate on the principle of additive color mixing through use of blue, green, and red dots (or vertical lines) on the picture screen. When viewed at a distance, the light from the closely spaced screen elements forms a continuous color image.

Whereas color television simulates different colors through *additive* mixture of blue, green, and red *lights*, color photography is based on the principle of *subtractive* color mixture using superimposed yellow, magenta, and cyan *dyes*. These three dye colors are termed the *subtractive primaries*, and each results from subtracting one of the additive primaries from white light. That is, yellow dye absorbs the blue component of white light. Magenta dye absorbs the green component of white light. Cyan dye absorbs the red component of white light.

The subtractive color-mixing process is illustrated in Plate 2b. This plate shows three circular filters being held in front of a source of white light. The filters contain yellow, magenta, and cyan dye. The yellow dye absorbs blue light from the white background and transmits green and red. The magenta dye absorbs green light and transmits blue and red. The cyan dye absorbs red light and transmits blue and green. The superimposition of magenta and cyan dyes results in the passage of only blue light from the background. This comes about since the magenta dye absorbs the green component of the white background and the cyan dye absorbs the red component. Superimposition of the yellow and cyan dyes results in the perception of green. Likewise, superimposition of yellow and magenta dyes results in the perception of red. Where all three dyes overlap, all light from the white background is absorbed and black results.

In color photography, various proportions of yellow, magenta, and cyan dye are superimposed to control the proportionate amount of blue, green, and red light that reaches the eye. Hence, the subtractive mixture of yellow, magenta, and cyan dyes on a photograph is used to control the additive mixture of blue, green, and red light reaching the eye of the observer. To accomplish this, color film is manufactured with three emulsion layers that are sensitive to blue, green, and red light but contain yellow, magenta, and cyan dye after processing.

Structure and Spectral Sensitivity of Color Film

The basic cross-sectional structure and spectral sensitivity of color film are shown in Figure 2.18. As shown in Figure 2.18a, the top film layer is sensitive to blue light, the second layer to green and blue light, and the third to red and blue light. Because these bottom two layers have blue sensitivity as well as the desired green and red sensitivities, a blue-absorbing filter layer is introduced between the first and second photosensitive layers. This filter layer blocks the passage of blue light beyond the blue-sensitive layer. This effectively results in selective sensitization of each of the film layers to the blue, green, and red primary colors. The yellow (blue-absorbing) filter layer has no permanent effect on the appearance of the film because it is dissolved during processing.

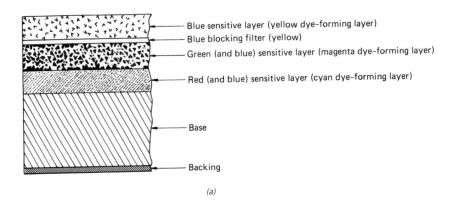

- Blue sensitive layer (yellow dye–forming layer)
- Blue blocking filter (yellow)
- Green (and blue) sensitive layer (magenta dye–forming layer)
- Red (and blue) sensitive layer (cyan dye–forming layer)
- Base
- Backing

(a)

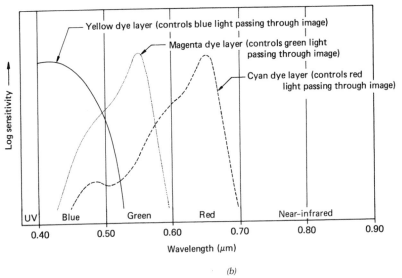

(b)

Figure 2.18 Structure and sensitivity of color film: (*a*) generalized cross section; (*b*) spectral sensitivities of the three dye layers. (Adapted from Eastman Kodak Company, 1992.)

From the standpoint of spectral sensitivity, the three layers of color film can be thought of as three black and white silver halide emulsions (Figure 2.18*b*). Again, the colors physically present in each of these layers after the film is processed are *not* blue, green, and red. Rather, after processing, the blue-sensitive layer contains yellow dye, the green-sensitive layer contains magenta dye, and the red-sensitive layer contains cyan dye (see Figure 2.18*a*). The amount of dye introduced in each layer is inversely related to the intensity of the corresponding primary light present in the scene photographed. When viewed in composite, the dye layers produce the visual sensation of the original scene.

The manner in which the three dye layers of color film operate is shown in Figure 2.19. For purposes of illustration, the original scene is represented schematically in (*a*) by a row of boxes that correspond to scene reflectance in four spectral bands: blue, green, red, and near IR. During exposure, the blue-sensitive layer is activated by the blue light, the green-sensitive layer is activated by the green light, and the red-sensitive layer is activated by the red light, as shown in (*b*). No layer is activated by the near-IR energy since the film is not sensitive to near-IR energy. During processing, dyes are introduced into each sensitivity layer in *inverse* proportion to the intensity of light recorded in each layer. Hence the more intense the exposure of the blue layer to blue light, the less yellow dye is introduced in the image and the more magenta and cyan dyes are introduced. This is shown in (*c*), where, for blue light, the yellow dye layer is clear and the other two layers contain magenta and cyan dyes. Likewise, green exposure results in the introduction of yellow and cyan dyes, and red exposure results in the introduction of yellow and magenta

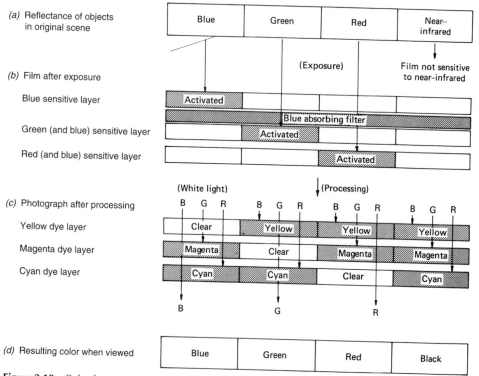

Figure 2.19   Color formation with color film. (Adapted from Eastman Kodak Company, 1992.)

dyes. When the developed image is viewed with a white light source (*d*), we perceive the colors in the original scene through the subtractive process. Where a blue object was present in the scene, the magenta dye subtracts the green component of the white light, the cyan dye subtracts the red component, and the image appears blue. Green and red are produced in an analogous fashion. Other colors are produced in accordance with the proportions of blue, green, and red present in the original scene.

The human eye is more sensitive to green wavelengths than to blue and red wavelengths. For this reason, some films (e.g., Fujicolor Nexia and Superia) employ a fourth, 'cyan-sensitive,' layer in order to improve the color fidelity of the film. This layer is physically located between the green-sensitive and red-sensitive film layers. The sensitivity curve for this layer lies beneath the shorter-wavelength portion of the green-sensitive layer's curve. Magenta dye is used for this 'cyan-sensitive' layer, in addition to the green-sensitive layer.

2.8 COLOR INFRARED FILM

The assignment of a given dye color to a given spectral sensitivity range is a film manufacturing parameter that can be varied arbitrarily. The color of the dye developed in any given emulsion layer need not bear any relationship to the color of light to which the layer is sensitive. Any desired portions of the photographic spectrum, including the near IR, can be recorded on color film with any color assignment.

In contrast to "normal" color film, *color IR* film is manufactured to record green, red, and the photographic portion (0.7 to 0.9 μm) of the near-IR scene energy in its three emulsion layers. The dyes developed in each of these layers are again yellow, magenta, and cyan. The result is a "false color" film in which blue images result from objects reflecting primarily green energy, green images result from objects reflecting primarily red energy, and red images result from objects reflecting primarily in the near-IR portion of the spectrum.

The basic structure and spectral sensitivity of color IR film are shown in Figure 2.20. (Note that there are some overlaps in the sensitivities of the layers.) The process by which the three primary colors are reproduced with such films is shown in Figure 2.21. Various combinations of the primary colors and complementary colors, as well as black and white, can also be reproduced on the film, depending on scene reflectance. For example, an object with a high reflectance in both green and near IR would produce a magenta image (blue plus red). It should be noted that most color IR films are designed to be used with a yellow (blue-absorbing) filter over the camera lens. As further described in Section 2.9, the yellow filter blocks the passage of any light having a wavelength below about 0.5 μm. This means that the blue (and

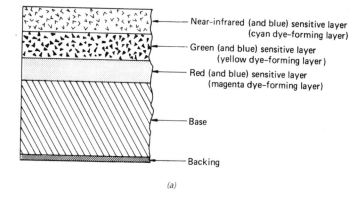

(a)

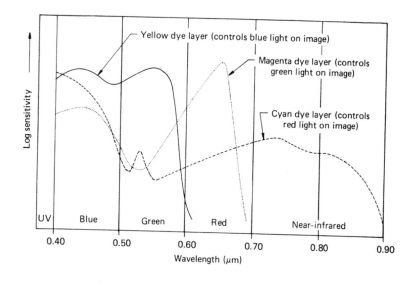

(b)

Figure 2.20 Structure and sensitivity of color IR film: (a) generalized cross section; (b) spectral sensitivities of the three dye layers. (Adapted from Eastman Kodak Company, 1992.)

UV) scene energy is not permitted to reach the film, a fact that aids in the interpretation of color IR imagery. If a yellow (blue-absorbing) filter were not used, it would be very difficult to ascribe any given image color to a particular ground reflectance because of the nearly equal sensitivity of all layers of the film to blue energy. The use of a blue-absorbing filter has the further advantage of improving haze penetration because the effect of Rayleigh scatter is reduced when the blue light is filtered out.

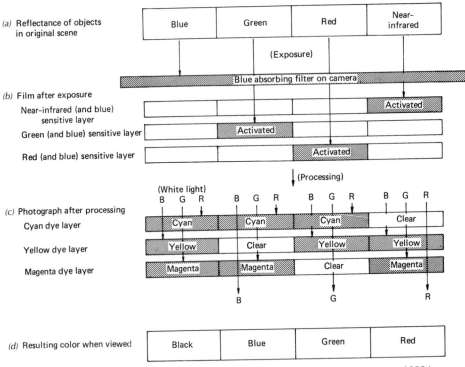

Figure 2.21   Color formation on color IR film. (Adapted from Eastman Kodak Company, 1992.)

Color IR film was developed during World War II to detect painted targets that were camouflaged to look like vegetation. Because healthy vegetation reflects IR energy much more strongly than it does green energy, it generally appears in various tones of red on color IR film. However, objects painted green generally have low IR reflectance. Thus, they appear blue on the film and can be readily discriminated from healthy green vegetation. Because of its genesis, color IR film has often been referred to as "camouflage detection film." With its vivid color portrayal of near-IR energy, color IR film has become an extremely useful film for resource analyses.

Plate 3 illustrates normal color (*a*) and color IR (*b*) aerial photographs of a portion of the University of Wisconsin–Madison campus. The grass, tree leaves, and football field reflect more strongly in the green than in the blue or red and thus appear green in the natural color photograph. The healthy grass and tree leaves reflect much more strongly in the IR than in the green or red and thus appear red in the color IR photograph. The football field has artificial turf that does not reflect well in the IR and thus does not appear red. The large rectangular gravel parking area adjacent to the natural grass practice fields appears a light brown in the normal color photograph and nearly white in the color IR photograph. This means it has a high reflectance in green, red, and IR. The red-roofed buildings appear a greenish-yellow on the color IR film, which means that they reflect highly in the red and also have some IR reflectance. The fact that the near-IR-sensitive layer of the film also has some sensitivity to red (Figure 2.20) also contributes to the greenish-yellow color of the red roofs when photographed on the color IR film.

Almost every aerial application of color IR photography deals with photographing *reflected sunlight*. The amount of energy *emitted* from the earth at ambient temperature (around 300 K) is insignificant in the range of 0.4 to 0.9 μm and hence cannot be photographed. This means that color IR film cannot, for example, be used to detect the temperature difference between two water bodies or between wet and dry soils. As explained in Chapter 5, electronic sensors (such as radiometers or thermal scanners) operating in the wavelength range 3 to 5 or 8 to 14 μm can be used to distinguish between temperatures of such objects.

The energy *emitted* from extremely hot objects such as flames from burning wood (forest fires or burning buildings) or flowing lava *can* be photographed on color and color IR film. Figure 2.22 shows blackbody radiation curves for earth features at an ambient temperature of 27°C (300 K) and flowing lava at 1100°C (1373 K). As calculated from Wien's displacement law (Eq. 1.5), the peak wavelength of the emitted energy is 9.7 μm for the earth features at 27°C and 2.1 μm for lava at 1100°C. When the spectral distribution of emitted energy is calculated, it is found that the energy emitted from the features

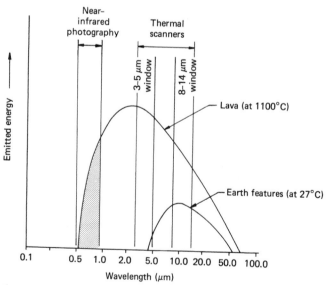

Figure 2.22   Blackbody radiation curves for earth surface features (at 27°C) and flowing lava (at 1100°C).

at 27°C is essentially zero over the range of photographic wavelengths. In the case of flowing lava at 1100°C, the emitted energy in the range of IR photography (0.5 to 0.9 μm) is sufficient to be recorded on photographic films.

Plate 4 shows normal color (*a*) and color IR (*b*) aerial photographs of flowing lava on the flank of Kilauea Volcano on the Island of Hawaii. Although the emitted energy can be seen as a faint orange glow on the normal color photograph, it is more clearly evident on the color IR film. The orange tones on the color IR photograph represent IR energy *emitted* from the flowing lava. The pink tones represent sunlight *reflected* from the living vegetation (principally tree ferns). Keep in mind that it is *only* when the temperature of a feature is extremely high that IR film will record energy emitted by an object. At all other times, the film is responding to *reflected* IR energy that is not directly related to the temperature of the feature.

2.9 FILTERS

Film type is only one variable that determines what information is recorded by a photographic remote sensing system. Equally important is the spectral makeup of the energy exposing the film. Through the use of filters, we can be selective about which wavelengths of energy reflected from a scene we allow to reach the film. Filters are transparent (glass or gelatin) materials that, by absorption or reflection, eliminate or reduce the energy reaching a film in selected portions of the photographic spectrum. They are placed in the optical path of a camera in front of the lens.

Aerial camera filters consist mainly of organic dyes suspended in glass or in a dried gelatin film. Filters are most commonly designated by *Kodak Wratten* filter numbers. They come in a variety of forms having a variety of spectral transmittance properties. The most commonly used spectral filters are *absorption filters*. As their name indicates, these filters absorb and transmit energy of selected wavelengths. A "yellow" filter, for example, absorbs blue energy incident upon it and transmits green and red energy. The green and red energy combine to form yellow—the color we would see when looking through the filter if it is illuminated by white light (see Plate 2*b*).

Absorption filters are often used in film–filter combinations that permit differentiation of objects with nearly identical spectral response patterns in major portions of the photographic spectrum. For example, two objects may appear to reflect the same color when viewed only in the visible portion of the spectrum but may have different reflection characteristics in the UV or near-IR region.

Figure 2.23 illustrates generalized spectral reflectance curves for natural grass and artificial turf, such as those shown in Plate 3. Because the artificial turf is manufactured with a green color to visually resemble natural grass, the reflectance in blue, green, and red is similar for both surfaces. However, the

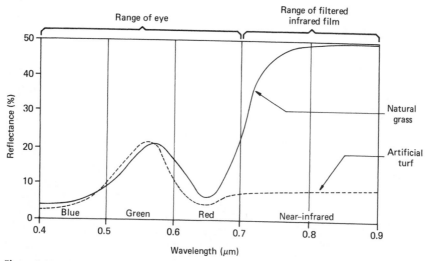

Figure 2.23 Generalized spectral reflectance curves for natural grass and artificial turf.

natural grass reflects very highly in the near IR whereas the artificial turf does not. If we wish to distinguish between natural grass and artificial turf using black and white photography, we can photograph the scene using black and white IR-sensitive film with an absorption filter over the camera lens that blocks all wavelengths shorter than 0.7 μm. Figure 2.24 illustrates the result of such photography. Figure 2.24*a* shows the scene as photographed on panchromatic film in which the natural grass and artificial turf have a similar photographic tone. Figure 2.24*b* shows the scene photographed on black and white IR film using a filter transmitting only wavelengths longer than 0.7 μm. In this case, the natural grass has a very light photographic tone (high IR reflectance) and the artificial turf a very dark photographic tone (low IR reflectance). The filter used in such photography, which selectively absorbs energy below a certain wavelength, is referred to as a *short wavelength blocking* filter or a *high pass* filter.

When one is interested in sensing the energy in only an isolated narrow portion of the spectrum, a *bandpass* filter may be used. Wavelengths above and below a specific range are blocked by such a filter. The spectral transmittance curves for a typical high pass filter and a bandpass filter are illustrated in Figure 2.25. Several high pass and bandpass filters may be used simultaneously to selectively photograph various wavelength bands on separate film images. This results in *multiband imaging*, which we describe in Section 2.14.

There is a large selection of filters from which to choose for any given application. Manufacturers' literature describes the spectral transmittance properties of each available type (e.g., Eastman Kodak Company, 1990). It should be noted that *low pass* absorption filters are not available. *Interference*

Figure 2.24 Simultaneous oblique aerial photographs showing the effect of filtration on discrimination of ground objects. On panchromatic film (*a*) natural grass and artificial turf have a similar tone. When scene energy is filtered such that only wavelengths longer than 0.7 μm are incident on black and white IR film (*b*), the natural grass has a very light tone and the artificial turf a very dark tone.

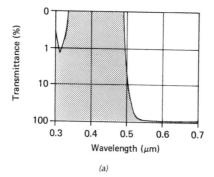

(a)

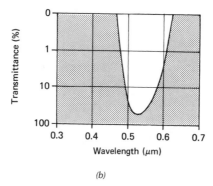

(b)

Figure 2.25 Typical transmission curves for filter types commonly used in aerial photography: (a) typical high pass filter (Kodak Wratten No. 12); (b) typical bandpass filter (Kodak Wratten No. 58). (Adapted from Eastman Kodak Company, 1992.)

filters must be used when short wavelength transmittance is desired. These filters reflect rather than absorb unwanted energy. They are also used when extremely narrow bandpass characteristics are desired.

Panchromatic aerial film is usually exposed through a yellow (blue-absorbing) filter to reduce the effects of atmospheric haze. Black and white IR-sensitive aerial film can be exposed through any of several filters. Typically, a yellow filter (which also transmits IR wavelengths) is used for forestry purposes and a red (which also transmits IR energy) or IR-only filter is used when delineation of water bodies is desired. Normal color film is usually exposed through a UV-absorbing (haze) filter and color IR film through a yellow filter, as shown in Figure 2.26.

Antivignetting filters are often used to improve the uniformity of exposure throughout an image. As described in Section 2.4, there is a geometrically based decrease in illumination with increasing distance from the center of a photograph. To negate the effect of this illumination *falloff*, antivignetting filters are designed to be strongly absorbing in their central area and progressively transparent in their circumferential area. To reduce the number of filters used, and thereby the number of between-filter reflections

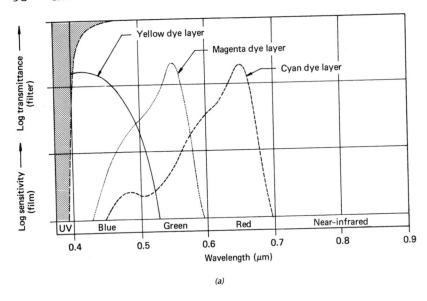

(a)

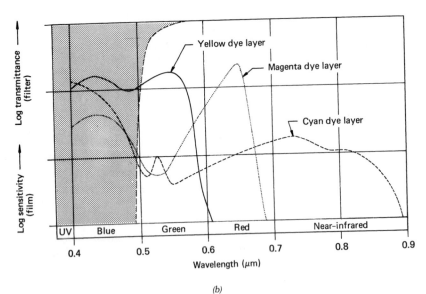

(b)

Figure 2.26 Spectral sensitivities for typical color and color IR film–filter combinations: (a) color film with UV-absorbing (haze) filter; (b) color IR film with Kodak Wratten™ No. 12 (yellow) filter. (Adapted from Eastman Kodak Company, 1990 and 1992.)

possible, antivignetting features are often built into haze and other absorption filters.

A final note on filtering techniques in aerial photography is that color films (particularly IR-sensitive films) are somewhat sensitive to aging. This causes their color layers to often "go out of balance." For example, the sensitivity of the IR-sensitive layer of a film might decrease with age relative to the other two layers. Such a film might still be exposed with satisfactory results if a *color-compensating* filter is used.

When using filters, it is often necessary to increase exposure to compensate for radiation absorption by the filter. Hence, filter manufacturers publish *filter factors*, or multiplying factors, to express the number of times by which an exposure must be increased for a given filter. Published filter factors are intended only as approximate guidelines, as actual factors vary for different exposure conditions.

2.10   AERIAL FILM CAMERAS

Aerial photographs can be made with virtually any type of camera. Many successful applications have employed aerial photographs made from light aircraft with handheld 35-mm cameras. For example, the photographs in Plates 3, 4, and 9 were made in this manner. The simplicity and low cost of purchase and operation of 35-mm cameras make them ideal sensors for small-area analysis. (The size of images taken with a 35-mm system is 24 × 36 mm; the width of the film is 35 mm.) Seventy-millimeter cameras are also used in many applications. (The size of images made with these systems is 55 × 55 mm.) Most aerial photographic remote sensing endeavors, however, entail the use of aerial photography made with precision-built aerial cameras. These cameras are specifically designed to expose a large number of photographs in rapid succession with the ultimate in geometric fidelity.

There are many different models of aerial film cameras currently in use. Here, we discuss *single-lens frame* cameras and *panoramic* cameras. In Section 2.12, we discuss digital cameras used for aerial photography.

Single-Lens Frame Cameras

Single-lens frame cameras are by far the most common cameras in use today. They are used almost exclusively in obtaining aerial photographs for remote sensing in general and photogrammetric mapping purposes in particular. *Mapping* cameras (often referred to as *metric* or *cartographic* cameras) are single-lens frame cameras designed to provide extremely high geometric image quality. They employ a low distortion lens system held in a fixed position relative to the plane of the film. The film format size (the nominal size of each image) is commonly a square 230 mm on a side. The total width of the film used is 240 mm, and the film magazine capacity ranges up to film lengths of 120 m. A frame of imagery is acquired with each opening of the camera shutter, which is generally tripped automatically at a set frequency by an electronic device called an *intervalometer*. Figure 2.27 illustrates a typical aerial mapping camera and its associated gyro-stabilized suspension mount.

Although mapping cameras typically use film with a 230 × 230 mm image format, other cameras with different image sizes have been built. For example, a special-purpose camera built for NASA called the *Large Format Camera* (LFC) had a 230 × 460 mm image format. It was used on an experimental basis to photograph the earth from the space shuttle (Doyle, 1985). (See Figures 3.3 and 4.22 for examples of LFC photography.)

Recall that for an aerial camera the distance between the center of the lens system and the film plane is equal to the focal length of the lens. It is at this fixed distance that light rays coming from an effectively infinite distance away from the camera come to focus on the film. (Most mapping cameras cannot be focused for use at close range.) For mapping purposes, 152-mm-focal-length lenses are most widely used. Lenses with 90 and 210 mm focal

Figure 2.27 Z/I Imaging RMK TOP Aerial Survey Camera System. Camera operation is computer driven and can be integrated with a GPS unit. (Courtesy Z/I Imaging Corporation.)

lengths are also used for mapping. Longer focal lengths, such as 300 mm, are used for very high altitude applications. Frame camera lenses are somewhat loosely termed as being either (1) *normal angle* (when the angular field of view of the lens system is up to 75°), (2) *wide angle* (when the field of view is 75° to 100°), and (3) *superwide angle* (when the field of view is greater than 100°) (angle measured along image diagonal).

Figure 2.28 illustrates the principal components of a single-lens frame mapping camera. The *lens cone assembly* includes the *lens, filter, shutter,* and *diaphragm.* The lens is generally composed of multiple lens elements that gather the light rays from a scene and bring them to focus in the *focal plane.* The filter serves any of the various functions enumerated in the previous section. The shutter and diaphragm (typically located between lens elements) control film exposure. The shutter controls the duration of exposure (from $\frac{1}{100}$ to $\frac{1}{1000}$ sec) while the diaphragm forms an aperture that can be varied in size. The camera *body* typically houses an electrical film drive

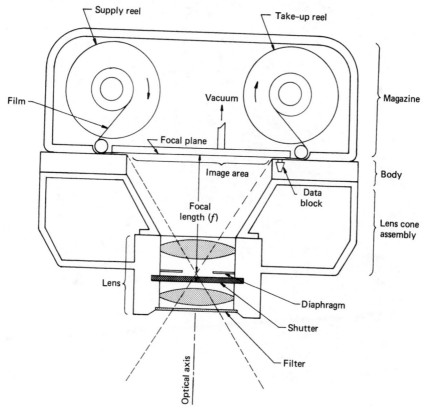

Figure 2.28 Principal components of a single-lens frame mapping camera.

mechanism for advancing the film, flattening the film during exposure, cocking the shutter, and tripping the shutter. The *camera magazine* holds the film supply and takeup reels, the film-advancing mechanism, and the film-flattening mechanism. Film flattening during exposure is often accomplished by drawing the film against a vacuum plate lying behind the focal plane. The focal plane is the plane in which the film is exposed. The *optical axis* of the camera is perpendicular to the film plane and extends through the center of the lens system.

Figure 2.29 Vertical aerial photograph taken with a 230 × 230-mm precision mapping film camera showing Langenburg, Germany. Note the camera fiducial marks on each side of image. Data blocks (on left of image) record image identification, clock, level bubble, and altimeter. Frame number is recorded in lower left corner of image. Scale 1:13,200. (Courtesy Carl Zeiss.)

During the time a frame camera shutter is opened for exposure of a photograph, aircraft motion causes the image to blur. To negate this effect, many frame cameras have built-in *image motion compensation*. This works by moving the film across the focal plane at a rate just equal to the rate of image movement. The camera system illustrated in Figure 2.27 incorporates this capability.

Shown in Figure 2.29 is a *vertical photograph* made with a mapping camera whose optical axis was directed as nearly vertical as possible at the instant of exposure. Note the appearance of the four *fiducial marks* at the middle of the image sides. (As illustrated in Figure 3.12, some mapping cameras incorporate corner fiducials.) These marks define the frame of reference for spatial measurements made from such aerial photos (explained in Chapter 3). Lines connecting opposite fiducial marks intersect at a photograph's *principal point*. As part of the manufacturer's calibration of a mapping camera, the camera focal length, the distances between fiducial marks, and the exact location of the principal point are precisely determined.

It should be noted that there are many single-lens frame cameras that are strictly *reconnaissance* cameras, as opposed to mapping cameras. These cameras come in a wide variety of configurations and are not described in any detail here. Most are designed to faithfully record image detail without necessarily providing the geometric fidelity of mapping cameras. However, to acquire high quality color photographs, these cameras must have color-corrected lenses that focus all colors at the same image plane. Many reconnaissance cameras have been designed for optimum focusing when black and white photographs are taken through a minus blue filter. Such cameras are generally not acceptable for color work because the blue image light that they record will be out of focus and degrade image quality.

Panoramic Cameras

Another major type of film camera we consider is the panoramic camera. This camera views only a comparatively narrow angular field at any given instant through a narrow slit. Ground areas are covered by either rotating the camera lens or rotating a prism in front of the lens. Figure 2.30 illustrates the design using lens rotation.

In Figure 2.30, the terrain is scanned from side to side, transverse to the direction of flight. The film is exposed along a curved surface located at the focal distance from the rotating lens assembly, and the angular coverage of the camera can extend from horizon to horizon. The exposure slit moves along the film as the lens rotates, and the film is held fixed during a given exposure. After one scan is completed, the film is advanced for the next exposure.

Panoramic cameras incorporating the rotating prism design contain a fixed lens and a flat film plane. Scanning is accomplished by rotating the

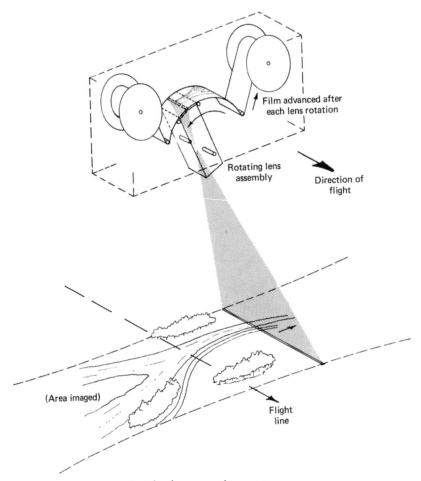

Film advanced after
each lens rotation

Rotating lens
assembly

Direction of
flight

(Area imaged)

Flight
line

Figure 2.30   Operating principle of a panoramic camera.

prism in front of the lens, yielding imagery geometrically equivalent to that of the rotating lens camera.

Figure 2.31 illustrates the pictorial detail and large area of coverage characteristic of panoramic photography. The distortions inherent in panoramic imaging are also apparent in the figure. Areas near the two ends of the photograph are compressed. This scale variation, called *panoramic distortion*, is a result of the cylindrical shape of the focal plane and the nature of scanning. Also, *scan positional distortion* is introduced in panoramic imaging due to forward motion of the aircraft during the time a scan is made.

Typical of rotating prism panoramic cameras is the *optical bar camera* used by NASA in high altitude flights for reconnaissance purposes. This camera incorporates a 610-mm-focal-length lens, has a total field of view of 120° (60° to each side of the flight line), has a film capacity of 2000 m, and is typi-

Figure 2.31 Panoramic photograph with 180° scan angle. Note image detail, large area of coverage, and geometric distortion. (Courtesy USAF Rome Air Development Center.)

cally flown at an altitude of 19,800 m. This yields extremely broad ground coverage, extending 34.3 km to each side of the flight path. As shown in Figure 2.32, the camera can also be used to obtain stereoscopic coverage over the same area. Such cameras have been used extensively for high altitude aerial reconnaissance and were used to photograph more than half the area of the moon during NASA's Apollo missions.

Compared to frame cameras, panoramic cameras cover a much larger ground area. With their narrower lens field of view, panoramic cameras can produce images with greater detail than frame images. Hence, panoramic images yield a broad, yet detailed view of the ground. These factors make panoramic cameras ideal sensors in large-area photographic analyses; *however*, panoramic photographs have the disadvantage that they lack the geometric fidelity of frame camera images. Also, atmospheric effects vary greatly in different portions of the image, because the distance from the camera to the ground in different parts of the scene varies much more with panoramic cameras than with frame cameras, especially when a 180° scan angle is used.

Panoramic photography has been used extensively by the U.S. Forest Service (USFS) and the U.S. Environmental Protection Agency (EPA). The USFS has used panoramic cameras for photographic interpretation purposes such as forest pest damage detection and planning the associated timber salvage operations. The principal advantages of the panoramic camera in such applications are its high image resolution and large area of coverage. The principal disadvantages are its unusual image format of 115 × 1500 mm and the continuously changing photo scale.

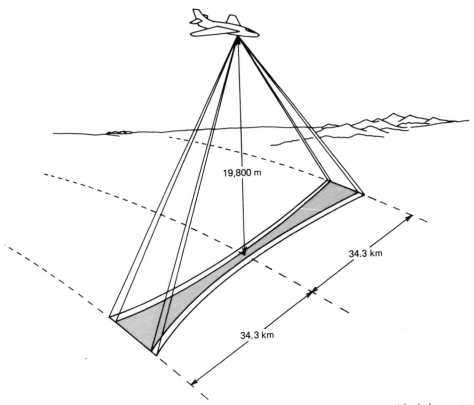

19,800 m

34.3 km

34.3 km

Figure 2.32 Typical ground coverage pattern using the optical bar panoramic camera (shaded area represents area of image overlap). (Adapted from ITEK Optical Systems drawing.)

The EPA has operated panoramic cameras from a device called the *Enviro-Pod*. The device is strapped underneath an aircraft and typically contains two panoramic cameras that record on 70-mm film. One camera is aimed vertically and the other is mounted in a forward-looking orientation. The forward-looking orientation permits the photographing of some objects that are obscured in the vertical photographs (e.g., barrels under a shed roof). The Enviro-Pod was frequently used to obtain high resolution images of industrial pollutants, hazardous waste sites, emergency episodes, and other activities of environmental consequence. The Intelligence Reconnaissance Imagery System III (IRIS III) is another optical imagery system that uses a high-resolution panoramic camera. This system incorporates a 600-mm-focal-length lens and is flown in the U-2R aircraft at altitudes above 20,000 m for various surveillance and reconnaissance purposes. Similarly, SPIN-2 satellite photographs (Section 6.16), taken from an altitude of 220 km, are also acquired using a panoramic camera.

2.11   FILM RESOLUTION

Spatial resolution is an expression of the optical quality of an image produced by a particular camera system. Resolution is influenced by a host of parameters, such as the resolving power of the film and camera lens used to obtain an image, any uncompensated image motion during exposure, the atmospheric conditions present at the time of image exposure, the conditions of film processing, and so on. Some of these elements are quantifiable. For example, we can measure the resolving power of a film by photographing a standard test chart. Such a chart is shown in Figure 2.33. It consists of groups

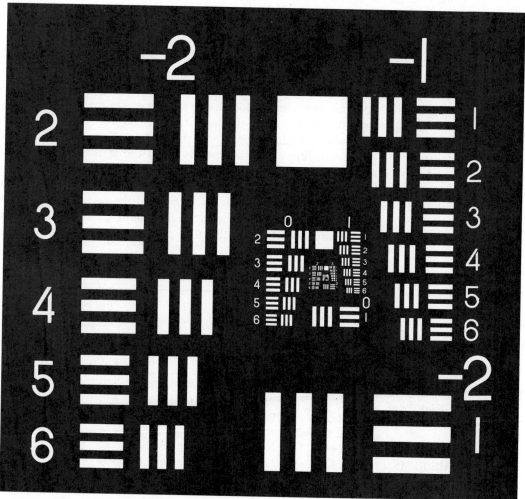

Figure 2.33   Resolving power test chart. (Courtesy Teledyne-Gurley Co.)

of three parallel lines separated by spaces equal to the width of the lines. Successive groups systematically decrease in size within the chart. The resolving power of a film is the reciprocal of the center-to-center distance (in millimeters) of the lines that are just "distinguishable" in the test chart image when viewed under a microscope. Hence, film resolving power is expressed in units of lines per millimeter. Film resolving power is sometimes referred to in units of "line pairs" per millimeter. In this case, the term *line pair* refers to a line (white) and a space (black) of equal width, as shown in Figure 2.33. The terms *lines per millimeter* and *line pairs per millimeter* refer to the same line spacing and can be used interchangeably. Film resolving power is specified at a particular contrast ratio between the lines and their background. This is done because resolution is very strongly influenced by contrast. Typical aerial film resolutions are shown in Table 2.2. Note the significant difference in film resolution between the 1000 : 1 and 1.6 : 1 contrast ratios.

An alternative method of determining film resolution that eliminates the subjectivity of deciding when lines are just "distinguishable" is the construction of a film's *modulation transfer function*. In this method, a scanning microdensitometer (Section 2.5) is used to scan across images of a series of "square-wave" test patterns similar to the one shown in Figure 2.34a. An ideal film would exactly record not only the brightness variation (modulation) of the test pattern but also the distinct edges in the pattern. For actual films, the fidelity of the film recording process depends upon the spatial frequency of the pattern. For test patterns with a small number of lines per millimeter, the maximum and minimum brightness values as measured from the film image (Figure 2.34b) might correspond exactly with those of the test pattern. At the spatial frequency of this test pattern, the film's modulation transfer is said to be 100 percent. However, note that the test pattern edges are somewhat

TABLE 2.2   Resolution of Selected Kodak Aerial Films

Film Name	Film Number	Film Type	Film Resolution (line pairs/mm)	
			1000 : 1 Contrast Ratio	1.6 : 1 Contrast Ratio
PLUS-X AEROGRAPHIC	2402	Panchromatic (negative)	130	55
AERECON High Altitude	3409	Panchromatic (negative)	630	320
Infrared AEROGRAPHIC	2424	Black and white infrared (negative)	125	50
AEROCOLOR III Negative	2444	Normal color (negative)	125	80
AEROCHROME III MS	2427	Normal color (positive)	100	80
AEROCHROME III Infrared	1443	Color infrared (positive)	100	63

Source: Eastman Kodak Company website (www.kodak.com).

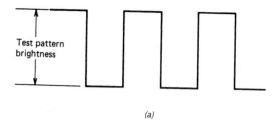

(a)

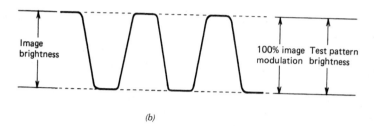

(b)

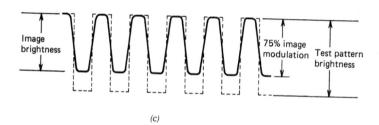

(c)

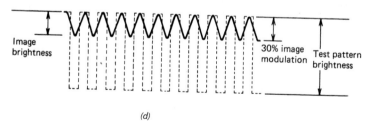

(d)

Figure 2.34 (a) Square-wave test pattern. (b) Modulation transfer of image of test pattern shown in (a). (c, d) Modulation transfer of images of test patterns having higher spatial frequency. [Note in (b) that 100 percent modulation occurs, but the image shows a reduction in edge sharpness as compared with the test pattern.] In (c), edge sharpness is further reduced, and, in addition, modulation transfer is reduced to 75 percent of that of the test pattern. In (d), further sharpness is lost and modulation transfer is reduced to 30 percent. (Adapted from Wolf and Dewitt, 2000.)

rounded on the film image. As the line width and spacing of the test pattern are reduced, density scans across the film image of the test pattern will produce both reduced modulations and increased edge rounding. This is illustrated in Figures 2.34c and d (showing 75 and 30 percent modulation transfer, respectively). By measuring film densities across many such patterns of progressively higher spatial frequency, a complete curve for the modulation transfer function can be constructed (Figure 2.35). Again, this curve expresses the fidelity with which images of features over a range of different sizes or spatial frequencies can be recorded by a given film.

The resolution, or modulation transfer, of any given film is primarily a function of the size distribution of the silver halide grains in the emulsion. In general, the higher the granularity of a film, the lower its resolving power. However, films of higher granularity are generally more sensitive to light, or *faster*, than those having lower granularity. Hence, there is often a trade-off between film "speed" and resolution.

The resolving power of any particular camera–film *system* can be measured by flying over and photographing a large bar target array located on the ground. The imagery thus obtained incorporates the image degradation realized in flight resulting from such factors as atmospheric effects and residual image motion during exposure (including that due to camera vibrations). The

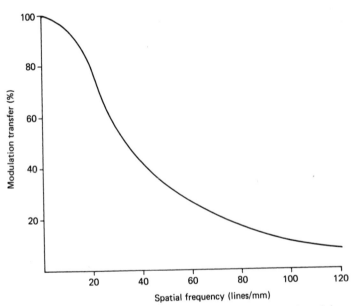

Figure 2.35   Curve of modulation transfer function (MTF). (Adapted from Wolf and Dewitt, 2000).

advantage to this is that we can begin to judge the *dynamic* spatial resolution of the total photographic system instead of the *static* resolution of any one of its components.

We might use the results of such a dynamic resolution test to compare various systems, but the numbers involved are difficult to interpret in a practical sense. Our interest in measuring a system's resolution goes beyond determining the ability of a system to record distinct images of small, nearly contiguous objects of a given shape on a test chart. We are interested not only in object *detection* but also in object *recognition* and *identification*. Hence, "spatial resolution" defies precise definition. At the detection level, the objective is to discern separate objects discretely. At the recognition level, we attempt to determine what objects are—for example, trees versus row crops. At the identification level, we more specifically identify objects—for example, oak trees versus corn.

The effects of scale and resolution can be combined to express image quality in terms of a *ground resolution distance* (GRD). This distance extrapolates the dynamic system resolution on a film to a ground distance. We can express this as

$$\text{GRD} = \frac{\text{reciprocal of image scale}}{\text{system resolution}} \tag{2.13}$$

For example, a photograph at a scale of 1 : 50,000 taken with a system having a dynamic resolution of 40 lines/mm would have a ground resolution distance of

$$\text{GRD} = \frac{50,000}{40} = 1250 \text{ mm} = 1.25 \text{ m}$$

This result assumes that we are dealing with an original film at the scale at which it was exposed. Enlargements would show some loss of image definition in the printing and enlarging process.

In summary, the ground resolution distance provides a framework for comparing the expected capabilities of various images to record spatial detail. However, this and any other measure of spatial resolution must be used with caution because many unpredictable variables enter into what can and cannot be detected, recognized, or identified on an aerial photograph.

2.12 ELECTRONIC IMAGING

The previous sections of this chapter have dealt with traditional photography, an imaging process that uses photographic film as the recording and storage medium for images. Here and in the next section we describe electronic image acquisition. Table 2.3 outlines how photographic and electronic

TABLE 2.3   Comparison of Photographic versus Electronic Image Processing

Characteristic	Photographic Processing	Electronic Image Processing
Data capture	Silver halide film in a camera	Photosensitive solid-state devices
Data storage	Photographic film or prints	Magnetic, optical, and solid-state media
Data manipulation	Chemical developing and optical printing	Digital image processing
Data transmission	Mail, delivery service, fax	Telemetry, telephone lines, computer networks
Softcopy display	Projected slides or movies	Computer monitors, television, projection video
Hardcopy display	Silver halide prints	Dye sublimation, inkjet, thermal, and laser printers

Source: Adapted from Khosla, 1992.

imaging differ in their means of image capture, storage, manipulation, transmission, and display.

Electronic imaging instruments typically use one- or two-dimensional detector arrays of light-sensing solid-state devices (photodiodes) for image acquisition, with each *photosite* (position in the array) sensing one pixel in the image field. Most often, there is one photodiode per photosite. Either CCD (charge-coupled device) or CMOS (complementary metal–oxide–semiconductor) image sensors are used. CMOS sensors are less widely used than CCD sensors because they typically have had less light sensitivity, produced data with greater "noise," and required substantially more image processing time. Newer CMOS sensors are overcoming these limitations. When electromagnetic energy strikes the surface of either type of photosite, electronic charges are produced, with the magnitude of the charges being proportional to the scene brightness. Many of the CCD and CMOS image sensors designed for use in remote sensing are capable of differentiating a wider range of scene brightness values than most photographic films.

Charge-coupled device image sensors are monochromatic. To obtain full-color data, each photosite of the CCD is typically covered with a blue, green, or red filter. Usually, photosites are square, with alternating blue-, green-, and red-sensitive sites arranged in a *Bayer pattern* (Figure 2.36). Half of the filters in this pattern are green; the remainder are blue or red, in equal numbers. To assign blue, green, and red values to each photosite, the two missing colors at each photosite are interpolated from surrounding photosites of the same color (see Section 7.2 and Appendix C for a discussion of resampling schemes for such interpolation). The resulting data set is a two-dimensional array of discrete pixels, with each pixel having three DNs representing the scene brightness in each spectral band of sensing. The *color depth*, or *quantization level*, of CCD sensors is typically 8 to 12 bits (256 to 4096 gray levels

G	B	G	B	G	B	G	B	G	B
R	G	R	G	R	G	R	G	R	G
G	B	G	B	G	B	G	B	G	B
R	G	R	G	R	G	R	G	R	G
G	B	G	B	G	B	G	B	G	B
R	G	R	G	R	G	R	G	R	G
G	B	G	B	G	B	G	B	G	B
R	G	R	G	R	G	R	G	R	G
G	B	G	B	G	B	G	B	G	B
R	G	R	G	R	G	R	G	R	G

Figure 2.36 Bayer pattern of blue, green, and red CCD filters. Note that one-half of these filters are green, one-quarter are blue, and one-quarter are red.

per band). A Bayer pattern of yellow, magenta, and cyan filters can also be used (in this case, half the filters are yellow; the remainder are magenta or cyan, in equal numbers). This configuration results in more energy reaching each photosite (providing an increased signal-to-noise ratio), but resampling computations to arrive at red, green, and blue values for each photosite are more complex and time consuming. A combination of yellow, magenta, cyan, and green filters (these filters are present in equal numbers) can also be used. Also, photosites need not be square; some CCDs utilize octagonal-shaped photosites. Under development (as of 2002) are multilayer CCDs where each photosite has three color-sensitive layers (blue, green, and red), akin to color films.

Charge-coupled device photosites need not be limited to a single photodiode sensor; CCDs are available with two photodiodes per photosite, with a larger primary photodiode with high sensitivity and a smaller secondary photodiode with lower sensitivity. It is claimed that this system can capture four times the range of scene brightness values than CCDs with only one photodiode per photosite. This should provide for both the capture of greater detail in highlight (very bright) areas, as well as a greater ability to resolve detail in darker (often shadow) areas.

Also available is a three-layer **CMOS** sensor that has three photodetectors (blue, green, and red) at every pixel location. This is based on the natural property of silicon to absorb different wavelengths of light at different depths. By placing photodetectors at different depths in the **CMOS** sensor, blue, green, and red energy can be sensed independently. Conceptually, this should result in sharper images, better color, and freedom from color artifacts resulting from interpolation (resampling) that are common in Bayer-pattern CCD sensors.

Digital cameras use a camera body and lens but record image data with CCD or CMOS sensors rather than with film. In turn, the electrical signals generated by these detectors are stored digitally, typically using media such as flash memory or CD and DVD media. While this process is not "photography" in the traditional sense (images are not recorded directly onto photographic film), it is often referred to as "digital photography." Two-dimensional array sizes of digital cameras for aerial imaging that use 35-mm camera bodies typically range from about 2000 × 3000 pixels to about 3000 × 4500 pixels. A digital camera's sensitivity to light is measured using the same ISO scale that is used by film cameras. Typical values range from ISO 100 to ISO 1600. The higher sensitivity levels (larger ISO numbers) allow for photographing at shorter exposure times, or in reduced-light conditions, but result in greater noise in the data.

As an example, Figure 2.37 illustrates the Kodak DCS Pro 14n three-band digital camera that uses a Nikon camera body and has a 3048-pixel row by 4560-pixel column CMOS sensor 24 × 36 mm in size (same size as 35-mm film images), with a color depth of 12-bits per band (4096 gray levels per

Figure 2.37 Kodak DCS Pro 14n digital camera. (Courtesy Eastman Kodak Company.)

band). Kodak DCS cameras have been used by astronauts to photograph the earth from several space shuttle missions.

As mentioned above, 35-mm camera format digital cameras typically use a minimum size sensor of around 2000 × 3000 (6,000,000) pixels. No single number can represent the number of pixels photographic film would have—*if it had pixels*. Furthermore, the pixels in the CCD array used in this camera are uniform in size and shape and are arranged in a systematic geometric pattern, whereas the "pixels" in a photographic film (silver halide grain clusters) are irregular in size, shape, and spatial distribution. However, a reasonable equivalent number is around 6 million pixels for a frame of 35-mm film. Thus, the current resolving power of small-format digital imaging systems compares favorably with that of photographic film of similar format and scale.

Digital cameras for aerial photography that use 70-mm-format camera bodies typically have a 4096 × 4096 pixel array. Many of these cameras use "filmback" replacements for 70-mm-format cameras such as the Hasselblad, Mamiya, and Rollei cameras. Because of the array size, data volumes for these cameras are larger than for digital camera systems using 35-mm format camera bodies.

Figure 2.38 is an example of a larger-format digital camera than the 35- and 70-mm-format cameras just described. It shows the Z/I Imaging Digital Mapping Camera (DMC), which consists of eight synchronously operating CCD-based digital cameras. Four of these are panchromatic cameras that have a 7000 × 4000 pixel array and a focal length of 120 mm. In postprocessing of the image data, the four pan images are mosaicked and reprojected through a common virtual perspective center to create a single-frame image having approximately 14,000 × 8000 pixels. The other four cameras are single band (blue, green, red, and near IR) systems, each incorporating a 3000 × 2000 pixel array and a 25-mm-focal-length lens. The ground footprint for these cameras is approximately equivalent to that of the mosaicked pan image. The resulting images can be combined (three at a time) into "normal color" and/or "color infrared" images, as illustrated (with another imaging system) in Figure 2.40 and Plate 7 (Section 2.14). In addition, the color bands can be co-registered with the higher-resolution panchromatic images using techniques explained in Section 7.6.

Very large two-dimensional arrays can be formed by using a one-dimensional linear CCD detector array that records successive lines of data as the aircraft or spacecraft moves along its flight or orbit path (see Section 5.3). An example of this is the 12,000-pixel CCD linear array associated with the ADS40 Airborne Digital Sensor described in Section 5.5.

Plate 5 shows examples of "normal color" and "color infrared" digital camera images, acquired by the Emerge Direct Digital Imagery system. This three-band system has an array size of 4079 × 4092 pixels and can sense in

Figure 2.38 Z/I Imaging Digital Mapping Camera. (Courtesy Z/I Imaging Corporation.)

either blue, green, and red bands (normal color) or green, red, and near IR bands (color infrared). Ground resolution cell sizes for data collection by this system are typically 0.15 to 1.0 m per pixel, depending principally on flying height. This system has provided digital orthorectified mosaic imagery products to a broad range of users for a variety of **GIS**, mapping, and land management applications.

Digital cameras are typically used for the type of aerial imaging previously done using small-format (35- and 70-mm) film cameras. Relatively inexpensive camera platforms can be used for small-format digital cameras, including ultralight and light aircraft, remote control model aircraft, hot-air blimps, tethered balloons, and kites. At the other extreme, Kodak DCS digital cameras have also been used to obtain images of the earth from the U.S. space shuttle. Figure 2.39 is an astronaut "photograph" from a flying height of 367 km using a Kodak DCS 460 digital camera. It shows several snow-covered volcanoes in the Aleutian Islands. Near the center of the image, smoke can be seen rising from 1730-m-high Mt. Cleveland on Chuginadak Island. Mt. Cleveland is one of the most active volcanoes in the Aleutian Island chain

Figure 2.39 Astronaut "photograph" from a flying height of 367 km using a Kodak DCS 460 digital camera showing several snow-covered volcanoes in the Aleutian Island chain. Note the smoke near the center of the image rising from 1730-m-high Mt. Cleveland, one of the most active volcanoes in the chain. Scale 1 : 270,000. (NASA photograph.)

and produced a major eruption of volcanic ash just 7 weeks after this image was exposed.

The use of small-format digital cameras provides several advantages over the use of 35- and 70-mm film cameras, including rapid turnaround time (images are available for viewing during the flight and immediately afterward) and an inherently computer-compatible format (photographic films need to be scanned to produce computer-compatible data). Image exposures can be optimized in flight—intended targets can be viewed, images acquired, histograms analyzed, and aperture, shutter speed, or ISO adjusted to provide an optimum exposure range. Digital camera data are stored on a reusable medium; the memory devices used for storing digital images can be erased and reused essentially indefinitely. Also, digital images can be copied repeatedly with no loss of image quality unless the data have been excessively compressed. New "wavelet" compression schemes have been shown to produce excellent results with compression ratios of up to 20 : 1 for digital aerial imagery.

The exposure latitude of digital cameras exceeds that of most films, and, as stated earlier, the resolving power of small-format (35- and 70-mm) digital images is comparable to that of photographic film images of similar format and scale. Another advantage of digital cameras over film cameras is their greater suitability for scientific work, as they can be more easily calibrated radiometrically using targets in the laboratory or field. The CCDs used for digital imaging are linear in response to energy across their entire dynamic range, whereas film response is linear only in the "straight-line portion" of the characteristic curve (Section 2.5). Also, most digital cameras used for aerial imaging have the capability of recording positional data obtained from a GPS receiver along with the image data recorded by the camera.

In their computer-compatible format, digital images can be readily processed and displayed using the digital image processing techniques explained in Chapter 7, including various image enhancement and classification techniques.

The applications in which electronic imaging systems (including both digital frame cameras and aerial video systems) have been successfully operated are numerous. Aerial electronic imaging has been applied in a variety of ways in the resource domains of forestry, range management, agriculture, water resources, geology, and environmental assessment. More specific examples include analysis of hazardous waste sites, detection of soil conditions, land use/land cover mapping, wild rice mapping, wildlife censusing, wildlife habitat studies, trout stream monitoring, right-of-way monitoring, water quality studies, precision agriculture assessments, wetland mapping, crop condition assessment, detection of forest insect and disease problems, irrigation mapping, and detection of frost damage in citrus groves, to mention but a few.

2.13   AERIAL VIDEOGRAPHY

Aerial videography is a form of electronic imaging whereby analog or digital video signals are recorded on magnetic tape or on CD or DVD media. Cameras used for aerial videography can have many configurations, including single-band cameras, multiband cameras, or multiple single-band cameras and can sense in the visible, near-IR, and mid-IR wavelengths (not all cameras have this full range).

Shuttered cameras, with exposure times as short as $\frac{1}{10,000}$ sec, are used for both analog and digital aerial videography. An exposure time of $\frac{1}{1,000}$ sec is typically used, which essentially eliminates image motion.

Analog video recording that follows the NTSC RS-170 standard (used principally in North America and Japan) uses signals conforming to an industry specification of 485 horizontal lines per image frame. The NTSC signals have an aspect ratio (frame width divided by frame height) of 4 : 3.

Image resolution in the vertical direction is set by the number of discrete lines per image frame (485 image lines in the case of NTSC signals). Image resolution in the horizontal direction is defined in terms of "TV lines" (TVLs). Although analog video signals do not have discrete pixels, it is useful to think of horizontal resolution in terms of pixels per line. The number of TVLs is the number of pixels contained in a horizontal distance equal to the frame height. The most widely used analog formats for aerial videography are the Super-VHS and Hi-8 formats, with a resolution of about 400 TVLs. Because of the 4 : 3 aspect ratio, the number of pixels in one frame of a Hi-8 recording would be about 535 $\times$ 485, or about 260,000 pixels. This is only about $\frac{1}{50}$ as many pixels per frame as a digital frame camera such as the Kodak DCS Pro 14n.

Digital video recording can follow one of many standards. The primary consumer digital camcorder formats (as of 2002) are Mini-DV, Digital8, MICROMV, and DVD, all with a resolution of about 500 TVLs. There are several DVCPRO formats: DVCPRO, DVCPRO50, DVCPRO P, and DVCPRO HD. They differ in scanning method (interlaced versus progressive), video format (image size), and image data compression. As an example, the DVCPRO format has a resolution of 800 $\times$ 485 pixels (388,000 pixels per frame). Other digital video recording formats exist, and more are being developed. This is an improvement over analog video recording, but still only about $\frac{1}{30}$ as many pixels per frame as a high-end 35-mm-format digital frame camera. High definition television (HDTV) has an aspect ratio of 16 : 9. Digital recordings of HDTV signals contain 1920 $\times$ 1080 pixels, or about 2.0 million pixels per frame.

Both analog and digital video recordings can be viewed on conventional television equipment through playback of the video tape in a video cassette recorder (VCR). Slow-speed and freeze-frame playback can be used to aid in visual image analysis. Digital video data are in a format that can be readily displayed and processed in a microcomputer. Analog video data can also be viewed and analyzed digitally by using a *video capture card* (digitizer) in a microcomputer.

The use of aerial videography has several advantages when compared with small-format (35- and 70-mm) aerial film photography. As with digital frame cameras, video images can be viewed in the aircraft at the time of data acquisition and are available for analysis immediately after a flight. The resulting imagery is also inexpensive—material costs are typically much lower than for 35-mm photographs for the same area of coverage. Another advantage is that video cassettes have an audio track, which means that verbal comments about specific features or locations can be recorded in synchronization with the imagery. In addition, GPS equipment can be used to record the latitude, longitude, and elevation of the aircraft directly onto the video image at the time of data acquisition.

Ground objects sensed by aerial videography are viewed from multiple vantage points as they pass under the aircraft. Thus, there are multiple view angle–sun angle combinations that can be analyzed.

Aerial video recording has the capability of recording more than 200,000 frames on a single tape (2 hours of flying time), which makes it especially well suited to imaging linear features such as roads, power lines, and rivers.

The principal disadvantage of video recording is the lower spatial resolution in comparison with 35- and 70-mm aerial film cameras and digital frame cameras. Also, aerial video imagery cannot be easily used for stereoscopic viewing, and video tapes can be somewhat cumbersome to index and handle when subsequent viewing of discrete image segments (or single frames) is desired (this is especially true of analog video recordings). Aerial video data recorded on CD or DVD media are much less cumbersome to index and retrieve.

Plate 6 shows two examples of color IR video composite images compared with color IR photographs. In this plate, (a) and (b) are images of a cotton field infested with harvester ants. Although the spatial resolution of the video image (a) is lower than the photograph (b), most of the ant mounds (white spots) can easily be detected in the video image. Shown in (c) and (d) are images of a cotton field with saline soils. The barren to sparsely vegetated saline areas can generally be distinguished as readily in the video as in the photographic image. The color tones of the video composite images are comparable to the color IR film tones in both scenes.

2.14 MULTIBAND IMAGING

Earlier in this chapter, we saw that both normal color and color IR films were three-layer systems. Normal color film has layers individually sensitive to blue, green, and red energy, and color IR film (used with an appropriate filter) has layers individually sensitive to green, red, and near-IR energy. Upon processing, these three-layer films produce color images because of the dye layers that control the amount of blue, green, and red energy that form the final photographic image.

Three individual bands from many types of sensing systems (e.g., digital cameras, video cameras, and various airborne and satellite multispectral scanners) can be displayed simultaneously to produce color images.

Multiband images are images sensed simultaneously from essentially the same geometric vantage point but in different bands of the electromagnetic energy spectrum. In the case of *multiband photography*, different parts of the spectrum are sensed with different film–filter combinations. Multiband digital camera images and video images are also typically exposed onto the camera's CCD or CMOS sensor(s) through different filters. Electro-optical sensors such as the Landsat Thematic Mapper (discussed in Chapter 6) typically sense in at least several bands of the electromagnetic spectrum. Hyperspectral sensors (Chapter 5) may sense in hundreds of very narrow portions of the spectrum.

The "best" combination of multiband images for discriminating a given scene varies with the spectral response patterns for the objects of interest within that scene. The "taking apart" of object reflectances (and/or emittances) through multiband imaging normally yields enhanced contrast between different terrain feature types and between different conditions of the same feature type. To optimize this contrast, band–color combinations are chosen for the specific features of interest in spectral regions where the maximum spectral reflectance differences are known, or are anticipated, to exist.

Multiband images are typically viewed by selecting three of the available bands and displaying one band as blue, one band as green, and one band as red through the use of some additive color device, typically a computer monitor. (Computer monitors, like color televisions, have three separate graphic planes that control independently the intensity of blue, green, and red light at each position on the screen.)

Figure 2.40 shows the images of four individual bands acquired by the ADAR System 5500, a multispectral digital aerial camera system. The standard configuration for this system is to employ four CCD sensors, capturing data in the blue, green, red, and near-IR spectral regions. Ground resolution cell sizes used for ADAR applications are typically 0.5 to 3 m per pixel, depending on the flying height. In addition to storing digital image data, this system continually monitors a GPS receiver, recording latitude, longitude, and elevation data to provide the location of the aircraft at the time each image is acquired. ADAR data have been used in a variety of applications, including natural resource management, forestry, wetland monitoring, precision agriculture, and urban planning.

Plate 7 shows "normal color" and "color IR" images that were prepared from the individual band images shown in Figure 2.40 by combining these image bands, as shown in Table 2.4. Note that, as is typical of multiband data in this spectral range, there is a great deal of correlation among the reflections in the blue, green, and red spectral bands and a large difference in reflections between the three visible-wavelength bands and the near-IR band. For example, the water is much darker toned in the near-IR band, and the trees are much lighter toned in the near-IR band. There are, however, sufficient differences among reflections in the blue, green, and red bands to produce a color composite image with colors that resemble a normal color photograph.

Multiband viewing is not restricted to visible and near-IR bands. Multiband viewing may also include mid-IR and thermal-IR bands, as illustrated later. Regardless of the number and wavelength bands of the images, only three bands are selected for viewing at one time, with one band displayed as blue, one band as green, and one band as red.

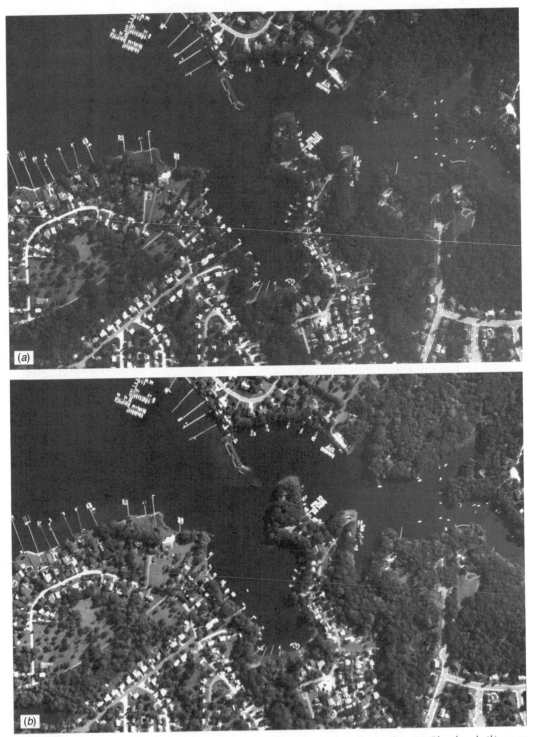

Figure 2.40 Multiband digital camera images, Chesapeake Bay, MD, early October. (*a*) Blue band, (*b*) green band, (*c*) red band, and (*d*) near IR band. Scale 1 : 9000. (Courtesy Positive Systems, Inc.)

Figure 2.40 *(Continued)*

TABLE 2.4   Band–Color Combinations Used to Produce Normal Color and Color Infrared Images

To Simulate a Normal Color Image		To Simulate a Color Infrared Image	
EM Band	Display Color	EM Band	Display Color
Blue	Blue	Green	Blue
Green	Green	Red	Green
Red	Red	Near IR	Red

2.15   CONCLUSION

As we indicated earlier in this chapter, aerial photography has historically been the most widely used form of remote sensing due to its general availability, geometric integrity, versatility, and economy. However, as with any other sensing system, aerial photography has certain limitations and requirements. Airphotos are often difficult to obtain, handle, store, calibrate, and interpret. Current technological trends seem to indicate that more and more inherently digital recording systems will be used in many applications where aerial photographs have traditionally been employed.

SELECTED BIBLIOGRAPHY

Ahmad, A., and J.H. Chandler, "Photogrammetric Capabilities of the Kodak DC40, DCS420 and DCS460 Digital Cameras," *Photogrammetric Record*, vol. 16, no. 94, 1999, pp. 601–615.

American Society for Photogrammetry and Remote Sensing (ASPRS), *15th Biennial Workshop on Videography & Color Photography in Resource Assessment*, ASPRS, Bethesda, MD, 1995.

American Society for Photogrammetry and Remote Sensing (ASPRS), "The First North American Symposium on Small Format Aerial Photography," *Technical Papers*, ASPRS, Bethesda, MD, 1997a.

American Society for Photogrammetry and Remote Sensing (ASPRS), *16th Biennial Workshop on Videography & Color Photography in Resource Assessment*, ASPRS, Bethesda, MD, 1997b.

American Society for Photogrammetry and Remote Sensing (ASPRS), *Corona Between the Sun and the Earth: The First NRO Reconnaissance Eye in Space*, ASPRS, Bethesda, MD, 1997c.

American Society of Photogrammetry (ASP), *Manual of Color Aerial Photography*, ASP, Falls Church, VA, 1968.

American Society of Photogrammetry (ASP), *Manual of Photogrammetry*, 4th ed., ASP, Falls Church, VA, 1980.

American Society of Photogrammetry (ASP), *Manual of Remote Sensing*, 2nd ed., ASP, Falls Church, VA, 1983.

Baker, S., "San Francisco in Ruins: The 1906 Aerial Photographs of George R. Lawrence," *Landscape*, vol. 30, no. 2, 1989, pp. 9–14.

Canadian Journal of Remote Sensing, Special Issue on Aerial Optical Remote Sensing, vol. 21, no. 3, 1995.

Doyle, F.J., "The Large Format Camera on Shuttle Mission 41-G," *Photogrammetric Engineering and Remote Sensing*, vol. 51, no. 2, 1985, pp. 200–203.

Drake, S., "Visual Interpretation of Vegetation Classes from Airborne Videography: An Evaluation of Observer Proficiency with Minimal Training," *Photogrammetric Engineering and Remote Sensing*, vol. 62, no. 8, 1996, pp. 969–978.

Eastman Kodak Company, *Aerial Imaging—Online Publications*, Available over the Internet at http://www.kodak.com/US/en/government/aerial/technicalPubs/.

Eastman Kodak Company, *Applied Infrared Photography*, Eastman Kodak Company, Rochester, NY, 1981.

Eastman Kodak Company, *Kodak Photographic Filters Handbook*, Eastman Kodak Company, Rochester, NY, 1990.

Eastman Kodak Company, *Kodak Data for Aerial Photography*, 6th ed., Eastman Kodak Company, Rochester, NY, 1992.

Edirisinghe, A., G.E. Chapman, and J.P. Louis, "Radiometric Corrections for Multispectral Airborne Video Imagery," *Photogrammetric Engineering and Remote Sensing*, vol. 67, no. 8, 2001, pp. 915–922.

Escobar, D.E., et al., "A Twelve-Band Airborne Digital Video Imaging System (ADVIS)," *Remote Sensing of Environment*, vol. 66, 1998, pp. 122–128.

Fritz, N.L., "Optimum Methods for Using Infrared-Sensitive Color Films," *Photogrammetric Engineering*, vol. 33, no. 10, 1967, pp. 1128–1138.

Fritz, N.L., "Filters: An Aid in Color-Infrared Photography," *Photogrammetric Engineering and Remote Sensing*, vol. 43, no. 1, 1977, pp. 61–72.

Gao, J., and S.M. O'Leary, "The Role of Spatial Resolution in Quantifying SSC from Airborne Remotely Sensed Data," *Photogrammetric Engineering and Remote Sensing*, vol. 63, no. 3, 1997, pp. 267–271.

Graham, R., *Digital Imaging*, Whittles Publishing, Caithness, Scotland, 1998.

Hess, L.L., et al., "Geocoded Digital Videography for Validation of Land Cover Mapping in the Amazon Basin," *International Journal of Remote Sensing*, vol. 23, no. 7, 2002, pp. 1527–1555.

Holopainen, M., and G. Wang, "The Calibration of Digitized Aerial Photographs for Forest Stratification," *International Journal of Remote Sensing*, vol. 19, no. 4, 1998, pp. 677–696.

Jacobson, R.E., *The Manual of Photography—Photographic and Digital Imaging*, 9th ed., Focal Press, Oxford, UK, 2000.

Khosla, R.P., "From Photons to Bits," *Physics Today*, vol. 45, no. 12, 1992, pp. 42–49.

King, D.J., "Airborne Multispectral Digital Camera and Video Sensors: A Critical Review of System Designs and Applications," *Canadian Journal of Remote Sensing*, vol. 21, no. 3, August 1995, pp. 245–273.

King, D., P. Walsh, and F. Ciuffreda, "Airborne Digital Frame Camera Imaging for Elevation Determina-

tion," *Photogrammetric Engineering and Remote Sensing*, vol. 60, no. 11, 1994, pp. 1321–1326.

Light, D.L., "Film Cameras or Digital Sensors? The Challenge Ahead for Aerial Imaging," *Photogrammetric Engineering and Remote Sensing*, vol. 62, no. 3, 1996, pp. 285–291.

Luman, D.E., C. Stohr, and L. Hunt, "Digital Reproduction of Historical Aerial Photographic Prints for Preserving a Deteriorating Archive," *Photogrammetric Engineering and Remote Sensing*, vol. 63, no. 10, 1997, pp. 1171–1179.

Luther, A.C., *Video Camera Technology*, Artech House, Boston, MA, 1998.

Luther, A.C., *Video Recording Technology*, Artech House, Boston, MA, 1999.

Milton, E.J., "Low-Cost Ground-Based Digital Infra-Red Photography," *International Journal of Remote Sensing*, vol. 23, no. 5, 2002, pp. 1001–1007.

Photogrammetric Engineering and Remote Sensing, Special Issue on Panoramic Photography and Forest Remote Sensing, vol. 48, no. 5, 1982.

Newhall, B., *Airborne Camera*, Hastings House, NY, 1969.

Ramsey, E.W., et al., "Mapping Chinese Tallow with Color-Infrared Photography," *Photogrammetric Engineering and Remote Sensing*, vol. 68, no. 3, 2002, pp. 251–255.

Robinson, A.H., et al., *Elements of Cartography*, 6th ed., Wiley, New York, 1995.

Slater, P.N., *Remote Sensing and Optical Systems*, Addison-Wesley, Reading, MA, 1980.

Tomer, M.D., J.L. Anderson, and J.A. Lamb, "Assessing Corn Yield and Nitrogen Uptake Variability with Digitized Aerial Infrared Photographs," *Photogrammetric Engineering and Remote Sensing*, vol. 63, no. 3, 1997, pp. 299–306.

Warner, W.S., R.W. Graham, and R.E. Read, *Small Format Aerial Photography*, American Society for Photogrammetry and Remote Sensing, Bethesda, MD, 1996.

Weigand, C.L., J.H. Everitt, and A.J. Richardson, "Comparison of Multispectral Video and SPOT-1 HRV Observations for Cotton Affected by Soil Salinity," *International Journal of Remote Sensing*, vol. 13, no. 8, 1992, pp. 1511–1525.

Wolf, P.R., and B. Dewitt, *Elements of Photogrammetry with Applications in GIS*, 3rd ed., McGraw-Hill Higher Education, New York, NY, 2000.

3 BASIC PRINCIPLES OF PHOTOGRAMMETRY

3.1 INTRODUCTION

Photogrammetry is the science and technology of obtaining spatial measurements and other geometrically reliable derived products from photographs. Photogrammetric analysis procedures can range from obtaining approximate distances, areas, and elevations using hardcopy photographic products, unsophisticated equipment, and simple geometric concepts to generating precise digital elevation models (DEMs), orthophotos, thematic GIS data, and other derived products through the use of digital raster images and relatively sophisticated analytical techniques.

We use the terms *digital* and *softcopy* photogrammetry interchangeably to refer to any photogrammetric operation involving the use of digital raster photographic image data rather than hardcopy images. Digital photogrammetry is changing rapidly and forms the basis for most current photogrammetric operations. However, the same basic geometric principles apply to traditional hardcopy (analog) and softcopy (digital) procedures. In fact, it is often easier to visualize and understand these principles in a hardcopy context and then extend them to the softcopy environment. This is the approach we adopt in this discussion. We also stress *aerial* photogrammetric techniques and procedures, but the same general principles hold for *space*-based operations.

Historically, the most common use of photogrammetry has been to produce hardcopy topographic maps. Today, photogrammetric procedures are used extensively to produce a range of GIS data products such as precise raster image backdrops for vector data and digital elevation models. Thematic data (in three dimensions) can also be extracted directly from photographs for inclusion in a GIS.

In this chapter, we introduce only the most basic aspects of the broad subject of photogrammetry. Our objective is to provide the reader with a fundamental understanding of how hardcopy photographs can be used to measure and map earth surface features and how softcopy systems work conceptually. We discuss the following photogrammetric activities.

1. **Determining the scale of a vertical photograph and estimating horizontal ground distances from measurements made on a vertical photograph.** The scale of a photograph expresses the mathematical relationship between a distance measured on the photo and the corresponding horizontal distance measured in a ground coordinate system. Unlike maps, which have a single constant scale, aerial photographs have a range of scales that vary in proportion to the elevation of the terrain involved. Once the scale of a photograph is known at any particular elevation, ground distances at that elevation can be readily estimated from corresponding photo distance measurements.

2. **Using area measurements made on a vertical photograph to determine the equivalent areas in a ground coordinate system.** Computing ground areas from corresponding photo area measurement is simply an extension of the above concept of scale. The only difference is that whereas ground distances and photo distances vary linearly, ground areas and photo areas vary as the square of the scale.

3. **Quantifying the effects of relief displacement on vertical aerial photographs.** Again unlike maps, aerial photographs in general do not show the true plan or top view of objects. The images of the tops of objects appearing in a photograph are displaced from the images of their bases. This is known as *relief displacement* and causes any object standing above the terrain to "lean away" from the principal point of a photograph radially. Relief displacement, like scale variation, precludes the use of aerial photographs directly as maps. However, reliable ground measurements and maps can be obtained from vertical photographs if photo measurements are analyzed with due regard for scale variations and relief displacement.

4. **Determination of object heights from relief displacement measurements.** While relief displacement is usually thought of as an image distortion that must be dealt with, it can also be used to estimate the heights of objects appearing on a photograph. As we later

illustrate, the magnitude of relief displacement depends on the flying height, the distance from the photo principal point to the feature, and the height of the feature. Because these factors are geometrically related, we can measure an object's relief displacement and radial position on a photograph and thereby determine the height of the object. This technique provides limited accuracy but is useful in applications where only approximate object heights are needed.

5. **Determination of object heights and terrain elevations by measurement of image parallax.** The previous operations are performed using vertical photos individually. Many photogrammetric operations involve analyzing images in the area of overlap of a stereopair. Within this area, we have two views of the same terrain, taken from different vantage points. Between these two views, the relative positions of features lying closer to the camera (at higher elevation) will change more from photo to photo than the positions of features farther from the camera (at lower elevation). This change in relative position is called *parallax*. It can be measured on overlapping photographs and used to determine object heights and terrain elevations.

6. **Use of ground control points.** The accuracy of photogrammetric measurements is usually premised on the use of *ground control points*. These are points that can be accurately located on the photograph and for which we have information on their ground coordinates and/or elevations (often through GPS observations). This information is used as "geometric ground truth" to calibrate photo measurements. For example, we commonly use ground control to determine the true (slightly tilted) angular orientation of a photograph, the flying height of a photograph, and the airbase of a pair of overlapping photographs (the distance between successive photo centers). This information is critical in a host of photogrammetric operations.

7. **Mapping with aerial photographs.** As mentioned previously, "mapping" from aerial photographs can take on numerous forms and can employ either hardcopy or softcopy approaches. Traditionally, topographic maps have been produced from hardcopy stereopairs in a device called a *stereoplotter*. With this type of instrument, the photographs are mounted in special projectors that can be mutually oriented to precisely correspond to the angular tilts present when the photographs were taken. Once oriented properly, the projectors recreate an accurate model of the terrain that, when viewed stereoscopically, can be used to plot a planimetric map having no relief distortions. In addition, topographic contours can be plotted on the map and the height of vertical features appearing in the model can be determined.

Whereas a stereoplotter is designed to transfer *map* information, without distortions, from stereo photographs, a similar device can be used to transfer *image* information, with distortions removed. The resulting undistorted image is called an *orthophotograph* (or *orthophoto*). Orthophotos combine the geometric utility of a map with the extra "real-world image" information provided by a photograph. The process of creating an orthophoto depends on the existence of a reliable DEM for the area being mapped. The DEMs are usually prepared photogrammetrically as well. In fact, *photogrammetric workstations* generally provide the integrated functionality for such tasks as generating DEMs, digital orthophotos, topographic maps, perspective views, and "fly-throughs," as well as the extraction of spatially referenced GIS data in two or three dimensions.

8. **Preparation of a flight plan to acquire aerial photography.** Whenever new photographic coverage of an area is to be obtained, a photographic flight mission must be planned. This process begins with selecting an image scale, camera lens and format size, and desired image overlap. The flight planner can then determine such geometric factors as the appropriate flying height, the distance between image centers, the direction and spacing of flight lines, and the number of images required to cover the study area. Based on these factors, a flight map and a list of specifications are prepared for the firm providing the photographic services.

Each of these photogrammetric operations is covered in separate sections in this chapter. We first discuss some general geometric concepts that are basic to these techniques.

3.2 BASIC GEOMETRIC CHARACTERISTICS OF AERIAL PHOTOGRAPHS

Geometric Types of Aerial Photographs

Aerial photographs are generally classified as either vertical or oblique. *Vertical photographs* are those made with the camera axis directed as vertically as possible. Vertical photography made with a single-lens frame camera is by far the most common type of aerial photography used in remote sensing applications. However, a "truly" vertical aerial photograph is rarely obtainable because of unavoidable angular rotations, or tilts, caused by the angular attitude of the aircraft at the instant of exposure. These unavoidable tilts cause slight (1° to 3°) unintentional inclination of the camera optical axis, resulting in the acquisition of *tilted photographs*.

Virtually all photographs are tilted. When tilted unintentionally and slightly, tilted photographs are usually referred to as being "vertical." For most elementary measurement applications, these photographs are treated as being vertical without introduction of serious error.

When aerial photographs are taken with an intentional inclination of the camera axis, *oblique photographs* result. *High oblique photographs* include an image of the horizon, and *low oblique photographs* do not. We limit our discussion in this chapter to the subject of vertical aerial photographs.

Taking Vertical Aerial Photographs

Most vertical aerial photographs are taken with frame cameras along *flight lines*, or *flight strips*. The line traced on the ground directly beneath the aircraft during acquisition of photography is called the *nadir line*. This line connects the image centers of the vertical photographs. Figure 3.1 illustrates the typical character of the photographic coverage along a flight line. Successive photographs are generally taken with some degree of *endlap*. Not only does this lapping ensure total coverage along a flight line, but an endlap of at least 50 percent is essential for total *stereoscopic coverage* of a project area. Stereoscopic coverage consists of adjacent pairs of overlapping vertical photographs called *stereopairs*. Stereopairs provide two different perspectives of the ground area in their region of endlap. When images forming a stereopair are viewed through a stereoscope, each eye psychologically occupies the vantage

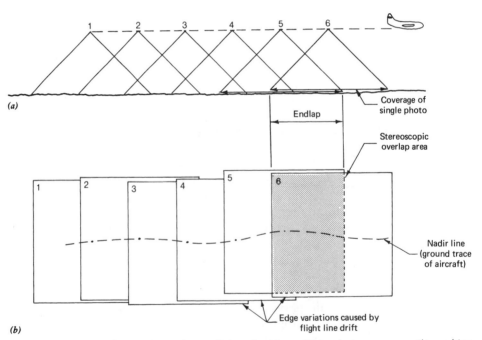

Figure 3.1 Photographic coverage along a flight strip: (*a*) conditions during exposure; (*b*) resulting photography.

Figure 3.2   Acquisition of successive photographs yielding a stereopair. (Courtesy Wild Heerbrugg, Inc.)

point from which the respective image of the stereopair was taken in flight. The result is the perception of a three-dimensional *stereomodel*. As pointed out in Chapter 4, most applications of aerial photographic interpretation entail the use of stereoscopic coverage and stereoviewing.

Successive photographs along a flight strip are taken at intervals that are controlled by the camera *intervalometer*, a device that automatically trips the camera shutter at desired times. The area included in the overlap of successive photographs is called the *stereoscopic overlap area*. Typically, successive photographs contain 55 to 65 percent overlap to ensure at least 50 percent endlap over varying terrain, in spite of unintentional tilt. Figure 3.2 illustrates the ground coverage relationship of successive photographs forming a stereopair having approximately a 60 percent stereoscopic overlap area.

The ground distance between the photo centers at the times of exposure is called the *air base*. The ratio between the air base and the flying height above ground determines the *vertical exaggeration* perceived by photo interpreters. The larger the *base–height ratio*, the greater the vertical exaggeration.

Figure 3.3 shows Large Format Camera photographs of Mt. Washington and vicinity, New Hampshire. These stereopairs illustrate the effect of varying the percentage of photo overlap and thus the base–height ratio of the photographs. These photographs were taken from a flying height of 364 km. The

Figure 3.3 Large Format Camera stereopairs, Mt. Washington and vicinity, New Hampshire, April 4, 1985; scale 1 : 800,000 (1.5 times enlargement from original image scale): (*a*) 0.30 base-height ratio; (*b*) 1.2 base-height ratio. (Courtesy NASA and ITEK Optical Systems.)

stereopair in (*a*) has a base–height ratio of 0.30. The stereopair in (*b*) has a base–height ratio of 1.2 and shows much greater apparent relief (greater vertical exaggeration) than (*a*).

Most project sites are large enough for multiple-flight-line passes to be made over the area to obtain complete stereoscopic coverage. Figure 3.4 illustrates how adjacent strips are photographed. On successive flights over the area, adjacent strips have a *sidelap* of approximately 30 percent. Multiple strips comprise what is called a *block* of photographs. Modern aerial surveys usually employ data from the aircraft's precise GPS navigation system to control flight line direction, flight line spacing, and photo exposure intervals.

(b)

Figure 3.3 *(Continued)*

A given photographic mission can entail the acquisition of literally hundreds of exposures. Quite often, a flight *index mosaic* is assembled by piecing together the individual photographs into a single continuous picture. This enables convenient visual reference to the area included in each image. Figure 3.5 illustrates such a mosaic.

Geometric Elements of a Vertical Photograph

The basic geometric elements of a vertical aerial photograph are depicted in Figure 3.6. Light rays from terrain objects are imaged in the plane of the film negative after intersecting at the camera lens exposure station, L. The negative is located behind the lens at a distance equal to the lens focal length, f.

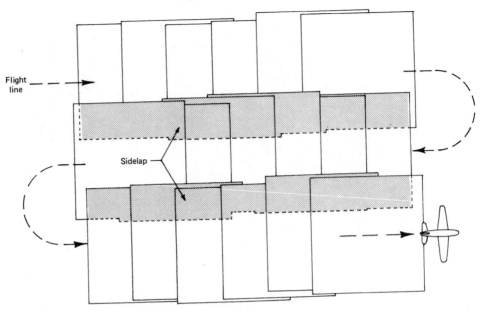

Figure 3.4 Adjacent flight lines over a project area.

Assuming the size of a paper print positive (or film positive) is equal to that of the negative, positive image positions can be depicted diagrammatically in front of the lens in a plane located at a distance f. This rendition is appropriate in that most photo positives used for measurement purposes are contact printed, resulting in the geometric relationships shown.

The x and y coordinate positions of image points are referenced with respect to axes formed by straight lines joining the opposite fiducial marks (see Figure 2.29) recorded on the positive. The x axis is arbitrarily assigned to the fiducial axis most nearly coincident with the line of flight and is taken as positive in the forward direction of flight. The positive y axis is located 90° counterclockwise from the positive x axis. Because of the precision with which the fiducial marks and the lens are placed in a metric camera, the photocoordinate origin, o, can be assumed to coincide exactly with the *principal point*, the intersection of the lens optical axis and the film plane. The point where the prolongation of the optical axis of the camera intersects the terrain is referred to as the *ground principal point*, O. Images for terrain points A, B, C, D, and E appear geometrically reversed on the negative at a', b', c', d', and e' and in proper geometric relationship on the positive at a, b, c, d, and e. (Throughout this chapter we refer to points on the image with lowercase letters and corresponding points on the terrain with uppercase letters.)

The xy photocoordinates of a point are the perpendicular distances from the xy coordinate axes. Points to the right of the y axis have positive x coordi-

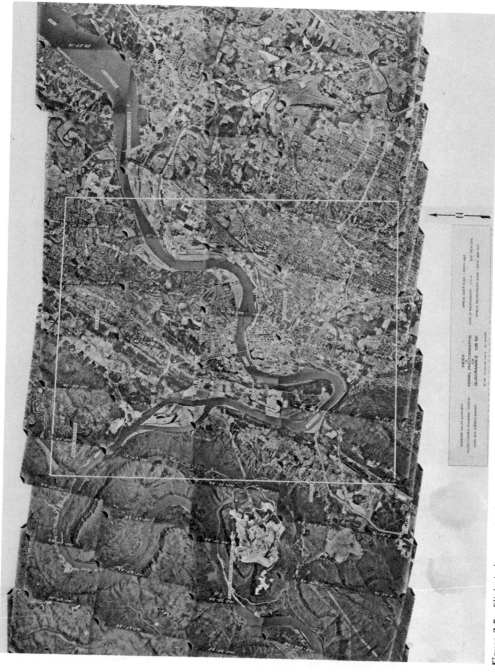

Figure 3.5 Flight index mosaic showing four east–west flight lines of aerial photography over Chattanooga, TN. Area outlined in white indicates coverage area of a single 1 : 24,000 quadrangle map. (Courtesy Mapping Services Branch, Tennessee Valley Authority.)

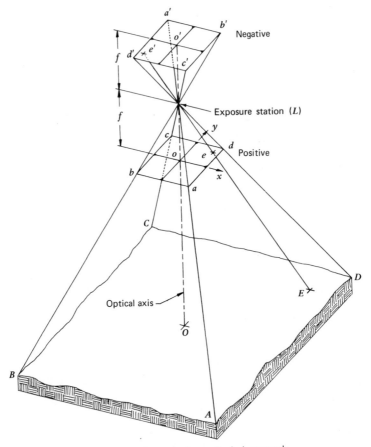

Figure 3.6   Basic geometric elements of a vertical photograph.

nates and points to the left have negative x coordinates. Similarly, points above the x axis have positive y coordinates and those below have negative y coordinates.

Photocoordinate Measurement

Measurements of photocoordinates may be obtained using any one of many measurement devices. These devices vary in their accuracy, cost, and availability. For rudimentary photogrammetric problems—where low orders of measurement accuracy are acceptable—a triangular *engineer's scale* or *metric scale* may be used. When using these scales, measurement accuracy is generally improved by taking the average of several repeated measurements. Measurements are also generally more accurate when made with the aid of a magnifying lens.

Photocoordinates can also be measured using a *coordinate digitizer* (see Figure 3.11). Such devices continuously display the *xy* positions of a spatial reference mark as it is positioned anywhere on the photograph. Another option for photocoordinate measurement is the use of a precision instrument called a *comparator*. A *monocomparator* can be used to measure very accurate coordinates on one photograph at a time; a *stereocomparator* can be used for making measurements on stereopairs.

In softcopy photogrammetric operations, individual points in a photograph are referenced by their row and column coordinates in the digital raster representation of the image. The relationship between the row and column coordinate system and the camera's fiducial axis coordinate system is determined through the development of a mathematical *coordinate transformation* between the two systems. This process requires that some points have their coordinates known in both systems. The fiducial marks are used for this purpose in that their positions in the focal plane are determined during the calibration of the camera, and they can be readily measured in the row and column coordinate system. (Appendix C contains a description of the mathematical form of the *affine coordinate transformation*, which is often used to interrelate the fiducial and row and column coordinate systems.)

Irrespective of what approach is used to measure photocoordinates, these measurements contain errors of varying sources and magnitudes. These errors stem from sources such as camera lens distortions, atmospheric refraction, earth curvature, failure of the fiducial axes to intersect at the principal point, and shrinkage or expansion of the photographic material on which measurements are made. Sophisticated photogrammetric analyses include corrections for all these errors. For simple measurements made on paper prints, such corrections are usually not employed because errors introduced by slight tilt in the photography will outweigh the effect of the other distortions.

3.3 PHOTOGRAPHIC SCALE

One of the most fundamental and frequently used geometric characteristics of aerial photographs is that of *photographic scale*. A photograph "scale," like a map scale, is an expression that states that one unit (any unit) of distance on a photograph represents a specific number of units of actual ground distance. Scales may be expressed as *unit equivalents*, *representative fractions*, or *ratios*. For example, if 1 mm on a photograph represents 25 m on the ground, the scale of the photograph can be expressed as 1 mm = 25 m (unit equivalents), or $\frac{1}{25,000}$ (representative fraction), or 1:25,000 (ratio).

Quite often the terms "large scale" and "small scale" are confused by those not working with expressions of scale on a routine basis. For example, which photograph would have the "larger" scale—a 1:10,000 scale photo covering several city blocks or a 1:50,000 photo that covers an entire city? The intuitive answer is often that the photo covering the larger "area" (the entire city) is the larger scale product. This is not the case. The larger scale product is the 1:10,000 image because it shows ground features at a larger, more detailed, size. The 1:50,000 scale photo of the entire city would render ground features at a much smaller, less detailed size. Hence, in spite of its larger ground coverage, the 1:50,000 photo would be termed the smaller scale product.

A convenient way to make scale comparisons is to remember that the same objects are smaller on a "smaller" scale photograph than on a "larger" scale photo. Scale comparisons can also be made by comparing the magnitudes of the representative fractions involved. (That is, $\frac{1}{50,000}$ is smaller than $\frac{1}{10,000}$.)

The most straightforward method for determining photo scale is to measure the corresponding photo and ground distances between any two points. This requires that the points be mutually identifiable on both the photo and a map. The scale S is then computed as the ratio of the photo distance d to the ground distance D,

$$S = \text{photo scale} = \frac{\text{photo distance}}{\text{ground distance}} = \frac{d}{D} \qquad (3.1)$$

EXAMPLE 3.1

Assume that two road intersections shown on a photograph can be located on a 1:25,000 scale topographic map. The measured distance between the intersections is 47.2 mm on the map and 94.3 mm on the photograph. (a) What is the scale of the photograph? (b) At that scale, what is the length of a fence line that measures 42.9 mm on the photograph?

Solution

(a) The ground distance between the intersections is determined from the map scale as

$$0.0472 \text{ m} \times \frac{25,000}{1} = 1180 \text{ m}$$

By direct ratio, the photo scale is

$$S = \frac{0.0943 \text{ m}}{1180 \text{ m}} = \frac{1}{12,513} \quad or \quad 1:12,500$$

(Note that because only three significant, or meaningful, figures were present in the original measurements, only three significant figures are indicated in the final result.)

(b) The ground length of the 42.9-mm fence line is

$$D = \frac{d}{S} = 0.0429 \text{ m} \div \frac{1}{12,500} = 536.25 \text{ m} \quad \text{or} \quad 536 \text{ m}$$

For a vertical photograph taken over flat terrain, scale is a function of the focal length f of the camera used to acquire the image and the flying height above the ground, H', from which the image was taken. In general,

$$\text{Scale} = \frac{\text{camera focal length}}{\text{flying height above terrain}} = \frac{f}{H'} \tag{3.2}$$

Figure 3.7 illustrates how we arrive at Eq. 3.2. Shown in this figure is the side view of a vertical photograph taken over flat terrain. Exposure station L is at

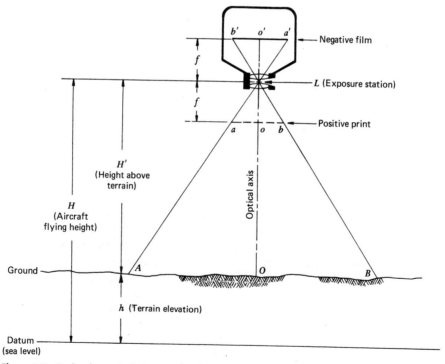

Figure 3.7 Scale of a vertical photograph taken over flat terrain.

an aircraft *flying height H* above some *datum*, or arbitrary base elevation. The datum most frequently used is mean sea level. If flying height H and the elevation of the terrain h are known, we can determine H' by subtraction ($H' = H - h$). If we now consider terrain points A, O, and B, they are imaged at points a', o', and b' on the negative film and at a, o, and b on the positive print. We can derive an expression for photo scale by observing similar triangles Lao and LAO, which are corresponding photo and ground distances. That is,

$$S = \frac{\overline{ao}}{\overline{AO}} = \frac{f}{H'} \tag{3.3}$$

Equation 3.3 is identical to our scale expression of Eq. 3.2. Yet another way of expressing these equations is

$$S = \frac{f}{H - h} \tag{3.4}$$

Equation 3.4 is the most commonly used form of the scale equation.

EXAMPLE 3.2

A camera equipped with a 152-mm-focal-length lens is used to take a vertical photograph from a flying height of 2780 m above mean sea level. If the terrain is flat and located at an elevation of 500 m, what is the scale of the photograph?

Solution

$$\text{Scale} = \frac{f}{H - h} = \frac{0.152 \text{ m}}{2780 \text{ m} - 500 \text{ m}} = \frac{1}{15,000} \quad \text{or} \quad 1:15,000$$

The most important principle expressed by Eq. 3.4 is that photo scale is a function of terrain elevation h. Because of the level terrain, the photograph depicted in Figure 3.7 has a constant scale. However, *photographs taken over terrain of varying elevation will exhibit a continuous range of scales associated with the variations in terrain elevation.* Likewise, tilted and oblique photographs have nonuniform scales.

EXAMPLE 3.3

Assume a vertical photograph was taken at a flying height of 5000 m above sea level using a camera with a 152-mm-focal-length lens. (a) Determine the photo scale at points A and B, which lie at elevations of 1200 and 1960 m. (b) What ground distance corresponds to a 20.1-mm photo distance measured at each of these elevations?

Solution

(a) By Eq. 3.4,

$$S_A = \frac{f}{H - h_A} = \frac{0.152 \text{ m}}{5000 \text{ m} - 1200 \text{ m}} = \frac{1}{25,000} \quad \text{or} \quad 1:25,000$$

$$S_B = \frac{f}{H - h_B} = \frac{0.152 \text{ m}}{5000 \text{ m} - 1960 \text{ m}} = \frac{1}{20,000} \quad \text{or} \quad 1:20,000$$

(b) The ground distance corresponding to a 20.1-mm photo distance is

$$D_A = \frac{d}{S_A} = 0.0201 \text{ m} \div \frac{1}{25,000} = 502.5 \text{ m} \quad \text{or} \quad 502 \text{ m}$$

$$D_B = \frac{d}{S_B} = 0.0201 \text{ m} \div \frac{1}{20,000} = 402 \text{ m}$$

Often it is convenient to compute an *average scale* for an entire photograph. This scale is calculated using the average terrain elevation for the area imaged. Consequently, it is exact for distances occurring at the average elevation and is approximate at all other elevations. Average scale may be expressed as

$$S_{\text{avg}} = \frac{f}{H - h_{\text{avg}}} \tag{3.5}$$

where h_{avg} is the average elevation of the terrain shown in the photograph.

The result of photo scale variation is geometric distortion. All points on a *map* are depicted in their true relative horizontal (planimetric) positions, but points on a *photo* taken over varying terrain are displaced from their true "map positions." This difference results because a map is a scaled *orthographic* projection of the ground surface, whereas a vertical photograph yields a *perspective* projection. The differing nature of these two forms of projection is illustrated in Figure 3.8. As shown, a map results from projecting vertical rays from ground points to the map sheet (at a particular scale). A photograph results from projecting converging rays through a common point within the camera lens. Because of the nature of this projection, any variations in terrain elevation will result in scale variation *and* displaced image positions.

On a map we see a top view of objects in their true relative horizontal positions. On a photograph, areas of terrain at the higher elevations lie

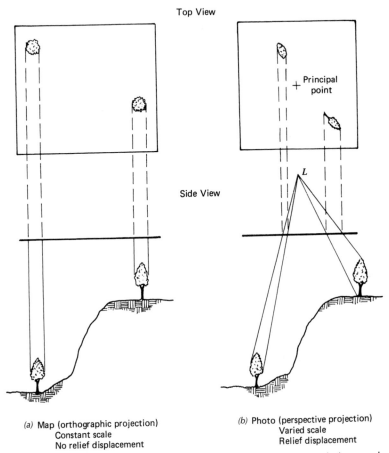

Top View

Side View

Principal point

L

(a) Map (orthographic projection)
Constant scale
No relief displacement

(b) Photo (perspective projection)
Varied scale
Relief displacement

Figure 3.8 Comparative geometry of (a) a map and (b) a vertical aerial photograph. Note differences in size, shape, and location of the two trees.

closer to the camera at the time of exposure and therefore appear larger than corresponding areas lying at lower elevations. Furthermore, the tops of objects are always displaced from their bases (Figure 3.8). This distortion is called *relief displacement* and causes any object standing above the terrain to "lean" away from the principal point of a photograph radially. We treat the subject of relief displacement in Section 3.6.

By now the reader should see that the only circumstance wherein an aerial photograph can be treated as if it were a map directly is in the case of a vertical photograph imaging uniformly flat terrain. This is rarely the case in practice and the image analyst must always be aware of the potential geometric distortions introduced by such influences as tilt, scale variation, and relief displacement. Failure to deal with these distortions will often lead, among other things, to a lack of geometric "fit" among image-derived and nonimage data sources in a GIS. However, if these factors are properly addressed photogrammetrically, extremely reliable measurements and map products can be derived from aerial photography.

3.4 GROUND COVERAGE OF AERIAL PHOTOGRAPHS

The ground coverage of a photograph is, among other things, a function of camera format size. For example, an image taken with a camera having a 230 × 230-mm format (on 240-mm film) has about 17.5 times the ground area coverage of an image of equal scale taken with a camera having a 55 × 55-mm format (on 70-mm film) and about 61 times the ground area coverage of an image of equal scale taken with a camera having a 24 × 36-mm format (on 35-mm film). As with photo scale, the ground coverage of photography obtained with any given format is a function of focal length and flying height above ground, H'. For a constant flying height, the width of the ground area covered by a photo varies inversely with focal length. Consequently, photos taken with shorter focal length lenses have larger areas of coverage (and smaller scales) than do those taken with longer focal length lenses. For any given focal length lens, the width of the ground area covered by a photo varies directly with flying height above terrain, with image scale varying inversely with flying height.

The effect that flying height has on ground coverage and image scale is illustrated in Figures 3.9a, b, and c. These images were all taken over Chattanooga, Tennessee, with the same camera type equipped with the same focal length lens but from three different altitudes. Figure 3.9a is a high altitude, small-scale image showing virtually the entire Chattanooga metropolitan area. Figure 3.9b is a lower altitude, larger scale image showing the ground area outlined in Figure 3.9a. Figure 3.9c is a yet lower altitude, larger scale image of the area outlined in Figure 3.9b. Note the trade-offs between the ground area covered by an image and the object detail available in each of the photographs.

3.5 AREA MEASUREMENT

The process of measuring areas using aerial photographs can take on many forms. The accuracy of area measurement is a function of not only the measuring device used, but also the degree of image scale variation due to relief in the terrain and tilt in the photography. Although large errors in

(a)

Figure 3.9 (a) Scale 1 : 210,000 vertical aerial photograph showing Chattanooga, TN. This figure is a 1.75× reduction of an original photograph taken with f = 152.4 mm from 18,300 m flying height. (NASA photograph.) (b) Scale 1 : 35,000 vertical aerial photograph providing coverage of area outlined in (a). This figure is a 1.75× reduction of an original photograph taken with f = 152.4 mm from 3050 m flying height. (c) Scale 1 : 10,500 vertical aerial photograph providing coverage of area outlined in (b). This figure is a 1.75× reduction of an original photograph taken with f = 152.4 mm from 915 m flying height. (Courtesy Mapping Services Branch, Tennessee Valley Authority.)

Figure 3.9 *(Continued)*

Figure 3.9 *(Continued)*

area determinations can result even with vertical photographs in regions of moderate to high relief, accurate measurements may be made on vertical photos of areas of low relief.

Simple scales may be used to measure the area of simply shaped features. For example, the area of a rectangular field can be determined by simply measuring its length and width. Similarly, the area of a circular feature can be computed after measuring its radius or diameter.

EXAMPLE 3.4

A rectangular agricultural field measures 8.65 cm long and 5.13 cm wide on a vertical photograph having a scale of 1:20,000. Find the area of the field at ground level.

Solution

$$\text{Ground length} = \text{photo length} \times \frac{1}{S} = 0.0865 \text{ m} \times 20,000 = 1730 \text{ m}$$

$$\text{Ground width} = \text{photo width} \times \frac{1}{S} = 0.0513 \times 20,000 = 1026 \text{ m}$$

$$\text{Ground area} = 1730 \text{ m} \times 1026 \text{ m} = 1,774,980 \text{ m}^2 = 177 \text{ ha}$$

The ground area of an irregularly shaped feature is usually determined by measuring the area of the feature on the photograph. The photo area is then converted to a ground area from the following relationship:

$$\text{Ground area} = \text{photo area} \times \frac{1}{S^2}$$

EXAMPLE 3.5

The area of a lake is 52.2 cm^2 on a 1:7500 vertical photograph. Find the ground area of the lake.

Solution

$$\text{Ground area} = \text{photo area} \times \frac{1}{S^2} = 0.00522 \text{ m}^2 \times 7500^2 = 293,625 \text{ m}^2 = 29.4 \text{ ha}$$

Numerous methods can be used to measure the area of irregularly shaped features on a photograph. One of the simplest techniques employs a transparent grid overlay consisting of lines forming rectangles or squares of known area. The grid is placed over the photograph and the area of a ground unit is estimated by counting grid units that fall within the unit to be measured. Perhaps the most widely used grid overlay is a *dot grid* (Figure 3.10). This grid, composed of uniformly spaced dots, is superimposed over the photo, and the dots falling within the region to be measured are counted. From knowledge of the dot density of the grid, the photo area of the region can be computed.

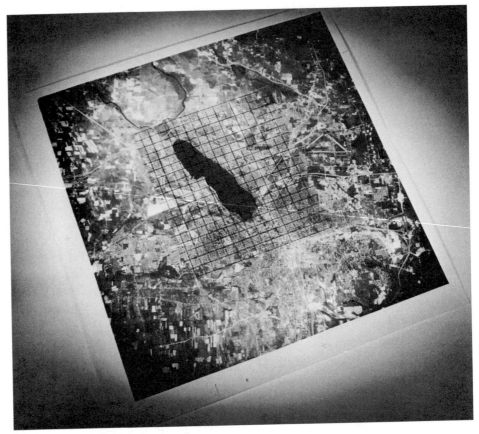

Figure 3.10 Transparent dot grid overlay.

EXAMPLE 3.6

A flooded area is covered by 129 dots on a 25-dot/cm^2 grid on a 1:20,000 vertical aerial photograph. Find the ground area flooded.

Solution

$$\text{Dot density} = \frac{1 \text{ cm}^2}{25 \text{ dots}} \times 20,000^2 = 16,000,000 \text{ cm}^2/\text{dot} = 0.16 \text{ ha/dot}$$

$$\text{Ground area} = 129 \text{ dots} \times 0.16 \text{ ha/dot} = 20.6 \text{ ha}$$

The dot grid is an inexpensive tool and its use requires little training. When numerous regions are to be measured, however, the counting proce-

Figure 3.11 Precision coordinate digitizer. (Courtesy Altek Corp.)

dure becomes quite tedious. An alternative technique is to use either a coordinate digitizer or digitizing tablet (Figure 3.11). These devices are typically interfaced with a computer such that area determination simply involves tracing around the boundary of the region of interest and the area can be read out directly. When photographs are available in softcopy format, area measurement often involves digitizing from a computer monitor using a mouse or other form of cursor control.

3.6 RELIEF DISPLACEMENT OF VERTICAL FEATURES

Characteristics of Relief Displacement

In Figure 3.8, we illustrated the effect of relief displacement on a photograph taken over varied terrain. In essence, an increase in the elevation of a feature causes its position on the photograph to be displaced radially outward from the principal point. Hence, when a vertical feature is photographed, relief displacement causes the top of the feature to lie farther from the photo center than its base. As a result, vertical features appear to lean away from the center of the photograph.

The pictorial effect of relief displacement is illustrated by the aerial photographs shown in Figure 3.12. These photographs depict the construction site of the Watts Bar Nuclear Plant adjacent to the Tennessee River. An operating coal-fired steam plant with its fan-shaped coal stockyard is shown in the

Figure 3.12 Vertical photographs of the Watts Bar Nuclear Power Plant Site, near Kingston, TN. In (a) the two plant cooling towers appear near the principal point and exhibit only slight relief displacement. The towers manifest severe relief displacement in (b). (Courtesy Mapping Services Branch, Tennessee Valley Authority.)

upper right of Figure 3.12a; the nuclear plant is shown in the center. Note particularly the two large cooling towers adjacent to the plant. In (a) these towers appear nearly in top view because they are located very close to the principal point of this photograph. However, the towers manifest some relief displacement as the top tower appears to lean somewhat toward the upper right and the bottom tower toward the lower right. In (b) the towers are

(b)

Figure 3.12   *(Continued)*

shown at a greater distance from the principal point. Note the increased relief displacement of the towers. We now see more of a "side view" of the objects since the images of their tops are displaced farther than the images of their bases. These photographs illustrate the radial nature of relief displacement and the increase in relief displacement with an increase in the radial distance from the principal point of a photograph.

The geometric components of relief displacement are illustrated in Figure 3.13, which shows a vertical photograph imaging a tower. The photograph is

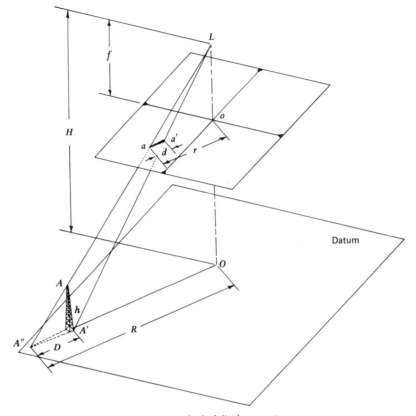

Figure 3.13 Geometric components of relief displacement.

taken from flying height H above datum. When considering the relief displacement of a vertical feature, it is convenient to arbitrarily assume a datum plane placed at the base of the feature. If this is done, the flying height H must be correctly referenced to this same datum, *not* mean sea level. Thus, in Figure 3.13 the height of the tower (whose base is at datum) is h. Note that the top of the tower, A, is imaged at a in the photograph whereas the base of the tower, A', is imaged at a'. That is, the image of the top of the tower is radially displaced by the distance d from that of the bottom. The distance d is the relief displacement of the tower. The equivalent distance projected to datum is D. The distance from the photo principal point to the top of the tower is r. The equivalent distance projected to datum is R.

We can express d as a function of the dimensions shown in Figure 3.13. From similar triangles $AA'A''$ and LOA'',

$$\frac{D}{h} = \frac{R}{H}$$

Expressing distances D and R at the scale of the photograph, we obtain

$$\frac{d}{h} = \frac{r}{H}$$

Rearranging the above equation yields

$$d = \frac{rh}{H} \tag{3.6}$$

where

d = relief displacement
r = radial distance on the photograph from the principal point to the displaced image point
h = height above datum of the object point
H = flying height above the same datum chosen to reference h

An analysis of Eq. 3.6 indicates mathematically the nature of relief displacement seen pictorially. That is, relief displacement of any given point increases as the distance from the principal point increases (this can be seen in Figure 3.12), and it increases as the elevation of the point increases. Other things being equal, it decreases with an increase in flying height. Hence, under similar conditions high altitude photography of an area manifests less relief displacement than low altitude photography. Also, there is no relief displacement at the principal point (since $r = 0$).

Object Height Determination from Relief Displacement Measurement

Equation 3.6 also indicates that relief displacement increases with the feature height h. This relationship makes it possible to indirectly measure heights of objects appearing on aerial photographs. By rearranging Eq. 3.6, we obtain

$$h = \frac{dH}{r} \tag{3.7}$$

To use Eq. 3.7, both the top and base of the object to be measured must be clearly identifiable on the photograph and the flying height H must be known. If this is the case, d and r can be measured on the photograph and used to calculate the object height h. (When using Eq. 3.7, it is important to remember that H must be referenced to the elevation of the base of the feature, not to mean sea level.)

EXAMPLE 3.7

For the photo shown in Figure 3.13, assume that the relief displacement for the tower at A is 2.01 mm, and the radial distance from the center of the photo to the top of the tower is 56.43 mm. If the flying height is 1220 m above the base of the tower, find the height of the tower.

Solution
By Eq. 3.7

$$h = \frac{dH}{r} = \frac{2.01 \text{ mm } (1220 \text{ m})}{56.43 \text{ mm}} = 43.4 \text{ m}$$

While measuring relief displacement is a very convenient means of calculating heights of objects from aerial photographs, the reader is reminded of the assumptions implicit in the use of the method. We have assumed use of truly vertical photography, accurate knowledge of the flying height, clearly visible objects, precise location of the principal point, and a measurement technique whose accuracy is consistent with the degree of relief displacement involved. If these assumptions are reasonably met, quite reliable height determinations may be made using single prints and relatively unsophisticated measuring equipment.

Correcting for Relief Displacement

In addition to calculating object heights, quantification of relief displacement can be used to correct the image positions of terrain points appearing in a photograph. Keep in mind that terrain points in areas of varied relief exhibit relief displacements as do vertical objects. This is illustrated in Figure 3.14. In this figure, the datum plane has been set at the average terrain elevation (not at mean sea level). If all terrain points were to lie at this common elevation, terrain points A and B would be located at A' and B' and would be imaged at points a' and b' on the photograph. Due to the varied relief, however, the position of point A is shifted radially outward on the photograph (to a), and the position of point B is shifted radially inward (to b). These changes in image position are the relief displacements of points A and B. Figure 3.14b illustrates the effect they have on the geometry of the photo. Because A' and B' lie at the same terrain elevation, the image line a'b' accurately represents the scaled horizontal length and directional orientation of the ground line AB. When the relief displacements are introduced, the resulting line ab has a considerably altered length and orientation.

Angles are also distorted by relief displacements. In Figure 3.14b, the horizontal ground angle ACB is accurately expressed by a'cb' on the photo. Due to the displacements, the distorted angle acb will appear on the photograph. Note that, because of the radial nature of relief displacements, angles about the origin of the photo (such as aob) will not be distorted.

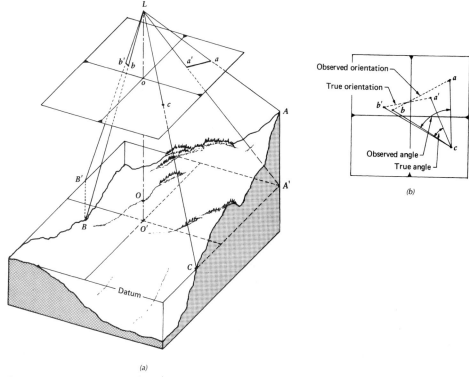

Figure 3.14 Relief displacement on a photograph taken over varied terrain: (a) displacement of terrain points; (b) distortion of horizontal angles measured on photograph.

Relief displacement can be corrected for by using Eq. 3.6 to compute its magnitude on a point-by-point basis and then laying off the computed displacement distances radially (in reverse) on the photograph. This procedure establishes the datum-level image positions of the points and removes the relief distortions, resulting in planimetrically correct image positions at datum scale. This scale can be determined from the flying height above datum ($S = f/H$). Ground lengths, directions, angles, and areas may then be directly determined from these corrected image positions.

EXAMPLE 3.8

Referring to the vertical photograph depicted in Figure 3.14, assume that the radial distance r_a to point A is 63.84 mm and the radial distance r_b to point B is 62.65 mm. Flying height H is 1220 m above datum, point A is 152 m above datum, and point B is 168 m below datum. Find the radial distance and direction one must lay off from points a and b to plot a' and b'.

Solution
By Eq. 3.6

$$d_a = \frac{r_a h_a}{H} = \frac{63.84 \text{ mm} \times 152 \text{ m}}{1220 \text{ m}} = 7.95 \text{ mm} \qquad \text{(plot inward)}$$

$$d_b = \frac{r_b h_b}{H} = \frac{62.65 \text{ mm} \times (-168 \text{ m})}{1220 \text{ m}} = -8.63 \text{ mm} \qquad \text{(plot outward)}$$

3.7 IMAGE PARALLAX

Characteristics of Image Parallax

Thus far we have limited our discussion to photogrammetric operations involving only single vertical photographs. Numerous applications of photogrammetry incorporate the analysis of stereopairs and use of the principle of *parallax*. The term *parallax* refers to the apparent change in relative positions of stationary objects caused by a change in viewing position. This phenomenon is observable when one looks at objects through a side window of a moving vehicle. With the moving window as a frame of reference, objects such as mountains at a relatively great distance from the window appear to move very little within the frame of reference. In contrast, objects close to the window, such as roadside trees, appear to move through a much greater distance.

In the same way that the close trees move relative to the distant mountains, terrain features close to an aircraft (i.e., at higher elevation) will appear to move relative to the lower elevation features when the point of view changes between successive exposures. These relative displacements form the basis for three-dimensional viewing of overlapping photographs. In addition, they can be measured and used to compute the elevations of terrain points.

Figure 3.15 illustrates the nature of parallax on overlapping vertical photographs taken over varied terrain. Note that the relative positions of points *A* and *B* change with the change in viewing position (in this case, the exposure station). Note also that the *parallax displacements occur only parallel to the line of flight*. In theory, the direction of flight should correspond precisely to the fiducial *x* axis. In reality, however, unavoidable changes in the aircraft orientation will usually slightly offset the fiducial axis from the flight axis. The true flight line axis may be found by first locating on a photograph the points that correspond to the image centers of the preceding and succeeding photographs. These points are called the *conjugate principal points*. A line drawn through the principal points and the conjugate principal points defines the

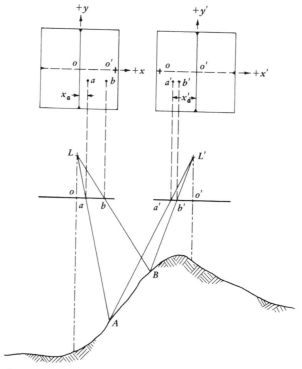

Figure 3.15 Parallax displacements on overlapping vertical photographs.

flight axis. As shown in Figure 3.16, all photographs except those on the ends of a flight strip normally have two sets of flight axes. This happens because the aircraft's path between exposures is usually slightly curved. In Figure 3.16, the flight axis for the stereopair formed by photos 1 and 2 is flight axis 12. The flight axis for the stereopair formed by photos 2 and 3 is flight axis 23.

The line of flight for any given stereopair defines a photocoordinate x axis for use in parallax measurement. Lines drawn perpendicular to the flight line and passing through the principal point of each photo form the photographic y axes for parallax measurement. The parallax of any point, such as A in Figure 3.15, is expressed in terms of the flight line coordinate system as

$$p_a = x_a - x_a'$$

(3.8)

where

p_a = parallax of point A

x_a = measured x coordinate of image a on the left photograph of the stereopair

x_a' = x coordinate of image a' on the right photograph

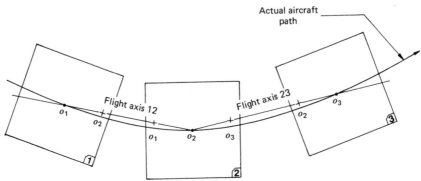

Figure 3.16 Flight line axes for successive stereopairs along a flight strip. (Curvature of aircraft path is exaggerated.)

The x axis for each photo is considered positive to the right of each photo principal point. This makes x'_a a negative quantity in Figure 3.15.

Object Height and Ground Coordinate Location from Parallax Measurement

Figure 3.17 shows overlapping vertical photographs of a terrain point, A. Using parallax measurements, we may determine the elevation at A and its ground coordinate location. Referring to Figure 3.17a, the horizontal distance between exposure stations L and L' is called B, the *air base*. The triangle in Figure 3.17b results from superimposition of the triangles at L and L' in order to graphically depict the nature of parallax p_a as computed from Eq. 3.8 algebraically. From similar triangles $La'_x a_x$ (Figure 3.17b) and $LA_x L'$ (Figure 3.17a)

$$\frac{p_a}{f} = \frac{B}{H - h_A}$$

from which

$$H - h_A = \frac{Bf}{p_a} \tag{3.9}$$

Rearranging yields

$$h_A = H - \frac{Bf}{p_a} \tag{3.10}$$

Also, from similar triangles $LO_A A_x$ and Loa_x,

$$\frac{X_A}{H - h_A} = \frac{x_a}{f}$$

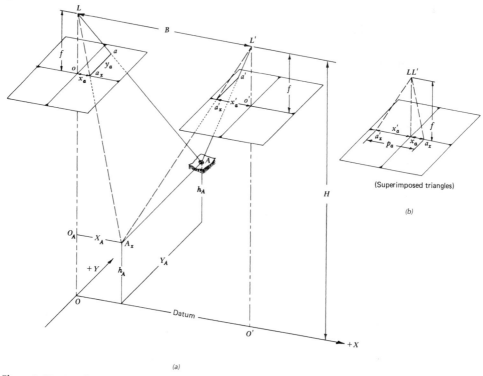

Figure 3.17   Parallax relationships on overlapping vertical photographs: (*a*) adjacent photographs forming a stereopair; (*b*) superimposition of right photograph onto left.

from which

$$X_A = \frac{x_a(H - h_A)}{f}$$

and substituting Eq. 3.9 into the above equation yields

$$X_A = B\frac{x_a}{p_a} \qquad (3.11)$$

A similar derivation using y coordinates yields

$$Y_A = B\frac{y_a}{p_a} \qquad (3.12)$$

Equations 3.10 to 3.12 are commonly known as the *parallax equations*. In these equations, X and Y are ground coordinates of a point with respect to an arbitrary coordinate system whose origin is vertically below the left exposure station and with positive X in the direction of flight; p is the parallax of the point in question; and x and y are the photocoordinates of the point on the left-hand photo. The major assumptions made in the derivation of these equations are that the photos are truly vertical and that they are taken from the same flying height. If these assumptions are sufficiently met, a complete survey of the ground region contained in the photo overlap area of a stereopair can be made.

EXAMPLE 3.9

The length of line AB and the elevation of its endpoints, A and B, are to be determined from a stereopair containing images a and b. The camera used to take the photographs has a 152.4-mm lens. The flying height was 1200 m (average for the two photos) and the air base was 600 m. The measured photographic coordinates of points A and B in the "flight line" coordinate system are $x_a = 54.61$ mm, $x_b = 98.67$ mm, $y_a = 50.80$ mm, $y_b = -25.40$ mm, $x_a' = -59.45$ mm, and $x_b' = -27.39$ mm. Find the length of line AB and the elevations of A and B.

Solution
From Eq. 3.8

$$p_a = x_a - x_a' = 54.61 - (-59.45) = 114.06 \text{ mm}$$

$$p_b = x_b - x_b' = 98.67 - (-27.39) = 126.06 \text{ mm}$$

From Eqs. 3.11 and 3.12

$$X_A = B\frac{x_a}{p_a} = \frac{600 \times 54.61}{114.06} = 287.27 \text{ m}$$

$$X_B = B\frac{x_b}{p_b} = \frac{600 \times 98.67}{126.06} = 469.63 \text{ m}$$

$$Y_A = B\frac{y_a}{p_a} = \frac{600 \times 50.80}{114.06} = 267.23 \text{ m}$$

$$Y_B = B\frac{y_b}{p_b} = \frac{600 \times (-25.40)}{126.06} = -120.89 \text{ m}$$

Applying the Pythagorean theorem yields

$$AB = [(469.63 - 287.27)^2 + (-120.89 - 267.23)^2]^{1/2} = 428.8 \text{ m}$$

From Eq. 3.10, the elevations of A and B are

$$h_A = H - \frac{Bf}{p_a} = 1200 - \frac{600 \times 152.4}{114.06} = 398 \text{ m}$$

$$h_B = H - \frac{Bf}{p_b} = 1200 - \frac{600 \times 152.4}{126.06} = 475 \text{ m}$$

In many applications, the *difference* in elevation between two points is of more immediate interest than is the actual value of the elevation of either point. In such cases, the change in elevation between two points can be found from

$$\Delta h = \frac{\Delta p \, H'}{p_a} \tag{3.13}$$

where

Δh = difference in elevation between two points whose parallax *difference* is Δp

H' = flying height above the lower point

p_a = parallax of the higher point

Using this approach in our previous example yields

$$\Delta h = \frac{12.00 \times 802}{126.06} = 77 \text{ m}$$

Note this answer agrees with the value computed above.

Parallax Measurement

To this point in our discussion, we have said little about how parallax measurements are made. In Example 3.9 we assumed that x and x' for points of interest were measured directly on the left and right photos, respectively. Parallaxes were then calculated from the algebraic differences of x and x', in accordance with Eq. 3.8. This procedure becomes cumbersome when many points are analyzed, since two measurements are required for each point.

Figure 3.18 illustrates the principle behind methods of parallax measurement that require only a single measurement for each point of interest. If the two photographs constituting a stereopair are fastened to a base with their flight lines aligned, the distance D remains constant for the setup, and the parallax of a point can be derived from measurement of the single distance d. That is, $p = D - d$. Distance d can be measured with a simple scale, *assuming a and a' are identifiable*. In areas of uniform photo tone, individual features may not be identifiable, making the measurement of d very difficult.

Employing the principle illustrated in Figure 3.18, a number of devices have been developed to increase the speed and accuracy of parallax measurement. These devices also permit parallax to be easily measured in areas

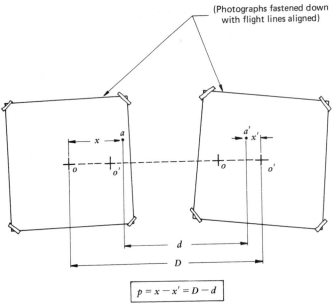

Figure 3.18 Alignment of a stereopair for parallax measurement.

of uniform photo tone. All employ stereoscopic viewing and the principle of the *floating mark*. This principle is illustrated in Figure 3.19. While viewing through a stereoscope, the image analyst uses a device that places small identical marks over each photograph. These marks are normally dots or crosses etched on transparent material. The marks—called *half marks*—are positioned over similar areas on the left-hand photo and the right-hand photo. The left mark is seen only by the left eye of the analyst and the right mark is seen only by the right eye. The relative positions of the half marks can be shifted along the direction of flight until they visually fuse together, forming a single mark that appears to "float" at a specific level in the stereomodel. The apparent elevation of the floating mark varies with the spacing between the half marks. Figure 3.19 illustrates how the fused marks can be made to float and can actually be set on the terrain at particular points in the stereomodel. Half-mark positions (a, b), (a, c), and (a, d) result in floating-mark positions in the model at B, C, and D.

A very simple device for measuring parallax is the *parallax wedge*. It consists of a transparent sheet of plastic on which are printed two converging lines or rows of dots (or graduated lines). Next to one of the converging lines is a scale that shows the horizontal distance between the two lines at each

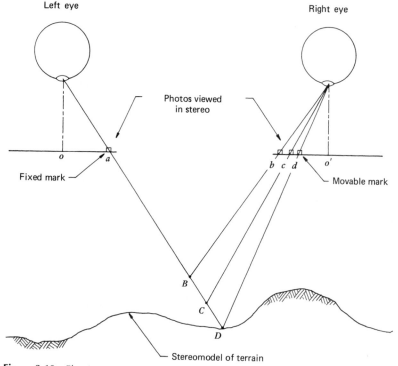

Figure 3.19 Floating-mark principle. (Note that only the right half mark is moved to change the apparent height of the floating mark in the stereomodel.)

point. Consequently, these graduations can be thought of as a series of distance d measurements as shown in Figure 3.18.

Figure 3.20 shows a parallax wedge set up for use. The wedge is positioned so that one of the converging lines lies over the left photo in a stereopair and one over the right photo. When viewed in stereo, the two lines fuse together over a portion of their length, forming a single line that appears to float in the stereomodel. Because the lines on the wedge converge, the floating line appears to slope through the stereoscopic image.

Figure 3.21 illustrates how a parallax wedge might be used to determine the height of a tree. In Figure 3.21a, the position of the wedge has been adjusted until the sloping line appears to intersect the top of the tree. A reading is taken from the scale at this point (58.55 mm). The wedge is then positioned such that the line intersects the base of the tree, and a reading is taken (59.75 mm). The difference between the readings (1.20 mm) is used to determine the tree height.

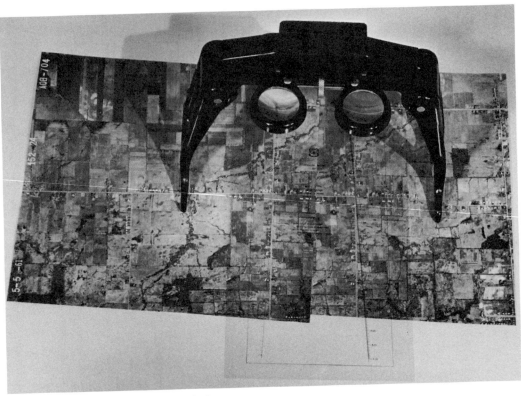

Figure 3.20 Parallax wedge oriented under lens stereoscope.

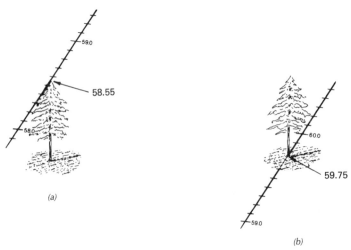

Figure 3.21 Parallax wedge oriented for taking a reading on (a) the top and (b) the base of a tree.

EXAMPLE 3.10

The flying height for an overlapping pair of photos is 1600 m above the ground and p_a is 75.60 mm. Find the height of the tree illustrated in Figure 3.21.

Solution
From Eq. 3.13

$$\Delta h = \frac{\Delta p\, H'}{p_a}$$

$$= \frac{1.20 \times 1600}{75.60} = 25 \text{ m}$$

Parallax measurement in softcopy photogrammetric systems usually involves some form of numerical *image correlation* to match points on the left photo of a stereopair to their conjugate images on the right photo. Figure 3.22

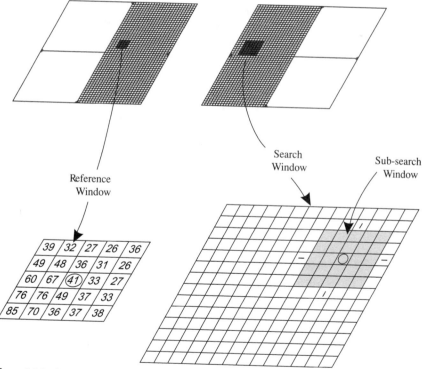

Search
Window

Sub-search
Window

Reference
Window

39	32	27	26	36
49	48	36	31	26
60	67	(41)	33	27
76	76	49	37	33
85	70	36	37	38

Figure 3.22   Principle of image matching. (After Wolf, 1983.)

illustrates the general concept of digital image matching. Shown is a stereo-pair of photographs with the pixels contained in the overlap area depicted (greatly exaggerated in size). A *reference window* on the left photograph comprises a local neighborhood of pixels around a fixed location. In this case the reference window is square and 5 × 5 pixels in size. (Windows vary in size and shape based on the particular matching technique.)

A *search window* is established on the right-hand photo of sufficient size and general location to encompass the conjugate image of the central pixel of the reference window. The initial location of the search window can be determined based on the location of the reference window, the camera focal length, and the size of the area of overlap. A subsearch "moving window" of pixels (Chapter 7) is then systematically moved pixel by pixel about the rows and columns of the search window and the numerical correlation between the digital numbers within the reference and subsearch windows at each location of the moving subsearch window is computed. The conjugate image is assumed to be at the location where the correlation is a maximum.

There are various types of algorithms that can be used to perform image matching. The details of these procedures are not as important as the general concept of locating the conjugate image point for all points on a reference image. The resulting photocoordinates can then be used in the various parallax equations earlier described (Eqs. 3.10 to 3.13). However, the parallax equations assume perfectly vertical photography and equal flying heights for all images. This simplifies the geometry and hence the mathematics of computing ground positions from the photo measurements. However, softcopy systems are not constrained to the above assumptions. Such systems employ mathematical models of the imaging process that readily handle variations in the flying height and attitude of each photograph. As we discuss in Section 3.9, the relationship among image coordinates, ground coordinates, the exposure station position, and angular orientation of each photograph is normally described by a series of *collinearity equations*. They are used in the process of analytical *aerotriangulation*, which involves determining the X, Y, Z, ground coordinates of individual points based on photocoordinate measurements.

3.8 GROUND CONTROL FOR AERIAL PHOTOGRAPHY

As we have indicated throughout, most remote sensing analyses involve some form of ground reference data. Photogrammetric operations are no exception. In fact, for most photogrammetric activities one form of ground reference data is often essential—ground control. *Ground control* refers to physical points on the ground whose ground positions are known with respect to some horizontal coordinate system and/or vertical datum. When mutually identifiable on the ground and on a photograph, ground control points can be used to establish the exact spatial position and orientation of a photograph relative to the ground at the instant of exposure.

Ground control points may be *horizontal control points, vertical control points,* or both. Horizontal control point positions are known planimetrically in some *XY* coordinate systems (e.g., a state plane coordinate system). Vertical control points have known elevations with respect to a level datum (e.g., mean sea level). A single point with known planimetric position and known elevation can serve as both a horizontal and vertical control point.

Historically, ground control has been established through ground-surveying techniques in the form of triangulation, trilateration, traversing, and leveling. Currently, the establishment of ground control is aided by the use of GPS procedures. The details of these and other more sophisticated surveying techniques used for establishing ground control are not important for the reader of this book to understand. It does warrant reiteration, however, that *accurate ground control is essential to virtually all photogrammetric operations* because photogrammetric measurements can only be as reliable as the ground control on which they are based. Measurements on the photo can be accurately extrapolated to the ground only when we know the location and orientation of the photograph relative to the ground at the instant of exposure. Ground control is required to determine these factors.

As we mentioned, ground control points must be clearly identifiable both on the ground and on the photography being used. Often, control points are selected and surveyed after photography has been taken, thereby ensuring that the points are identifiable on the image. Cultural features, such as road intersections, are often used as control points in such cases. If a ground survey is made prior to a photo mission, control points may be premarked with artificial targets to aid in their identification on the photography. Crosses that contrast with the background land cover make ideal control point markers. Their size is selected in accordance with the scale of the photography to be flown and their material form can be quite variable. In many cases, markers are made by simply painting white crosses on roadways. Alternatively, markers can be painted on contrasting sheets of Masonite, plywood, or heavy cloth.

Research is now extremely active and promising in developing means to minimize the effort needed to establish ground control in photogrammetric operations. One approach uses a pair of interrelated GPS receivers, one stationed at a ground control point, and the other in the aircraft transporting the camera. In this way, the precise location of each photograph, in the ground coordinate system, can be determined. At the same time, an *Inertial Measurement Unit (IMU)* is employed to determine the angular orientation or attitude of each exposure. With the location and attitude of each photograph known, there is no need for ground control (except at the base station).

3.9 MAPPING WITH AERIAL PHOTOGRAPHS

Stereoscopic Plotting Instruments

As we mentioned in the introductory section of this chapter, one of the most widespread uses of photogrammetry is in the preparation of topographic maps. *Stereoplotters* are precision analog instruments designed for this purpose. Two projectors are used that can be adjusted in their position and angular orientation to duplicate the exact relative position and orientation of the aerial camera at the instants the two photos of a stereopair were exposed. That is, camera tilts are precisely re-created in the projection process. Likewise, the base distance between exposures and differences in flying heights are simulated by adjusting the relative positions of the projectors.

Conceptually, the operating principle of a stereoplotter is quite simple. Each photograph in a stereopair is the result of rays projected from the terrain, through a lens, onto an image plane that has a particular position and attitude. In a stereoplotter, the direction of projection is simply reversed. We project rays from the photographs (in the same relative orientation in which they were taken) to form a greatly reduced scale model of the terrain in the overlap area. The model can be viewed and measured in three dimensions and can be projected orthographically to a map sheet. This process eliminates the perspective view distortions present when we attempt to map directly from a single photo. It also eliminates errors introduced by tilts and unequal flying heights when measuring parallax on stereopairs.

There are many different types of stereoplotters, but all of them are made up of three basic components:

1. A projection system (to create the terrain model).
2. A viewing system (to enable the instrument operator to see the model stereoscopically).
3. A measuring and tracing system (for measuring elevations in the model and tracing features onto a map sheet).

Figure 3.23 illustrates a *direct optical projection* plotter. Such systems directly project overlapping images onto a *tracing table* where the terrain model is viewed in stereo. The projectors can be both translated along and rotated about their x, y, and z axes. This permits the instrument operator to perform a *relative orientation* to re-create the position and angular orientation of the two photographs at the time of exposure. This is done by adjusting the projectors until all conjugate image points coincide in the y direction (at this point, only the elevation-caused x parallax remains). The two projectors are then adjusted in tandem to arrive at an *absolute orientation* of the projected model. This is accomplished by scaling and leveling the pair of projectors until the images of ground control points occupy their appropriate preplotted posi-

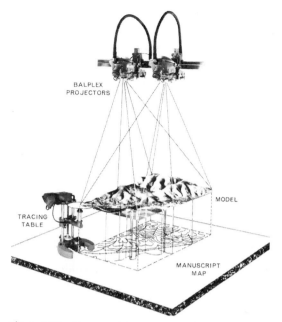

Figure 3.23 Stereomodel projected in a Balplex stereo-plotter. (Courtesy TBR Associates, Inc.)

tions on the map manuscript (Figure 3.23). Following these two orientation procedures, a geometrically correct model will be projected and a complete topographic map may be drawn.

In a direct projection plotter, both images in a stereopair are projected onto the same surface. In order to see stereo, the operator's eyes must view each image separately. This can be accomplished by using an *anaglyphic viewing system*, in which one photo is projected through a cyan filter and the other through a red filter. By viewing the model through eyeglasses having corresponding color filters, the operator's left eye will see only the left photo and the right eye will see only the right photo.

Anaglyphic viewing systems cannot be used with color photography. Even in the case of viewing black and white images, these systems reduce the brightness and resolution of the projected images. An alternative projection method employs a *polarized platen viewer (PPV)* system, which uses polarizing filters and eyeglass lenses instead of colored filters. *Stereo image alternator (SIA)* systems use shutters in the projectors to rapidly alternate the projection of the two photos. The operator views the model through a synchronized shutter system, causing the left eye and right eye to see the images from the corresponding projector only.

Portions of the stereomodel are projected onto a *tracing table platen*, which has a small point of light at its center. This point forms a floating mark whose elevation can be adjusted by raising or lowering the platen. Because the stereomodel has been absolutely oriented to ground control, the platen table height can be equated to terrain elevations. These elevations may be directly read from a height meter on the tracing platen.

Features are mapped planimetrically in the model by tracing them with the floating mark, while continuously raising and lowering the mark to maintain contact with the terrain. Relief displacement of the plotted positions is eliminated by this process. A pencil attached to the platen plots the feature being traced onto the map manuscript, which is located on the plotter table.

Contours are compiled by setting the platen height meter at a desired contour elevation and moving the floating mark along the terrain so that it just maintains contact with the surface of the model. For the novice, these plotting operations seem impossible to perform. It takes considerable training and experience to become proficient at the art of accurate topographic map preparation using a stereoplotter.

To this point, we have discussed direct optical projection plotters only. These were the earliest type of stereoplotter to be developed and were used extensively during the 1950s and 1960s. Another class of stereoplotter instruments receiving substantial use historically is the *mechanical* and *optical-mechanical* projection plotters. These systems simulate the direct projection of rays by mechanical or optical-mechanical means. The operator views the photographs in stereo through a binocular system and the viewing optics are connected to a measuring and tracing system by a precise mechanical linkage. Such a design increases the accuracy of the map production process by minimizing the optical distortions associated with direct project systems.

As digital computers became readily available, *analytical stereoplotters* were developed in order to determine the positions in a stereomodel through mathematical simulation of light ray projection. This was yet another increase in the accuracy and versatility of the stereoplotting process in that such systems have virtually no optical or mechanical limitations. They use hardcopy images and involve linking a comparator-type viewing and measuring system to a computer. The operator simply inputs the camera focal length and data about the calibrated positions of the fiducial marks (and lens distortions) into the computer. Then, under cursor control, the coordinates of the fiducial marks and several ground control points are measured and the computer performs the complete orientation of the stereomodel. Positions in the model can then be mathematically transformed into ground coordinates and elevations. Many systems provide for simultaneous viewing, in color, of the stereomodel and digitized line work (Figure 3.24). Modern softcopy systems employ the same analytical photogrammetric procedures but utilize digital rather than analog images.

Figure 3.24 Zeiss P3 Planicomp analytical plotter (background) with display terminal (foreground). (Courtesy Z/I Imaging Corporation.)

Orthophotos

As implied by their name, orthophotos are orthographic photographs. They do not contain the scale, tilt, and relief distortions characterizing normal aerial photographs. In essence, orthophotos are "photomaps." Like maps, they have one scale (even in varying terrain), and like photographs, they show the terrain in actual detail (not by lines and symbols). Hence, orthophotos give the resource analyst the "best of both worlds"—a product that can be readily interpreted like a photograph but one on which true distances, angles, and areas may be measured directly. Because of these characteristics, orthophotos make excellent base maps for compiling data to be input to a GIS or overlaying and editing data already incorporated in a GIS. Orthophotos also enhance the communication of spatial data, since data users can often relate better to an orthophoto than a conventional line and symbol map or display.

Analog orthophotos are generated from overlapping conventional photos in a process called *differential rectification*. The result of this process is elimination of photo-scale variation and image displacement resulting from relief

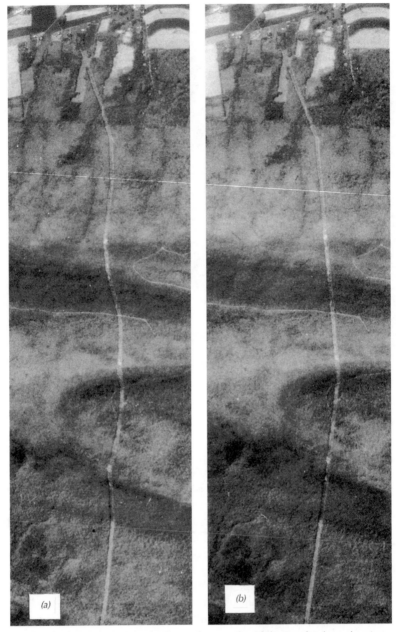

(a)

(b)

Figure 3.25 Portion of (*a*) a perspective photograph and (*b*) an orthophoto showing a power line clearing traversing hilly terrain. (Note the excessive crookedness of the power line clearing in the perspective photo that is eliminated in the orthophoto.) (Courtesy USGS.)

and tilt. Figure 3.25 illustrates the effect of this process. Figure 3.25*a* is a conventional (perspective) photograph of a power line clearing traversing a hilly forested area. The excessively crooked appearance of the clearing is due to relief displacement. Figure 3.25*b* is a portion of an orthophoto covering the same area. In this image, relief displacement has been removed and the true path of the power line is shown.

Analog orthophotos are prepared in instruments called *orthophotoscopes* in much the same manner as maps are prepared in direct optical projection stereoplotters. However, instead of plotting *selected* features in the stereomodel onto a base map, *all* points of the stereomodel are photographed onto an *orthophoto negative*, which is then used to print the orthophotograph.

The principle of operation of a direct optical projection orthophotoscope is shown in Figure 3.26. In operation, diapositives for a stereopair are relatively and absolutely oriented in the instrument as if normal map compilation were to commence. Unlike that of a normal stereoplotter, however, the floating mark for an orthophotoscope is a very small slit in a film holder containing

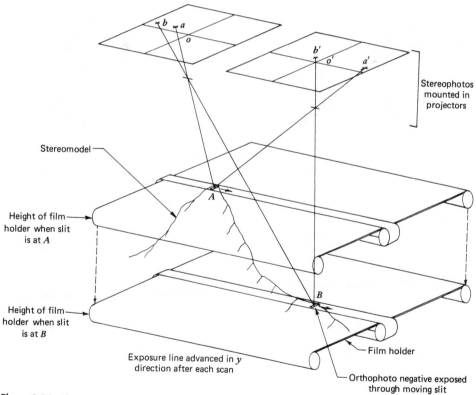

Figure 3.26 Operating principle of direct optical projection orthophotoscope.

the orthophoto negative. A small area of the negative is exposed to light through the slit that is continuously scanned across the film. At the end of each scan, the slit is moved over and scanned in reverse to expose an adjacent strip of film. In this way, the film is eventually exposed to the full stereomodel through the small slit. Along the way, the instrument operator controls the height of the film holder, keeping the slit just in contact with the terrain. Hence, point by point, the scale variation and relief displacement present in the original photography are removed by varying the projection distance. (All tilt distortions are eliminated in the process of orienting the stereomodel.)

Orthophotos alone do not convey topographic information. However, they can be used as base maps for contour line overlays prepared in a separate stereoplotting operation. The result of overprinting contour information on an orthophoto is a *topographic orthophotomap*. Much time is saved in the preparation of such maps because the instrument operator need not map the planimetric data in the map compilation process. Figure 3.27 illustrates a portion of a topographic orthophotomap.

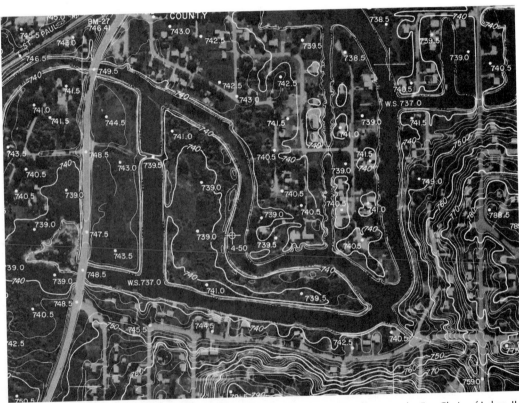

Figure 3.27 Portion of a 1 : 4800 topographic orthophotomap. Photography taken over the Fox Chain of Lakes, IL. (Courtesy Alster and Associates, Inc.)

The method of orthophoto preparation illustrated in Figure 3.26 is termed "on line" in that the terrain scanning and film exposure occur simultaneously. In "off-line" systems, the stereomodel is first scanned using a standard floating mark, instead of the film holder slit. During this process, the scan line elevation profiles are stored in digital form. At a later time, the digital profiles are read by instruments that automatically raise and lower the film holder while exposing a negative. An off-line system offers the advantage that the operator can vary the terrain scanning rate, devoting more time to complex topography. Also, the profiles can be rechecked, and mistakes can be corrected.

The digital terrain profiles generated in the orthophoto production process actually comprise a DEM of the terrain covered by the orthophoto. As with any other DEM, the data extracted from these profiles can be used in a broad range of GIS and digital image processing operations (Chapter 7). Also it should be noted that if a DEM already exists for an area, it can be used as the basis for creating an orthophoto. Hence, DEMs created from previous orthophoto production or from digitized maps often form the basis for producing new orthophotos.

Orthophotos may be viewed stereoscopically when they are paired with *stereomates*. These products are photographs made in an orthophoto instrument by *introducing* image parallax as a function of known terrain elevations obtained during the production of their corresponding orthophoto. Figure 3.28 illustrates an orthophoto and a corresponding stereomate that may be viewed stereoscopically. These products were generated as a part of a stereoscopic orthophotomapping program undertaken by the Canadian Forest Management Institute. The advantage of such products is that they combine the attributes of an orthophoto with the benefits of stereo observation. (Note that Figure 3.25 can also be viewed in stereo. This figure consists of an orthophoto and one of the two photos comprising the stereopair from which the orthophoto was produced.)

Photogrammetric Workstations

Photogrammetric workstations (Figure 3.29) involve integrated hardware and software systems for end-to-end spatial data capture, manipulation, analysis, storage, display, and output of softcopy images. Such systems not only incorporate the functionality of analytical stereoplotters but also provide for automated generation of DEMs, computation of digital orthophotos, preparation of perspective views (singly or in a series of fly-throughs), and capture of two-dimensional and three-dimensional data for direct entry into a GIS. Digital orthophotos in particular receive widespread use in modern spatial data systems because they make ideal image bases for image interpretation and feature encoding in a GIS environment. Not only do they greatly facilitate the

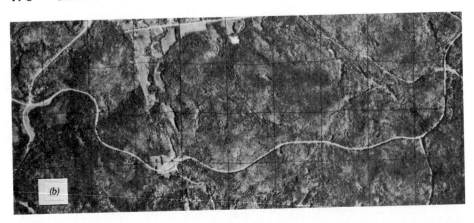

Figure 3.28 Stereo orthophotograph showing a portion of Gatineau Park, Canada: (a) An orthophoto and (b) a stereomate provide for three-dimensional viewing of the terrain. Measurements made from, or plots made on, the orthophoto have map accuracy. Forest-type information is overprinted on this scene along with a Universal Transverse Mercator (UTM) grid. Note that the UTM grid is square on the orthophoto but is distorted by the introduction of parallax on the stereomate. Scale 1 : 38,000. (Courtesy Forest Management Institute, Canadian Forestry Service.)

Figure 3.29 Z/I Imaging ImageStation 2002 digital photogrammetric workstation (shown with optional dual monitors). System employs an infrared emitter (on top of right monitor) and liquid-crystal glasses (left side of digitizing table) for stereoscopic viewing. (Courtesy Z/I Imaging Corporation.)

capture of many forms of GIS data, but they also provide a graphic base for displaying and updating such data.

Plate 8 is yet another example of the need for, and influence of, the distortion correction provided through the orthophoto production process. Shown in (*a*) is an original uncorrected color photograph taken over an area of high relief in Glacier National Park. The digital orthophoto corresponding to the uncorrected photograph in (*a*) is shown in (*b*). Note the locational errors that would be introduced if GIS data were developed from the uncorrected image. GIS analysts are encouraged to use digital orthophotos in their work whenever possible. A major federal source of such data in the United States is the U.S. Geological Survey (USGS) National Digital Orthophoto Program (NDOP), listed in Appendix B.

Figures 3.30 and 3.31 illustrate the visualization capability of sample output products from a photogrammetric workstation. Figure 3.30 shows a perspective view of a rural area located near Madison, Wisconsin. This image was created by draping digital orthophoto data over a DEM of the same area. Figure 3.31 shows a stereopair (*a*) and a perspective view (*b*) of the Clinical Science Center, located on the University of Wisconsin-Madison campus. The images of the various faces of the buildings shown in (*b*) were extracted from the original digitized aerial photograph in which that face was displayed with the greatest relief displacement in accordance with the direction of the perspective view. This process is done automatically from among all of the relevant photographs of the original block of aerial photographs covering the area of interest. Figures 3.30 and 3.31 show only a small sample of the possible output products from a photogrammetric workstation. It should also be noted that most of these systems also provide links to image processing software, making them amenable to the analysis of virtually any source of digital image data (e.g., high resolution satellite data).

Figure 3.30   Perspective view of a rural area generated digitally by draping orthophoto image data over a digital elevation model of the same area. (Courtesy University of Wisconsin-Madison, Environmental Remote Sensing Center, and NASA Affiliated Research Center Program.)

Figure 3.31 Vertical stereopair (a) covering the ground area depicted in the perspective view shown in (b). The image of each building face shown in (b) was extracted automatically from the photograph in which that face was shown with the maximum relief displacement in the original block of aerial photographs covering the area. (Courtesy University of Wisconsin-Madison, Campus Mapping Project.)

Before leaving the topic of photogrammetric workstations and softcopy photogrammetry in general, it is worth reiterating that the accuracy and flexibility of these systems is rooted in their use of digital data and the mathematical principles of analytical photogrammetry. Among the most important and common principles incorporated in the operation of these systems is the *collinearity condition*. As shown in Figure 3.32, collinearity is the condition in which the exposure station of any photograph, any object point in the ground coordinate system, and its photographic image all lie on a straight line. This condition holds irrespective of the angular tilt of a photograph. The possible angular rotations of the image plane from that of an equivalent vertical photograph are the ω, ϕ, and κ angles shown near the top of Figure 3.32.

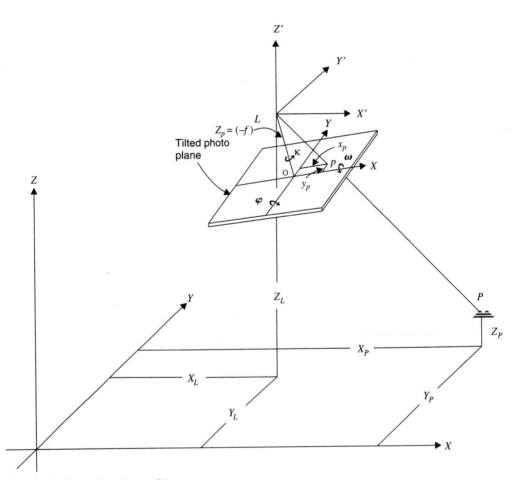

Figure 3.32 The collinearity condition.

The equations that express the collinearity condition are the *collinearity equations*. They describe the relationships among image coordinates, ground coordinates, the exposure station position, and angular orientation of a photograph as follows:

$$x_p = -f\left[\frac{m_{11}(X_P - X_L) + m_{12}(Y_P - Y_L) + m_{13}(Z_P - Z_L)}{m_{31}(X_P - X_L) + m_{32}(Y_P - Y_L) + m_{33}(Z_P - Z_L)}\right] \quad (3.14)$$

$$y_p = -f\left[\frac{m_{21}(X_P - X_L) + m_{22}(Y_P - Y_L) + m_{23}(Z_P - Z_L)}{m_{31}(X_P - X_L) + m_{32}(Y_P - Y_L) + m_{33}(Z_P - Z_L)}\right] \quad (3.15)$$

where

$$x_p, y_p = \text{image coordinates of any point } p$$
$$f = \text{focal length}$$
$$X_P, Y_P, Z_P = \text{ground coordinates of point } P$$
$$X_L, Y_L, Z_L = \text{ground coordinates of exposure station } L$$
$$m_{11}, \ldots, m_{33} = \text{coefficients of a } 3 \times 3 \text{ rotation matrix defined by the}$$
angles ω, ϕ, and κ that transforms the ground coordinate system to the image coordinate system

The above equations are nonlinear and contain nine unknowns: the exposure station position (X_L, Y_L, Z_L), the three rotation angles (ω, ϕ, and κ, which are imbedded in the m coefficients), and the three object point coordinates (X_P, Y_P, Z_P). (The equations are linearized using Taylor's theorem.)

The collinearity equations are at the heart of many softcopy photogrammetric operations. For example, if the location of the exposure station is known (X_L, Y_L, Z_L) as well as the rotation angles ω, ϕ, and κ, from which m_{11}, $\ldots$, m_{33} are known, then any position on the ground (X_P, Y_P, Z_P) can be located on the photo (at x_p, y_p). This is how DEMs are used to generate digital orthophotos. That is, at each ground position in the DEM (X_P, Y_P, Z_P) the associated image point can be computed through the collinearity equations as x_p, y_p. The brightness value of the image at that point is then inserted into an output array and the process is repeated for every line and column position in the DEM to form the entire digital orthophoto. A minor complication in this whole process is the fact that rarely will the photocoordinate value (x_p, y_p) computed for a given DEM cell be exactly centered over a pixel in the original digital input image. Accordingly, the process of *resampling* (Chapter 7 and Appendix C) is employed to determine the best brightness value to assign to each pixel in the orthophoto based on a consideration of the brightness values of a neighborhood of pixels surrounding each computed photocoordinate position (x_p, y_p).

The six parameters X_L, Y_L, Z_L and ω, ϕ, and κ comprise the *exterior orientation* of a photograph. A combination of GPS and IMU system observations can be used to determine the exterior orientation of an image if such equipment is available. If not, the six elements of exterior orientation can be determined through the process of *space resection*. In this process, photocoordinates for at least three ground control points are measured. In that Eqs. 3.14 and 3.15 yield two equations for each control point, three such points yield six equations that can be solved simultaneously for the six unknowns. If more than three control points are available, then more than six equations can be formed and a least squares solution for the unknowns is performed.

Other applications of collinearity include *relative orientation* of a pair of overlapping photographs. This is similar to the relative orientation of projectors in a stereoplotter. *Absolute orientation* is accomplished by performing a coordinate transformation between the relatively oriented model and ground control (again, similar to fitting preplotted ground control in analog stereoplotter map compilation).

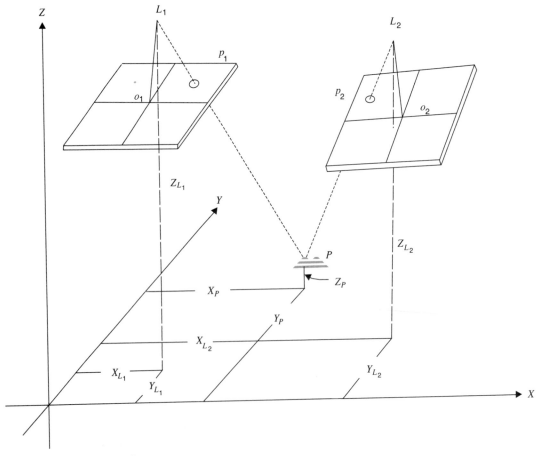

Figure 3.33 Space intersection.

Another important application of the collinearity equations is the process of *space intersection*. This is a procedure by which the X, Y, and Z coordinates of any point in the overlap of a stereopair of tilted photographs can be determined. As shown in Figure 3.33, space intersection makes use of the fact that corresponding rays from overlapping photos intersect at a unique point. Image correlation (Section 3.7) is generally used to match any given point with its conjugate image to determine the point of intersection.

Space intersection makes use of the fact that a total of four collinearity equations can be written for each point in the overlap area. Two of these equations relate to the point's x and y coordinates on the left-hand photo; two result from the x' and y' coordinates measured on the right-hand photo. If the exterior orientation of both photos is known, then the only unknowns in each equation are X, Y, and Z for the point under analysis. Given four equations for three unknowns, a least squares solution for the ground coordinates of each point can be performed. Systematic sampling throughout the overlap area in this manner forms the basis for DEM production with softcopy systems. While this process is highly automated, the image correlation involved is often far from perfect. This normally leads to the need to edit the resulting DEM, but the automated DEM compilation process is still a very useful one.

A final important function performed by photogrammetric workstations is *bundle adjustment*. This process is the extension of the principles described above to an entire block of photos simultaneously. In its most robust form, bundle adjustment provides for simultaneous mathematical treatment of photocoordinate measurements and ground control measurements, as well as treating the exterior orientation parameters of each photo in the block as unknowns. It can be thought of as simultaneous space resection and space intersection of all photos in a block at once. This process both minimizes the need for ground control and increases the precision of the resulting location of points in the ground coordinate system.

3.10 FLIGHT PLANNING

Frequently, the objectives of a photographic remote sensing project can only be met through procurement of new photography of a study area. These occasions can arise for many reasons. For example, photography available for a particular area could be outdated for applications such as land use mapping. In addition, available photography may have been taken in the wrong season. For example, photography acquired for topographic mapping is usually flown in the fall or spring to minimize vegetative cover. This photography will likely be inappropriate for applications involving vegetation analysis. Furthermore, existing photos could be at an inappropriate scale for the application at hand or they could have been taken with an unsuitable film type. Frequently, analysts who require color infrared coverage of an area will find that only black and white panchromatic photography is available. Highly specialized applications may require unusual film–filter combinations or exposure settings, making it highly unlikely that existing photography will be suitable.

When new photography is required, the interpreter is frequently involved in planning the flight. The interpreter soon learns firsthand that one of the most important parameters in an aerial mission is beyond the control of even the best planner—the weather. In most areas, only a few days of the year are ideal for aerial photography. In order to take advantage of clear weather, commercial aerial photography firms will fly many jobs in a single day, often at widely separated locations. Flights are usually scheduled between 10 a.m. and 2 p.m. for maximum illumination and minimum shadow. Overall, a great deal of time, effort, and expense go into the planning and execution of a photographic mission. In many respects, it is an art as well as a science.

Below, we discuss the geometric aspects of the task of flight planning. The parameters needed for this task are (1) the focal length of the camera to be used; (2) the film format size; (3) the photo scale desired; (4) the size of the area to be photographed; (5) the average elevation of the area to be photographed; (6) the overlap desired; (7) the sidelap desired; and (8) the ground speed of the aircraft to be used.

Based on the above parameters, the mission planner prepares computations and a flight map that indicate to the flight crew (1) the flying height above datum from which the photos are to be taken; (2) the location, direction, and number of flight lines to be made over the area to be photographed; (3) the time interval between exposures; (4) the number of exposures on each flight line; and (5) the total number of exposures necessary for the mission.

Flight plans are normally portrayed on a map for the flight crew. However, old photography, an index mosaic, or even a satellite image may be used for this purpose. The computations prerequisite to preparing a flight plan are given in the following example.

EXAMPLE 3.11

A study area is 10 km wide in the east–west direction and 16 km long in the north–south direction (see Figure 3.34). A camera having a 152.4-mm-focal-length lens and a 230-mm format is to be used. The desired photo scale is 1 : 25,000 and the nominal endlap and sidelap are to be 60 and 30 percent. Beginning and ending flight lines are to be positioned along the boundaries of the study area. The only map available for the area is at a scale of 1 : 62,500. This map indicates that the average terrain

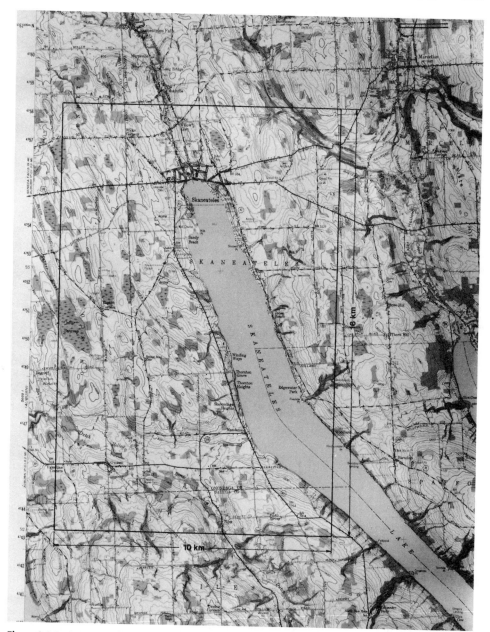

Figure 3.34 A 10 × 16-km study area over which photographic coverage is to be obtained.

elevation is 300 m above datum. Perform the computations necessary to develop a flight plan and draw a flight map.

Solution

(a) Use north–south flight lines. Note that using north–south flight lines minimizes the number of lines required and consequently the number of aircraft turns and realignments necessary. (Also, flying in a cardinal direction often facilitates the identification of roads, section lines, and other features that can be used for aligning the flight lines.)

(b) Find the flying height above terrain ($H' = f/S$) and add the mean site elevation to find flying height above mean sea level:

$$H = \frac{f}{S} + h_{avg} = \frac{0.1524 \text{ m}}{1/25,000} + 300 \text{ m} = 4110 \text{ m}$$

(c) Determine ground coverage per image from film format size and photo scale:

$$\text{Coverage per photo} = \frac{0.23 \text{ m}}{1/25,000} = 5750 \text{ m on a side}$$

(d) Determine ground separation between photos on a line for 40 percent advance per photo (i.e., 60 percent endlap):

$$0.40 \times 5750 \text{ m} = 2300 \text{ m between photo centers}$$

(e) Assuming an aircraft speed of 160 km/hr, the time between exposures is

$$\frac{2300 \text{ m/photo}}{160 \text{ km/hr}} \times \frac{3600 \text{ sec/hr}}{1000 \text{ m/km}} = 51.75 \text{ sec} \qquad \text{(use 51 sec)}$$

(f) Because the intervalometer can only be set in even seconds (this varies between models), the number is rounded off. By rounding down, at least 60 percent coverage is ensured. Recalculate the distance between photo centers using the reverse of the above equation:

$$51 \text{ sec/photo} \times 160 \text{ km/hr} \times \frac{1000 \text{ m/km}}{3600 \text{ sec/hr}} = 2267 \text{ m}$$

(g) Compute the number of photos per 16-km line by dividing this length by the photo advance. Add one photo to each end and round the number up to ensure coverage:

$$\frac{16,000 \text{ m/line}}{2267 \text{ m/photo}} + 1 + 1 = 9.1 \text{ photos/line} \qquad \text{(use 10)}$$

(h) If the flight lines are to have a sidelap of 30 percent of the coverage, they must be separated by 70 percent of the coverage:

$$0.70 \times 5750 \text{ m coverage} = 4025 \text{ m between flight lines}$$

(i) Find the number of flight lines required to cover the 10-km study area width by dividing this width by distance between flight lines (note: this division gives number of spaces between flight lines; add 1 to arrive at the number of lines):

$$\frac{10,000 \text{ m width}}{4025 \text{ m/flight line}} + 1 = 3.48 \qquad \text{(use 4)}$$

The adjusted spacing between lines for using four lines is

$$\frac{10,000 \text{ m}}{4 - 1 \text{ spaces}} = 3333 \text{ m/space}$$

(j) Find the spacing of flight lines on the map (1 : 62,500 scale):

$$3333 \text{ m} \times \frac{1}{62,500} = 53.3 \text{ mm}$$

(k) Find the total number of photos needed:

$$10 \text{ photos/line} \times 4 \text{ lines} = 40 \text{ photos}$$

(Note: The first and last flight lines in this example were positioned coincident with the boundaries of the study area. This provision ensures complete coverage of the area under the "better safe than sorry" philosophy. Often, a savings in film, flight time, and money is realized by experienced flight crews by moving the first and last lines in toward the middle of the study area.)

In computer-supported mission planning, the position of the exposure station for each photo would also be computed for use in the aircraft navigation system.

The above computations would be summarized on a flight map as shown in Figure 3.35. In addition, a set of detailed specifications outlining the material, equipment, and procedures to be used for the mission would be agreed upon prior to the mission. These specifications typically spell out the requirements and tolerances for flying the mission, the form and quality of the products to be delivered, and the ownership rights to the original images. Among other things, mission specifications normally include such details as mission timing, ground control requirements, camera calibration characteristics, film and filter type, exposure conditions, scale tolerance, endlap, sidelap, tilt and crab, photographic quality, GPS encoding, product indexing, and product delivery schedules.

Table 3.1 summarizes the flight planning parameters for various commonly used photo scales. For each scale, the table indicates the required flying height (above ground), the ground spacing between photos and between flight lines, and the lineal and areal coverage of a single photo. The table is based on a 230-mm (9-in.) format camera having a 152.4-mm- (6-in.-) focal-length lens.

JOB 1 - OWASCO

4 LINES NORTH-SOUTH
H = 4110m (MSL) (13,480 ft (MSL)
t = 51 sec FOR 60% ENDLAP
f = 152.4 mm
23 cm x 23 cm FORMAT
1:25000 CONTACT SCALE

Figure 3.35 Flight map for Example 3.11. (Lines indicate centers of each flight line to be followed.)

TABLE 3.1 Flight Planning Parameters for a 230-mm[a] (9-in.) Format Mapping Camera Equipped with a 152.4-mm- (6-in.-) Focal-Length Lens

Photo Scale (Ratio)	Photo Scale (in./ft)	Flying Height (ft)	Exposure Spacing (ft, 60% endlap)	Flight Line Spacing (ft, 30% sidelap)	Length (ft, one side)	Area (mi²)
				English Units		
1 : 1800	1″/150′	900	540	945	1,350	0.07
1 : 2400	1″/200′	1,200	720	1,260	1,800	0.12
1 : 3000	1″/250′	1,500	900	1,575	2,250	0.18
1 : 3600	1″/300′	1,800	1,080	1,890	2,700	0.26
1 : 4800	1″/400′	2,400	1,440	2,520	3,600	0.46
1 : 6000	1″/500′	3,000	1,800	3,150	4,500	0.73
1 : 7200	1″/600′	3,600	2,160	3,780	5,400	1.05
1 : 9600	1″/800′	4,800	2,880	5,040	7,200	1.86
1 : 12,000	1″/1000′	6,000	3,600	6,300	9,000	2.91
1 : 15,840	1″/1320′	7,920	4,752	8,316	11,880	5.06
1 : 20,000	1″/1667′	10,000	6,000	10,500	15,000	8.07
1 : 24,000	1″/2000′	12,000	7,200	12,600	18,000	11.6
1 : 40,000	1″/3333′	20,000	12,000	21,000	30,000	32.3
1 : 60,000	1″/5000′	30,000	18,000	31,500	45,000	72.6
1 : 80,000	1″/6667′	40,000	24,000	42,000	60,000	129.1
1 : 100,000	1″/8333′	50,000	30,000	52,500	75,000	201.8

Photo Scale (Ratio)	Photo Scale (mm/m)	Flying Height (m)	Exposure Spacing (m, 60% endlap)	Flight Line Spacing (m, 30% sidelap)	Length (m, one side)	Area (km²)
				SI Units		
1 : 1800	1 mm/1.8 m	274	165	288	411	0.17
1 : 2400	1 mm/2.4 m	366	219	384	549	0.30
1 : 3000	1 mm/3.0 m	457	274	480	686	0.47
1 : 3600	1 mm/3.6 m	549	329	576	823	0.68
1 : 4800	1 mm/4.8 m	732	439	768	1,097	1.20
1 : 6000	1 mm/6.0 m	914	549	960	1,372	1.88
1 : 7200	1 mm/7.2 m	1,097	658	1,152	1,646	2.71
1 : 9600	1 mm/9.6 m	1,463	878	1,536	2,195	4.82
1 : 12,000	1 mm/12.0 m	1,829	1,097	1,920	2,743	7.52
1 : 15,840	1 mm/15.8 m	2,414	1,448	2,535	3,621	13.1
1 : 20,000	1 mm/20.0 m	3,048	1,829	3,200	4,572	20.9
1 : 24,000	1 mm/24.0 m	3,658	2,195	3,840	5,486	30.1
1 : 40,000	1 mm/40.0 m	6,096	3,658	6,401	9,144	83.6
1 : 60,000	1 mm/60.0 m	9,144	5,486	9,601	13,716	188.1
1 : 80,000	1 mm/80.0 m	12,192	7,315	12,802	18,288	334.5
1 : 100,000	1 mm/100.0 m	15,240	9,144	16,002	22,860	522.6

[a]Calculations based on an actual film format of 228.6 × 228.6 mm.

3.11   CONCLUSION

As we have seen in this chapter, photogrammetry is a very large and rapidly changing subject. Historically, most photogrammetric operations were analog in nature, involving the physical projection and measurement of hardcopy images with the aid of precise optical or mechanical equipment. Today, mathematical models using softcopy data are common. Most softcopy systems are also amenable to the analysis of many different types of digital image data (e.g., airborne digital camera data, satellite data).

With links to various GIS and image processing software, softcopy workstations often represent highly integrated systems for spatial data capture, manipulation, analysis, storage, display, and output.

SELECTED BIBLIOGRAPHY

Afek, Y., and A. Brand, "Mosaicking of Orthorectified Aerial Images," *Photogrammetric Engineering and Remote Sensing*, vol. 64, no. 2, 1998, pp. 115–125.

Agouris, P., A. Stefanidis, and S. Gyftakis, "Differential Snakes for Change Detection in Road Segments," *Photogrammetric Engineering and Remote Sensing*, vol. 67, no. 12, 2001, pp. 1391–1399.

American Society for Photogrammetry and Remote Snsing (ASPRS), *Historical Development of Photogrammetric Methods and Instruments*, ASPRS, Bethesda, MD, 1989.

American Society for Photogrammetry and Remote Sensing (ASPRS), *Digital Photogrammetry: An Addendum to the Manual of Photogrammetry*, ASPRS, Bethesda, MD, 1996a.

American Society for Photogrammetry and Remote Sensing (ASPRS), *Close Range Photogrammetry and Machine Vision*, ASPRS, Bethesda, MD, 1996b.

American Society for Photogrammetry and Remote Sensing (ASPRS), *Digital Elevation Model Technologies and Applications: The DEM Users Manual*, ASPRS, Bethesda, MD, 2001.

American Society for Photogrammetry and Remote Sensing (ASPRS), *Manual of Photogrammetry*, 5th ed., ASPRS, Bethesda, MD, in press.

Arlinghaus, S.L., et al., *Practical Handbook of Digital Mapping Terms and Concepts*, Lewis Publishers, Boca Raton, FL, 1994.

Baltsavias, E.P., "A Comparison between Photogrammetry and Laser Scanning," *ISPRS Journal of Photogrammetry and Remote Sensing*, vol. 54, nos. 2/3, 1999, pp. 83–94.

Bryan, P.G., et al., "Digital Rectification Techniques for Architectural and Archaeological Presentation," *Photogrammetric Record*, vol. 16, no. 93, 1999, pp. 399–415.

Egels, Y., and M. Kasser, *Digital Photogrammetry*, Taylor & Francis, New York, 2001.

Falkner, E., *Aerial Mapping: Methods and Applications*, Lewis Publishers, Boca Raton, FL, 1994.

Fryer, J., and K. McIntosh, "Enhancement of Image Resolution in Digital Photogrammetry," *Photogrammetric Engineering and Remote Sensing*, vol. 67, no. 6, 2001, pp. 741–749.

Gauthier, R.P., M. Maloley, and K.B. Fung, "Land-Cover Anomaly Detection along Pipeline Rights-of-Way," *Photogrammetric Engineering and Remote Sensing*, vol. 67, no. 12, 2001, pp. 1377–1389.

Georgopoulos, A., et al., "Three Dimensional Visualization of the Built Environment Using Digital Orthophotography," *Photogrammetric Record*, vol. 15, no. 90, 1997, pp. 913–921.

Goad, C.C., and M. Yang, "A New Approach to Precision Airborne GPS Positioning for Photogrammetry," *Photogrammetric Engineering and Remote Sensing*, vol. 63, no. 9, 1997, pp. 1067–1077.

Graham, R., and A. Koh, *Digital Aerial Survey Theory and Practice*, CRC Press, Boca Raton, FL, 2002.

Greenfeld, J., "Evaluating the Accuracy of Digital Orthophoto Quadrangles (DOQ) in the Context of

Parcel-Based GIS," *Photogrammetric Engineering and Remote Sensing*, vol. 67, no. 2, 2001, pp. 199–205.

Gruen, A., "Tobago-A Semi Automated Approach for the Generation of 3-D Building Models," *ISPRS Journal of Photogrammetry and Remote Sensing*, vol. 53, no. 2, 1998, pp. 108–118.

Gyer, M.S., "Methods for Computing Photogrammetric Refraction Corrections for Vertical and Oblique Photographs," *Photogrammetric Engineering and Remote Sensing*, vol. 62, no. 3, 1996, pp. 301–310.

Habib, A.F., R. Uebbing, and K. Novak, "Automatic Extraction of Road Signs from Terrestrial Color Imagery," *Photogrammetric Engineering and Remote Sensing*, vol. 65, no. 5, 1999, pp. 597–601.

Habib, A.F., Y.R. Lee, and M. Morgan, "Surface Matching and Change Detection Using a Modified Hough Transformation for Robust Parameter Estimation," *Photogrammetric Record*, vol. 17, no. 98, 2001, pp. 303–315.

Habib, A., and D. Kelley, "Single-Photo Resection Using the Modified Hough Transform," *Photogrammetric Engineering and Remote Sensing*, vol. 67, no. 8, 2001, pp. 909–914.

Heipke, C., "Automation of Interior, Relative, and Absolute Orientation," *ISPRS Journal of Photogrammetry and Remote Sensing*, vol. 51, no. 1, 1997, pp. 1–19.

Hinz, A., and H. Heier, "The Z/I Imaging Digital Camera System," *Photogrammetric Record*, vol. 16, no. 96, 2000, pp. 929–936.

Hohle, J., "Experiences with the Production of Digital Orthophotos," *Photogrammetric Engineering and Remote Sensing*, vol. 62, no. 10, 1996, pp. 1189–1194.

Hu, Y., and C.V. Tao, "Updating Solutions of the Rational Function Model Using Additional Control Information," *Photogrammetric Engineering and Remote Sensing*, vol. 68, no. 7, 2002, pp. 715–723.

ISPRS Journal of Photogrammetry and Remote Sensing, Special Issue on Medical Imaging and Photogrammetry, vol. 56, nos. 5–6, 2002.

Kinn, G., "Direct Georeferencing in Digital Imaging Practice," *Photogrammetric Engineering and Remote Sensing*, vol. 68, no. 5, 2002, p. 399.

Konecny, G., *Photogrammetry*, Walter de Gruyter, New York, 2002a.

Konecny, G. *Geoinformation Remote Sensing, Photogrammetry and Geographical Information Systems*, Taylor & Francis, New York, 2002b.

Krupnik, A., and T. Schenk, "Experiments with Matching in the Object Space for Aerotriangulation," *ISPRS Journal of Photogrammetry and Remote Sensing*, vol. 52, no. 5, 1997, pp. 160–168.

Lee, J.E., and S.D. Johnson, "Expectancy of Cloudless Photographic Days in the Contiguous United States," *Photogrammetric Engineering and Remote Sensing*, vol. 51, no. 12, 1985, pp. 1883–1891.

Li, R., et al., "Photogrammetric Processing of High-Resolution Airborne and Satellite Linear Array Stereo Images for Mapping Applications," *International Journal of Remote Sensing*, vol. 23, no. 20, 2002, pp. 4451–4473.

Liang, T., and C. Heipke, "Automatic Relative Orientation of Aerial Images," *Photogrammetric Engineering and Remote Sensing*, vol. 62, no. 1, 1996, pp. 47–55.

Light, D.L., "The New Camera Calibration System at the U.S. Geological Survey," *Photogrammetric Engineering and Remote Sensing*, vol. 58, no. 2, 1992, pp. 185–188.

Light, D.L., "C-Factor for Softcopy Photogrammetry," *Photogrammetric Engineering and Remote Sensing*, vol. 65, no. 6, 1999, pp. 667–669.

Light, D., "An Airborne Direct Digital Imaging System," *Photogrammetric Engineering and Remote Sensing*, vol. 67, no. 11, 2001, pp. 1299–1305.

McIntosh, K., and A. Krupnik, "Integration of Laser-Derived DSMs and Matched Image Edges for Generating an Accurate Surface Model," *ISPRS Journal of Photogrammetry and Remote Sensing*, vol. 56, no. 3, 2002, pp. 167–176.

Mikhail, E.M., J.S. Bethel, and J.C. McGlone, *Introduction to Modern Photogrammetry*, Wiley, New York, 2001.

Mostafa, M.M.R., and K.P. Schwarz, "A Multi-Sensor System for Airborne Image Capture and Georeferencing," *Photogrammetric Engineering and Remote Sensing*, vol. 66, no. 12, 2000, pp. 1417–1423.

Photogrammetric Engineering and Remote Sensing, Special Issue on Softcopy Photogrammetry, vol. 62, no. 6, 1996.

Photogrammetric Engineering and Remote Sensing, Special Issue on Softcopy Photogrammetry, vol. 63, no. 8, 1997.

Rau, J.Y., N.Y. Chen, and L.C. Chen, "True Orthophoto Generation of Built-up Areas Using Multi-View Images," *Photogrammetric Engineering and Remote Sensing*, vol. 68, no. 6, 2002, pp. 581–588.

Sahar, L., and A. Krupnik, "Semiautomatic Extraction of Building Outlines from Large-Scale Aerial Images," *Photogrammetric Engineering and Remote Sensing*, vol. 65, no. 4, 1999, pp. 459–465.

Schenk, T., *Digital Photogrammetry*, TerraScience–Verlag, Laurelville, OH, 1999.

Schenk, T., "Object Recognition in Digital Photogrammetry," *Photogrammetric Record*, vol. 16, no. 95, 2000, pp. 743–762.

Tang, L., J. Braun, and R. Debitsch, "Automatic Aerotriangulation—Concept, Realization, and Results," *ISPRS Journal of Photogrammetry and Remote Sensing*, vol. 52, no. 3, 1997, pp. 122–131.

Tao, C.V., "Semi-Automated Object Measurement Using Multiple-Image Matching from Mobile Mapping Image Sequences," *Photogrammetric Engineering and Remote Sensing*, vol. 66, no. 12, 2000, pp. 1477–1485.

Tao, C.V., and Y. Hu, "A Comprehensive Study of the Rational Function Model for Photogrammetric Processing," *Photogrammetric Engineering and Remote Sensing*, vol. 67, no. 12, 2001, pp. 1347–1357.

Tao, C.V., and Y. Hu, "3D Reconstruction Methods Based on the Rational Function Model," *Photogrammetric Engineering and Remote Sensing*, vol. 68, no. 7, 2002, pp. 705–714.

Tommaselli, A.M.G., and C.L. Tozzi, "A Recursive Approach to Space Resection Using Straight Lines," *Photogrammetric Engineering and Remote Sensing*, vol. 62, no. 1, 1996, pp. 57–66.

Tu, Z.W., and R.X. Li, "A Framework for Automatic Recognition of Spatial Features from Mobile Mapping Imagery," *Photogrammetric Engineering and Remote Sensing*, vol. 68, no. 3, 2002, pp. 267–276.

Warner, W.S., R.W. Graham, and R.E. Read, *Small Format Aerial Photography*, American Society for Photogrammetry and Remote Sensing, Bethesda, MD, 1996.

Wewel, F., F. Scholten, and K. Gwinner, "High Resolution Stereo Camera (HRSC)-Multispectral 3D-Data Acquisition and Photogrammetric Data Processing," *Canadian Journal of Remote Sensing*, vol. 26, no. 5, 2000, pp. 466–474.

Wolf, P.R., Elements of Photogrammetry, 2nd ed., McGraw-Hill, New York, 1983.

Wolf, P.R., and B.A. Dewitt, *Elements of Photogrammetry: With Applications in GIS*, 3rd ed., McGraw-Hill, New York, 2000.

Zhou, G., et al., "Orthorectification of 1960s Satellite Photographs Covering Greenland," *IEEE Transactions on Geoscience and Remote Sensing*, vol. 40, no. 6, 2002, pp. 1257–1259.

Zhou, G.Q., J. Albertz, and K. Gwinner, "Extracting 3D Information Using Spatio-Temporal Analysis of Aerial Image Sequences," *Photogrammetric Engineering and Remote Sensing*, vol. 65, no. 7, 1999, pp. 823–832.

4 INTRODUCTION TO VISUAL IMAGE INTERPRETATION

4.1 INTRODUCTION

When we look at aerial and space images, we see various objects of different sizes and shapes. Some of these objects may be readily identifiable while others may not, depending on our own individual perceptions and experience. When we can identify what we see on the images and communicate this information to others, we are practicing *image interpretation*. The images contain raw image *data*. These data, when processed by a human interpreter's brain, become usable *information*.

As previously mentioned, aerial photography dates to the year 1858 and the balloon photographs of Nadar. Aerial photography did not receive much emphasis during the ensuing decades because the process was cumbersome and risky and the results uncertain. However, some scientists and inventors did recognize the potential value of aerial photographs in presenting a new view of the earth's surface. Military strategists also understood the potential of this medium for the remote acquisition of military information. During World War I, aerial photography was established as an operational military reconnaissance tool.

After World War I, many nonmilitary applications appeared. The experience gained by World War I pilots in taking pictures from the air convinced quite a number of them that they could put their newly acquired skills to

193

work on such civilian applications as agricultural surveys, timber surveys, and mineral exploration. The U.S. Department of Agriculture's Agricultural Stabilization and Conservation Service (USDA-ASCS) began photographing selected counties of the United States on a repetitive basis in 1937.

Images from space with various levels of detail have been available since the 1960s. Earlier images were of low resolution and were often at oblique angles. Since the advent of the Landsat satellite program in the 1970s and the SPOT satellite program in the 1980s (see Chapter 6), near-vertical images with resolutions useful for earth resources mapping have been available.

In this chapter, we explore the applications of image interpretation to the solution of a variety of problems in different fields. *Aerial photographic image interpretation* is emphasized in this chapter, but it should be noted that most of the principles described here also apply to the interpretation of space images. Space images are often smaller in scale than aerial images and are typically not analyzed stereoscopically.

Because image interpretation is best learned through the experience of viewing hundreds of remotely sensed images according to the requirements of specific fields of application, we cannot hope to train our readers in image interpretation. Here, we will simply present many potential applications of image interpretation and illustrate a few with sample images.

In the next two sections, we present the fundamentals of image interpretation and describe basic image interpretation equipment. In Sections 4.4 through 4.16 we treat the application of image interpretation to a variety of application areas—land use/land cover mapping, geologic and soil mapping, agriculture, forestry, rangeland, water resources, urban and regional planning, wetland mapping, wildlife ecology, archaeology, environmental assessment, disaster assessment, and landform identification and evaluation.

4.2 FUNDAMENTALS OF VISUAL IMAGE INTERPRETATION

Aerial and space images contain a detailed record of features on the ground at the time of data acquisition. An image interpreter systematically examines the images and, frequently, other supporting materials such as maps and reports of field observations. Based on this study, an interpretation is made as to the physical nature of objects and phenomena appearing in the images. Interpretations may take place at a number of levels of complexity, from the simple recognition of objects on the earth's surface to the derivation of detailed information regarding the complex interactions among earth surface and subsurface features. Success in image interpretation varies with the training and experience of the interpreter, the nature of the objects or phenomena being interpreted, and the quality of the images being utilized. Generally, the most capable image interpreters have keen powers of observation coupled with imagination and a great deal of patience. In addition, it is important that the interpreter have a thorough understanding of the phenomenon being studied as well as knowledge of the geographic region under study.

Elements of Image Interpretation

Although most individuals have had substantial experience in interpreting "conventional" photographs in their daily lives (e.g., newspaper photographs), the interpretation of aerial and space images often departs from everyday image interpretation in three important respects: (1) the portrayal of features from an overhead, often unfamiliar, perspective; (2) the frequent use of wavelengths outside of the visible portion of the spectrum; and (3) the depiction of the earth's surface at unfamiliar scales and resolutions (Campbell, 2002). While these factors may be insignificant to the experienced image interpreter, they can represent a substantial challenge to the novice image analyst! A systematic study of aerial and space images usually involves several basic characteristics of features shown on an image. The exact characteristics useful for any specific task and the manner in which they are considered depend on the field of application. However, most applications consider the following basic characteristics, or variations of them: shape, size, pattern, tone (or hue), texture, shadows, site, association, and resolution (Olson, 1960).

Shape refers to the general form, configuration, or outline of individual objects. In the case of stereoscopic images, the object's *height* also defines its shape. The shape of some objects is so distinctive that their images may be identified solely from this criterion. The Pentagon building near Washington, D.C., is a classic example. All shapes are obviously not this diagnostic, but every shape is of some significance to the image interpreter.

Size of objects on images must be considered in the context of the image scale. A small storage shed, for example, might be misinterpreted as a barn if size were not considered. Relative sizes among objects on images of the same scale must also be considered.

Pattern relates to the spatial arrangement of objects. The repetition of certain general forms or relationships is characteristic of many objects, both natural and constructed, and gives objects a pattern that aids the image interpreter in recognizing them. For example, the ordered spatial arrangement of trees in an orchard is in distinct contrast to that of forest tree stands.

Tone (or *hue*) refers to the relative brightness or color of objects on an image. Figure 1.9 showed how relative photo tones could be used to distinguish between deciduous and coniferous trees on black and white infrared photographs. Figure 4.50c (in Section 4.16) shows a striking pattern of light-toned and dark-toned soils where the tonal patterns vary according to the drainage conditions of the soil (the lighter toned areas are topographically higher and drier; the darker toned areas are lower and wetter). Without tonal differences, the shapes, patterns, and textures of objects could not be discerned.

Texture is the frequency of tonal change on an image. Texture is produced by an aggregation of unit features that may be too small to be discerned individually on the image, such as tree leaves and leaf shadows. It is a product of their individual shape, size, pattern, shadow, and tone. It determines the overall visual "smoothness" or "coarseness" of image features. As the scale of the image is reduced, the texture of any given object or area becomes progressively finer and ultimately disappears. An interpreter can often distinguish between features with similar reflectances based on their texture differences. An example would be the smooth texture of green grass as contrasted with the rough texture of green tree crowns on medium-scale airphotos. Another example can be seen in Figure 4.19, which shows the contrasting textures of two tree species.

Shadows are important to interpreters in two opposing respects: (1) the shape or outline of a shadow affords an impression of the profile view of objects (which aids interpretation) and (2) objects within shadows reflect little light and are difficult to discern on an image (which hinders interpretation). For example, the shadows cast by various tree species or cultural features (bridges, silos, towers, etc.) can definitely aid in their identification on airphotos. Also, the shadows resulting from subtle variations in terrain elevations, especially in the case of low sun angle images, can aid in assessing natural topographic variations that may be diagnostic of various geologic landforms. As a general rule, images are more easily interpreted when shadows fall toward the observer. This is especially true when images are examined monoscopically, where relief cannot be seen directly, as in stereoscopic images. In Figure 4.1*a*, a large ridge with numerous side valleys can be seen in the center of the image. When this image is inverted (i.e., turned such that the shadows fall away from the observer), as in (*b*), the result is a confusing image that almost seems to have a valley of sorts running through the center of the image (from bottom to top). This arises because one "expects" light sources to generally be above objects (ASPRS, 1997b, p. 73).

Site refers to topographic or geographic location and is a particularly important aid in the identification of vegetation types. For example, certain tree species would be expected to occur on well-drained upland sites, whereas other tree species would be expected to occur on poorly drained lowland sites. Also, various tree species occur only in certain geographic areas (e.g., redwoods occur in California, but not in Indiana).

Association refers to the occurrence of certain features in relation to others. For example, a Ferris wheel might be difficult to identify if standing in a field near a barn but would be easy to identify if in an area recognized as an amusement park.

Resolution depends on many factors, but it always places a practical limit on interpretation because some objects are too small or have too little contrast with their surroundings to be clearly seen on the image.

Figure 4.1 Photograph illustrating the effect of shadow direction on the interpretability of terrain. Island of Kauai, Hawaii, mid-January. Scale 1 : 48,000. (a) Shadows falling toward observer, (b) same image turned such that shadows are falling away from observer. (USDA–ASCS panchromatic photograph.)

Other factors, such as image scale, image color balance, and condition of images (e.g., torn or faded photographic prints) also affect the success of image interpretation activities.

Many of the above elements of airphoto interpretation are illustrated in Figure 4.2. Figure 4.2*a* is a nearly full-scale copy of a 240-mm airphoto that was at an original scale of 1 : 28,000. Parts (*b*) through (*e*) of Figure 4.2 show four scenes extracted and enlarged from this airphoto. Among the land cover types in Figure 4.2*b* are water, trees, suburban houses, grass, a divided highway, and a drive-in theater. Most of the land cover types are easily identified in this figure. The drive-in theater could be difficult for inexperienced interpreters to identify, but a careful study of the elements of image interpretation leads to its identification. It has a unique *shape* and *pattern*. Its *size* is consistent with a drive-in theater (note the relative size of the cars on the highway and the parking spaces of the theater). In addition to the curved rows of the parking area, the *pattern* also shows the projection building and the screen. The identification of the screen is aided by its *shadow*. It is located in *association* with a divided highway, which is accessed by a short roadway.

Many different land cover types can be seen in Figure 4.2*c*. Immediately noticeable in this photograph, at upper left, is a feature with a superficially

Figure 4.2 Aerial photographic subscenes illustrating the elements of image interpretation, Detroit Lakes area, Minnesota, mid-October. (*a*) Reduced slightly (97.4%) from original photographic of scale 1:28,000; (*b*) and (*c*) enlarged about 6.5 times to a scale of 1:4300; (*d*) enlarged to a scale of 1:15,000; (*e*) enlarged to a scale of 1:24,000. North is to the bottom of the page. (Courtesy KBM, Inc.)

Figure 4.2 *(Continued)*

similar appearance to the drive-in theater. Careful examination of this feature, and the surrounding grassy area, leads to the conclusion that this is a baseball diamond. The trees that can be seen in numerous places in the photograph are casting shadows of their trunks and branches because the mid-October date of this photograph is a time when deciduous trees are in a leaf-off condition. Seen in the right one-third of the photograph is a residential area. Running top to bottom through the center of the image is a commercial area with buildings that have a larger size than the houses in the residential area and large parking areas surrounding these larger buildings.

Figure 4.2 *(Continued)*

Figure 4.2*d* shows two major linear features. Near the top of the photograph is a divided highway. Running diagonally from upper left to lower right is an airport runway 1390 m long (the scale of this figure is 1:15,000, and the length of this linear feature is 9.28 cm at this scale). The terminal area for this airport is near the top center of Figure 4.2*d*.

Figure 4.2*e* illustrates natural versus constructed features. The water body at *a* is a natural feature, with an irregular shoreline and some surrounding wetland areas (especially visible at the narrow end of the lake). The water body at *b* is part of a sewage treatment plant; the shoreline of this feature has unnaturally straight sections in comparison with the water body shown at *a*.

Image Interpretation Strategies

As previously mentioned, the image interpretation process can involve various levels of complexity, from a simple direct recognition of objects in the scene to the inference of site conditions. An example of direct recognition would be the identification of a highway interchange. Assuming the interpreter has some experience with the vertical perspective of aerial and space images, recognition of a highway interchange should be a straightforward process. On the other hand, it may often be necessary to infer, rather than directly observe, the characteristics of features based on their appearance on images. In the case of a buried gas pipeline, for example, the actual pipeline cannot be seen, but there are often changes at the ground surface caused by the buried pipeline that are visible on aerial and space images. Soils are typically better drained over the pipeline because of the sand and gravel used for backfill, and the presence of a buried pipeline can often be inferred by the appearance of a light-toned linear streak across the image. Also, the interpreter can take into account the probability of certain ground cover types occurring on certain dates at certain places. Knowledge of the crop development stages (*crop calendar*) for an area would determine if a particular crop is likely to be visible on a particular date. For example, corn, peas, and winter wheat would each have a significant vegetative ground cover on different dates. Likewise, in a particular growing region, one crop type may be present over a geographic area many times larger than that of another crop type; therefore, the probability of occurrence of one crop type would be much greater than another.

In a sense, the image interpretation process is like the work of a detective trying to put all the pieces of evidence together to solve a mystery. For the interpreter, the mystery might be presented in terms of trying to understand why certain areas in an agricultural field look different from the rest of that field. At the most general level, the interpreter must recognize the area under study as an agricultural field. Beyond this, consideration might be made as to whether the crop present in the field is a row crop (e.g., corn) or a continuous cover crop (e.g., alfalfa). Based on the crop calendar and regional growing conditions, a decision might be made that the crop is indeed corn, rather than another row crop, such as soybeans. Furthermore, it might be noted that the anomalously appearing areas in the field are associated with areas of slightly higher topographic relief relative to the rest of the field. With knowledge of the recent local weather conditions, the interpreter might infer that the anomalously appearing areas are associated with drier soil conditions and the corn in these areas is likely drought stressed. Hence, the interpreter uses the process of *convergence of evidence* to successively increase the accuracy and detail of the interpretation.

Image Interpretation Keys

The image interpretation process can often be facilitated through the use of *image interpretation keys*. Keys can be valuable training aids for novice interpreters and provide useful reference or refresher materials for more experienced interpreters. An image interpretation key helps the interpreter evaluate the information presented on aerial and space images in an organized and consistent manner. It provides guidance about the correct identification of features or conditions on the images. Ideally, a key consists of two basic parts: (1) a collection of annotated or captioned images (preferably stereopairs) illustrative of the features or conditions to be identified and (2) a graphic or word description that sets forth in some systematic fashion the image recognition characteristics of those features or conditions. Two general types of image interpretation keys exist, differentiated by the method of presentation of diagnostic features. A *selective key* contains numerous example images with supporting text. The interpreter selects the example that most nearly resembles the feature or condition found on the image under study.

An *elimination key* is arranged so that the interpretation proceeds step by step from the general to the specific and leads to the elimination of all features or conditions except the one being identified. Elimination keys often take the form of *dichotomous keys* where the interpreter makes a series of choices between two alternatives and progressively eliminates all but one possible answer. Figure 4.3 shows a dichotomous key prepared for the identification of fruit and nut crops in the Sacramento Valley, California. The use of elimination keys can lead to more positive answers than selective keys but may result in erroneous answers if the interpreter is forced to make an uncertain choice between two unfamiliar image characteristics.

As a generalization, keys are more easily constructed and more reliably utilized for cultural feature identification (houses, bridges, roads, water towers) than for vegetation or landform identification. However, a number of keys have been successfully employed for agricultural crop identification and tree species identification. *Such keys are normally developed and used on a region-by-region and season-by-season basis in that the appearance of vegetation can vary widely with location and season.*

Wavelengths of Sensing

The band(s) of the electromagnetic energy spectrum selected for aerial and space imaging affects the amount of information that can be interpreted from the images. Numerous examples of this are scattered throughout this book.

Temporal Aspects of Image Interpretation

The temporal aspects of natural phenomena are important for image interpretation because such factors as vegetative growth and soil moisture vary during the year. For crop identification, more positive results can be achieved

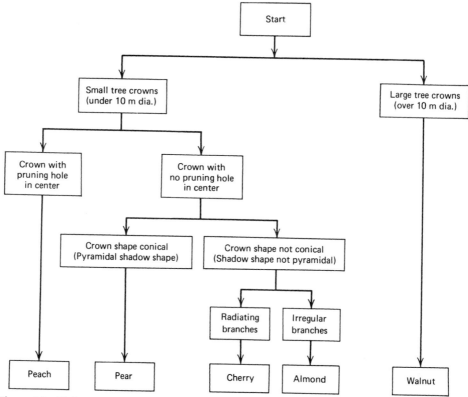

Figure 4.3 Dichotomous airphoto interpretation key to fruit and nut crops in the Sacramento Valley, CA, designed for use with 1 : 6000 scale panchromatic aerial photographs. (Adapted from American Society of Photogrammetry, 1983. Copyright © 1975, American Society of Photogrammetry. Reproduced with permission.)

by obtaining images at several times during the annual growing cycle. Observations of local vegetation emergence and recession can aid in the timing of image acquisition for natural vegetation mapping. In addition to seasonal variations, weather can cause significant short-term changes. Because soil moisture conditions may change dramatically during the day or two immediately following a rainstorm, the timing of image acquisition for soil studies is very critical.

Image Scale

Image scale affects the level of useful information that can be extracted from aerial and space images. Typical scales and areas of coverage are shown in

Table 4.1 for some of the more commonly available sources of aerial photographs. Although terminology with regard to airphoto scale has not been standardized, we can consider that *small-scale* airphotos have a scale of 1:50,000 or smaller, *medium-scale* airphotos have a scale between 1:12,000 and 1:50,000, and *large-scale* airphotos have a scale of 1:12,000 or larger.

In the case of nonphotographic sensors (including digital cameras), images do not have a scale per se; rather, they have a certain ground resolution cell size (e.g., 10×10 or 20×20 m square per pixel), as illustrated in Figures 1.19 and 1.20, and can be reproduced at various scales.

In the figure captions of this book, we have stated the scale of many images—including photographic, multispectral, and radar images—so that the reader can develop a feel for the degree of detail that can be extracted from images of varying scales.

As generalizations, the following statements can be made about the appropriateness of various image scales for resource studies. Small-scale images are used for reconnaissance mapping, large-area resource assessment, and general resource management planning. Medium-scale images are used for

TABLE 4.1   Typical Aerial Photograph Scales and Areas of Coverage for 240-mm Format Film (230×230-mm Image)

Photo Scale	Area per Frame (km)	Comments
1:130,000 or 1:120,000	29.9×29.9 or 27.6×27.6	NASA high altitude photography, 152-mm-focal-length lens
1:80,000	18.4×18.4	National High Altitude Photography (NHAP) program, 152-mm-focal-length lens, panchromatic film
1:65,000 or 1:60,000	14.9×14.9 or 13.8×13.8	NASA high altitude photography, 305-mm-focal-length lens
1:58,000	13.3×13.3	NHAP program, 210-mm-focal-length lens, color infrared film
1:40,000	9.2×9.2	National Aerial Photography Program (NAPP); U.S. Geological Survey (USGS) and USDA current mapping photography programs
1:24,000	5.5×5.5	Photography flown to match scale of USGS $7\frac{1}{2}'$ quadrangle maps
1:20,000	4.6×4.6	Typical scale of older (archival) USGS and USDA mapping photography
1:15,840	3.6×3.6	Traditional U.S. Forest Service (USFS) photography (4 in./mile)
1:6000	1.4×1.4	Typical EPA photography for emergency response and intensive analysis of hazardous waste sites

the identification, classification, and mapping of such features as tree species, agricultural crop type, vegetation community, and soil type. Large-scale images are used for the intensive monitoring of specific items such as surveys of the damage caused by plant disease, insects, or tree blowdown. Large-scale images are also used by the EPA for emergency response to hazardous waste spills and for the intensive site analysis of hazardous waste sites.

Conducted between 1980 and 1987, the National High-Altitude Photography (NHAP) program was a federal multiagency activity coordinated by the U.S. Geological Survey (USGS). It provided nationwide photographic coverage at a nominal scale of 1 : 58,000 using color infrared film and of 1 : 80,000 using panchromatic film exposed simultaneously. Stereoscopic photographs were typically taken from an aircraft altitude of 12,200 m above mean terrain, with a sun angle of at least 30° to minimize shadows, and on days with no cloud cover and minimal haze. The NHAP flight lines were flown in a north–south direction and were positioned to correspond to the centerline of each map sheet included in the national USGS 7.5′ quadrangle series. NHAP I began in 1980 and provided nationwide coverage under the "leaf-off" conditions of spring and fall. NHAP II began in 1985 with photography under "leaf-on" conditions. The leaf-off conditions are preferable for land cover mapping when it is important to be able to see as much detail as possible underneath trees. Leaf-on conditions are preferred for vegetation mapping. In 1987, the program name was changed to the National Aerial Photography Program (NAPP) in recognition of modifications in the user requirements and flight specifications (the flying height is no longer "high altitude").

The objectives of NAPP are to provide a complete photographic database of the 48 conterminous United States and to provide 5- to 7-year updates to the database. The first three NAPP photo acquisition cycles were 1987 to 1991, 1992 to 1996, and 1997 to 2003. The NAPP specifications call for 1 : 40,000 scale cloud-free black and white or color infrared photography with stereoscopic coverage. Spatial resolution of the NAPP photos is 1 to 2 m. Vegetation status (leaf-on, leaf-off) depends on requirements for the respective state involved.

As with NHAP coverage, NAPP photographs are taken along north–south flight lines positioned with respect to the USGS 7.5′ quadrangle maps covering the United States. Two flight lines (each incorporating five photographs) are included over each map area. Thus, every other NAPP photo is nominally centered on one-quarter of a 7.5′ quadrangle.

The NAPP photos have proven to be an extremely valuable ongoing source of medium-scale images supporting a wide range of applications. NAPP photos are the primary source of aerial photographs used in the production of 1-m resolution digital orthophotos for the National Digital Orthophoto Program (NDOP).

Information on the availability of NHAP and NAPP photographs can be obtained from the USGS EROS Data Center (see Appendix B).

Approaching the Image Interpretation Process

There is no single, "right" way to approach the image interpretation process. The specific image products and interpretation equipment available will, in part, influence how a particular interpretation task is undertaken. Beyond these factors, the specific goals of the task will determine the image interpretation process employed. Many applications simply require the image analyst to identify and count various discrete objects occurring in a study area. For example, counts may be made of such items as motor vehicles, residential dwellings, recreational watercraft, or animals. Other applications of the interpretation process often involve the identification of anomalous conditions. For example, the image analyst might survey large areas looking for such features as failing septic systems, sources of water pollution entering a stream, areas of a forest stressed by an insect or disease problem, or evidence of sites having potential archaeological significance.

Many applications of image interpretation involve the delineation of discrete areal units throughout images. For example, the mapping of land use, soil types, or forest types requires the interpreter to outline the boundaries between areas of one type versus another. Such tasks can be problematic when the boundary is not a discrete edge, but rather a "fuzzy edge" or gradation from one type of area to another, as is common with natural phenomena such as soils and natural vegetation.

Two extremely important issues must be addressed before an interpreter undertakes the task of delineating separate areal units on aerial or space images. The first is the definition of the *classification system* or criteria to be used to separate the various categories of features occurring in the images. For example, in mapping land use the interpreter must fix firmly in mind what specific characteristics determine if an area is "residential," "commercial," or "industrial." Similarly, the forest type mapping process must involve clear definition of what constitutes an area to be delineated in a particular species, height, or crown density class.

The second important issue in delineation of discrete areal units on images is the selection of the *minimum mapping unit* (MMU) to be employed in the process. This refers to the smallest size areal entity to be mapped as a discrete area. Selection of the MMU will determine the extent of detail conveyed by an interpretation. This is illustrated in Figure 4.4. In (a), a small MMU results in a much more detailed interpretation than does the use of a large MMU, as illustrated in (b).

Once the classification system and MMU have been determined, the interpreter can begin the process of delineating boundaries between feature types.

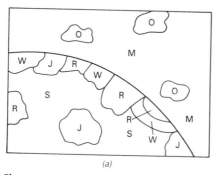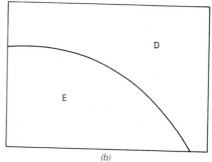

Figure 4.4 Influence of minimum mapping unit size on interpretation detail. (*a*) Forest types mapped using a small MMU: O, oak; M, maple; W, white pine; J, jack pine; R, red pine; S, spruce. (*b*) Forest types mapped using a large MMU: D, deciduous; E, evergreen.

Experience suggests that it is advisable to delineate the most highly contrasting feature types first and to work from the general to the specific. For example, in a land use mapping effort it would be better to separate "urban" from "water" and "agriculture" before separating more detailed categories of each of these feature types based on subtle differences.

In certain applications, the interpreter might choose to delineate *photomorphic regions* as part of the delineation process. These are regions of reasonably uniform tone, texture, and other image characteristics. When initially delineated, the feature type identity of these regions may not be known. Field observations or other ground truth can then be used to verify the identity of each region. Regrettably, there is not always a one-to-one correspondence between the appearance of a photomorphic region and a mapping category of interest. However, the delineation of such regions often serves as a stratification tool in the interpretation process and can be valuable in applications such as vegetation mapping (where photomorphic regions often correspond directly to vegetation classes of interest).

Image Preparation and Viewing

Before undertaking any visual image interpretation task, there are several other factors the image analyst should consider. These range from collecting any relevant collateral sources of information (e.g., maps, field reports, other images) to identifying what viewing equipment is available. Good lighting and access to equipment yielding a range of image magnifications are essential. Beyond this, the interpreter will also want to be sure the images to be viewed are systematically labeled and indexed so cross-referencing to other data sources (e.g., maps) is facilitated. Boundary delineations might be made directly on the images or interpretations might be made directly in a digital format if the required equipment is available. Often, delineations are made on either a clear acetate or Mylar overlay affixed to the images. In such cases, it is important to mark a number of points (e.g., fiducial marks, road intersections) on the overlay to be used to ensure proper registration of the overlay to the image during interpretation (and if the overlay and image are separated and then need to be reregistered).

When the interpretation involves multiple overlapping photographs along a flight line or series of flight lines, the interpreter should first delineate the *effective areas* for the photo coverage before commencing the interpretation. The effective area is typically defined as the central area on each photograph bounded by lines bisecting the area of overlap with every adjacent photograph. Interpreting only within these areas ensures that the entire ground area included in the photos is covered, but with no duplication of interpretation effort. Likewise, because the effective area of a photograph includes all areas closer to the center of that photograph than to the center of any other, it is the area in which objects can be viewed with the least relief displacement. This minimizes the effect of topographic displacement when data interpreted from the individual photos are transferred to a composite base map.

Effective areas can be established by drawing lines on one photo of a stereopair which approximately bisect the endlap and sidelap and transferring three or four points stereoscopically to the adjacent photo (usually at the high and low points of the terrain) along the original line. The transferred points are then connected with straight lines. In areas of high relief, the transferred line will not be straight (due to relief displacement).

Sometimes, effective areas are delineated on every other photograph, rather than on each photograph. In this case, photos without effective areas are used for stereoscopic viewing but are not used for mapping purposes. The advantage of mapping on every photo is the minimization of relief displacement. The disadvantages include the need to delineate, interpret, and transfer twice as many effective areas. In any case, the interpreter should make certain that interpretations crossing the boundaries between effective areas match both spatially and in terms of the identification of the interpreted unit. That is, interpretation polygons spanning more than one effective area must have one interpretation label and the portions in each photo must "line up" with one another.

4.3 BASIC VISUAL IMAGE INTERPRETATION EQUIPMENT

Visual image interpretation equipment generally serves one of several fundamental purposes: viewing images, making measurements on images, performing image interpretation tasks, and transferring interpreted information to base maps or digital databases. Basic equipment for viewing images and transferring interpreted information is described here. Equipment involved in performing measuring and mapping tasks was described in Chapter 3.

The airphoto interpretation process typically involves the utilization of stereoscopic viewing to provide a three-dimensional view of the terrain. Some space images are also analyzed stereoscopically. The stereo effect is possible because we have binocular vision. That is, since we have two eyes that are slightly separated, we continually view the world from two slightly different perspectives. Whenever objects lie at different distances in a scene, each eye sees a slightly different view of the objects. The differences between the two views are synthesized by the mind to provide depth perception. Thus, the two views provided by our separated eyes enable us to see in three dimensions.

When aerial photographs overlap, they also provide two views taken from separated positions. By viewing the left photograph of a pair with the left eye and the right photo with the right eye, we obtain a three-dimensional view of the terrain surface. A *stereoscope* facilitates the stereoviewing process. This book contains many *stereopairs*, or *stereograms*, which can be viewed in three dimensions using a lens stereoscope such as shown in Figure 4.5. An average separation of about 58 mm between common points has been used in the stereograms in this book. The exact spacing varies somewhat because of the different elevations of the points. The original vertical aerial photographs with about 60 percent endlap have been cropped and are mounted here with 100 percent overlap.

Figure 4.5 Simple lens stereoscope.

Figure 4.6 Stereoscopic vision test. (Courtesy Carl Zeiss, Inc.)

Figure 4.6 can be used to test stereoscopic vision. When this diagram is viewed through a stereoscope, the rings and other objects should appear to be at varying distances from the observer. Your stereovision ability can be evaluated by filling in Table 4.2 (answers are in the second part of the table). People whose eyesight is very weak in one eye may not have the ability to see in stereo. This will preclude three-dimensional viewing of the stereograms in this book. However, many people with essentially monocular vision have become proficient photo interpreters. In fact, many forms of interpretation involve monocular viewing with such basic equipment as handheld magnifying glasses or tube magnifiers (2 × to 10 × lenses mounted in a transparent stand).

Some people will be able to view the stereograms in this book without a stereoscope. This can be accomplished by holding the book about 25 cm from your eyes and allowing the view of each eye to drift into a straight-ahead viewing position (as when looking at objects at an infinite distance) while still maintaining focus on the stereogram. When the two images have fused into one, the stereogram will be seen in three dimensions. Most persons will find stereoviewing without proper stereoscopes to be a tiring procedure, producing "eyestrain." It is, however, a useful technique to employ when stereoscopes are not available.

Several types of stereoscopes are available, utilizing lenses or a combination of lenses, mirrors, and prisms.

Lens stereoscopes, such as the one shown in Figure 4.5, are portable and comparatively inexpensive. Most are small instruments with folding legs. The lens spacing can usually be adjusted from about 45 to 75 mm to accommodate individual eye spacings. Lens magnification is typically 2 power but may be adjustable. The principal disadvantage of small lens stereoscopes is that the images must be quite close together to be positioned properly underneath the lenses. Because of this, the interpreter cannot view the entire stereoscopic area of 240-mm aerial photographs without raising the edge of one of the photographs.

Mirror stereoscopes use a combination of prisms and mirrors to separate the lines of sight from each of the viewer's eyes. Such stereoscopes typically have little or no magnification in their normal viewing mode. Binoculars can be fitted to the eyepieces to provide a magnification of 2 to 4 power, with a resulting decrease in field of view. With a mirror stereoscope using little or no

TABLE 4.2 Stereovision Test for Use with Figure 4.6

PART I

Within the rings marked 1 through 8 are designs that appear to be at different elevations. Using "1" to designate the highest elevation, write down the depth order of the designs. It is possible that two or more designs may be at the same elevation. In this case, use the same number for all designs at the same elevation.

Ring 1		**Ring 6**	
Square	(2)	Lower left circle	( )
Marginal ring	(1)	Lower right circle	( )
Triangle	(3)	Upper right circle	( )
Point	(4)	Upper left circle	( )
		Marginal ring	( )
Ring 7		**Ring 3**	
Black flag with ball	( )	Square	( )
Marginal ring	( )	Marginal ring	( )
Black circle	( )	Cross	( )
Arrow	( )	Lower left circle	( )
Tower with cross	( )	Upper center circle	( )
Double cross	( )		
Black triangle	( )		
Black rectangle	( )		

PART II

Indicate the relative elevations of the rings 1 through 8.

( )( )( )( )( )( )( )( )

Highest   Lowest

PART III

Draw profiles to indicate the relative elevations of the letters in the words "prufungstafel" and "stereoskopisches sehen."

P R U F U N G S T A F E L   S T E R E O S K O P I S C H E S   S E H E N

(Answers to Stereovision Test on next page)

TABLE 4.2 *(Continued)*

PART I

Ring 1

		Ring 6	
Square	(2)	Lower left circle	(4)
Marginal ring	(1)	Lower right circle	(5)
Triangle	(3)	Upper right circle	(1)
Point	(4)	Upper left circle	(3)
		Marginal ring	(2)

Ring 7

		Ring 3	
Black flag with ball	(5)	Square	(4)
Marginal ring	(1)	Marginal ring	(2)
Black circle	(4)	Cross	(3)
Arrow	(2)	Lower left circle	(1)
Tower with cross	(7)	Upper center circle	(5)
Double cross	(2)		
Black triangle	(3)		
Black rectangle	(6)		

PART II

(7) (6) (5) (1) (4) (2)[a] (3)[a] (8)
Highest   Lowest

PART III

PRUFUNGSTAFEL   STEREOSKOPISCHES   SEHEN

[a]Rings 2 and 3 are at the same elevation.

magnification, the interpreter can view all or most of the stereoscopic portion of a 240-mm stereopair without moving either the photographs or the stereoscope. This type of stereoscope has the disadvantage that it is too large for easy portability and is much more costly than simple lens stereoscopes.

Zoom stereoscopes have a continuously variable magnification, typically 2.5 to 10 power. Zoom stereoscopes are expensive precision instruments, typically with a very high lens resolution.

Either paper prints or transparencies can be viewed using a stereoscope. Paper prints are more convenient to handle, more easily annotated, and better suited to field use; transparencies have better spatial resolution and color fidelity. An interpreter would generally use a simple lens or mirror stereoscope with paper prints and a more elaborate viewer such as the zoom stereoscope with transparencies. Transparencies are placed on a *light table* (Figure 4.7) for viewing because the light source must come from behind the transparency. The spectral characteristics of the transparency and light table

Figure 4.7 Light table and zoom stereoscope. Richards MIM 4 table with motorized film advance and Bausch & Lomb model 240 stereoscope. (Courtesy The Richards Corporation.)

lamps should be balanced for optimum viewing conditions. Light tables typically have bulbs with a "color temperature" around 3500 K, which means that the spectral distribution of their light output is similar to that of a blackbody heated to 3500 K. The color temperature of "noon daylight" is about 5500 K; tungsten bulbs used for indoor lighting have a color temperature of about 3200 K.

Once information has been interpreted from aerial and space images, it is frequently transferred to a base map. When the base map is not at the same scale as the image, special optical devices may be used in the transfer process. Some of these devices use high precision opaque projectors to enlarge or reduce the image data to the scale of the map. Other devices employ viewing systems that optically superimpose a view of an image and a map. By adjusting the magnification of the two views, the image can be matched to the scale of the map.

An example of the latter device was the Bausch & Lomb Stereo Zoom Transfer Scope, which has been replaced by the Thales-Optem *Digital Transfer Scope* (*DTS*) shown in Figure 4.8. The DTS combines interpretation of hardcopy images and superimposition of a map directly with a geographic information system on a personal computer. It allows the operator to simultaneously view both a vector format digital base map, displayed on a

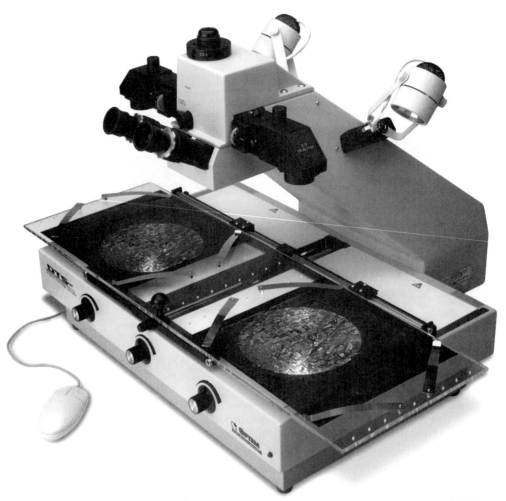

Figure 4.8 Thales-Optem Wide-Field Digital Transfer Scope. (Courtesy Thales-Optem International, Inc.)

liquid-crystal display (LCD) monitor within the instrument, and a stereopair of hardcopy images (paper or film). The DTS allows interpretation of aerial or satellite images in stereo (or mono) through the eyepieces. Through a combination of the zooming optical system (7:1) and software, this device can accommodate a large disparity of image and map scales. The software also "warps" (reshapes, scales, and transforms) the base map data to match the scale and geographic distortions of the hardcopy image. Image interpretation (map editing, feature delineation, and annotation) is then performed in the transformed image space. After the interpretation, the original map data and the information compiled from interpretation are then transformed back into the original, geometrically correct map space. The DTS can also superimpose two images (e.g., a raster image displayed on the internal monitor and hard-copy image) for change detection analysis. A digital camera can be attached to the built-in camera port of the DTS (the black cylinder located on the top of the instrument head) for capturing the superimposition of a digital map and hardcopy image for use in publications or further analysis.

4.4 LAND USE/LAND COVER MAPPING

A knowledge of land use and land cover is important for many planning and management activities and is considered an essential element for modeling and understanding the earth as a system. Land cover maps are presently being developed from local to national to global scales. The use of panchro-matic, medium-scale aerial photographs to map land use has been an ac-cepted practice since the 1940s. More recently, small-scale aerial photographs and satellite images have been utilized for land use/land cover mapping.

The term *land cover* relates to the type of feature present on the surface of the earth. Corn fields, lakes, maple trees, and concrete highways are all exam-ples of land cover types. The term *land use* relates to the human activity or economic function associated with a specific piece of land. As an example, a tract of land on the fringe of an urban area may be used for single-family housing. Depending on the level of mapping detail, its *land use* could be de-scribed as urban use, residential use, or single-family residential use. The same tract of land would have a *land cover* consisting of roofs, pavement, grass, and trees. For a study of the socioeconomic aspects of land use plan-ning (school requirements, municipal services, tax income, etc.), it would be important to know that the use of this land is for single-family dwellings. For a hydrologic study of rainfall-runoff characteristics, it would be important to know the amount and distribution of roofs, pavement, grass, and trees in this tract. Thus, a knowledge of both land use and land cover can be important for land planning and land management activities.

The USGS devised a land use and land cover classification system for use with remote sensor data in the mid-1970s (Anderson et al., 1976). The basic concepts and structure of this system are still valid today. A number of more recent land use/land cover mapping efforts follow these basic concepts and, although their mapping units may be more detailed or more specialized, and they may use more recent remote sensing systems as data sources, they still follow the basic structure originally set forth by the USGS. In the remainder of this section, we first explain the USGS land use and land cover classifica-tion system, then describe some ongoing land use/land cover mapping efforts in the United States and elsewhere.

Ideally, land use and land cover information should be presented on sepa-rate maps and not intermixed as in the USGS classification system. From a practical standpoint, however, it is often most efficient to mix the two systems when remote sensing data form the principal data source for such mapping activities. While land cover information can be directly interpreted from appropriate remote sensing images, information about human activity on the land (land use) cannot always be inferred directly from land cover. As an example, extensive recreational activities covering large tracts of land are not particularly amenable to interpretation from aerial photographs or satellite images. For instance, hunting is a common and pervasive recreational use occurring on land that would be classified as some type of forest, range, wetland, or agricultural land during either a ground survey or image interpretation. Thus, additional information sources are needed to supplement the land cover data. Supplemental information is also necessary for determining the use of such lands as parks, game refuges, or water conservation districts that may have land uses coincident with administrative boundaries not usually identifiable on remote sensor images. Recognizing that some information cannot be derived from remote sensing data, the USGS system is based on categories that can be reasonably interpreted from aerial or space imagery.

The USGS land use and land cover classification system was designed according to the following criteria: (1) the minimum level of interpretation accuracy using remotely sensed data should be at least 85 percent, (2) the accuracy of interpretation for the several categories should be about equal, (3) repeatable results should be obtainable from one interpreter to another and from one time of sensing to another, (4) the classification system should be applicable over extensive areas, (5) the categorization should permit land use to be inferred from the land cover types, (6) the classification system should be suitable for use with remote sensor data obtained at different times of the year, (7) categories should be divisible into more detailed subcategories that can be obtained from large-scale imagery or ground surveys, (8) aggregation of categories must be possible, (9) comparison with future land use and land cover data should be possible, and (10) multiple uses of land should be recognized when possible.

It is important to note that these criteria were developed prior to the widespread use of satellite imagery and computer-assisted classification techniques. While most of the 10 criteria have withstood the test of time, experience has shown that the first two criteria regarding overall and per class consistency and accuracy are not always attainable when mapping land use and land cover over large, complex geographic areas. In particular, when using computer-assisted classification methods, it is frequently not possible to map consistently at a single level of the USGS hierarchy. This is typically due to the occasionally ambiguous relationship between land cover and spectral response and the implications of land use on land cover.

The basic USGS land use and land cover classification system for use with remote sensor data is shown in Table 4.3. The system is designed to use four

TABLE 4.3   USGS Land Use/Land Cover Classification System for Use with Remote Sensor Data

Level I	Level II
1 Urban or built-up land	11 Residential
	12 Commercial and service
	13 Industrial
	14 Transportation, communications, and utilities
	15 Industrial and commercial complexes
	16 Mixed urban or built-up land
	17 Other urban or built-up land
2 Agricultural land	21 Cropland and pasture
	22 Orchards, groves, vineyards, nurseries, and ornamental horticultural areas
	23 Confined feeding operations
	24 Other agricultural land
3 Rangeland	31 Herbaceous rangeland
	32 Shrub and brush rangeland
	33 Mixed rangeland
4 Forest land	41 Deciduous forest land
	42 Evergreen forest land
	43 Mixed forest land
5 Water	51 Streams and canals
	52 Lakes
	53 Reservoirs
	54 Bays and estuaries
6 Wetland	61 Forested wetland
	62 Nonforested wetland
7 Barren land	71 Dry salt flats
	72 Beaches
	73 Sandy areas other than beaches
	74 Bare exposed rock
	75 Strip mines, quarries, and gravel pits
	76 Transitional areas
	77 Mixed barren land
8 Tundra	81 Shrub and brush tundra
	82 Herbaceous tundra
	83 Bare ground tundra
	84 Wet tundra
	85 Mixed tundra
9 Perennial snow or ice	91 Perennial snowfields
	92 Glaciers

"levels" of information, two of which are detailed in Table 4.3. A multilevel system has been devised because different degrees of detail can be obtained from different aerial and space images, depending on the sensor system and image resolution.

The USGS classification system also provides for the inclusion of more detailed land use/land cover categories in Levels III and IV. Levels I and II, with classifications specified by the USGS (Table 4.3), are principally of inter-

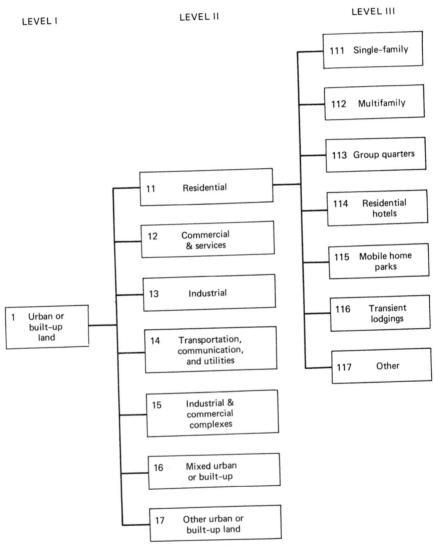

Figure 4.9 An example aggregation of land use/land cover types.

est to users who desire information on a nationwide, interstate, or statewide basis. Levels III and IV can be utilized to provide information at a resolution appropriate for regional (multicounty), county, or local planning and management activities. Again, as shown in Table 4.3, Level I and II categories are specified by the USGS. It is intended that Levels III and IV be designed by the local users of the USGS system, keeping in mind that the categories in each level must aggregate into the categories in the next higher level. Figure 4.9 illustrates a sample aggregation of classifications for Levels III, II, and I.

Table 4.4 lists representative image interpretation formats for the four USGS land use and land cover classification levels (Jensen, 2000). Level I was originally designed for use with low to moderate resolution satellite data such as Landsat Multispectral Scanner (MSS) images. (See Chapter 6 for a description of the Landsat satellites and the other satellite systems mentioned below.) Image resolutions of 20 to 100 m are appropriate for this level of mapping.

Level II was designed for use with small-scale aerial photographs. Image resolutions of 5 to 20 m are appropriate for this level of mapping. The most widely used image type for this level has been high altitude color infrared photographs. However, small-scale panchromatic aerial photographs (Figure 4.10), Landsat Thematic Mapper (TM) and Enhanced Thematic Mapper Plus (ETM+) data, SPOT satellite data, and Indian Remote Sensing Satellite (IRS) data are also representative data sources for many Level II mapping categories.

The general relationships shown in Table 4.4 are not intended to restrict users to particular data sources or scales, either in the original imagery or in the final map products. For example, Level I land use/land cover information, while efficiently and economically gathered over large areas by the Landsat MSS, could also be interpreted from conventional medium-scale photographs or compiled from a ground survey. Conversely, some of the Level II categories have been accurately interpreted from Landsat MSS data.

TABLE 4.4   Representative Image Interpretation Formats for Various Land Use/Land Cover Classification Levels

Land Use/Land Cover Classification Level	Representative Format for Image Interpretation
I	Low to moderate resolution satellite data (e.g., Landsat MSS data)
II	Small-scale aerial photographs; moderate resolution satellite data (e.g., Landsat TM data)
III	Medium-scale aerial photographs; high resolution satellite data (e.g., IKONOS data)
IV	Large-scale aerial photographs

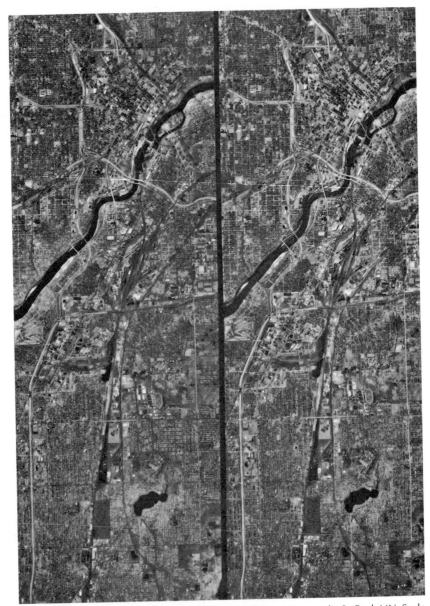

Figure 4.10 Small scale panchromatic aerial photographs, Minneapolis-St. Paul, MN. Scale 1 : 94,000. (North to right.) Stereopair. (Courtesy Mark Hurd Aerial Surveys, Inc.)

For mapping at Level III, substantial amounts of supplemental information, in addition to that obtained from medium-scale images, may need to be acquired. At this level, a resolution of 1 to 5 m is appropriate. Both aerial photographs and high resolution satellite data can be used as data sources at this level.

Mapping at Level IV also requires substantial amounts of supplemental information, in addition to that obtained from aerial images. At this level, a resolution of 0.25 to 1.0 m is appropriate. Large-scale aerial photographs are often the most appropriate form of remotely sensed data for this level of mapping.

The USGS definitions for Level I classes are set forth in the following paragraphs. This system is intended to account for 100 percent of the earth's land surface (including inland water bodies). Each Level II subcategory is explained in Anderson et al. (1976) but is not detailed here.

Urban or *built-up land* is composed of areas of intensive use with much of the land covered by structures. Included in this category are cities; towns; villages; strip developments along highways; transportation, power, and communication facilities; and areas such as those occupied by mills, shopping centers, industrial and commercial complexes, and institutions that may, in some instances, be isolated from urban areas. This category takes precedence over others when the criteria for more than one category are met. For example, residential areas that have sufficient tree cover to meet *forest land* criteria should be placed in the urban or built-up land category.

Agricultural land may be broadly defined as land used primarily for production of food and fiber. The category includes the following uses: cropland and pasture, orchards, groves and vineyards, nurseries and ornamental horticultural areas, and confined feeding operations. Where farming activities are limited by soil wetness, the exact boundary may be difficult to locate and *agricultural land* may grade into *wetland*. When wetlands are drained for agricultural purposes, they are included in the *agricultural land* category. When such drainage enterprises fall into disuse and if wetland vegetation is reestablished, the land reverts to the *wetland* category.

Rangeland historically has been defined as land where the potential natural vegetation is predominantly grasses, grasslike plants, forbs, or shrubs and where natural grazing was an important influence in its precivilization state. Under this traditional definition, most of the rangelands in the United States are in the western range, the area to the west of an irregular north–south line that cuts through the Dakotas, Nebraska, Kansas, Oklahoma, and Texas. Rangelands also are found in additional regions, such as the Flint Hills (eastern Kansas), the southeastern states, and Alaska. The historical connotation of rangeland is expanded in the USGS classification to include those areas in the eastern states called brushlands.

Forest land represents areas that have a tree-crown areal density (crown closure percentage) of 10 percent or more, are stocked with trees capable of producing timber or other wood products, and exert an influence on the climate or water regime. Lands from which trees have been removed to less than 10 percent crown closure but that have not been developed for other uses are also included. For example, lands on which there are rotation cycles of clearcutting and blockplanting are part of the forest land category. Forest land that is extensively grazed, as in the southeastern United States, would also be included in this category because the dominant cover is forest and the dominant activities are forest related. Areas that meet the criteria for forest land and also urban and built-up land are placed in the latter category. Forested areas that have wetland characteristics are placed in the *wetland* class.

The *water* category includes streams, canals, lakes, reservoirs, bays, and estuaries.

The *wetland* category designates those areas where the water table is at, near, or above the land surface for a significant part of most years. The hydrologic regime is such that aquatic or hydrophytic vegetation is usually established, although alluvial and tidal flats may be nonvegetated. Examples of wetlands include marshes, mudflats, and swamps situated on the shallow margins of bays, lakes, ponds, streams, and artificial impoundments such as reservoirs. Included are wet meadows or perched bogs in high mountain valleys and seasonally wet or flooded basins, playas, or potholes with no surface water outflow. Shallow water areas where aquatic vegetation is submerged are classified as *water* and are not included in the *wetland* category. Areas in which soil wetness or flooding is so short-lived that no typical wetland vegetation is developed belong in other categories. Cultivated wetlands such as the flooded fields associated with rice production and developed cranberry bogs are classified as *agricultural land*. Uncultivated wetlands from which wild rice, cattails, and so forth are harvested are retained in the *wetland* category, as are wetlands grazed by livestock. Wetland areas drained for any purpose belong to the other land use/land cover categories such as urban or built-up land, agricultural land, rangeland, or forest land. If the drainage is discontinued and wetland conditions resume, the classification will revert to *wetland*. Wetlands managed for wildlife purposes are properly classified as *wetland*.

Barren land is land of limited ability to support life and in which less than one-third of the area has vegetation or other cover. This category includes such areas as dry salt flats, beaches, bare exposed rock, strip mines, quarries, and gravel pits. Wet, nonvegetated barren lands are included in the wetland category. Agricultural land temporarily without vegetative cover because of cropping season or tillage practices is considered *agricultural land*. Areas of intensively managed forest land that have clear-cut blocks evident are classified as *forest land*.

Tundra is the term applied to the treeless regions beyond the geographic limit of the boreal forest and above the altitudinal limit of trees in high mountain ranges. In North America, tundra occurs primarily in Alaska and northern Canada and in isolated areas of the high mountain ranges.

Perennial snow or *ice* areas occur because of a combination of environmental factors that cause these features to survive the summer melting season. In so doing, they persist as relatively permanent features on the landscape.

As noted above, some parcels of land could be placed into more than one category, and specific definitions are necessary to explain the classification priorities. This comes about because the USGS land use/land cover classification system contains a mixture of land activity, land cover, and land condition attributes.

Several land use/land cover mapping efforts that have been undertaken in the United States and elsewhere use the USGS land use/land cover classification system, or variations thereof. A representative subset of these efforts is summarized below (new initiatives continue to develop given the importance of such data).

The USGS land use/land cover classification system was used to prepare Level I and II land use/land cover maps for most of the conterminous United States and Hawaii at a scale of 1:250,000. For most categories, a minimum mapping unit of 16 ha was used. A limited number of maps are available at a scale of 1:100,000. Digital land use/land cover data have been compiled from these maps by the USGS in vector format and are available for the conterminous United States and Hawaii. Polygons delineating "natural" areas have a minimum size of 4 ha with a minimum feature width of 400 m. The minimum size of polygons representing cultural features (e.g., urban areas, highways) is also 4 ha, with a minimum feature width of 200 m. The data are also available in raster (grid-cell) format. The digital data use the Universal Transverse Mercator (UTM) coordinate system and can be transformed into other map projections. In digital form, the land use/land cover data can be combined with other data types in a GIS. These data are available from the USGS EROS Data Center (see Appendix B) at no cost when downloaded over the Internet.

Traditionally, land cover mapping activities in the United States, as well as a variety of natural resource mapping activities, have used topographic base maps produced by the USGS, at scales generally ranging from 1:24,000 to 1:250,000, for their base maps. Over time, mapping activities have become increasingly digitally based. In response to this, the USGS will be providing, on the Internet, current, accurate, and nationally consistent digital data and topographic maps derived from those data. The resulting product, *The National Map*, will be a seamless, continuously maintained set of geographic base data that will serve as a foundation for integrating, sharing, and using other data easily and consistently. The National Map will include the following: (1) high resolution digital orthorectified imagery that will provide some of the feature information now symbolized on topographic maps; (2) high resolution surface elevation data, including bathymetry, to derive contours for primary series topographic maps and to support production of accurate orthorectified imagery; (3) vector feature data for hydrography, transportation (roads, railways, and waterways), structures, government unit boundaries, and publicly owned land boundaries; (4) geographic names for physical and cultural features to support the U.S. Board on Geographic Names, and other names, such as those for highways and streets; and, (5) extensive land cover data. Data will be seamlessly and consistently classified, enabling users to extract data and information for irregular geographic areas, such as counties or drainage basins. For further information, see http://nationalmap.usgs.gov.

The USGS and the EPA are cooperating on a North American Landscape Characterization Project (NALC) that has completed production of a standardized digital data set for the contiguous 48 states and Mexico that contains Landsat MSS satellite data (Chapter 6) from the 1970s, 1980s, and 1990s. This data set allows researchers to inventory land use and land cover and to conduct land cover change analysis using a standardized data set.

The U.S. Department of Agriculture's Natural Resources Conservation Service (NRCS) conducts a statistically based National Resources Inventory (NRI) at 5-year intervals. The NRI is an inventory of land use and land cover, soil erosion, prime farmland, wetlands, and other natural resource characteristics of nonfederal rural land in the United States. It provides a record of the nation's conservation accomplishments and future program needs. Specifically, it provides valuable information concerning the effect of legislative actions on protecting land from erosion and slowing the rate of loss of wetlands, wildlife habitat diversity, and prime agricultural land. The NRI data are also used by federal and state agencies in regional and state environmental planning to determine the magnitude and location of soil and water resource problems. With each 5-year cycle of data gathering, the NRI places increasing emphasis on remote sensing data acquisition and computer-based data analysis.

As part of NASA's Earth Observing System (EOS) Pathfinder Program, the USGS, the University of Nebraska-Lincoln, and the European Commission's Joint Research Centre are involved with a Global Land Cover Characterization study that has generated a 1-km resolution global land cover database for use in a wide range of environmental research and modeling applications. The global land cover characteristics database was developed on a continent-by-continent basis and each continental database contains unique elements based on the geographical aspects of the specific continent. The initial remote sensing data source for this study was Advanced Very High Resolution Radiometer (AVHRR) data (Chapter 6) spanning April 1992 through March 1993. These global land cover data are available from the USGS EROS Data Center (see Appendix B) at no cost when downloaded over the Internet.

The USGS National Gap Analysis Program (GAP) is a state-, regional-, and national-level program established to provide map data, and other information, about natural vegetation, distribution of native vertebrate species, and land ownership. Gap analysis is a scientific method for identifying the degree to which native animal species and natural communities are represented in our present-day mix of conservation lands. Those species and communities not adequately represented in the existing network of conservation lands constitute conservation "gaps." The GAP data sets are produced at a nominal scale of 1 : 100,000. A minimum mapping unit (MMU) of 30 × 30 m (the resolution of the Landsat TM data typically used for data acquisition and analysis) is generally used. (Some GAP data sets use a larger MMU.)

The Multi-Resolution Land Characteristics (MRLC) Consortium was established as an interagency initiative to provide a consistent national Landsat TM data set and, ultimately, a 30-m national land cover classification for the conterminous United States. The original members of the MLRC were the USGS, the EPA, the National Oceanic and Atmospheric Administration (NOAA), and the U.S. Forest Service (USFS). Later joining the MLRC were the National Aeronautics and Space Administration (NASA) and the Bureau of Land Management (BLM). This consortium of agencies developed a common set of requirements and specifications for Landsat TM data acquisition and data processing. The initial result of MRLC was a multitemporal geocoded Landsat TM data set for the 48 conterminous states, with dates of data acquisition centered on 1992. The ultimate purpose of this data set was the development of the land cover database needed by each of the participating projects and programs. Several complementary land cover databases are being developed from the MRLC coverage. For additional information, see the MRLC website at http://www.epa.gov/mlrc.

Land use and land cover mapping is also being addressed by various groups outside of the United States. For example, the Land Cover Working Group of the Asian Association on Remote Sensing is involved with the preparation of a 1-km-resolution land cover database of Asia based on AVHRR satellite data.

The Coordination of Information on the Environment (CORINE) land cover initiative is led by the European Environment Agency. The CORINE land cover database provides a pan-European inventory of biophysical land cover, using a 44-class nomenclature. It is made available on a 250 by 250 m grid database. CORINE land cover is a key database for integrated environmental assessment across Europe. For additional information, see http://www.eea.eu.int.

The Africover project of the United Nations Food and Agriculture Organization has as its goal the establishment, for the whole of Africa, of a digital geo-referenced database on land cover, the *Multipurpose Africover Database for Environmental Resources*, at a 1 : 200,000 scale (1 : 100,000 for small countries and specific areas). This project has been prepared in response to a number of national requests for assistance on the implementation of reliable and geo-referenced information on natural resources (e.g., "early warning" of potential agricultural problems, forest and rangeland monitoring, watershed management, wildlife monitoring, natural resource planning, production statistics, biodiversity and climate change assessment) at subnational, national, and regional levels. The land cover information is mainly derived from visual interpretation of recent high resolution satellite images that have been enhanced digitally. Additional information about the Africover database is available at http://www.africover.org.

4.5   GEOLOGIC AND SOIL MAPPING

The earth has a highly complex and variable surface whose topographic relief and material composition reflect the bedrock and unconsolidated materials that underlie each part of the surface as well as the agents of change that have acted on them. Each type of rock, each fracture or other effect of internal movement, and each erosional and depositional feature bear the imprint of the processes that produced them. Persons who seek to describe and explain earth materials and structures must understand geomorphological principles and be able to recognize the surface expressions of the various materials and structures. Through the processes of visual image interpretation and geologic and soil mapping, these materials and structures can be identified and evaluated. Geologic and soil mapping will always require a considerable amount of field exploration, but the mapping process can be greatly facilitated through the use of visual image interpretation. Here, we briefly describe the application of visual image interpretation to geologic and soil mapping. Section 4.16 provides a more detailed coverage of this application and contains stereoscopic aerial photographs illustrating visual image interpretation for landform identification and evaluation.

Geologic Mapping

The first aerial photographs taken from an airplane for geologic mapping purposes were used to construct a mosaic covering Bengasi, Libya, in 1913. In general, the earliest uses of airphotos were simply as base maps for geologic data compilation, especially as applied to petroleum exploration. Some interpretive use of aerial photographs began in the 1920s. Since the 1940s, the interpretive use of airphotos for geologic mapping and evaluation has been widespread.

Geologic mapping involves the identification of landforms, rock types, and rock structures (folds, faults, fractures) and the portrayal of geologic units and structures on a map or other display in their correct spatial relationship with one another. Mineral resource exploration is an important type of geologic mapping activity. Because most of the surface and near-surface mineral deposits in accessible regions of the earth have been found, current emphasis is on the location of deposits far below the earth's surface or in inaccessible regions. Geophysical methods that provide deep penetration into the earth are generally needed to locate potential deposits and drill holes are required to confirm their existence. However, much information about potential areas for mineral exploration can be provided by interpretation of surface features on aerial photographs and satellite images.

Interpretation of satellite images and aerial photographs is a critical preliminary step for most field-based geologic mapping projects, especially in remote regions. Such imagery provides an efficient, comparatively low-cost means of targeting key areas for the much more expensive ground-based field surveys. Images are used to locate areas where rocks are exposed at the surface and are thus accessible to the geologist for study, and to trace key geologic units across the landscape. Images also allow the geologist to make important distinctions between landforms, relate them to the geologic processes that formed them, and thus interpret the geologic history of the area. At the most utilitarian level, images can be indispensable to the geologist in navigation and in determining how to access an area of interest, whether by road or trail, by river, or by air. (The terrain and vegetation characteristics will dictate how a geologist will be able to reach the rocks and features that must be evaluated and mapped.)

Satellite images provide geologists several advantages over aerial photographs: (1) they provide the advantage of large-area or synoptic coverage, which allows an analyst to examine in single scenes (or in mosaics) the geological portrayal of the earth on a regional basis; (2) they provide the ability to analyze multispectral bands quantitatively in terms of numbers (DNs), which permits the application of computer processing routines to discern and enhance certain compositional properties of earth materials; and, (3) they provide the capability of merging different types of remote sensing data products (e.g., reflectance images with radar or thermal imagery) or combining them with topographic elevation data and other kinds of information bases (e.g., thematic maps, geophysical measurements, chemical sampling surveys), which enables new solutions to determining interrelations among various natural properties of earth phenomena. These new space-driven approaches are beginning to revolutionize the ways in which geologists conduct their field studies, and they have proven to be indispensable techniques for improving the geologic mapping process and carrying out practical exploration for mineral and energy resources on an extensive scale.

Multistage image interpretation is typically utilized in geologic studies. The interpreter may begin by making interpretations of satellite images at scales of 1:250,000 to 1:1,000,000, then examining high altitude stereoscopic aerial photographs at scales from 1:58,000 to 1:130,000. For detailed mapping, stereoscopic aerial photographs at scales as large as 1:20,000 may be utilized.

Small-scale mapping often involves the mapping of *lineaments*, regional linear features that are caused by the linear alignment of morphological features, such as streams, escarpments, and mountain ranges, and tonal features that in many areas are the surface expressions of fractures or fault zones. Increased moisture retention within the fractured material of a fracture or fault zone is often manifested as distinctive vegetation or small impounded bodies known as "sag ponds." Major lineaments can range from a few to hundreds of kilometers in length. Note the clear expression of the linear Garlock and San Andreas faults in Figure 4.11*a*, a small-scale satellite image that covers an

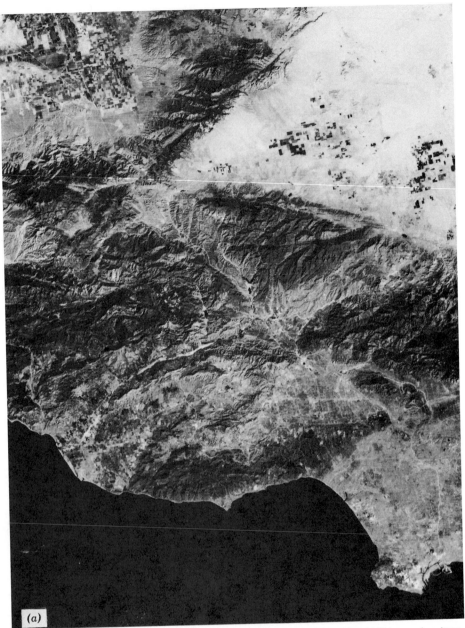

(a)

Figure 4.11 Extensive geologic features visible on satellite imagery. (a) Landsat MSS image, Los Angeles, CA, and vicinity. Scale 1 : 1,000,000. (b) Map showing major geologic faults and major earthquake sites. For updated information on southern California earthquakes, see http://www.scecdc.scec.org/clickmap.html. (Adapted from Williams and Carter, 1976.)

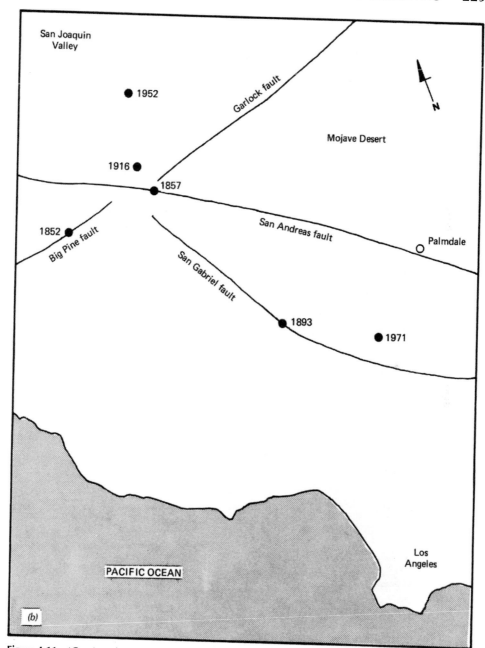

Figure 4.11   (*Continued*)

area 127 by 165 km in size. The mapping of lineaments is also important in mineral resource studies because many ore deposits are located along fracture zones.

Several factors influence the detection of lineaments and other topographic features of geologic significance. One of the most important is the angular relationship between the feature and the illumination source. In general, features that trend parallel to the illumination source are not detected as readily as those that are oriented perpendicularly. Moderately low illumination angles are preferred for the detection of subtle topographic linear features. An example of low sun angle photography, taken under wintertime conditions with snow-covered ground, is shown in Figure 4.12. This figure shows an early space station (Skylab) photograph of northwestern Wyoming and eastern Idaho. The light-toned area at the upper left is the Snake River Plain of Idaho, an extensive area of volcanic flood basalt. The dark, tree-covered area at the

Figure 4.12 Low oblique photograph (from Skylab), eastern Idaho and northwestern Wyoming (including Yellowstone and Grand Teton National Parks), late afternoon, midwinter, scale approximately 1 : 5,000,000 at photo center, black and white reproduction of color original. (NASA image.)

upper right is Yellowstone National Park. Below and to the left of Yellowstone Park are the Grand Teton mountains and "Jackson Hole." The tilted sedimentary rocks of the Wyoming Range and Salt River Range can be seen in the lower left quadrant, and the glaciated Wind River Range at lower right. Side-looking radar images (see Chapter 8) often provide a relatively low angle illumination that accentuates such topographic features.

Many interpreters use a *Ronchi grid* for lineament mapping. The Ronchi grid is a diffraction grating commonly ruled with 78 lines/cm that enhances or suppresses linear features on an image. When an image is viewed through the grid, with the grid held near the eye, linear features parallel to the grid appear diffused and suppressed, and linear features perpendicular to the grid are enhanced.

Many geologists believe that reflection in spectral bands around 1.6 and 2.2 μm is particularly important for mineral exploration and lithologic mapping. These bands cannot be photographed, but they can be sensed with various multispectral and hyperspectral scanners (see Chapters 5 and 6). Also, the examination of multiple narrow bands in the thermal infrared spectral region shows great promise in discriminating rock and mineral types.

Although monoscopic viewing is often suitable for lineament mapping, *lithologic mapping*, the mapping of rock units, is greatly enhanced by the use of stereoscopic viewing. As outlined in Section 4.16, the process of rock unit identification and mapping involves the stereoscopic examination of images to determine the topographic form (including drainage pattern and texture), image tone, and natural vegetative cover of the area under study. In unvegetated areas, many lithologic units are distinguishable on the basis of their topographic form and spectral properties. In vegetated areas, identification is much more difficult because the rock surface is obscured, and some of the more subtle aspects of changes in vegetative cover must be considered.

Because some 70 percent of the earth's land surface is covered with vegetation, a geobotanical approach to geologic unit discrimination is important. The basis of *geobotany* is the relationship between a plant's nutrient requirements and two interrelated factors—the availability of nutrients in the soil and the physical properties of the soil, including the availability of soil moisture. The distribution of vegetation often is used as an indirect indicator of the composition of the underlying soil and rock materials. A geobotanical approach to geologic mapping using remotely sensed images suggests a cooperative effort among geologists, soil scientists, and field-oriented botanists, each of whom should be familiar with remote sensing. An especially important aspect of this approach is the identification of vegetation anomalies related to mineralized areas. Geobotanical anomalies may be expressed in a number of ways: (1) anomalous distribution of species and/or plant communities, (2) stunted or enhanced growth and/or anomalously sparse or dense ground cover, (3) alteration of leaf pigment and/or physiographic processes that produce leaf color changes, and (4) anomalous changes in the phenologic cycle, such as early foliage change or senescence in the fall, alteration of flowering periods, and/or late leaf flush in the spring. Such vegetation anomalies are best identified by analyzing images acquired several times during the year, with emphasis placed on the growing period, from leaf flush in the spring to fall senescence. Using this approach, "normal" vegetation conditions can be established, and anomalous conditions can be more readily identified.

Soil Mapping

Detailed soil surveys form a primary source of resource information about an area. Hence, they are used heavily in such activities as comprehensive land use planning. Understanding soil suitability for various land use activities is essential to preventing environmental deterioration associated with misuse of land. In short, if planning is to be an effective tool for guiding land use, it must be premised on a thorough inventory of the natural resource base; soil data are an essential facet of such inventories.

Detailed soil surveys are the product of an intensive study of soil resources by trained scientists. The delineation of soil units has traditionally utilized airphoto interpretation coupled with extensive field work. Soil scientists traverse the landscape on foot, identify soils, and delineate soil boundaries. This process involves the field examination of numerous soil profiles (cross sections) and the identification and classification of soil units. The soil scientist's experience and training are relied on to evaluate the relationship of soils to vegetation, geologic parent material, landform, and landscape position. Airphoto interpretation has been utilized since the early 1930s to facilitate the soil mapping process. Typically, panchromatic aerial photographs at scales ranging from 1 : 15,840 to 1 : 40,000 have been used as mapping bases.

Agricultural soil survey maps have been prepared for portions of the United States by the USDA since about the year 1900. Most of the soil surveys published since 1957 contain soil maps printed on a photomosaic base at a scale of 1 : 24,000, 1 : 20,000, or 1 : 15,840. Beginning in the mid-1980s, soil survey map information for many counties has been made available both as line maps and as digital files that can be incorporated into geographic information systems. The original purpose of these surveys was to provide technical assistance to farmers and ranchers for cropland and grazing operations. Soil surveys published since 1957 contain information about the suitability of each mapped soil unit for a variety of uses. They contain information for such purposes as estimating yields of common agricultural crops; evaluating rangeland suitability; determining woodland productivity; assessing wildlife habitat conditions; judging suitability for various recreational uses; and determining suitability for various developmental uses, such as highways, local streets and roads, building foundations, and septic tank absorption fields.

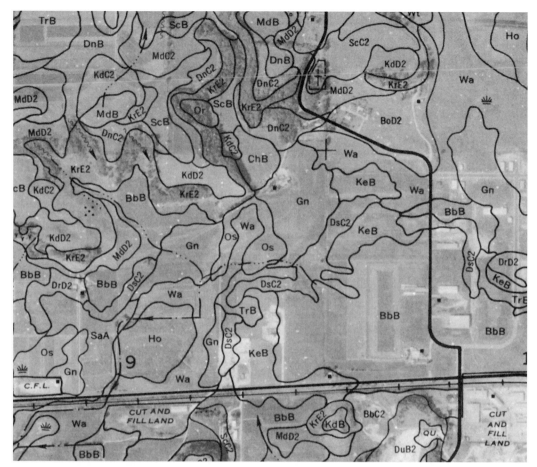

Figure 4.13 Portion of a USDA–ASCS soil map, Dane County, WI. Original scale 1:15,840 (4 in. = 1 mile). (From U.S. Department of Agriculture, 1977.)

The USDA Natural Resources Conservation Service (formerly the Soil Conservation Service) provides soil survey maps in digital form for many areas of the United States. Since 1994 it has provided nationwide detailed soil information by means of the *National Soil Information System*, an online soil attribute database system, available at http://nasis.nrcs.usda.gov.

A portion of a 1:15,840 scale USDA soil map printed on a photomosaic base is shown as Figure 4.13. Table 4.5 shows a sampling of the kind of soil information and interpretations contained in USDA soil survey reports. This map and table show that the nature of soil conditions and, therefore, the appropriateness of land areas for various uses can vary greatly over short distances. As with soil map data, much of the interpretive soil information (such as shown in Table 4.5) is available as computer-based files.

TABLE 4.5 Soil Information and Interpretation for Five Soils Shown in Figure 4.13

Map Unit (Figure 4.13)	Soil Name	Soil Description	Depth to Groundwater Table (cm)	Predicted Corn Yield (kg/ha)	Predicted Degree of Limitations for Use As		
					Septic Tank Absorption Fields	Dwellings with Basements	Sites for Golf Course Fairways
BbB	Batavia silt loam, gravelly substratum, 2–6% slope	100–200 cm silt over stratified sand and gravel	>150	8700	Moderate	Slight	Slight
Ho	Houghton muck, 0–2% slope	Muck at least 150 cm deep	0–30	8100 (when drained)	Very severe	Very severe	Severe
KrE2	Kidder soils, 20–35% slope	About 60 cm silt over sandy loam glacial till	>150	Not suited	Severe	Severe	Severe
MdB	McHenry silty loam, 2–6% slope	25–40 cm silt over sandy loam glacial till	>150	7000	Slight	Slight	Slight
Wa	Wacousta silty clay loam, 0–2% slope	Silty clay loam and silt loam glacial lakebed materials	0–30	7000	Very severe	Very severe	Severe

Source: From U.S. Department of Agriculture, 1977.

As described in Section 1.4, the reflection of sunlight from bare (unvegetated) soil surfaces depends on many interrelated factors, including soil moisture content, soil texture, surface roughness, the presence of iron oxide, and the organic matter content. A unit of bare soil may manifest significantly different image tones on different days, depending especially on its moisture content. Also, as the area of vegetated surfaces (e.g., leaves) increases during the growing season, the reflectance from the scene is more the result of vegetative characteristics than the soil type.

Plate 9 illustrates the dramatically different appearance of one agricultural field, approximately 15 ha in size, during one growing season. Except for a small area at the upper right, the entire field is mapped as one soil type by the USDA (map unit BbB, as shown in Figure 4.13 and described in Table 4.5). The soil parent materials in this field consist of glacial meltwater deposits of stratified sand and gravel overlain by 45 to 150 cm of loess (wind-deposited silt). Maximum relief is about 2 m and slope ranges from 0 to 6 percent. This field was planted to corn (*Zea mays* L.) in May and harvested in November.

Plates 9a, b, and c illustrate the change in surface moisture patterns visible on the cultivated soil over a span of 48 hours in early summer. During this period, the corn plants were only about 10 cm tall, and consequently most of the field surface was bare soil. The area received about 2.5 cm of rain on June 29. On June 30, when the photo in Plate 9a was exposed, the moist soil had a nearly uniform surface tone. By July 2 (9c), distinct patterns of dry soil surface (light image tone) could be differentiated from areas of wet soil surface (darker image tones). The dry areas have relatively high infiltration capacity and are slight mounds of 1 to 2 m relief. These topographic highs have very gentle slopes. Rainfall that does not infiltrate into the soil on these areas runs off onto lower portions of the landscape. These lower areas remain wet longer because they have relatively low infiltration capacity and receive runoff from the higher areas in addition to their original increment of rainfall.

Plates 9d, e, and f illustrate changes in the appearance of the corn crop during the growing season. By August 11 (9d), the corn had grown to a height of 2 m. Vegetation completely covered the soil surface and the field had a very uniform appearance. However, by September 17 (9e), distinct tonal patterns were again evident. Very little rain fell on this field during July, August, and early September, and growth of the corn during this period was dependent on moisture stored in the soil. In the dry areas, shown in light tan-yellow, the leaves and stalks of the corn were drying out and turning brown. In the wetter areas of pink and red photo colors, the corn plants were still green and continuing to grow. Note the striking similarity of the pattern of wet and dry soils in (9c) versus the "green" and brown areas of corn in (9e). The pattern seen in the September photograph (9e) persists in the October photograph (9f); however, there are larger areas of dry corn in October.

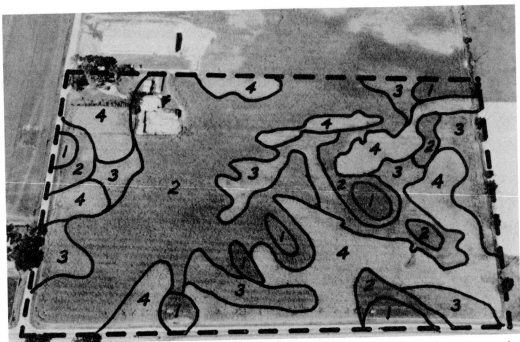

Figure 4.14 Oblique aerial photograph, September 17, with overlay showing four levels of soil moisture (see also Plate 9*e*), Dane County, WI. Scale approximately 1 : 3300 at photo center.

Based on these photographs, a soil scientist was able to divide the soil moisture conditions in this field into four classes, as shown in Figure 4.14. Field inspection of selected sites in each of the four units produced the information in Table 4.6. Note that the corn yield is more than 50 percent greater in unit 2 than in unit 4.

This sequence of photographs taken during one growing season illustrates that certain times of the year are better suited to image acquisition for soil mapping (and crop management) purposes than others. In any given region and season, the most appropriate dates will vary widely, depending on

TABLE 4.6 Selected Characteristics of the Four Soil Units Shown in Figure 4.14

Characteristic	Unit 1	Unit 2	Unit 3	Unit 4
Thickness of silt over sand and gravel	At least 150 cm	105–135 cm	90–120 cm	45–105 cm
Soil drainage class (see Section 4.16)	Somewhat poorly drained	Moderately well drained	Moderately well to well drained	Well drained
Average corn yield, kg/ha)	Not sampled	9100	8250	5850

many factors, including temperature, rainfall, elevation, vegetative cover, and soil infiltration characteristics.

4.6   AGRICULTURAL APPLICATIONS

When one considers the components involved in studying the worldwide supply and demand for agricultural products, the applications of remote sensing in general are indeed many and varied. The scope of the physical, biological, and technological problems facing modern agriculture is an extremely broad one that is intimately related with worldwide problems of population, energy, environmental quality, climate, and weather. These factors are in turn influenced by human values and traditions and economic, political, and social systems. We make no attempt here to look at the "big picture" of how remote sensing is used in agriculture. Instead, we consider the direct application of visual image interpretation in three selected areas: crop-type classification, "precision farming," and crop management in general.

Crop-type classification (and area inventory) through visual image interpretation is based on the premise that specific crop types can be identified by their spectral response patterns and image texture. Successful identification of crops requires a knowledge of the developmental stages of each crop in the area to be inventoried. This information is typically summarized in the form of a *crop calendar* that lists the expected developmental status and appearance of each crop in an area throughout the year. Because of changes in crop characteristics during the growing season, it is often desirable to use images acquired on several dates during the growing cycle for crop identification. Often, crops that appear very similar on one date will look quite different on another date, and several dates of image acquisition may be necessary to obtain unique spectral response patterns from each crop type. When photographic methods are employed, the use of color and color infrared films provides advantages over the use of panchromatic film because of the increased spectral information of the color materials. Also, stereoscopic coverage provides the advantage of being able to use plant height in the discrimination process.

When only broad classes of crops are to be inventoried, single-date panchromatic photography may be sufficient. Table 4.7 shows a dichotomous airphoto interpretation key developed for the identification of major crop and land cover types in agricultural areas of California using medium-scale panchromatic aerial photographs. This tabular style of dichotomous key is an alternative format to the style shown in Figure 4.3. This generalized classification scheme does not attempt to distinguish among various types of vine and bush crops, row crops, or continuous cover crops. When specific crop types are to be inventoried, a more detailed interpretation key employing multidate aerial imaging, using color and/or color infrared film, or multispectral imagery may be required.

TABLE 4.7   Dichotomous Airphoto Interpretation Key for Identification of Major Crop and Land Cover Types in Agricultural Areas of California for Use with Summertime Panchromatic Aerial Photographs at a Scale of 1:15,000

1. Vegetation or soil clearly discernible on photographs	See 2
1. Vegetation and soil either absent or largely obscured by artificial structures, bare rock, or water	Nonproductive lands
2. Cultivation pattern absent; field boundaries irregularly shaped	See 3
2. Cultivation pattern present; field boundaries regularly shaped	See 5
3. Trees present, covering most of ground surface	Timberland
3. Trees absent or widely scattered; ground surface covered by low-lying vegetation	See 4
4. Crowns of individual plants discernible; texture coarse and mottled	Brushland
4. Crowns of individual plants not discernible; texture fine	Grassland
5. Crop vegetation absent	Fallow
5. Crop vegetation present	See 6
6. Crowns of individual plants clearly discernible	See 7
6. Crowns of individual plants not clearly discernible	See 8
7. Alignment and spacing of individual trees at intervals of 6 m or more	Orchards
7. Alignment and spacing of individual plants at intervals of 3 m or more	Vine and bush crops
8. Rows of vegetation clearly discernible, usually at intervals of 0.5–1.5 m	Row crops
8. Rows of vegetation not clearly discernible; crops forming a continuous cover before reaching maturity	See 9
9. Evidence of use by livestock present; evidence of irrigation from sprinklers or ditches usually conspicuous	Irrigated pasture crops
9. Evidence of use by livestock absent; evidence of irrigation from sprinklers or ditches usually inconspicuous or absent; bundles of straw or hay and harvesting marks frequently discernible	Continuous cover crops (small grains, hay, etc.)

Source: From National Research Council, 1970.

Figure 4.15 illustrates some of the factors important to visual image interpretation for crop identification. The photographs shown in this figure are black and white reproductions of color infrared originals covering the same area on two different dates during the same growing season. The area shown is part of an experimental farm located in southern Minnesota. The crops present are alfalfa, corn, sunflowers, soybeans, and wheat. The photograph

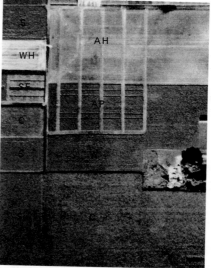

Figure 4.15 Black and white reproductions of large-scale multidate color infrared photographs of various agricultural crops in southern Minnesota (scale 1:3600): (*a*) June 25; (*b*) August 17. Crops shown in (*b*) are alfalfa (A), alfalfa plots (AP), harvested alfalfa (AH), corn (C), soybeans (S), sunflowers (SF), and harvested wheat (WH). (Courtesy University of Minnesota Remote Sensing Laboratory.)

shown in (*a*) was taken during the early summer; the stereopair shown in (*b*) was taken in the late summer. Collectively, these images demonstrate the importance of date of image acquisition, photo tone and texture, and stereoscopic coverage in the crop discrimination process.

Precision farming or "precision crop management (PCM)," has been defined as an *information- and technology-based agricultural management system to identify, analyze, and manage site-soil spatial and temporal variability within fields for optimum profitability, sustainability, and protection of the environment* (Robert, 1997). An essential component of precision farming, or PCM, is the use of *variable rate technology* (*VRT*). This refers to the application of various crop production inputs on a location-specific basis. Variable rate equipment has been developed for a range of materials, including herbicide, fertilizer, water, insecticide, and seeds. The within-field position of such equipment is typically determined through the use of GPS technology. In turn, the GPS guidance system is often linked to a GIS, which provides the "intelligence" necessary to determine the rate at which a given material should be applied at a given location within a field. The data used to determine application rates for various materials usually come from a range of sources (e.g., remote sensing images, detailed soil mapping, historical yield records, equipment-mounted sensors).

Historically, agricultural fields have been managed as if they were homogeneous units. That is, one set of management practices would be used for the entire field. For example, one fertilizer rate or one herbicide rate would be used across the entire field. These rates would tend to be chosen for the most productive soil occurring in the field, meaning many portions of the field would be overtreated or undertreated. The goal of PCM is to target application rates to the location-specific conditions occurring within a field and thereby maximize profitability and minimize energy waste and surface and ground water pollution.

Plate 9 illustrated how variable the soil-site conditions can be within a single field in terms of the surface moisture patterns observable prior to corn crop development (9*a*-*c*) and following an extended period with little rainfall as the corn crop matured (9*e*).

Figure 4.16 illustrates another cornfield that manifests a great deal of spatial variability in terms of crop development. Multiple factors contribute to the anomalous tones occurring across this field. Among these are planter skips (small bright rectangles), drought stress (irregular bright areas), weeds (difficult to see in this black and white reproduction), and water erosion (narrow dark ditch trending from lower center toward upper right).

Whereas PCM is a relatively recent development, the utility of image interpretation in *crop management* in general has been documented for many years. For example, large-scale images have proven useful for documenting deleterious conditions due to crop disease, insect damage, plant stress from other causes, and disaster damage. The most successful applications have uti-

Figure 4.16 Anomalous growth conditions within a cornfield. Black and white rendering of a color composite acquired with 1 m ground resolution cell size per pixel, east-central Indiana, early August. (Courtesy Purdue University Laboratory for Applications of Remote Sensing and Emerge. Copyright © 1998.)

lized large-scale color infrared aerial photographs taken on various dates. In addition to "stress detection," such photographs can provide many other forms of information important to crop management. Table 4.8 lists the kinds of information potentially available from large-scale color infrared aerial photographs obtained at different times in the growing season.

Some of the plant diseases that have been detected using visual image interpretation are southern corn leaf blight, bacterial blight of field beans, potato wilt, sugar beet leaf spot, stem rust of wheat and oats, late blight fungus of potatoes, fusarium wilt and downy mildew in tomato and watermelon, powdery mildew on lettuce and cucumber, powdery mildew in barley, root rotting fungus in cotton, vineyard *Armillaria mellea* soil fungus, pecan root rot, and coconut wilt. Some types of insect damage that have been detected are aphid infestation in corn fields, phylloxera root feeding damage to vineyards, red mite damage to peach tree foliage, and plant damage due to fire ants, harvester ants, leaf cutting ants, army worms, and grasshoppers. Other types of plant damage that have been detected include those from moisture stress, iron deficiency, nitrogen deficiency, excessive soil salinity, wind and water erosion, rodent activity, road salts, air pollution, and cultivator damage.

Visual image interpretation for crop condition assessment is a much more difficult task than visual image interpretation for crop type and area inventory. Ground reference data are essential, and in most studies to date,

TABLE 4.8 **Typical Crop Management Information Potentially Obtainable from Large-Scale Color Infrared Aerial Photographs**

Preplanting
Study variations in soil surface moisture, texture, and organic content in bare fields. Monitor residue and check conditions of terraces, grass waterways, and other surface features.

Plowing/Planting
Determine plowing and planting progress, poorly or excessively drained areas, runoff and erosion problems, and tile line locations.

Emergence
Detect delayed emergence and low plant density, looking for insect, disease, or weather problems, planting failure due to malfunctioning equipment, human error in planting, and effectiveness of preemergent herbicides. Determine necessary remedial measures (such as replanting).

Mid-growing Season
Check on stand growth and development through the growing season, looking for evidence of plant loss or damage due to adverse moisture conditions, misapplication of chemicals, insects, diseases, eroded topsoil, nitrogen deficiencies, and problems in irrigation distribution. Monitor effectiveness of herbicide treatment and drainage.

Preharvest
Check stand condition and acreage to be harvested, looking for lodging, significant weed infestations, or other potential problems for harvesting operations. Check for uniformity of ripening.

Postharvest
Determine total area harvested. Check field cover in harvested areas for weed and volunteer regrowth, erosion, and soil moisture problems.

As Required
Document special situations such as flooding, drought, frost, fire, hail storms, tornadoes, hurricanes, or other problems.

Source: Adapted from Baber and Flowerday, 1979.

comparisons have been made between healthy and stressed vegetation growing in adjacent fields or plots. Under these conditions, interpreters might discriminate between finer differences in spectral response than would be possible in a noncomparative analysis—that is, the level of success would be lower if they did not know a stress existed in an area. It would also be more difficult to differentiate among the effects of disease, insect damage, nutrient deficiencies, or drought, from variations caused by plant variety, plant maturity, planting rate, or background soil color differences. Because many stress effects are most apparent during dry spells, images should not be acquired too soon after rainy weather. (Note that Plate 6 shows color infrared aerial photographs and multiband video images of cotton fields with harvester ant damage and poor growth in areas with saline soils.)

In addition to crop damage due to disease, insects, and various other stresses, crop damage resulting from such disasters as flooding, drought, frost, fire, hailstorms, tornados, and hurricanes can be assessed by visual image interpretation.

Until the early 2000s, the overwhelming majority of applications of image interpretation to precision crop management involved airborne remote sensing as opposed to the use of satellite images. The reason for this is that the satellite systems available to that point lacked the spatial resolution, frequency of coverage, and data delivery timeliness required for crop management. With the increased availability of high resolution satellite data collected by pointable systems with very rapid data supply rates (such systems are covered in Chapter 6), the use of satellite data has greatly increased.

4.7   FORESTRY APPLICATIONS

Forestry is concerned with the management of forests for wood, forage, water, wildlife, and recreation. Because the principal raw product from forests is wood, forestry is especially concerned with timber management, maintenance and improvement of existing forest stands, and fire control. Forests of one type or another cover nearly a third of the world's land area. They are distributed unevenly and their resource value varies widely.

Visual image interpretation provides a feasible means of monitoring many of the world's forest conditions. We will be concerned principally with the application of visual image interpretation to tree species identification, studying harvested areas, timber cruising, and the assessment of disease and insect infestations.

The visual image interpretation process for *tree species identification* is generally more complex than for agricultural crop identification. A given area of forest land is often occupied by a complex mixture of many tree species, as contrasted with agricultural land where large, relatively uniform fields typically are encountered. Also, foresters may be interested in the species composition of the "forest understory," which is often blocked from view on aerial and satellite images by the crowns of the large trees.

Tree species can be identified on aerial and satellite images through the process of elimination. The first step is to eliminate those species whose presence in an area is impossible or improbable because of location, physiography, or climate. The second step is to establish which groups of species do occur in the area, based on a knowledge of the common species associations and their requirements. The final stage is the identification of individual tree species using basic image interpretation principles.

The image characteristics of shape, size, pattern, shadow, tone, and texture, as described in Section 4.2, are used by interpreters in tree species identification. Individual tree species have their own characteristic crown *shape* and *size*. As illustrated in Figures 4.17 and 4.18, some species have rounded crowns,

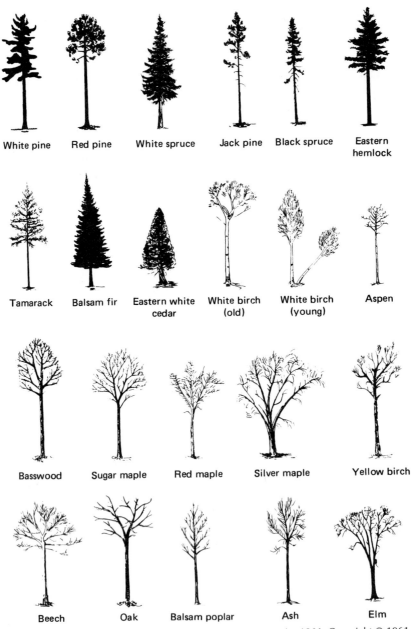

Figure 4.17 Silhouettes of forest trees. (From Sayn-Wittgenstein, 1961. Copyright © 1961, American Society of Photogrammetry. Reproduced with permission.)

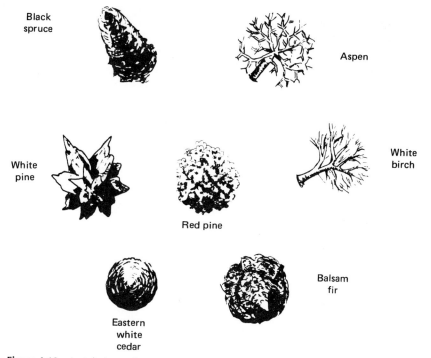

Figure 4.18 Aerial views of tree crowns. Note that most of these trees are shown with radial displacement. (From Sayn-Wittgenstein, 1961. Copyright © 1961, American Society of Photogrammetry. Reproduced with permission.)

some have cone-shaped crowns, and some have star-shaped crowns. Variations of these basic crown shapes also occur. In dense stands, the arrangement of tree crowns produces a *pattern* that is distinct for many species. When trees are isolated, *shadows* often provide a profile image of trees that is useful in species identification. Toward the edges of aerial images, relief displacement can afford somewhat of a profile view of trees. *Image tone* depends on many factors, and it is not generally possible to correlate absolute tonal values with individual tree species. Relative tones on a single image, or a group of images, may be of great value in delineating adjacent stands of different species. Variations in crown *texture* are important in species identification. Some species have a tufted appearance, others appear smooth, and still others look billowy. As mentioned in Section 4.2, image texture is very scale dependent.

Figure 4.19 illustrates how the above-described image characteristics can be used to identify tree species. A pure stand of black spruce (outlined area) surrounded by aspen is shown in Figure 4.19. Black spruce are coniferous trees with very slender crowns and pointed tops. In pure stands, the canopy is regular in pattern and the tree height is even or changes gradually with the quality of the site. The crown texture of dense black spruce stands is carpetlike

Figure 4.19 Aerial photographs of black spruce (outlined area) and aspen, Ontario, Canada. Scale 1:15,840. Stereopair. (From Zsilinszky, 1966. Courtesy Victor G. Zsilinszky, Ontario Centre for Remote Sensing.)

in appearance. In contrast, aspen are deciduous trees with rounded crowns that are more widely spaced and more variable in size and density than the spruce trees. The striking difference in image texture between black spruce and aspen is apparent in Figure 4.19.

The process of tree species identification using visual image interpretation is not as simple as might be implied by the straightforward examples shown in these figures. Naturally, the process is easiest to accomplish when dealing with pure, even-aged stands. Under other conditions, species identification can be as much of an art as a science. Identification of tree species has, however, been very successful when practiced by skilled, experienced interpreters. Field visitation is virtually always used to aid the interpreter in the type map compilation process.

The extent to which tree species can be recognized on aerial photographs is largely determined by the scale and quality of the images, as well as the variety and arrangement of species on the image. The characteristics of tree form, such as crown shape and branching habit, are heavily used for identification on large-scale images. The interpretability of these characteristics becomes progressively less as the scale is decreased. Eventually, the characteristics of individual trees become so indistinct that they are replaced by overall stand characteristics in terms of image tone, texture, and shadow pattern. On images at extremely large scales (such as 1:600), most species can be recognized almost entirely by their morphological characteristics. At this scale, twig structure, leaf arrangement, and crown shape are important clues to species recognition. At scales of 1:2400 to 1:3000, small and medium branches are still visible and individual crowns can be clearly distinguished. At 1:8000, individual trees can still be separated, except when growing in dense stands, but it is not always possible to describe crown shape. At 1:15,840 (Figure 4.19), crown shape can still be determined from tree shadows for large trees growing in the open. At scales smaller than 1:20,000, individual trees generally cannot be recognized when growing in stands, and stand tone and texture become the important identifying criteria (Sayn-Wittgenstein, 1961). This is particularly true when satellite data sources are employed.

Historically, the format most widely used for tree species identification has been black and white photographic paper prints at a scale of 1:15,840 to 1:24,000. Black and white infrared paper prints are especially valuable in separating evergreen from deciduous types. However, color and color infrared films, as well as digital frame cameras and video cameras and also high resolution multiband satellite images, are being used with increasing frequency.

It is difficult to develop visual image interpretation keys for tree species identification because individual stands vary considerably in appearance depending on age, site conditions, geographic location, geomorphic setting, and other factors. However, a number of elimination keys have been developed for use with aerial photographs that have proven to be valuable interpretive tools when utilized by experienced image interpreters. Tables 4.9 and 4.10 are examples of such keys.

Phenological correlations are useful in tree species identification. Changes in the appearance of trees in the different seasons of the year sometimes enable discrimination of species that are indistinguishable on single dates. The most obvious example is the separation of deciduous and evergreen trees that is easily made on images acquired when the deciduous foliage has fallen. This distinction can also be discerned on spring images acquired shortly after the flushing of leaves or on fall images acquired after the trees have turned color. For example, in the summer, panchromatic and color photographs show little difference in tone between deciduous and evergreen trees (Figure 1.9a). Differences in tones are generally quite striking, however, on summer color infrared and black and white infrared photographs (Figure 1.9b).

In spring images, differences in the time at which species leaf out can provide valuable clues for species recognition. For example, trembling aspen and white birch consistently are among the first trees to leaf out, while the oaks, ashes, and large-tooth aspen are among the last. These two groups could be distinguished on images acquired shortly after trembling aspen and white birch have leafed out. Tone differences between hardwoods, which are small during the summer, become definite during the fall, when some species turn yellow and others red or brown. The best species distinctions in the fall are obtained on images acquired when fall coloring is at its peak, rather than when some trees have lost their leaves (Sayn-Wittgenstein, 1961).

TABLE 4.9 Airphoto Interpretation Key for the Identification of Hardwoods in Summer

1. Crowns compact, dense, large	
2. Crowns very symmetrical and very smooth, oblong or oval; trees form small portion of stand	Basswood
2. Crowns irregularly rounded (sometimes symmetrical), billowy, or tufted	
3. Surface of crown not smooth, but billowy	Oak
3. Crowns rounded, sometimes symmetrical, smooth surfaced	Sugar maple,[a] beech[a]
3. Crowns irregularly rounded or tufted	Yellow birch[a]
1. Crowns small or, if large, open or multiple	
6. Crowns small or, if large, open and irregular, revealing light-colored trunk	
7. Trunk chalk white, often forked; trees tend to grow in clumps	White birch
7. Trunk light, but not white, undivided trunk reaching high into crown, generally not in clumps	Aspen
6. Crown medium sized or large; trunk dark	
8. Crown tufted or narrow and pointed	
9. Trunk often divided, crown tufted	Red maple
9. Undivided trunk, crown narrow	Balsam poplar
8. Crowns flat topped or rounded	
10. Crowns medium sized, rounded; undivided trunk; branches ascending	Ash
10. Crowns large, wide; trunk divided into big spreading branches	
11. Top of crown appears pitted	Elm
11. Top of crown closed	Silver maple

Source: From Sayn-Wittgenstein, 1961. Copyright © 1961, American Society of Photogrammetry. Reproduced with permission.

[a]A local tone-key showing levels 4 and 5 is usually necessary to distinguish these species.

Harvested areas are clearly visible on many aerial and satellite images. Figure 4.20 shows a satellite image illustrating timber harvesting in the northwestern United States. Here the darker toned areas are dense stands of Douglas fir and the lighter toned areas are recently cleared areas consisting of tree stumps, shrubs, and various grasses, in areas where essentially all trees have been removed during the harvesting operations. Mottled, intermediate toned areas have been replanted with Douglas fir trees and are at an intermediate growth stage. (For other examples of satellite images showing timber harvesting, see Chapters 6 and 8.)

Visual image interpretation is used extensively for "timber cruising." The primary objective of such operations is to determine the volume of timber that might be harvested from an individual tree or (more commonly) a stand

TABLE 4.10 Airphoto Interpretation Key for the Identification of Conifers

1. Crowns small; if large, then definitely cone shaped	
Crowns broadly conical, usually rounded tip; branches not prominent	Cedar
Crowns narrow, often cylindrical; trees frequently grow in swamps	Swamp-type black spruce
Crowns conical, deciduous, very light toned in fall, usually associated with black spruce	Tamarack
Crowns narrowly conical, very symmetrical, top pointed; branches less prominent than in white spruce	Balsam fir
Crowns narrowly conical; top often appears obtuse on photograph (except northern white spruce); branches more prominent than in balsam fir	White spruce, black spruce (except swamp type)
Crowns irregular, sometimes with pointed top; have thinner foliage and smoother texture than spruce and balsam fir	Jack pine
1. Crowns large and spreading, not narrowly conical; top often not well defined	
2. Crowns very dense, irregular or broadly conical	
Individual branches very prominent; crown usually irregular	White pine
Individual branches rarely very prominent; crown usually conical	Eastern hemlock
2. Crowns open, oval (circular in plain view)	Red pine

Source: From Sayn-Wittgenstein, 1961. Copyright © 1961, American Society of Photogrammetry. Reproduced with permission.

of trees. To be successful, image-based timber cruising requires a highly skilled interpreter working with both aerial or satellite and ground data. Image measurements on individual trees or stands are statistically related to ground measurements of tree volume in selected plots. The results are then extrapolated to large areas. The image measurements most often used are (1) tree height or stand height, (2) tree-crown diameter, (3) density of stocking, and (4) stand area.

The height of an individual tree, or the mean height of a stand of trees, is normally determined by measuring relief displacement or image parallax. The task of measuring tree-crown diameters is no different from obtaining other distance measurements on images. Ground distances are obtained from image distances via the scale relationship. The process is expedited by the use of special-purpose overlays similar to dot grids. Overlays are also used to measure the density of stocking in an area in terms of the crown closure or percentage of the ground area covered by tree crowns. Alternatively, some measure of the number of individual crowns per unit area may be made. The accuracy of these measurements is influenced by such factors as the band(s)

Figure 4.20   Satellite image (Landsat TM band 5, mid-IR), Cascade Mountains, Oregon, mid-April. Scale 1:130,000.

in which the image is sensed, the season of image acquisition, and the amount of shadow in the images.

Once data on individual trees or stands are extracted from images, they are statistically related (using multiple regression) with ground data on timber volume to prepare *volume tables*. The volume of *individual* trees is normally determined as a function of species, crown diameter, and height. This method of timber volume estimation is practical only on large-scale images and is normally used to measure the volume of scattered trees in open areas. More frequently, *stand volumes* are of interest. Stand volume tables are normally based on combinations of species, height, crown diameter, and crown closure (Table 4.11).

Visual image interpretation has been used in many instances to survey forest and urban shade tree damage from disease and insect infestations as well as from other causes. A variety of image bands and scales have been utilized for damage surveys. Although panchromatic photographs have often been used, the most successful surveys have typically used medium- or large-scale color and color infrared photographs (digital frame cameras and video cameras can also be used, as well as high resolution multiband satellite images). Some types of tree disease damage due to bacteria, fungus, virus, and other

TABLE 4.11 Estimated Volume of Kentucky Hardwood Stands

Average Stand Height (m)	Average Crown Diameter (m)	Estimated Volume (m³/ha) at Selected Crown Closures								
		15%	25%	35%	45%	55%	65%	75%	85%	95%
9	3–4	21	26	30	33	36	40	44	49	54
12	3–4	25	30	35	39	42	46	49	53	56
15	3–4	28	33	39	44	49	54	58	63	68
18	3–4	39	47	55	61	67	72	78	84	90
21	3–4	63	75	85	93	98	103	107	112	117
9	5–6	24	28	31	35	38	43	48	52	57
12	5–6	28	31	35	40	45	50	55	59	64
15	5–6	31	37	42	47	52	58	64	70	76
18	5–6	42	51	59	66	73	77	80	84	87
21	5–6	70	80	91	98	105	108	112	115	119
24	5–6	105	114	122	128	133	138	142	147	152
12	7–8	35	44	52	59	66	72	78	84	90
15	7–8	42	52	63	70	77	83	89	94	100
18	7–8	63	73	84	89	94	99	104	108	113
21	7–8	94	103	112	117	122	127	132	136	141
24	7–8	122	133	143	149	154	159	163	168	173
27	7–8	155	165	175	180	185	190	195	199	204
30	7–8	190	200	210	215	220	224	227	231	234
12	9+	59	72	84	89	94	99	104	108	113
15	9+	73	84	94	100	105	110	114	119	124
18	9+	91	101	110	115	120	125	130	135	140
21	9+	119	129	138	145	150	155	160	165	170
24	9+	150	159	168	175	182	186	190	195	200
27	9+	182	190	200	205	210	215	220	225	230
30	9+	213	222	231	236	241	245	248	252	255
33	9+	252	259	266	271	276	281	286	290	295

Source: Adapted from U.S. Department of Agriculture, 1978.

agents that have been detected using visual image interpretation are ash dieback, beech bark disease, Douglas fir root rot, Dutch elm disease, maple dieback, oak wilt, and white pine blister rust. Some types of insect damage that have been detected are those caused by the balsam wooly aphid, black-headed budworm, Black Hills bark beetle, Douglas fir beetle, gypsy moth larva, pine butterfly, mountain pine beetle, southern pine beetle, spruce budworm, western hemlock looper, western pine beetle, and white pine weevil. Other types of forest damage that have been detected include those resulting from air pollution (e.g., ozone, sulfur dioxide, "smog"), animals (e.g., beaver, deer, porcupine), fire, frost, moisture stress, soil salinity, nutrient imbalance, and storms.

In this discussion we have highlighted the application of visual image interpretation to tree species identification, studying harvested areas, timber cruising, and forest damage assessment. However, the forest management applications of visual image interpretation extend far beyond the scope of these four activities. Additional applications include such tasks as forest land appraisal, timber harvest planning, monitoring logging and reforestation, planning and assessing applications of herbicides and fertilizer in forest stands, assessing plant vigor and health in forest nurseries, mapping "forest fuels" to assess fire potential, planning fire suppression activities, assessing potential slope failures and soil erosion, planning forest roads, inventorying forest recreation resources, censusing wildlife and assessing wildlife habitat, and monitoring vegetation regrowth in fire lanes and power line rights-of-way.

Again, the success of virtually all of the above applications is premised on the existence of high quality reference data to aid in the interpretation. The use of aerial and satellite images and "conventional" ground methods of observation and measurement are typically closely intertwined. For example, timber volume inventories are basically premised on extensive ground measurement in sample plots (of tree volumes), but images are used to stratify the area to be inventoried and establish the location of these plots (typically based on interpreted tree type, stand area, and stocking density information). Thus, the image interpretation process complements, rather than replaces, the field activities.

4.8 RANGELAND APPLICATIONS

Rangeland has historically been defined as land where the potential natural vegetation is predominantly grasses, grasslike plants, forbs, or shrubs and where animal grazing was an important influence in its precivilization state. Rangelands not only provide forage for domestic and wild animals, they also represent areas potentially supporting land uses as varied as intensive agriculture, recreation, and housing.

Rangeland management utilizes rangeland science and practical experience for the purpose of the protection, improvement, and continued welfare of the basic rangeland resources, including soils, vegetation, endangered plants and animals, wilderness, water, and historical sites.

Rangeland management places emphasis on the following: (1) determining the suitability of vegetation for multiple uses, (2) designing and implementing vegetation improvements, (3) understanding the social and economic effects of alternative land uses, (4) controlling range pests and undesirable vegetation, (5) determining multiple-use carrying capacities, (6) reducing or eliminating soil erosion and protecting soil stability, (7) reclaiming soil and vegetation on disturbed areas, (8) designing and controlling livestock grazing systems, (9) coordinating rangeland management activities with other resource managers, (10) protecting and maintaining environmental quality, (11) mediating land use conflicts, and (12) furnishing information to policymakers (Heady and Child, 1994).

Given the expanse and remoteness of rangelands and the diversity and intensity of pressures upon them, visual image interpretation has been shown to be a valuable range management tool. A physical-measurement-oriented list of range management activities that have some potential for being accomplished by image interpretation techniques includes: (1) inventory and classification of rangeland vegetation, (2) determination of carrying capacity of rangeland plant communities, (3) determination of the productivity of rangeland plant communities, (4) condition classification and trend monitoring, (5) determination of forage and browse utilization, (6) determination of range readiness for grazing, (7) kind, class, and breed of livestock using a range area, (8) measurement of watershed values, including measurements of erosion, (9) making wildlife censuses and evaluations of rangelands for wildlife habitat, (10) evaluating the recreational use of rangelands, (11) judging and measuring the improvement potential of various range sites, and (12) implementing intensive grazing management systems (Tueller, 1996). Table 4.12 outlines the appropriate rangeland management uses of imagery of various scales.

TABLE 4.12 Appropriate Rangeland Management Uses of Aerial and Satellite Imagery of Various Scales

Imagery Scale	Rangeland Management Uses
1:100 to 1:500	Species identification, including grasses and seedlings, identification and measurement of erosion features, forage production estimates, rodent activities in the surface soil, assessment of the amounts of other surface features such as litter, and wildlife habitat assessment.
1:600 to 1:2000	Species measurements, erosion estimates over larger land areas, condition and trend assessment, production and utilization estimates, and wildlife habitat assessment.
1:5000 to 1:10,000	Detailed vegetation mapping, condition and trend assessment, production and utilization estimates, and wildlife habitat assessment.
1:15,000 to 1:30,000	Vegetation mapping at the habitat-type or ecological site level, allotment management planning, and planning for multiple use, including wildlife habitat assessment.
1:30,000 to 1:80,000	Planning for range management, vegetation and soil unit mapping on a pasture or allotment basis, and multiple-use planning, including wildlife habitat mapping.
1:100,000 to 1:2,500,000	Synoptic views for planning rangeland use and mapping large vegetation zones covering large areas such as entire mountain ranges.

Source: Adapted from Tueller, 1996.

4.9   WATER RESOURCE APPLICATIONS

Whether for irrigation, power generation, drinking, manufacturing, or recreation, water is one of our most critical resources. Visual image interpretation can be used in a variety of ways to help monitor the quality, quantity, and geographic distribution of this resource. In this section, we are concerned principally with the use of visual image interpretation in water pollution detection, lake eutrophication assessment, and flood damage estimation. Before describing each of these applications, let us review some of the basic properties of the interaction of sunlight with clear water.

In general, most of the sunlight that enters a clear water body is absorbed within about 2 m of the surface. The degree of absorption is highly dependent on wavelength. Near-infrared wavelengths are absorbed in only a few tenths of a meter of water, resulting in very dark image tones of even shallow water bodies on near-infrared images. Absorption in the visible portion of the spectrum varies quite dramatically with the characteristics of the water body under study. From the standpoint of the imaging of bottom details through clear water, the best light penetration is achieved between the wavelengths of 0.48 and 0.60 μm. Penetration of up to about 20 m in clear, calm ocean water has been reported in this wavelength band (Jupp et al., 1984; Smith and Jensen, 1998). Although blue wavelengths penetrate well, they are extensively scattered and an "underwater haze" results. Red wavelengths penetrate only a few meters.

The analysis of underwater features is often permitted by using imaging systems sensitive to at least the wavelengths of 0.48 to 0.60 μm. For example, excellent photographs of bottom details in clear ocean water have been obtained using both normal color and color infrared photography. White sand bottoms under clear ocean water will appear blue-green using normal color film and blue using color infrared film (with a yellow filter). Bottom details are somewhat sharper using color infrared film because the blue wavelengths are filtered out and, thus, the effects of "underwater haze" are minimized. With such photography, the color infrared film becomes essentially a two-layer film because there is almost no near-infrared reflection from the water and, therefore, virtually no image on the infrared-sensitive film layer.

Figure 4.21 illustrates the penetration of different wavelengths of sunlight into clear ocean water. The upper part of the photographs shows an exposed coral reef (varying amounts are exposed in the different frames due to wave action). The high infrared reflectance from the exposed coral results from the presence of algae that live in a symbiotic relationship with the coral. Most of the underwater reef consists of coral whose uppermost surfaces come to within about 0.3 m of the water surface. The keyhole-shaped area in the photo center has water depths ranging from very shallow near the dry white sand beach at lower right to a maximum of about 2 m near the center of the round part of the keyhole (upper left part of the photos).

Water Pollution Detection

All naturally occurring water contains some impurities. Water is considered polluted when the presence of impurities is sufficient to limit its use for a given domestic and/or industrial purpose. Not all pollutants are the result of human activity. Natural sources of pollution include such things as minerals leached from soil and decaying vegetation. When dealing with water pollution, it is appropriate to consider two types of sources: point and nonpoint. *Point sources* are highly localized, such as industrial outfalls. *Nonpoint sources*, such as fertilizer and sediment runoff from agricultural fields, have large and dispersed source areas.

Each of the following categories of materials, when present in excessive amounts, can result in water pollution: (1) organic wastes contributed by domestic sewage and industrial wastes of plant and animal origin that remove oxygen from the water through decomposition; (2) infectious agents contributed by domestic sewage and by certain kinds of industrial wastes that may transmit disease; (3) plant nutrients that promote nuisance growths of aquatic plant life such as algae and water weeds; (4) synthetic-organic chemicals such as detergents and pesticides resulting from chemical technology that are toxic to aquatic life and potentially toxic to humans; (5) inorganic chemical and mineral substances resulting from mining, manufacturing processes, oil plant operations, and agricultural practices that interfere with natural stream purification, destroy fish and aquatic life, cause excessive hardness of water supplies, produce corrosive effects, and in general add to the cost of water treatment; (6) sediments that fill streams, channels, harbors, and reservoirs, cause abrasion of hydroelectric power and pumping equipment, affect the fish and shellfish population by blanketing fish nests, spawn, and food supplies, and increase the cost of water treatment; (7) radioactive pollution resulting from the mining and processing of radioactive ores, the use of refined radioactive materials, and fallout following nuclear testing; and (8) temperature increases that result from the use of water for cooling purposes by steam electric power plants and industries and from impoundment of water in reservoirs, have harmful effects on fish and aquatic life, and reduce the capacity of the receiving water to assimilate wastes.

It is rarely possible to make a positive identification of the type and concentration of a pollutant by visual image interpretation alone. However, it is possible to use visual image interpretation to identify the point at which a discharge reaches a body of water and to determine the general dispersion characteristics of its plume. In some instances, such as the case of sediment

Figure 4.21 Black and white copies of color and color infrared photographs, Hanauma Bay, Island of Oahu, Hawaii: (*a*) normal color film (0.40–0.70 μm); (*b*) color infrared film with a Wratten No. 15 filter (0.50–0.90 μm); (*c*) color infrared film with a Wratten No. 29 filter (0.60–0.90 μm); (*d*) color infrared film with a Wratten No. 87 filter (0.74–0.90 μm).

Figure 4.21 *(Continued)*

suspended in water, it is possible to make valid observations about sediment concentrations using quantitative radiometry coupled with the laboratory analysis of selective water samples.

Sediment pollution is often clearly depicted on aerial and space images. Figure 4.22 is a space photograph showing the silt-laden waters of the Po River discharging into the Adriatic Sea. The seawater has a low reflectance of sunlight, similar to that for "Water (Clear)" shown in Figure 1.10. The spectral response pattern of the suspended solids resembles that of "Dry Bare Soil (Gray-Brown)" shown in Figure 1.10. Because the spectral response pattern of the suspended materials is distinct from that of the natural seawater, these two materials can be readily distinguished on the photograph.

When point-source pollutants—such as domestic and industrial wastes—enter natural water bodies, there is typically a dispersed plume similar to that shown in Figure 4.22. If pollutants have reflectance characteristics different from the water bodies, their mixing and dispersal can be traced on aerial photographs. Aerial photographs have been successfully used in the enforcement of antipollution laws. In such cases, it is normally mandatory that reference water samples be collected from within the plume and outside the plume coincident with the time of aerial photography. The aerial photographs can be used as evidence in court cases to establish the source of the pollutant samples collected. However, extreme care must be taken to follow the legal rules of evidence pertaining to photographic exhibits.

Materials that form films on the water surface, such as oil films, can also be detected through the use of aerial and satellite images. Oil enters the world's water bodies from a variety of sources, including natural seeps, municipal and industrial waste discharges, urban runoff, and refinery and shipping losses and accidents. Thick *oil slicks* have a distinct brown or black color. Thinner *oil sheens* and *oil rainbows* have a characteristic silvery sheen or iridescent color banding but do not have a distinct brown or black color. The principal reflectance differences between water bodies and oil films in the photographic part of the spectrum occur between 0.30 and 0.45 μm. Therefore, the best results are obtained when the imaging systems are sensitive to these wavelengths.

Normal color or ultraviolet aerial photography is often employed for the detection of oil films on water. Oil slicks on seawater can also be detected using imaging radar because of the dampening effect of oil slicks on waves (Chapter 8).

Lake Eutrophication Assessment

Water quality in inland lakes is often described in terms of *trophic state* (nutritional state). A lake choked with aquatic weeds or a lake with extreme-nuisance algal blooms is called a *eutrophic* (nutrient-rich) lake. A lake with

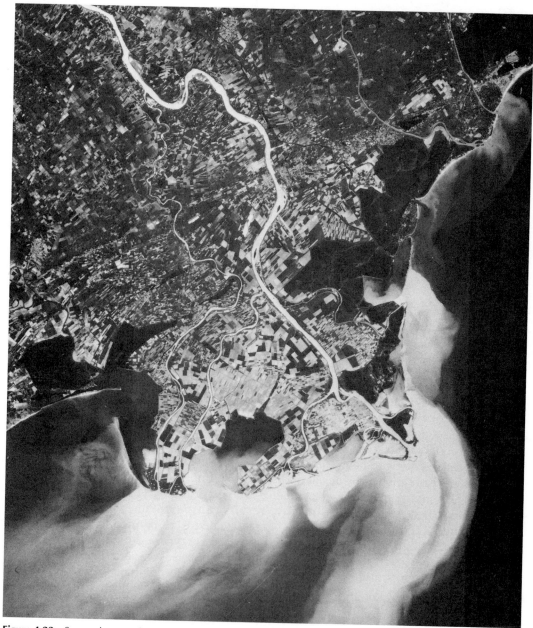

Figure 4.22 Space photograph (taken with the Large Format Camera) showing the silt-laden Po River discharging its sediments into the Adriatic Sea. Scale 1:350,000. (Courtesy NASA and ITEK Optical Systems.)

very clear water is called an *oligotrophic* (low nutrient, high oxygen) lake. The general process by which lakes age is referred to as *eutrophication*. Eutrophication is a natural process expressed in terms of geologic time. However, when influenced by human activity, the process is greatly accelerated and may result in "polluted" water conditions. Such processes are termed *cultural eutrophication* and are intimately related to land use/land cover.

What constitutes an unacceptable degree of eutrophication is a function of who is making the judgment. Most recreational users of water bodies prefer clear water free of excessive *macrophytes* (large aquatic plants) and *algae*. Swimmers, boaters, and water skiers prefer lakes relatively free of submersed macrophytes, while persons fishing for bass and similar fish generally prefer some macrophytes. Large concentrations of blue-green algae have an unpleasant odor that is offensive to most people, especially during "blooms," or periods following active algal growth. Green algae tend to be less bothersome, unless present in large quantities.

The use of visual image interpretation coupled with selective field observations is an effective technique for mapping aquatic macrophytes. Macrophyte community mapping can be aided by the use of image interpretation keys. More detailed information regarding total plant biomass or plant density can often be achieved by utilizing quantitative technique (Chapter 7). Airphoto interpretation has been used to economically plan and monitor operations such as mechanical harvesting or chemical treatment of weeds.

Concentrations of free-floating algae are a good indicator of a lake's trophic status. Excessive concentrations of blue-green algae are especially prevalent under eutrophic conditions. Seasonally, blooms of blue-green algae occur during warm water conditions in late summer, whereas diatoms are more common in the cold water of spring and fall. Green algae are typically present at any point in the seasonal cycle of lakes. Because the different broad classes of algae have somewhat different spectral response patterns, they can be distinguished by aerial and space imaging. However, the wavelengths corresponding to peak reflectance of blue-green and green algae are often close together, and the most positive results can be obtained using narrow band multiband photography with filters selected to maximize the differences between the spectral response of the two algae types or by using multispectral or hyperspectral scanners (Chapter 5) with at least several bands in the 0.45- to 0.60-μm wavelength range. Algae blooms floating on or very near the water surface also reflect highly in the near infrared, as illustrated in Figure 4.23. (See Plate 31*b* for an additional example of algae blooms.)

Flood Assessment

The use of aerial and satellite images for *flood assessment* is illustrated in Figures 4.24 to 4.27. Such images can help determine the extent of flooding and

Figure 4.23   Algae bloom on Lake Mendota, Madison, WI. Black and white copy of a color infrared photograph, early July. Approximate scale 1 : 6000. The algae form the light-toned swirling patterns. The curvilinear dark-tone band through the center of the photograph is the path of a motorized boat that has churned the water and brought clearer water to the surface.

the need for federal disaster relief funds, when appropriate, and can be utilized by insurance agencies to assist in assessing the monetary value of property loss.

Figure 4.24 is a multidate sequence of photographs showing river flooding and its aftereffects. Figure 4.24*a* is a late-summer aerial photograph showing the normal appearance of the Pecatonica River as it meanders through cropland in southern Wisconsin. Figure 4.24*b* shows the same area near the peak of a flood whose severity is expected only once each 100 years. The floodwater is about 3 m deep in the area at the center of this photograph. On the day before this photograph was taken, more than 150 mm of rain fell in a 2.5-hr period on the Pecatonica River watershed, which encompasses roughly 1800 km^2 above this area. Figure 4.24*c* shows the same area 3 weeks after flooding. The soils in the flooded area are moderately well drained to poorly drained silt loam alluvial soils that are high in fertility and in moisture-supplying capacity. The darkest soil tones in Figure 4.24*c* correspond to the poorly drained areas that are still quite wet 3 weeks after flooding. The widespread crop damage can be clearly seen on this photograph. Figure 4.24*d*

Figure 4.24 Black and white copies of panchromatic and color infrared aerial photographs showing flooding and its aftereffects, Pecatonica River near Gratiot, WI: (*a*) USDA–ASCS panchromatic photograph. Scale 1 : 9000. (*b*) Oblique color infrared photograph, June 30. (*c*) Oblique color infrared photograph, July 22. (*d*) Oblique color infrared photograph, August 11. The flying height for photos (*b*) to (*d*) was 1100 m.

Figure 4.24   *(Continued)*

Figure 4.25 Satellite images (composite of Landsat TM near- and mid-infrared bands) of the Missouri River near St. Louis. Scale 1 : 570,000. (*a*) Normal flow conditions, July 1988. (*b*) Flooding of a severity expected only once every 100 years, July 1993.

was taken 6 weeks after the flooding. Although the soil moisture conditions have returned to normal, the widespread crop damage is still very evident on this photograph. The streaked pattern of light-toned lines in the right-hand part clearly shows the direction of river flow at the time of flooding. Note that each light-toned streak is just downstream from a tree or group of trees and aligned with the direction of flow (the direction of flood water flow was from left to right in the right-hand part of this photograph).

Figure 4.25 illustrates the appearance of a flooding river from a satellite perspective. Here we see composite images of the near- and mid-infrared bands of Landsat TM data (Chapter 6), bands in which water appears very dark as contrasted with the surrounding vegetation. In (a) we see the Missouri River under low flow conditions, with a river width of about 500 m. In (b) we see a major flood in progress, with a river width of about 3000 m. This is the well-known 1993 flood, a flood with a severity expected only once every 100 years, that occurred on the Missouri and Mississippi Rivers.

Figures 4.26 and 4.27 illustrate new lakes formed by the overflow of the Nile River. First spotted in 1998 by astronauts aboard the Space Shuttle *Discovery*, the Toshka Lakes have fluctuated in size and shape through alternating dry spells and monsoons. Together, the lakes hold about one-quarter of the Nile's total water supply. The future existence of these lakes is unknown. Figure 4.26 illustrates the location of these lakes relative to Lake Nasser, the lake formed by the Aswan High Dam on the Nile River. Figure 4.27 illustrates how these lakes grew in size and number during the 2-year period beginning November 1999.

Other Selected Applications

A knowledge of *groundwater location* is important for both water supply and pollution control analysis. The identification of topographic and vegetation indicators of groundwater and the determination of the location of *groundwater discharge areas* (springs and seeps) can assist in the location of potential well sites. Also, it is important to be able to identify *groundwater recharge zones* in order to protect these areas (via zoning restrictions) from activities that would pollute the groundwater supply. Available image interpretation techniques cannot be used directly to map the depth to water in a groundwater system. However, vegetation types have been successfully used as indicators of approximate depth to groundwater. Estimates of *groundwater use* have also been made based on the interpretation of crop type, area, and irrigation method.

Additional water resource applications of visual image interpretation include hydrologic watershed assessment, riparian vegetation mapping, reservoir site selection, shoreline erosion studies, fish habitat surveys, snow cover mapping, floodplain and shoreland zoning compliance, and survey of recreational use of lakes and rivers.

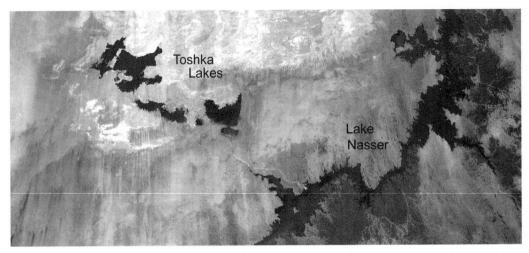

Figure 4.26 MODIS image of newly formed Toshka Lakes in southern Egypt. Approximate scale 1:3,000,000. September 14, 2002. (Courtesy NASA.)

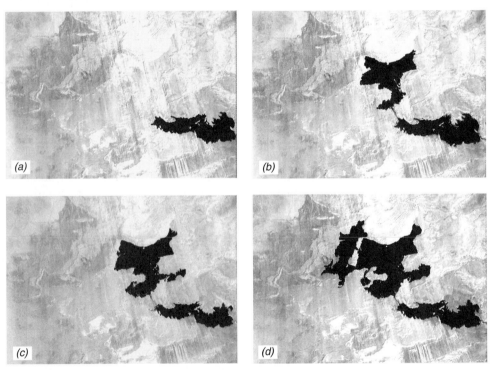

Figure 4.27 Landsat-7 ETM+ images of newly formed Toshka Lakes in southern Egypt. Approximate scale 1:2,000,000. (a) November 9, 1999, (b) January 12, 2000, (c) August 23, 2000, and (d) December 16, 2001. (Courtesy NASA.)

4.10   URBAN AND REGIONAL PLANNING APPLICATIONS

Urban and regional planners require nearly continuous acquisition of data to formulate governmental policies and programs. These policies and programs might range from the social, economic, and cultural domain to the context of environmental and natural resource planning. The role of planning agencies is becoming increasingly more complex and is extending to a wider range of activities. Consequently, there is an increased need for these agencies to have timely, accurate, and cost-effective sources of data of various forms. Several of these data needs are well served by visual image interpretation. A key example is land use/land cover mapping, as discussed in Section 4.4. Another example is the use of image interpretation to contribute data for land use suitability evaluation purposes, as outlined in Section 4.16. Here we discuss the utility of visual image interpretation in population estimation, housing quality studies, traffic and parking studies, site selection processes, and urban change detection.

Population estimates can be indirectly obtained through visual image interpretation. Traditionally, the procedure has been to use medium- to large-scale aerial photographs to estimate the number of dwelling units of each housing type in an area (single-family, two-family, multiple-family) and then multiply the number of dwelling units by the average family size per dwelling unit for each housing type. The identification of housing types is based on such criteria as size and shape of buildings, yards, courts, and driveways.

Images from the *Defense Meteorological Satellite Program (DMSP)* (Chapter 6) have been used to examine the earth's urban development from space. One of the DMSP's scanners, the *Operational Linescan System (OLS)*, is sensitive enough to detect low levels of visible and near-IR energy at night. With this sensor, it is possible to detect clouds illuminated by moonlight, plus lights from cities, towns, industrial sites, gas flares, and ephemeral events such as fires and lightning-illuminated clouds (Elvidge et al., 1997). Figure 4.28 is a global example of DMSP nighttime image data (a composite of hundreds of individual scenes). Figure 4.29 shows DMSP data of the eastern United States. The brightness of the light patterns in Figure 4.29 corresponds closely with the population distribution of the eastern United States, as seen in Figure 4.30. Care must be taken when attempting to compare data from different parts of the world. The brightest areas of the earth are the most urbanized, but not necessarily the most populated (compare western Europe with India and China). Nevertheless, DMSP data provide a valuable data source for tracking the growth of the earth's urbanization over time.

Visual image interpretation can also assist in *housing quality studies*. Many environmental factors affecting housing quality can be readily interpreted from aerial and satellite images, whereas others (such as the interior condition of buildings) cannot be directly interpreted. A reasonable estimate of housing quality can usually be obtained through statistical analysis of a limited, carefully selected set of environmental quality factors. Environmental factors that are

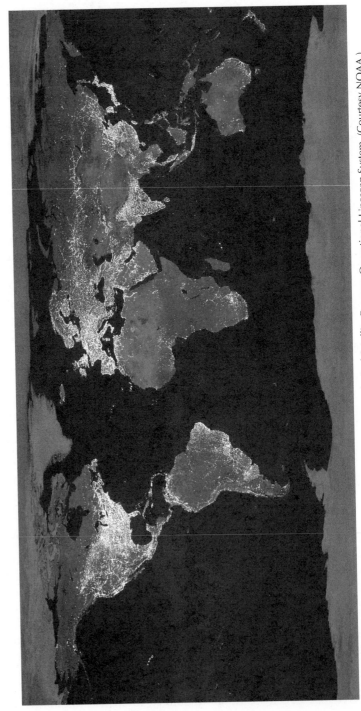

Figure 4.28 Earth lights at night as sensed by the Defense Meteorological Satellite Program Operational Linescan System. (Courtesy NOAA.)

268

interpretable from aerial photographs and that have been found to be useful in housing quality studies include house size, lot size, building density, building setback, street width and condition, curb and sidewalk condition, driveway presence/absence, garage presence/absence, vegetation quality, yard and open space maintenance, proximity to parkland, and proximity to industrial land use. Large-scale panchromatic photography has typically been used for housing quality studies. However, large- to medium-scale color infrared film has been shown to be superior in evaluating vegetation condition (lawns, shrubs, and trees).

Visual image interpretation can assist in *traffic and parking studies*. Traditional on-the-ground vehicle counts show the number of vehicles passing a few selected points over a period of time. An aerial image shows the distribution of vehicles over space at an instant of time. Vehicle spacings—and thus areas of congestion—can be evaluated by viewing such photographs. Average vehicle speeds can be determined when the image scale and time interval between exposures of overlapping photographs are known. The number and spatial distribution of vehicles parked in open-air lots and streets can be inventoried from aerial images. Not all vehicles in urban areas are visible on aerial images, however. Vehicles in tunnels and enclosed parking will obviously not be visible. In an area of tall buildings, streets near the edges of photographs may be hidden from view because of the radial relief displacement of the buildings. In addition, it may be difficult to discern vehicles in shadow areas on films of high contrast.

Visual image interpretation can assist in various location and siting problems, such as transportation route location, sanitary landfill site selection, power plant siting, commercial site selection, and transmission line location. The same general decision-making process is followed in each of these selection processes. First the factors to be assessed in the route/site selection process are determined. Natural and cultural features plus various economic, social, and political factors are considered. Then data files containing information on these factors are assembled and alternative routes/sites are analyzed and the final route/site is selected. Visual image interpretation has been used to collect much of the natural and cultural data dealing with topography, geology, soils, potential construction materials, vegetation, land use, wetland location, historical/archaeological sites, and natural hazards (earthquakes, landslides, floods, volcanoes, and tsunami). Various methods for obtaining such natural and cultural data through visual image interpretation are described elsewhere in this chapter. The task of analyzing the data is greatly facilitated by the use of a GIS.

Urban change detection mapping and analysis can be facilitated through the interpretation of multidate aerial and satellite images, such as the photographs shown in Figure 4.31, which illustrate the changes in an urban fringe area over a period of 53 years. The 1937 photograph (*a*) shows the area to be entirely agricultural land. The 1955 photograph (*b*) shows that a "beltline" highway has been constructed across the top of the area and that a gravel pit has begun operation in a glacial outwash plain at lower left. The 1968 photograph (*c*) shows that commercial development has begun at upper left and that extensive single-family housing development has begun at lower right. A

Figure 4.29 Defense Meteorological Satellite Program nighttime image of the eastern United States, 1981, 0.4- to 1.1-μm band. (Courtesy U.S. Air Force and NOAA.)

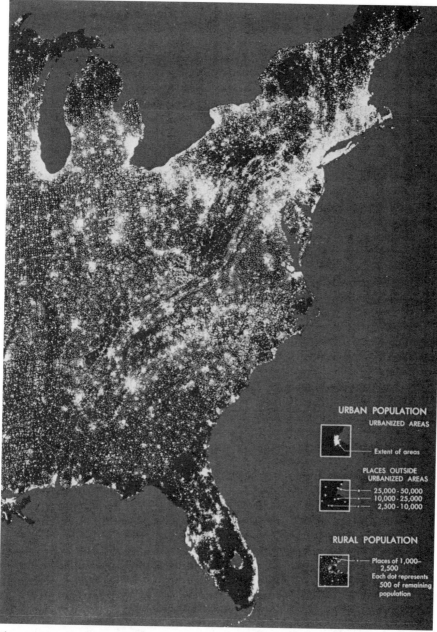

Figure 4.30 Map showing eastern U.S. population distribution in 1970. Population is proportional to brightness (see map legend). (U.S. Bureau of Census map.)

Figure 4.31 Multidate aerial photographs illustrating urban change, southwest Madison, WI (scale 1:20,000): (*a*) 1937; (*b*) 1955; (*c*) 1968; (*d*) 1990. [(*a–c*) USDA–ASCS photos. (*d*) Courtesy Dane County Regional Planning Commission.]

Figure 4.31   *(Continued)*

school has been constructed at lower center and the gravel pit continues operation. The 1990 photograph (d) shows that the commercial and single-family development has continued. Multiple-family housing units have been constructed at left. The gravel pit site is now a city park that was a sanitary landfill site for a number of years between the dates of photographs (c) and (d).

Plate 10 illustrates the use of satellite data for urban change detection. These Landsat images (Chapter 6) dramatically illustrate population growth in the Las Vegas metropolitan area (the fastest growing metropolitan area in the United States) over a 28-year period. The estimated population in the area shown was 300,000 in 1972 (a), and 1,425,000 in 2000 (b).

Visual image interpretation for urban change detection and analysis can be facilitated through the use of a Digital Transfer Scope (Section 4.3) as an aid in comparing images of two different dates or an image with a map.

4.11 WETLAND MAPPING

The value of the world's wetland systems has gained increased recognition. Wetlands contribute to a healthy environment in many ways. They act to retain water during dry periods, thus keeping the water table high and relatively stable. During periods of flooding, they act to reduce flood levels and to trap suspended solids and attached nutrients. Thus, streams flowing into lakes by way of wetland areas will transport fewer suspended solids and nutrients to the lakes than if they flow directly into the lakes. The removal of such wetland systems because of urbanization or other factors typically causes lake water quality to worsen. In addition, wetlands are important feeding, breeding, and drinking areas for wildlife and provide a stopping place and refuge for waterfowl. As with any natural habitat, wetlands are important in supporting species diversity and have a complex and important food web. Scientific values of wetlands include a record of biological and botanical events of the past, a place to study biological relationships, and a place for teaching. It is especially easy to obtain a feel for the biological world by studying a wetland. Other human uses include low intensity recreation and esthetic enjoyment.

Accompanying the increased interest in wetlands has been an increased emphasis on inventorying. The design of any particular wetland inventory is dependent on the objectives to be met by that inventory. Thus, a clearly defined purpose must be established before the inventory is even contemplated. Wetland inventories may be designed to meet the general needs of a broad range of users or to fulfill a very specific purpose for a particular application. Multipurpose and single-purpose inventories are both valid ways of obtaining wetland information, but the former minimizes duplication of effort. To perform a wetlands inventory, a classification system must be devised that will provide the information necessary to the inventory users. The system should be based primarily on enduring wetland characteristics so that the inventory does not become outdated too quickly, but the classification should also accommodate user information requirements for ephemeral wetland characteristics. In addition, the inventory system must provide a detailed description of specifically what is considered to be a wetland. If the wetland definition used for various "wetland maps" is not c learly stated, then it is not possible to tell if apparent wetland changes noted between maps of different ages result from actual wetland changes or are due to differences in concepts of what is considered a wetland.

At the federal level in the United States, four principal agencies are involved with wetland identification and delineation: (1) the Environmental Protection Agency, (2) the Army Corps of Engineers, (3) the Natural Resources Conservation Service, and (4) the Fish and Wildlife Service. The Environmental Protection Agency is concerned principally with water quality, the Army Corps of Engineers is concerned principally with navigable water issues that may be related to wetlands, the Natural Resources Conservation Service is concerned principally with identifying and mapping wetlands, and the Fish and Wildlife Service is principally interested in the use of wetlands for wildlife habitat. In 1989, these four agencies produced a *Federal Manual for Identifying and Delineating Jurisdictional Wetlands* (Federal Interagency Committee for Wetland Delineation, 1989), which provides a common basis for identifying and delineating wetlands. There is general agreement on the three basic elements for identifying wetlands: (1) hydrophytic vegetation, (2) hydric soils, and (3) wetland hydrology. *Hydrophytic vegetation* is defined as macrophytic plant life growing in water, soil, or substrate that is at least periodically deficient in oxygen as a result of excessive water content. *Hydric soils* are defined as soils that are saturated, flooded, or ponded long enough during the growing season to develop anaerobic (lacking free oxygen) conditions in the upper part. In general, hydric soils are flooded, ponded, or saturated for 1 week or more during the period when soil temperatures are above biologic zero (5°C) and usually support hydrophytic vegetation. *Wetland hydrology* refers to conditions of permanent or periodic inundation, or soil saturation to the surface, at least seasonally, hydrologic conditions that are the driving forces behind wetland formation. Numerous factors influence the wetness of an area, including precipitation, stratigraphy, topography, soil permeability, and plant cover. All wetlands typically have at least a seasonal abundance of water that may come from direct precipitation, overbank flooding, surface water runoff resulting from precipitation or snow melt, groundwater discharge, or tidal flooding.

Color infrared photography has been the preferred film type for wetlands image interpretation. It provides interpreters with a high level of contrast in image tone and color between wetland and nonwetland environments, and moist soil spectral reflectance patterns contrast more distinctively with less moist soils on color infrared film than on panchromatic or normal color films. Other multiband image types (e.g., multispectral scanners, hyperspectral scanners) can also be used, but should include at least one visible band and one near-infrared band.

An example of wetland mapping is shown in Figures 4.32 and 4.33. Figure 4.32 is a 6.7× enlargement of a color infrared airphoto that was used for

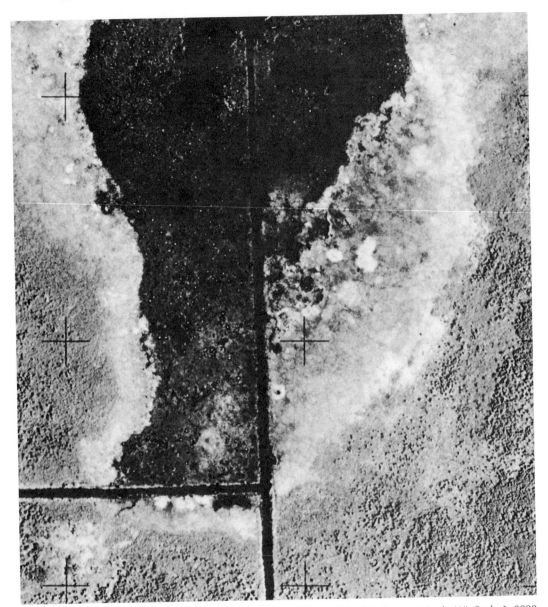

Figure 4.32   Black and white copy of a color infrared aerial photograph of Sheboygan Marsh, WI. Scale 1 : 9000 (enlarged 6.7 times from 1 : 60,000). Grid ticks appearing in image are from a reseau grid included in camera focal plane. (NASA image.)

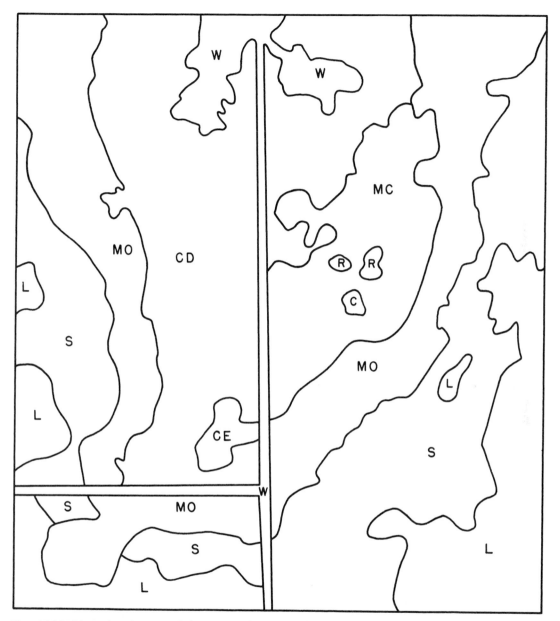

Figure 4.33 Vegetation classes in Sheboygan Marsh (scale 1 : 9000): W = open water, D = deep water emergents, E = shallow water emergents, C = cattail (solid stand), O = sedges and grasses, R = reed canary grass (solid stand), M = mixed wetland vegetation, S = shrubs, L = lowland conifer forest.

TABLE 4.13   Airphoto Interpretation Key to Vegetation Classes in Sheboygan Marsh for Use with Late-Spring 1:60,000 Color Infrared Film

Map Symbol (Figure 4.33)	Class Definition and Airphoto Interpretation Key
W	*Open water:* Areas of open water produce a dark blue image. The dark color and uniform smooth texture of the open water are in distinct contrast with the lighter tones of the surrounding vegetation.
D	*Deep water emergents:* These exist in water depths of 0.15–0.45 m or more and consist predominantly of cattail (*Typha latifolia, T. augustifolia*), burreed (*Sparganium eurycarpum*), and sometimes reedgrass (*Phragmites communis*). These species, when interspersed with water, form an image made up of a dull bluish color with soft texture, a tone produced by background reflectance of water blending with the vegetation reflectance. This subcommunity is sometimes interspersed with shallow water emergents.
E	*Shallow water emergents:* These consist of a mixture of such wetland species as cattail (*T. latifolia, T. augustifolia*), arrowhead (*Saggitaria latifolia*), water plantain (*Alisma plantago-aquatica*), burreed (*S. eurycarpum*), and several sedge species (*Carex lacustris, C. rostrata, C. stricta, C. aquatilis*) in water depths of 0.15 m or less. A medium bluish tone is produced that is lighter than the deep water areas.
C	*Cattail-solid stand:* This consists of solid stands of cattail (*T. latifolia, T. augustifolia*) that appear as mottled white patches in water ranging in depth from 0.10 to 0.75 m.
O	*Sedges and grasses:* The main components of a sedge meadow, sedges (*C. lacustris, C. rostrata, C. stricta, C. aquatilis*) and grasses (*Spartina* sp, *Phragmites* sp, *Calamagrostis* sp) are generally interspersed with small depressions of shallow water that together produce a continuous pattern of bluish water color intermixed with small white blotches.
R	*Reed canary grass-solid stand:* Reed canary grass appears as a uniform vegetation type that produces a bright white tone on the image. Reed canary grass occurs in small irregular patches and as linear features along stream banks. It is often difficult to differentiate from sedges and grasses because of the almost identical tones produced. Large areas of the species that were planted for marsh hay often retain their unnatural rectangular boundaries.
M	*Mixed wetland vegetation:* This consists primarily of sedges (*C. rostrata, C. stricta, C. lacustris*), forbs (marsh dock, *Rumex brittanica*; marsh bellflower, *Campanula aparinoides*; and marsh bedstraw, *Galium trifidum*), grasses (bluejoint, *Calamagrostis canadensis*), and cord grass (*Sparganium* sp). This community produces an interlacing pattern of magenta tones, light blues, and white colors, indicating the mixture of the component species.
S	*Shrubs:* This consists of buttonbush (*Cephalanthus occidentalis* L.), alder (*Alnus rugosa*), willow (*Salix interior, S. petiolaris, S. bebbiana*), and red osier dogwood (*Cornus stolonifera*). Shrubby areas have an intense magenta tone with coarse texture.
L	*Lowland conifer forest:* This consists, at this site, primarily of tamarack (*Larix laricina*) and white cedar (*Thuja occidentalis*) that display a deep mauve tone with considerable texture.

wetland vegetation mapping at an original scale of 1 : 60,000. The vegetation classification system and airphoto interpretation key are shown in Table 4.13. The wetland vegetation map (Figure 4.33) shows the vegetation in this scene grouped into nine classes. The smallest units mapped at the original scale of 1 : 60,000 are a few distinctive stands of reed canary grass and cattails about $\frac{1}{3}$ ha in size. Most of the units mapped are much larger.

Another example of wetland mapping can be seen in Plate 28, which illustrates multitemporal data merging as an aid in mapping invasive plant species, in this case reed canary grass.

At the federal level, the U.S. Fish and Wildlife Service is responsible for a National Wetlands Inventory that produces information on the characteristics, extent, and status of the nation's wetlands and deep water habitats. This information is used by federal, state, and local agencies, academic institutions, the U.S. Congress, and the private sector. As of 2002, the National Wetlands Inventory had mapped 90 percent of the lower 48 states, and 34 percent of Alaska. About 44 percent of the lower 48 states, and 13 percent of Alaska had been digitized. Congressional mandates require the production of status and trends reports at 10-year intervals. Updates to map and digital coverage are available on the Internet at: http//www.nwi.fws.gov.

4.12   WILDLIFE ECOLOGY APPLICATIONS

The term *wildlife* refers to animals that live in a wild, undomesticated state. *Wildlife ecology* is concerned with the interactions between wildlife and their environment. Related activities are *wildlife conservation* and *wildlife management*. Two aspects of wildlife ecology for which visual image interpretation can most readily provide useful information are wildlife habitat mapping and wildlife censusing.

A *wildlife habitat* provides the necessary combination of climate, substrate, and vegetation that each animal species requires. Within a habitat, the functional area that an animal occupies is referred to as its *niche*. Throughout evolution, various species of animals have adapted to various combinations of physical factors and vegetation. The adaptations of each species suit it to a particular habitat and rule out its use of other places. The number and type of animals that can be supported in a habitat are determined by the amount and distribution of food, shelter, and water in relation to the mobility of the animal. By determining the food, shelter, and water characteristics of a particular area, general inferences can be drawn about the ability of that area to meet the habitat requirements of different wildlife species. Because these requirements involve many natural factors, the image interpretation techniques described elsewhere in this chapter for mapping land cover, soil, forests, wetlands, and water resources are applicable to wildlife habitat analysis. Also, delineation of the "edges" between various landscape features is an important aspect of habitat analysis. Often, the interpreted habitat characteristics are incorporated in GIS-based modeling of the relationship between the habitat and the number and behavior of various species.

Figure 4.34 illustrates wildlife habitat mapping. This figure shows the Sheboygan Marsh, which was also shown in Figure 4.32 for the purpose of illustrating wetland vegetation mapping. In Figure 4.34, the nine vegetation classes shown in Figure 4.33 have been grouped into five wildlife habitat types, as follows: (1) *open water*, (2) *aquatic vegetation* (cattail, burreed, and reed grass), (3) *sedge meadow* (sedges and grasses), (4) *shrubs* (alder, willow, and dogwood), and (5) *lowland conifer forest* (tamarack and white cedar). Each of these five habitat types supports a significantly different population of mammals, birds, and fish. For example, a careful examination of the "aquatic vegetation" habitat area of Figure 4.34 on the original color infrared transparency (1 : 60,000) reveals that there are more than 100 white spots on the photograph, each surrounded by a dark area. Each of these white spots is a muskrat hut. Within the area of this photograph, muskrat huts can be found only in the area identified as aquatic vegetation habitat.

Wildlife censusing can be accomplished by ground surveys, aerial visual observations, or aerial imaging. Ground surveys rely on statistical sampling techniques and are often tedious, time consuming, and inaccurate. Many of the wildlife areas to be sampled are often nearly inaccessible on the ground. Aerial visual observations involve attempting to count the number of individuals of a species while flying over a survey area. Although this can be a low cost and relatively rapid type of survey, there are many problems involved. Aerial visual observations require quick decisions on the part of the observer regarding numbers, species composition, and percentages of various age and sex classes. Aggregations of mammals or birds may be too large for accurate counting in the brief time period available. In addition, low-flying aircraft almost invariably disturb wildlife, with much of the population taking cover before being counted.

Vertical aerial photography has been the best method of accurately censusing many wildlife populations. If the mammals or birds are not disturbed by the aircraft, the airphotos will permit very accurate counts to be undertaken. In addition, normal patterns of spatial distributions of individuals within groups will be apparent. Aerial photographs provide a permanent record that can be examined any number of times. Prolonged study of the photographs may reveal information that could not have been otherwise understood.

A variety of mammals and birds have been successfully censused using vertical aerial photography, including moose, elephants, whales, elk, sheep,

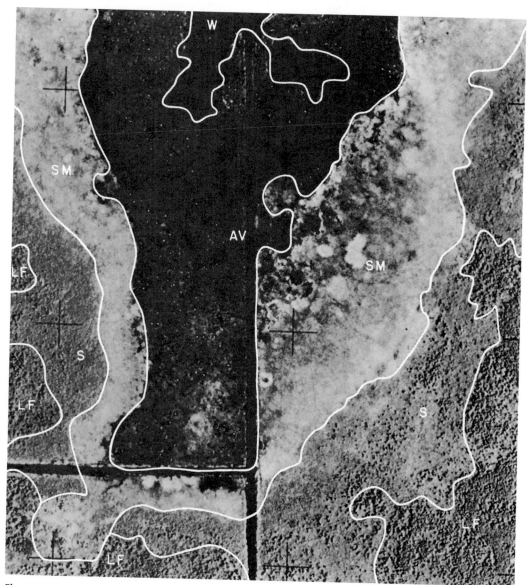

Figure 4.34 Wildlife habitat types in Sheboygan Marsh (scale 1 : 9000): W = open water, AV = aquatic vegetation, SM = sedge meadow, S = shrubs, LF = lowland conifer forest.

deer, antelope, sea lions, caribou, beavers, seals, geese, ducks, flamingos, gulls, oyster catchers, and penguins. Vertical aerial photography obviously cannot be used to census all wildlife populations. Only those that frequent relatively open areas during daylight hours can be counted. (Thermal scanning can be used to detect large animals in open areas.)

Wildlife censusing also requires that individual animals be large enough to be resolved on the photographs. A scale not smaller than 1:8000 is recommended for large mammals such as elk, whereas scales as large as 1:3000 should be used for smaller mammals such as sheep, deer, and antelope. A critical factor is the tonal contrast between the animal and its surroundings. For example, flocks of snow geese (large white birds) can be identified at a scale of 1:12,000 against a dark background. Individual birds are identifiable at scales of 1:6000 and larger.

In Figure 4.35, snow geese appear as white dots against a darker water background in the Bosque del Apache National Wildlife Refuge in New Mexico, where as many as 45,000 snow geese visit during their migrations.

Dark-colored wildlife species often can be discerned better in the winter against a snow or ice background than in the summer with a soil, vegetation, or water background. This is also the time of year when many species tend to band together and leaves have fallen from deciduous trees, making censusing possible even in certain kinds of forests. Special film–filter combinations can be selected to maximize the contrast.

The counting of individual animals on aerial images may present a problem when large numbers are present. Transparent grid overlays are often used as an aid in estimating numbers. Aerial images can also be used to stratify population densities (individuals per unit area) for use in stratified sampling techniques. Alternatively, aerial images not already in digital form can be digitized (via image scanning) and digital image processing techniques used to automatically "count" individuals.

Figure 4.36 shows a prairie dog colony on a plateau in South Dakota. Prairie dogs feed upon grasses and broad-leaved plants and construct burrows with mounded entrances. They disturb the ground in the vicinity of the colony, making the area susceptible to invasion by plants that exist in disturbed areas. The lighter toned area on the plateau in the center of the photograph is covered by such vegetation (mostly forbs) and the surrounding darker toned area is covered by native prairie grass. Each white spot in this lighter toned area is the bare soil associated with one prairie dog mound.

Figure 4.37 shows a large group of beluga whales (small white whales) that have congregated in an arctic estuarine environment principally for the purpose of calving. At the image scales shown here, it is possible to determine the number and characteristics of individual whales and to measure their lengths. On the full 230 × 230-mm aerial photograph from which Figure 4.37 was rephotographed, a total of about 1600 individual whales were counted. At

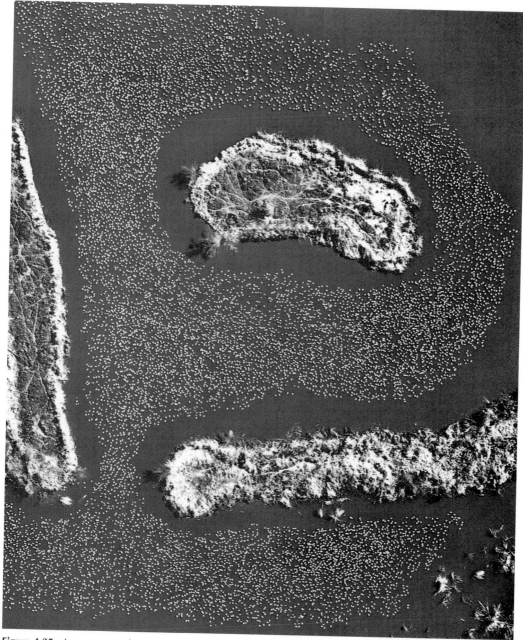

Figure 4.35 Large group of snow geese on water, Bosque del Apache National Wildlife Refuge, NM (black and white copy of a portion of a 230 × 230 mm color infrared aerial photograph). Late November. Approximate scale 1:1000. (Courtesy Kodak Aerial Services and U.S. Fish and Wildlife Service.)

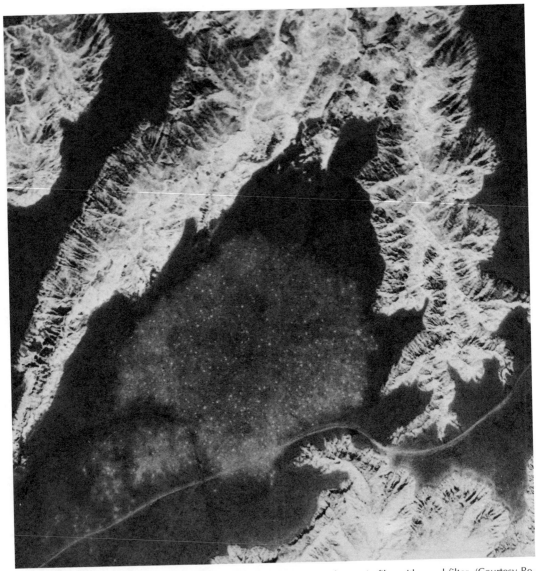

Figure 4.36 Prairie dog colony, Cuny Table, SD. Scale 1 : 9000. Panchromatic film with a red filter. (Courtesy Remote Sensing Institute, South Dakota State University.)

Figure 4.37   Large group of beluga whales, Cunningham Inlet, Somerset Island, northern Canada (black and white copy of photograph taken with Kodak Water Penetration Color Film, SO-224): (a) 1:2,400; (b) 1:800. (b) is a 3 times enlargement of the lower left portion of (a). (Courtesy J. D. Heyland, Metcalfe, Ontario.)

the original film scale of 1 : 2000, the average adult length was measured as 4 m and the average calf length was measured as 2 m. Numerous adults with calves can be seen, especially in the enlargement (Figure 4.37*b*). "Bachelor groups" of eight and six males can be seen at the lower left and lower right of Figure 4.37*b*.

4.13   ARCHAEOLOGICAL APPLICATIONS

Archaeology is concerned with the scientific study of historic or prehistoric peoples by analysis of the remains of their existence, especially those remains that have been discovered through earth excavation.

The earliest archaeological investigations dealt with obvious monuments of earlier societies. The existence of these sites was often known from histori-

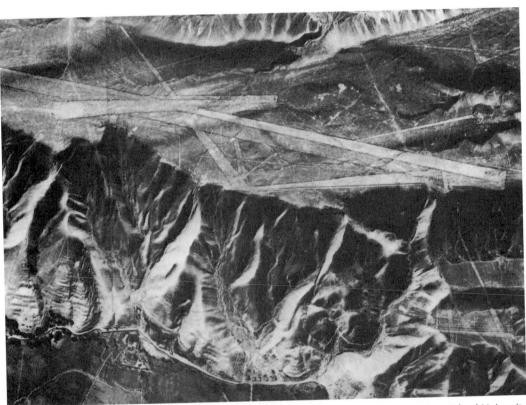

Figure 4.38   Vertical photomosaic showing Nazca Lines, Peru. (From Kosok, 1965. Courtesy Long Island University Press.)

cal accounts. Visual image interpretation has proven particularly useful in locating sites whose existence has been lost to history. Both surface and subsurface features of interest to archaeologists have been detected using visual image interpretation.

Surface features include visible ruins, mounds, rock piles, and various other surface markings. Examples of visible ruins are rock structures such as Stonehenge (England), castles (throughout Europe), and Indian dwellings in the southwestern United States. Examples of mounds are the bird-shaped and serpent-shaped Indian mounds of the Midwestern United States. Examples of rock structures are the various medicine wheels such as the Bighorn Medicine Wheel in Wyoming and the Moose Mountain Medicine Wheel in Saskatchewan. Other surface markings include Indian pictographs and the ancient Nazca Lines in Peru.

Figure 4.38 shows the Nazca Lines. They are estimated to have been made between 1300 and 2200 years ago and cover an area of about 500 km^2. Many geometric shapes have been found, as well as narrow straight lines that extend for as long as 8 km. They were made by clearing away literally millions of rocks to expose the lighter toned ground beneath. The cleared rocks were piled around the outer boundaries of the "lines." These markings were first noticed from the air during the 1920s. At that time, it was hypothesized that they formed a gigantic astronomical calendar, a belief still held by some scientists. The definite reason for their construction remains unknown.

Subsurface archaeological features include buried ruins of buildings, ditches, canals, and roads. When such features are covered by agricultural fields or native vegetation, they may be revealed on aerial or satellite images by tonal anomalies resulting from subtle differences in soil moisture or crop growth. On occasion, such features have been revealed by ephemeral differences in frost patterns.

Figure 4.39 shows the site of the ancient city of Spina on the Po River delta in Italy. Spina flourished during the fifth century B.C. and later became a "lost" city whose very existence was doubted by many. An extensive search for Spina ended in 1956 when it was identified on aerial photographs by an Italian archaeologist. Ancient Spina was a city of canals and waterways. The dark-toned linear features in Figure 4.39 are areas of dense vegetation growing in wet soils at the former location of the canals. The lighter toned rectangular areas are sparse vegetation over sand and the rubble of brick foundations. The light-toned linear features that run diagonally across this photograph are present-day drainage ditches.

The sites of more than a thousand Roman villas have been discovered in northern France through the use of 35-mm aerial photography. The buildings were destroyed in the third century A.D., but their foundation materials remain in the soil. In Figure 4.40, we see the villa foundation because of

Figure 4.39 Oblique aerial photograph showing site of the ancient city of Spina, Italy. (Courtesy Fotoaerea Valvassori, Ravenna, Italy.)

differences in crop vigor. The area shown in this figure has recently been converted from pasture to cropland. In the early years following such conversion, farmers apply little or no fertilizer to the fields. The cereal crops over the foundation materials are light toned owing to both the lack of fertilizer and a period of drought prior to the date of photography. The crops are darker toned over the remainder of the field. The main building (in the foreground) was 95 × 60 m.

Figure 4.41, an IKONOS image (Chapter 6) of the pyramids and sphinx at Giza, Egypt, shows the level of detail that can be seen in high resolution space images of archaeological sites. Shown at *a* is the Great Pyramid of Khufu. At

Figure 4.40 Oblique 35-mm airphoto of a cereal crop field in northern France. Differences in crop vigor reveal the foundation of a Roman villa. (Photograph by R. Agache. From Wilson, 1975. Courtesy The Council for British Archaeology, London.)

b is the Pyramid of Khafre; at *c* is the Pyramid of Menkaure, and just left of *d* is the Great Sphinx. The numerous small rectangular features left and right of the Great Pyramid are flat-topped funerary structures called "mastabas." The pyramids and sphinx were built about 4500 years ago. The length of each side of the Great Pyramid averages 230 m at its base, and the four sides are accurately oriented in the four cardinal directions (north, east, south, and west). The Great Sphinx is 73 m long, is carved out of limestone bedrock, and faces the rising sun.

Figure 4.41 IKONOS image, pyramids and sphinx at Giza, Egypt. Scale 1:12,600. North is to the bottom of the page. (Courtesy Space Imaging.)

4.14 ENVIRONMENTAL ASSESSMENT

Many human activities produce potentially adverse environmental effects. Examples include the construction and operation of highways, railroads, pipelines, airports, industrial sites, power plants, and transmission lines; subdivision and commercial developments; sanitary landfill and hazardous waste disposal operations; and timber harvesting and strip mining operations.

Dating as far back as 1969, the *National Environmental Policy Act* (NEPA) established as national policy the creation and maintenance of conditions that encourage harmony between people and their environment and minimize environmental degradation. This act requires that *environmental impact statements* be prepared for any federal action having significant impact on the environment. The key items to be evaluated in an environmental impact statement are (1) the environmental impact of the proposed action; (2) any adverse environmental effects that cannot be avoided should the action be implemented; (3) alternatives to the proposed action; (4) the relationship between local short term uses of the environment and the maintenance and enhancement of long term productivity; and (5) any irreversible and irretrievable commitments of resources that would be involved in the proposed action should it be implemented. Since the passage of NEPA, many other federal and state laws have been passed with environmental assessment as a primary component.

Environmental assessments involve, at a minimum, comprehensive inventory of physiographic, geologic, soil, cultural, vegetative, wildlife, watershed, and airshed conditions. Such assessments will typically draw on expertise of persons from many areas such as civil engineering, forestry, landscape architecture, land planning, geography, geology, archaeology, environmental economics, rural sociology, ecology, seismology, soils engineering, pedology, botany, biology, zoology, hydrology, water chemistry, aquatic biology, environmental engineering, meteorology, air chemistry, and air pollution engineering. Many of the remote sensing and image interpretation techniques set forth in this book can be utilized to assist in the conduct of such assessments. Overall, the applications of remote sensing in environmental monitoring and assessment are virtually limitless, ranging from environmental impact assessment to emergency response planning, landfill monitoring, permitting and enforcement, and natural disaster mitigation, to name but a few.

Effectively "real-time" imaging is used in such applications as responding to the spillage of hazardous materials. Such images are used to determine the extent and location of visible spillage and release, vegetation damage, and threats to natural drainage and human welfare. On the other hand, historical images are often used to conduct intensive site analyses of waste sites, augmenting these with current images when necessary. These analyses may include characterizing changes in surface drainage conditions through time; identifying the location of landfills, waste treatment ponds, and lagoons and their subsequent burial and abandonment; and detecting and identifying drums containing waste materials. Also, image interpretation may be used to help locate potential sites for drilling and sampling of hazardous wastes.

Figure 4.42 shows a plume of chlorine gas resulting from a train derailment. When such incidents occur, there is immediate need to assess downwind susceptibility of human exposure and the other potential impacts of the event. When available, remote sensing imagery and numerous forms of GIS

Figure 4.42 Low altitude, oblique aerial photograph of chlorine spill resulting from train derailment near Alberton, MT. (Courtesy RMP Systems.)

data are used in conjunction with ancillary ground information (such as wind speed and direction) to develop emergency evacuation and response plans in the general vicinity of the incident. In the specific case of Figure 4.42, the GIS and remote sensing initiated for immediate response was also very useful and important for the longer term monitoring of the spill site. Thus, the GIS established for the response was also useful in monitoring and change detection, as would be the case with many similar applications.

Figure 4.43 shows an oil spill in Bull Run, near Manassas, Virginia. Several thousand barrels of oil entered a nearby creek and then flowed into the connecting river and a reservoir. Several oil containment booms, clearly visible in Figure 4.43, were placed across the run to entrap the oil and facilitate cleanup efforts. Aerial photographs taken over the next few days followed the path of the oil spill movement. The photographs guided the on-scene coordinator in selecting locations for containment booms and pinpointing areas of oil accumulation. Subsequent photographs verified the success of the containment and cleanup of the spill. An example of a radar image showing an oil slick in the Mediterranean Sea can be seen in Figure 8.44.

Large-scale airphotos have been used in such applications as the identification of failing septic systems. The principal manifestations of septic system

Figure 4.43 Aerial photograph of an oil spill on Bull Run, near Manassas, Virginia. Note the series of booms placed across the river to entrap the oil (the flow of water is from right to left in this photograph). (Courtesy EPA Environmental Photographic Interpretation Center.)

failure are typically the upward or lateral movement of partially treated or untreated wastewater toward the soil surface. As the effluent moves upward and approaches the ground surface, the large amount of nutrients in the effluent causes enhanced growth in the vegetation directly above it. When the effluent reaches the surface, the overabundance of nutrients, coupled with an imbalance in the soil's air–water ratio, causes the vegetation to become stressed and eventually die. Finally, the effluent surfaces and either stands on the ground surface or flows downslope, often manifesting the same growth–stress–death pattern as it moves. Both normal color and color infrared photographs at a scale of around 1:8000 have been used for the detection of such situations (Evans, 1982). Open areas can be photographed throughout much of the year. Areas with sparse tree cover should be pho-tographed during early spring (after grasses have emerged but before tree leaves have appeared) or late fall (after tree leaves have dropped). Areas of dense tree cover may be impossible to analyze using airphoto interpretation at any time.

An analysis of the photo characteristics of color, texture, site, and asso-ciation, along with collateral soil information, is important for the identifi-cation of failing septic systems. Stereoscopic viewing is also important because it allows for the identification of slope, relief, and direction of sur-face drainage.

4.15 NATURAL DISASTER ASSESSMENT

Many forms of natural and human-induced disasters have caused loss of life, property damage, and damage to natural features. A variety of remote sensing systems can be used to detect, monitor, and respond to natural disasters, as well as assess disaster vulnerability. Here, we discuss wildfires, severe storms, floods, volcanic eruptions, dust and smoke, earthquakes, shoreline erosion, and landslides. NASA, NOAA, and the USGS maintain websites devoted to natural hazards (Appendix B).

Wildfires

Wildfires are a serious and growing hazard over much of the world. They pose a great threat to life and property, especially when they move into popu-lated areas. Wildfires are a natural process, and their suppression is now rec-ognized to have created greater fire hazards than in the past. Wildfire suppression has also disrupted natural plant succession and wildlife habitat in many areas.

Plate 11 shows a wildfire burning near the city of Los Alamos, New Mex-ico, as imaged by Landsat-7 (Chapter 6). Los Alamos is located on a series of mesas in the upper-right quadrant of the image. Plate 11a is a composite of bands 1, 2, and 3 (displayed as blue, green, and red, respectively), result-ing in a "normal color" image. Plate 11b is a composite of bands 2, 4, and 7 (again displayed as blue, green, and red) and is a "false color image." The normal color image shows the smoke plumes well but does not reveal much specific information about the areas that are burning at the time of image acquisition. In the false color image (b), the hottest actively burning areas appear bright red-orange in color, visible because of the large amount of en-ergy emitted from the fires in the 2.08- to 2.35-μm wavelength sensitivity range of band 7 (Figure 2.22). The darker red and near-black-toned areas are recently burned areas. The nearby darker green areas are unburned forest lands.

Severe Storms

Severe storms can take many forms, including tornados and cyclones. Tornadoes are rotating columns of air, usually with a funnel-shaped vortex several hundred meters in diameter, whirling destructively at speeds up to about 500 km/hr. Tornadoes occur most often in association with thunderstorms during the spring and summer in the midlatitudes of both the Northern and Southern Hemispheres. Cyclones are atmospheric systems characterized by the rapid, inward circulation of air masses about a low-pressure center, accompanied by stormy, often destructive, weather. Cyclones circulate counterclockwise in the Northern Hemisphere, and clockwise in the Southern Hemisphere. Hurricanes are severe tropical cyclones originating in the equatorial regions of the Atlantic Ocean or Caribbean Sea. Typhoons are tropical cyclones occurring in the western Pacific or Indian Oceans.

Tornado intensity is commonly estimated by analyzing damage to structures and then correlating it with the wind speeds required to produce such destruction. Tornado intensity is most often determined using the *Fujita Scale*, or *F-Scale* (Table 4.14). Although very few (about 2 percent) of all tornadoes reach F4 and F5 intensities, they account for about 65 percent of all deaths.

TABLE 4.14 Fujita Scale (F-Scale) of Tornado Intensity

F-Scale Value	Wind Speed (km/hr)	Tornado Intensity	Description of Damage	Examples
F0	64–116	Weak	Light	Branches broken, shallow-rooted trees pushed over
F1	117–181	Weak	Moderate	Surfaces peeled off roofs, mobile homes pushed off foundations or overturned
F2	182–253	Strong	Considerable	Roofs torn off frame houses, large trees snapped or uprooted
F3	254–332	Strong	Severe	Roofs and some walls torn off well-constructed houses, trains overturned, heavy cars lifted off ground and thrown
F4	333–418	Violent	Devastating	Well-constructed houses leveled, cars thrown.
F5	419–512	Violent	Incredible	Strong frame houses lifted off foundations and carried a considerable distance to disintegration, automobile-sized missiles flung through the air more than 100 m

Source: 2002 Encyclopedia Britannica online (http://www.britannica.com).

Plate 12 is a large-scale digital camera image showing a portion of the aftermath of an F4 tornado that struck Hayesville, Kansas. This tornado was responsible for 6 deaths, 150 injuries, and over $140 million in property damage.

Plate 32 contains "before" and "after" Landsat-7 satellite (Chapter 6) images of the damage caused by an F3 tornado that struck Burnett County, Wisconsin. This storm resulted in 3 deaths, 8 serious injuries, complete destruction of 180 homes and businesses, and damage to 200 others. Figure 4.44 is a large-scale aerial photograph showing the "blowdown" of trees by this tornado. Tornado damage is also shown in Figure 8.31, a radar image of a forested area in northern Wisconsin.

Hurricane intensity is most often determined using the Saffir–Simpson Hurricane Scale, which rates hurricane intensity on a scale from 1 to 5, and is used to give an estimate of potential property damage and flooding along the coast from a hurricane landfall. The strongest hurricanes are "Category 5" hurricanes, with winds greater than 249 km/hr. Category 5 hurricanes typically have a storm surge 5.5 m above normal sea level (this value varies widely with ocean bottom topography) and result in complete roof failure on many residences, and some complete building failures. Many shrubs, trees, and signs are blown down. Severe and extensive window and door damage can occur, as can complete destruction of mobile homes. Also, there is typically major damage to the lower floors of all structures located less than 5 m above sea level and within 500 m of the shoreline. Before the 1940s, many hurricanes went undetected. At present every hurricane is detected and tracked by satellite imaging. Figure 4.45 is a MODIS satellite (Chapter 6) image of an offshore hurricane that peaked as a Category 5 hurricane. A typical hurricane structure can be seen, with a counterclockwise flow of air, and a center "eye," with reduced wind speed and rainfall. Side-looking radar satellite images from Radarsat (Chapter 8) have also been used to monitor hurricanes.

Floods

Over-the-bank river flooding delivers valuable topsoil and nutrients to farmland and brings life to otherwise infertile regions of the world, such as the Nile River Valley. On the other hand, flash floods and large "100-year" floods are responsible for more deaths than tornadoes or hurricanes and cause great amounts of property damage.

Various examples of flooding can be found elsewhere in this book. Figure 4.24 is a multidate sequence of low altitude aerial photographs showing river flooding and its aftereffects. Figure 4.25 illustrates the appearance of a flooding river from a satellite perspective. Figures 4.26 and 4.27 illustrate new lakes formed by the overflow of the Nile River, again from a satellite perspective. And, Figure 8.48 shows flooding of Canada's Red River as imaged by the Radarsat satellite.

Figure 4.44 Aerial photograph showing the destruction of hundreds of trees by an F3 tornado that struck Burnett County, WI, on June 18, 2001. Scale 1:8100. (Courtesy Burnett County Land Information Office, University of Wisconsin-Madison Environmental Remote Sensing Center, and NASA Regional Earth Science Applications Center Program.)

Figure 4.45 Hurricane Herman off Baja California, September 2, 2002, as imaged by MODIS. (Courtesy NASA.)

Volcanic Eruptions

Volcanic eruptions are one of earth's most dramatic and violent agents of change. Eruptions often force people living near volcanoes to abandon their land and homes, sometimes forever. Volcanic activity in the last 300 years has killed more than 250,000 people, destroyed entire cities and forests, and severely disrupted local economies. Volcanoes can present a major hazard to those who live near them, for a variety of reasons: (1) pyroclastic eruptions can smother large areas of the landscape with hot ash, dust, and smoke within a span of minutes to hours; (2) red hot rocks spewed from the mouth of a volcano can ignite fires in nearby forests and towns, while rivers of molten lava can consume almost anything in their paths as they reshape the landscape; (3) heavy rains or rapidly melting summit snowpacks can trigger lahars, sluices of mud that can flow for miles, overrunning roads and villages; and, (4) large plumes of ash and gas ejected high into the atmosphere can influence climate, sometimes on a global scale (from USGS and NASA Natural Hazards websites—see Appendix B).

Plate 38 shows an extensive volcanic terrain in central Africa. Numerous lava flows can be seen on the slopes of Nyamuragiro volcano, which dominates the lower portion of the image. To the upper right of Nyamuragiro volcano is Nyiragongo volcano, which erupted in 2002 with loss of life and great property damage in and around the city of Goma, located on the shore of Lake Kivu, which can be seen at the right edge of Plate 38.

Plate 13 is a Landsat-5 (Chapter 6) image showing the eruption of 3350-m-high Mt. Etna, Italy, Europe's most active volcano. The bright red-orange areas on the volcano's flanks are flowing lava and fissures containing molten lava. This image is a composite of Landsat bands 2 (sensitive to green energy), 5 (mid-IR), and 7 (mid-IR). In this color composite, band 2 is displayed as blue, band 5 as green, and band 7 as red. Because the molten lava emits very little energy in green wavelengths, and a great deal of energy in the mid-IR (Figure 2.22), it appears in Plate 13 with a red-orange color. (Note that Plate 4b showed flowing lava as photographed with color infrared film.) The bright, puffy clouds near the volcano were formed from water vapor released during the eruption. A thick plume of airborne ash can be seen blowing from the volcano, to the southeast (north is to the top of Plate 13).

Dust and Smoke

Aerosols are small particles suspended in the air. Some occur naturally, originating from volcanoes, dust storms, and forest and grassland fires. Human activities, such as the burning of fossil fuels, prescribed fires, and the alteration of natural land surface cover (e.g., slash-and-burn activities), also generate aerosols. Many human-produced aerosols are small enough to be inhaled, so they can present a serious health hazard around industrial centers, or even hundreds of miles downwind. Additionally, thick dust or smoke plumes severely limit visibility and can make it hazardous to travel by air or road. Examples of smoke and volcanic ash can be seen in Plates 11 and 13, previously discussed.

Dust plumes have been observed in many arid regions around the globe. They can be extensive and travel great distances. For example, using satellite images, dust plumes originating near the west coast of Africa have been observed reaching the east coast of South America. Figure 4.46 shows an example of dust plumes over Baja California, blowing in a southwesterly direction, as imaged by the SeaWiFS satellite (Chapter 6).

Earthquakes

Earthquakes occur in many parts of the world and can cause considerable loss of life and property damage. Figure 4.11 illustrates extensive geologic features visible on satellite imagery that can be correlated with major geologic faults and major earthquake sites. Figure 6.29 shows the trace of ground cracks created during an earthquake that are clearly evident on a postearthquake satellite image. For information on earthquake hazards, see the USGS Earthquake Hazards Program website (Appendix B).

Shoreline Erosion

Driven by rising sea and lake levels, large storms, flooding, and powerful ocean waves, erosion wears away the beaches and bluffs along the world's shorelines. Bluff erosion rates vary widely, depending on geologic setting, waves, and weather. The erosion rate for a bluff can be regular over the years, or it can change from near zero for decades to several meters in a matter of seconds. Remote sensing studies of shoreline erosion have used a variety of platforms, from cameras in microlite aircraft to satellite data. Historical data using aerial photographs dating back to the 1930s, as well as more recent satellite data, can be used to document shoreline erosion over time. Also, recent studies of shoreline erosion have used Lidar data (Chapter 8).

Landslides

Landslides are mass movements of soil or rock down slopes and a major natural hazard because they are widespread. Globally, landslides cause an estimated 1000 deaths per year and great property damage. They commonly occur in conjunction with other major natural disasters, such as earthquakes,

Figure 4.46 Dust plumes over Baja California as imaged by the SeaWiFS satellite, February 2002. Scale 1:8,600,000. (Courtesy NASA.)

floods, and volcanic eruptions. Landslides can also be caused by excessive precipitation or human activities, such as deforestation or developments that disturb natural slope stability. They do considerable damage to infrastructure, especially highways, railways, waterways, and pipelines.

Historically, aerial photographs have been used extensively to characterize landslides and to produce landslide inventory maps, particularly because of their stereo-viewing capability and high spatial resolution. High resolution satellite data (Chapter 6), such as IKONOS, and the stereo data from SPOT-4, have proven useful for mapping large landslides, and the multi-incidence, stereo, and high resolution capabilities of Radarsat (Chapter 8) are also proving useful for landslide studies. Radar interferometry techniques (Chapter 8) have also been used in landslide studies in mountainous areas (Singhroy et al., 1998).

For additional information on landslide hazards, see the USGS Geologic Hazards—Landslides website (Appendix B).

4.16 PRINCIPLES OF LANDFORM IDENTIFICATION AND EVALUATION

Various terrain characteristics are important to soil scientists, geologists, geographers, civil engineers, urban and regional planners, landscape architects, real estate developers, and others who wish to evaluate the suitability of the terrain for various land uses. Because terrain conditions strongly influence the capability of the land to support various species of vegetation, an understanding of image interpretation for terrain evaluation is also important for botanists, conservation biologists, foresters, wildlife ecologists, and others concerned with vegetation mapping and evaluation.

The principal terrain characteristics that can be estimated by means of visual image interpretation are bedrock type, landform, soil texture, site drainage conditions, susceptibility to flooding, and depth of unconsolidated materials over bedrock. In addition, the slope of the land surface can be estimated by stereo image viewing and measured by photogrammetric methods.

Space limits the image interpretation process described to the assessment of terrain characteristics that are visible on medium-scale stereoscopic aerial photographs. Similar principles apply to nonphotographic and spaceborne sources.

Soil Characteristics

The term *soil* has specific scientific connotations to different groups involved with soil surveying and mapping. For example, engineers and agricultural soil scientists each have a different concept of soils and use a different terminology in describing soils. Most engineers consider all unconsolidated earth material lying above bedrock to be "soil." Agricultural soil scientists regard soil as a material that develops from a geologic parent material through the natural process of weathering and contains a certain amount of organic material and other constituents that support plant life. For example, a 10-m-thick deposit of glacial till over bedrock might be extensively weathered and altered to a depth of 1 m. The remaining 9 m would be relatively unaltered. An engineer would consider this a soil deposit 10 m thick lying over bedrock. A soil scientist would consider this a soil layer 1 m thick lying over glacial till parent material. We use the soil science (pedological) concept of soil in this chapter.

Through the processes of weathering, including the effects of climate and plant and animal activity, unconsolidated earth materials develop distinct layers that soil scientists call *soil horizons*. The top layer is designated the *A horizon* and called the *surface soil*, or *topsoil*. It can range from about 0 to 60 cm in thickness and is typically 15 to 30 cm. The A horizon is the most extensively weathered horizon. It contains the most organic matter of any horizon and has had some of its fine-textured particles washed down into lower horizons. The second layer is designated the *B horizon* and called the *subsoil*. It can range from 0 to 250 cm in thickness and usually is 45 to 60 cm. The B horizon contains some organic matter and is the layer of accumulation for the fine-textured particles washed down from the A horizon. The portion of the soil profile occupied by the A and B horizons is called the *soil* (or *solum*) by soil scientists. The *C horizon* is the underlying geologic material from which the A and B horizons have developed and is called the *parent material* (or *initial material*). The concept of soil profile development into distinct horizons is vitally important for agricultural soils mapping and productivity estimation as well as for many developmental uses of the landscape.

There are three principal origins of soil materials. *Residual soils* are formed in place from bedrock by the natural process of weathering. *Transported soils* are formed from parent materials that have been transported to their present location by wind, water, and/or glacial ice. *Organic soils* (muck and peat) are formed from decomposed plant materials in a very wet environment, typically in shallow lakes or areas with a very high groundwater table.

Soils consist of combinations of solid particles, water, and air. Particles are given size names, such as gravel, sand, silt, and clay, based on particle size. Particle size terminology is not standardized for all disciplines and several classification systems exist. Typical particle size definitions for engineers and agricultural soil scientists are shown in Table 4.15. For our purposes, the differences in particle size definitions between engineers and soil scientists for gravel, sand, silt, and clay are relatively unimportant. We use the soil science definition because it has a convenient system for naming combinations of particle sizes.

We consider materials containing more than 50 percent silt and clay to be *fine textured* and materials containing more than 50 percent sand and gravel to be *coarse textured*.

TABLE 4.15 Soil Particle Size Designations

Soil Particle Size Name	Soil Particle Size (mm)	
	Engineering Definition	Agricultural Soil Science Definition
Gravel	2.0–76.2	2.0–76.2
Sand	0.074–2.0	0.05–2.0
Silt	0.005–0.074	0.002–0.05
Clay	Below 0.005	Below 0.002

Soils have characteristic drainage conditions that depend on surface runoff, soil permeability, and internal soil drainage. We use the USDA soil drainage classification system (U.S. Department of Agriculture, 1997) for soils in their natural condition, with the seven soil drainage classes described as follows. (1) *Very poorly drained.* Natural removal of water from the soil is so slow that the water table remains at or near the surface most of the time. Soils of this drainage class usually occupy level or depressed sites and are frequently ponded. (2) *Poorly drained.* Natural removal of water from the soil is so slow that it remains wet for a large part of the time. The water table is commonly at or near the ground surface during a considerable part of the year. (3) *Somewhat poorly drained.* Natural removal of water from the soil is slow enough to keep it wet for significant periods, but not all the time. (4) *Moderately well drained.* Natural removal of water from the soil is somewhat slow so that the soil is wet for a small but significant part of the time. (5) *Well drained.* Natural removal of water from the soil is at a moderate rate without notable impedance. (6) *Somewhat excessively drained.* Natural removal of water from the soil is rapid. Many soils of this drainage class are sandy and very porous. (7) *Excessively drained.* Natural removal of water from the soil is very rapid. Excessively drained soils may be on steep slopes, very porous, or both.

Land Use Suitability Evaluation

Terrain information can be used to evaluate the suitability of land areas for a variety of land uses. Our emphasis is on suitability for developmental purposes, principally urban and suburban land uses.

The topographic characteristics of an area are one of the most important determinants of the suitability of an area for development. For subdivision development, slopes in the 2 to 6 percent range are steep enough to provide for good surface drainage and interesting siting and yet flat enough so that few significant site development problems will be encountered provided the soil is well drained. Some drainage problems may be encountered in the 0 to 2 percent range, but these can be readily overcome unless there is a large expanse of absolutely flat land with insufficient internal drainage. The site plan in the 6 to 12 percent range may be more interesting than in the 2 to 6 percent range but will be more costly to develop. Slopes over 12 percent present problems in street development and lot design and also pose serious problems when septic tanks are used for domestic sewage disposal. Severe limitations to subdivision development occur on slopes over 20 percent. For industrial park and commercial sites, slopes of not more than 5 percent are preferred.

The soil texture and drainage conditions also affect land use suitability. Well-drained, coarse-textured soils present few limitations to development. Poorly drained or very poorly drained, fine-textured soils can present severe limitations. Shallow groundwater tables and poor soil drainage conditions cause problems in septic tank installation and operation, in basement and foundation excavation, and in keeping basements water free after construction. In general, depths to the water table of at least 2 m are preferred. Depths of 1 to 2 m may be satisfactory where public sewage disposal is provided and buildings are constructed without basements.

Shallow depths to bedrock cause problems in septic tank installation and maintenance, in utility line construction, in basement and foundation excavation, and in street location and construction, especially when present in combination with steep slopes. Depths to bedrock over 2 m are preferred. Sites with a depth to bedrock of 1 to 2 m are generally unsatisfactory, but the development of these areas may be feasible in some cases. These sites are generally unsatisfactory where septic tank sewage disposal is to be provided. Also, additional excavation costs are involved where basements and public sewage disposal facilities are to be constructed. A depth to bedrock of less than 1 m presents serious limitations to development and is an unsatisfactory condition in almost all cases of land development.

Slope stability problems occur with certain soil-slope conditions. Although we will not discuss techniques for slope stability analysis using image interpretation, it should be mentioned that numerous areas of incipient landslide failure have been detected by image interpretation.

Despite the emphasis here on land development, it must be recognized that many land areas are worthy of preservation in their natural state because of outstanding topographic or geologic characteristics or because rare or endangered plant or animal species occupy those areas. The potential alteration of the hydrology of an area must also be kept in mind. In addition, the maintenance of prime agricultural land for agricultural rather than developmental use must be an important consideration in all land use planning decisions. Similar concerns also apply to the preservation of wetland systems.

Elements of Image Interpretation for Landform Identification and Evaluation

Image interpretation for landform identification and evaluation is based on a systematic observation and evaluation of key elements that are studied stereoscopically. These are topography, drainage pattern and texture, erosion, image tone, and vegetation and land use.

Topography

Each landform and bedrock type described here has its own characteristic topographic form, including a typical size and shape. In fact, there is often a distinct topographic change at the boundary between two different landforms.

With vertical photographs having a normal 60 percent overlap, most individuals see the terrain exaggerated in height about three or four times. Consequently, slopes appear steeper than they actually are. The specific amount of vertical exaggeration observed in any given stereopair is a function of the geometric conditions under which the photographs are viewed and taken.

Drainage Pattern and Texture

The drainage pattern and texture seen on aerial and space images are indicators of landform and bedrock type and also suggest soil characteristics and site drainage conditions.

Six of the most common drainage patterns are illustrated in Figure 4.47. The *dendritic drainage pattern* is a well-integrated pattern formed by a main stream with its tributaries branching and rebranching freely in all directions and occurs on relatively homogeneous materials such as horizontally bedded sedimentary rock and granite. The *rectangular drainage pattern* is basically a dendritic pattern modified by structural bedrock control such that the tributaries meet at right angles and is typical of flat-lying massive sandstone formations with a well-developed joint system. The *trellis drainage pattern* consists of streams having one dominant direction, with subsidiary directions of drainage at right angles, and occurs in areas of folded sedimentary rocks. The *radial drainage pattern* is formed by streams that radiate outward from a central area as is typical of volcanoes and domes. The *centripetal drainage pattern* is the reverse of the radial drainage pattern (drainage is directed toward a central point) and occurs in areas of limestone sinkholes, glacial kettle holes, volcanic craters, and other depressions. The *deranged drainage pattern* is a disordered pattern of aimlessly directed short streams, ponds, and wetland areas typical of ablation glacial till areas.

The previously described drainage patterns are all "erosional" drainage patterns resulting from the erosion of the land surface; they should not be confused with "depositional" drainage features that are remnants of the mode of origin of landforms such as alluvial fans and glacial outwash plains.

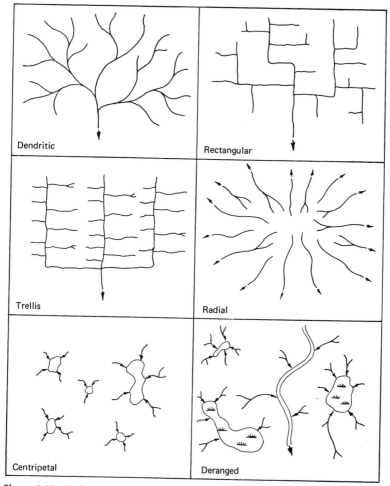

Figure 4.47 Six basic drainage patterns.

Coupled with drainage pattern is drainage texture. Figure 4.48 shows *coarse-textured* and *fine-textured* drainage patterns. Coarse-textured patterns develop where the soils and rocks have good internal drainage with little surface runoff. Fine-textured patterns develop where the soils and rocks have poor internal drainage and high surface runoff. Also, fine-textured drainage patterns develop on soft, easily eroded rocks, such as shale, whereas coarse-textured patterns develop on hard, massive rocks, such as granite.

Erosion

Gullies are small drainage features that may be as small as a meter wide and a hundred meters long. Gullies result from the erosion of unconsolidated material

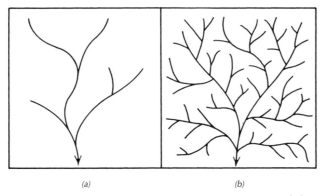

(a) (b)

Figure 4.48 Illustrative drainage patterns: (a) coarse-textured dendritic pattern; (b) fine-textured dendritic pattern.

by runoff and develop where rainfall cannot adequately percolate into the ground but instead collects and flows across the surface in small rivulets. These initial rivulets enlarge and take on a particular shape characteristic of the material in which they are formed. As illustrated in Figures 4.49 and 4.50, short gullies with V-shaped cross sections tend to develop in sand and gravel; gullies with U-shaped cross sections tend to develop in silty soils; and long gullies with gently rounded cross sections tend to develop in silty clay and clay soils.

Image Tone

The term *image tone* refers to the "brightness" at any point on an aerial or space image. The absolute value of the image tone depends not only on certain terrain characteristics but also on image acquisition factors such as film–filter combination (or the bands used for multispectral or hyperspectral scanning), exposure, and photographic/data processing. Image tone also depends on meteorological and climatological factors such as atmospheric haze, sun angle, and cloud shadows. Because of the effect of these non-terrain-related factors, image interpretation for terrain evaluation must rely on an analysis of *relative* tone values, rather than absolute tone values. Relative

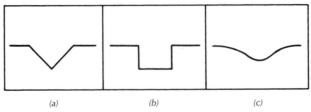

(a) (b) (c)

Figure 4.49 Illustrative gully cross sections: (a) sand and gravel; (b) silt; (c) silty clay or clay.

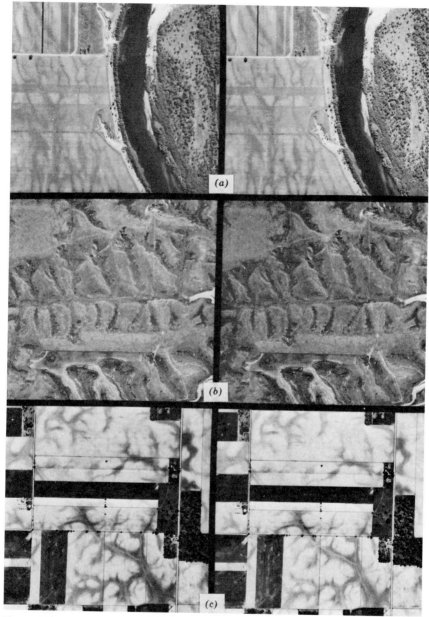

Figure 4.50 Stereopairs illustrating basic gully shapes: (*a*) sand and gravel terrace, Dunn County, WI; (*b*) loess (wind-deposited silt), Buffalo County, NE; (*c*) silty clay loam glacial till, Madison County, IN. Scale 1 : 20,000. (USDA–ASCS panchromatic photos.)

tone values are important because they often form distinct image patterns that may be of great significance in image interpretation.

The effect of terrain conditions on relative image tone can be seen in Figure 4.50c. In the case of bare soils (nonvegetated soils), the lighter toned areas tend to have a topographically higher position, a coarser soil texture, a lower soil moisture content, and a lower organic content. Figure 4.50c shows a striking tonal pattern often seen on fine-textured glacial till soils. The tonal differences are caused by differences in sunlight reflection due principally to the varying moisture content of the soil. The lighter toned areas are somewhat poorly drained silt loam soils on rises to 1 m above the surrounding darker toned areas of very poorly drained silty clay loam soils. The degree of contrast between lighter and darker toned bare soils varies depending on the overall moisture conditions of the soil, as illustrated in Plate 9.

The sharpness of the boundary between lighter and darker toned areas is often related to the soil texture. Coarser textured soils will generally have sharper gradations between light and dark tones while finer textured soils will generally have more gradual gradations. These variations in tonal gradients result from differences in capillary action occurring in soils of different textures.

Our discussion of image interpretation for terrain evaluation relates primarily to the use of panchromatic film because this film type has historically received the most use for this purpose. Subtle differences in soil and rock colors can be detected using multiple bands in the visible part of the spectrum (e.g., using color film rather than using panchromatic film), and subtle differences in soil moisture and vegetation vigor can be detected using at least one near-infrared band (e.g., color infrared film). Because there is a wide variety of soil and vegetation colors possible on color and color infrared films, it is not possible to consider them all here. Therefore, our discussion of image tone will describe tone as the shades of gray seen on panchromatic photographs. Persons working with color or color infrared photographs (or other sensors such as multispectral or hyperspectral scanners or side-looking radar) of specific geographic regions at specific times of the year can work out their own criteria for image tone evaluation following the principles outlined in this section.

Vegetation and Land Use

Differences in natural or cultivated vegetation often indicate differences in terrain conditions. For example, orchards and vineyards are generally located on well-drained soils, whereas truck farming activities often take place on highly organic soils such as muck and peat deposits. In many cases, however, vegetation and land use obscure differences in terrain conditions and the interpreter must be careful to draw inferences only from meaningful differences in vegetation and land use.

The Image Interpretation Process

Through an analysis of the elements of image interpretation (topography, drainage pattern and texture, erosion, image tone, vegetation, and land use), the image interpreter can identify different terrain conditions and can determine the boundaries between them. Initially, image interpreters will need to consider carefully each of the above elements individually and in combination in order to estimate terrain conditions. After some experience, these elements are often applied subconsciously as the interpreter develops the facility to recognize certain recurring image patterns almost instantaneously. In complex areas, the interpreter should not make snap decisions about terrain conditions but should carefully consider the topography, drainage pattern and texture, erosion, image tone, vegetation, and land use characteristics exhibited on the aerial and space images.

In the remainder of this section, we examine several of the principal bedrock types common on the earth's surface. For each of these, we consider geologic origin and formation, soil and/or bedrock characteristics, implications for land use planning, and image identification using the elements of image interpretation for terrain evaluation. Our illustrations are limited to occurrences in the United States. We emphasize the recognition of clear-cut examples of various bedrock types. In nature, there are many variations to each type. Interpreters working in specific localities can use the principles set forth here to develop their own image interpretation keys.

In cases where distinctions in image appearance must be made for different climatic situations, we will speak of "humid" and "arid" climates. We will consider *humid climates* to occur in areas that receive 50 cm or more rainfall per year and *arid climates* to occur in areas that receive less than 50 cm/year rainfall. In the United States, farming without irrigation is generally feasible in areas with a rainfall of about 50 cm/year or more. Areas receiving less than 50 cm/year rainfall typically require irrigation for farming.

Even the most searching and capable image analysis can benefit from field verification as the image interpretation process is seldom expected to stand alone. The image interpreter should consult existing topographic, geologic, and soil maps and should conduct a selective field check. The principal benefits of image interpretation for terrain evaluation should be a savings in time, money, and effort. The use of image interpretation techniques can allow for terrain mapping during periods of unsuitable weather for field mapping and can provide for more efficient field operations.

In order to illustrate the process of image interpretation for landform identification and evaluation, we will consider the terrain characteristics and image identification of several common bedrock types. Specifically, we treat the analysis of selected sedimentary and igneous rocks. The first three editions of this book treated the subject of landform identification and evaluation in greater detail by including discussions of aeolian landforms, glacial landforms, fluvial landforms, and organic soils (the first and second editions contain the most detailed coverage).

Sedimentary Rocks

The principal sedimentary rock types to be considered are sandstone, shale, and limestone. Sedimentary rocks are by far the most common rock type exposed at the earth's surface and extend over approximately 75 percent of the earth's land surface (igneous rocks extend over approximately 20 percent and metamorphic rocks over about 5 percent).

Sedimentary rocks are formed by the consolidation of layers of sediments that have settled out of water or air. Sediments are converted into coherent rock masses by lithification, a process that involves cementation and compaction by the weight of overlying sediments.

Clastic sedimentary rocks are rocks containing discrete particles derived from the erosion, transportation, and deposition of preexisting rocks and soils. The nature of the constituent particles and the way in which they are bound together determine the texture, permeability, and strength of the rocks. Clastic sedimentary rocks containing primarily sand-sized particles are called *sandstone*, those containing primarily silt-sized particles are called *siltstone*, and those containing primarily clay-sized particles are called *shale*.

Limestone has a high calcium carbonate content and is formed from chemical or biochemical action. The distinction between the two methods of formation is as follows. Chemically formed limestone results from the precipitation of calcium carbonate from water. Biochemically formed limestone results from chemical processes acting on shells, shell fragments, and plant materials.

The principal sedimentary rock characteristics that affect the appearance of the terrain on aerial and space images are *bedding, jointing,* and *resistance to erosion.*

Sedimentary rocks are typically stratified or layered as the result of variations in the depositional process. The individual strata or layers are called *beds.* The top and bottom of each bed have more or less distinct surfaces, called bedding planes, that delineate the termination of one bed and the beginning of another with somewhat different characteristics. Individual beds may range in thickness from a few millimeters to many meters. Beds in their initial condition usually are nearly horizontal but may be tilted to any angle by subsequent movements of the earth's crust.

Joints are cracks through solid bodies of rock with little or no movement parallel to joint surfaces. Joints in sedimentary rocks are primarily perpendicular to bedding planes and form plane surfaces that may intersect other joint planes. Several systematic joints constitute a joint set, and when two or more sets are recognized in an area, the overall pattern is called a joint system. Because joints are planes of weakness in rocks, they often form surfaces clearly visible on aerial and space images, especially in the case of sandstone. Streams often follow joint lines and may zig-zag from one joint line to another.

The *resistance to erosion* of sedimentary rocks depends on rock strength, permeability, and solubility. Rock strength depends principally on the strength of the bonding agent holding the individual sediment particles together and on the thickness of the beds. Thick beds of sandstone cemented by quartz are very strong and may be used as building materials. Thin beds of shale are often so weak that they can be shattered by hand into flakes and plates. Rock permeability refers to the ability of the rock mass to transmit water and depends on the size of the pore spaces between sediment particles and on the continuity of their connections. Sandstone is generally a very permeable rock. Shale is usually quite impermeable and water moves principally along joint planes rather than in sediment void spaces. Limestones high in calcium carbonate are soluble in water and may dissolve under the action of rainfall and ground water movement.

Here, we describe the characteristics of sandstone, shale, and limestone with a horizontally bedded attitude. The first and second editions of this book include a discussion of interbedded sedimentary rocks (both horizontally bedded and tilted).

Sandstone

Sandstone deposits commonly occur in beds a few meters thick interbedded with shale and/or limestone. Here we are concerned primarily with sandstone formations about 10 m or more in thickness.

Sandstone *bedding* is often prominent on images, especially when the sandstone beds occur over softer, more easily eroded formations such as shale. *Jointing* is prominent, with a joint system consisting of two or three dominant directions. The *resistance to erosion* varies, depending on the strength of the cementing agent. Sandstone cemented with iron compounds and silica is typically very strong, whereas sandstone cemented with carbonates is generally quite weak. Since sandstone is very permeable, most rainfall percolates downward through the rock rather than becoming erosion-producing surface runoff. Sandstone cemented with carbonates may weaken as percolating water dissolves the cementing agent.

In arid areas, there is seldom a residual soil cover over sandstone because any weathered sand particles are removed by wind erosion. In humid areas, the depth of residual soil cover depends on the strength of the cementing agent but is commonly less than 1 m and seldom more than 2 m. The residual soil texture in humid areas depends on the particle size of the sandstone and on the strength of the cementing agent. Weakly cemented sandstone weathers to sand while residual soils formed from strongly cemented sandstone may contain some silt and clay. Residual soils are typically well drained to excessively drained sand, loamy sand, and sandy loam.

Areas of massive sandstone beds with a residual soil cover are commonly undeveloped because of a combination of their typically rugged topography and shallow depths to bedrock. Buried sandstone strata are often an excellent source of groundwater for both individual homeowners and municipalities. Well-cemented sandstone rock is often used as building stone for residential construction.

Image Identification of Horizontally Bedded Sandstone

Topography: Bold, massive, relatively flat-topped hills with nearly vertical or very steep hillsides. *Drainage:* Coarse-textured, joint-controlled, modified dendritic pattern; often a rectangular pattern caused by perpendicular directions of joint sets. *Erosion:* Few gullies; V-shaped if present in residual soil. *Photographic image tone:* Generally light toned due to light rock color and excellent internal drainage of both residual soil and sandstone rock. Reddish sandstone in arid areas may photograph with a somewhat dark tone on panchromatic film. A dense tree cover over sandstone in humid areas generally appears dark, but in this case the interpreter is looking at the tree canopy rather than the soil or rock surface. *Vegetation and land use:* Sparse vegetation in arid areas. Commonly forested in humid areas as residual soil is too well drained to support crops. In a humid climate, flat-topped sandstone ridges with a loess cover are often farmed. *Other:* Sandstone is sometimes mistakenly identified as granite.

Figure 4.51 shows horizontally bedded sandstone in an arid climate interbedded with a few thin shale beds. The bedding can best be seen by inspecting the valley walls where the deeply incised stream cuts across the terrain. The direction of the major joint set is nearly vertical on the page; a secondary direction is perpendicular to the major joint set. These joint sets only partially control the direction of flow of the major stream but strongly influence the direction of secondary drainage.

Shale

Deposits of shale are common throughout the world as both thick deposits and thin deposits interbedded with sandstone and limestone. Shale *bedding* is very extensive with beds typically 1 to 20 cm in thickness. Bedding is not al-

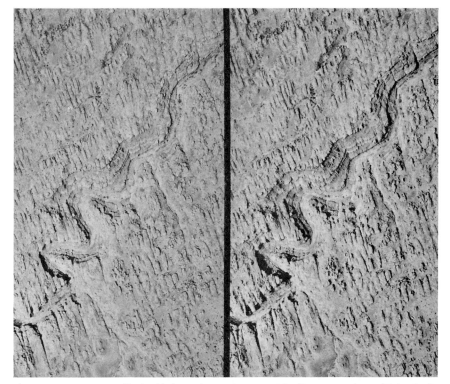

Figure 4.51 Horizontally bedded sandstone in an arid climate, southern Utah. Scale 1 : 20,000. (USGS panchromatic photos.)

ways visible on aerial and space images. However, if beds with a distinct difference in color or resistance to erosion are present or if shale is interbedded with sandstone or limestone, bedding may be seen. The effect of *jointing* is not always strong enough to alter the surface drainage system into a significantly joint-controlled pattern. The *resistance to erosion* is low, compared with other sedimentary rock types. Since shale is relatively impermeable, most rainfall runs off the ground surface, causing extensive erosion.

The depth of residual soil cover is generally less than 1 m and seldom more than 2 m. The residual soil is high in silt and clay, with textures typically silty loam, silty clay loam, silty clay, and clay. Internal soil drainage is typically moderately well drained or poorer, depending on soil texture and on soil and rock structure.

Although the topography in shale areas is generally favorable to urban development, the soil drainage and depth to bedrock conditions may limit residential development by causing problems in basement excavation and in septic tank installation and maintenance. The groundwater supply is extremely variable in shale bedrock. If the shale is strongly jointed, groundwater may be available. In many cases, however, it will be necessary to drill through the shale into an underlying water-bearing stratum.

Image Identification of Horizontally Bedded Shale

Topography: In an arid climate, minutely dissected terrain with steep stream/gully side slopes resulting from rapid surface runoff associated with short duration heavy rainfall. In a humid climate, gently to moderately sloping, softly rounded hills. *Drainage:* A dendritic pattern with gently curving streams; fine textured in arid climates and medium to fine textured in humid climates. *Erosion:* Gullies in residual soil have gently rounded cross sections. *Photographic image tone:* Varies widely, generally dark toned compared with sandstone and limestone. Differences in image tone may outline bedding. *Vegetation and land use:* Arid areas usually barren, except for desert vegetation. Humid areas intensively cultivated or heavily forested. *Other:* Shale is sometimes mistakenly identified as loess.

Figure 4.52 shows horizontally bedded shale in an arid climate. A comparison with Figure 4.51 illustrates the contrast in bedding, jointing, and resistance to erosion between shale and sandstone.

Limestone

Limestone consists mainly of calcium carbonate, which is soluble in water. Limestone that contains a significant amount of calcium carbonate and magnesium carbonate (or calcium magnesium carbonate) is called dolomitic limestone, or dolomite, and is less soluble in water. Limestone occurs throughout the world. For example, an area of very soluble limestone occurs in the United States in a region spanning portions of Indiana, Kentucky, and Tennessee.

Limestone *bedding* is generally not prominent on images unless the limestone is interbedded with sandstone or shale. *Jointing* is strong and determines the location of many of the pathways for subsurface drainage. However, jointing is generally not prominent on images of limestone in a humid climate. The *resistance to erosion* varies, depending on the solubility and jointing of the rock. Since calcium carbonate is soluble in water, many limestone areas have been severely eroded by rainfall and groundwater action.

The ground surface in areas of soluble limestone in humid climates is typically dotted with literally thousands of roughly circular depressions called *sinkholes.* They form when surface runoff drains vertically through the rock

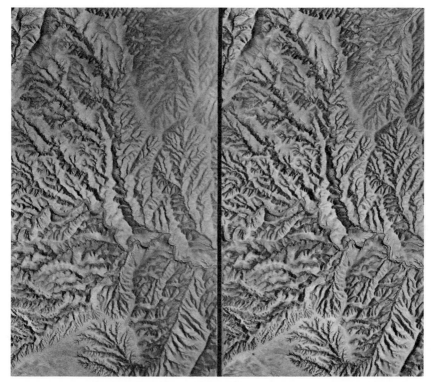

Figure 4.52 Horizontally bedded shale in an arid climate, Utah. Scale 1 : 26,700. (USGS panchromatic photos.)

along joint planes and the intersections of joint planes, gradually enlarging the underground drainageways by solution and causing the ground surface to collapse and form sinkholes.

There is generally only a shallow residual soil cover over limestone in arid areas where limestone often caps ridges and plateaus. In humid areas, the depth of residual soil cover is extremely variable and depends on the amount of solution weathering. Generally, residual soil depth ranges from 2 to 4 m for soluble limestone (which typically occurs as valleys or plains) and is somewhat less for dolomite (which may cap ridges and plateaus). The residual soil in humid areas contains a great deal of clay. Soil textures of clay, silty clay, clay loam, and silty clay loam are common. Soils are often well drained, except in sinkhole bottoms, due to soil structure and solution openings in the underlying rock. If these soils are extensively disturbed by human activity—such as subdivision development—soil drainage can become very poor.

Although limestone areas may be generally satisfactory for urban development, there are limiting characteristics that must be carefully considered. Because the residual soils contain a great deal of clay, they are relatively poor foundation soils. Often, it will be necessary to locate foundations directly on bedrock for proper building support. Although the soils are well drained in a natural condition, there may be problems with septic tank operation because of a low percolation rate in the disturbed soil. Groundwater may be difficult to locate and may be very "hard." In addition, effluent from septic tanks can often contaminate the groundwater. There is usually considerable variation in topography, depth to bedrock, and soil drainage conditions, requiring careful soil exploration and mapping before development proceeds. Sinkhole collapse under heavy loads such as construction equipment, highways, and airport runways is a serious problem in some limestone areas.

Image Identification of Horizontally Bedded Limestone

This discussion refers to soluble limestone in humid climates. *Topography:* A gently rolling surface broken by numerous roughly circular sinkholes that are typically 3 to 15 m in depth and 5 to 50 m in diameter. *Drainage:* Centripetal drainage into individual sinkholes. Very few surface streams. Surface streams from adjacent landforms or rock types may disappear underground via sinkholes when streams reach the limestone. *Erosion:* Gullies with gently rounded cross sections develop in the fine-textured residual soil. *Photographic image tone:* Mottled tone due to extensive sinkhole development. *Vegetation and land use:* Typically farmed, except for sinkhole bottoms that are often wet or contain standing water a portion of the year. *Other:* Limestone with extensive sinkhole development might be mistakenly identified as ablation till. Dolomitic limestone is more difficult to identify than soluble limestone. It is generally well drained and has subtle sinkholes.

Figure 4.53 shows horizontally bedded soluble limestone in a humid climate. Note the extensive sinkhole development (up to 40 sinkholes per square kilometer are present) and the complete lack of surface streams. The residual soils here are well-drained silty clay loam and silty clay 1.5 to 3 m deep over limestone bedrock.

Igneous Rocks

Igneous rocks are formed by the cooling and consequent solidification of magma, a molten mass of rock material. Igneous rocks are divided into two groups: intrusive and extrusive. *Intrusive igneous rocks* are formed when

Figure 4.53 Horizontally bedded soluble limestone in a humid climate, Harrison County, IN. Scale 1 : 20,000. (USDA–ASCS panchromatic photos.)

magma does not reach the earth's surface but solidifies in cavities or cracks it has made by pushing the surrounding rock apart or by melting or dissolving it. *Extrusive igneous rocks* are formed when magma reaches the ground surface.

Intrusive igneous rocks commonly occur in large masses in which the molten magma has cooled very slowly and solidified into large crystals. The crystal grains interlock closely to produce a dense, strong rock that is free of cavities. Erosion of overlying materials exposes intrusive igneous rocks.

Extrusive igneous rocks occur as various volcanic forms, including various types of lava flows, cones, and ash deposits. These rocks have cooled more rapidly than intrusive rocks and consequently have smaller crystals.

Intrusive Igneous Rocks

Intrusive igneous rocks range from granite, a light-colored, coarse-grained rock consisting principally of quartz and feldspar, to gabbro, a dark-colored, coarse-grained rock consisting principally of ferromagnesian minerals and feldspar. There are many intrusive igneous rocks intermediate between granite and gabbro in composition, such as granodiorite and diorite. We consider only the broad class of intrusive igneous rocks called *granitic rocks*, a term used to describe any coarse-grained, light-colored, intrusive igneous rock.

Granitic rocks occur as massive, *unbedded* formations such as the Sierra Nevada Mountains and the Black Hills of South Dakota. They are often strongly fractured into a series of irregularly oriented *joints* as a result of cooling from a molten state and/or pressure relief as overburden is eroded. Granitic rocks have a high *resistance to erosion*. As they weather, they tend to break or peel in concentric sheets through a process called exfoliation.

In arid areas, the depth of residual soil cover over granitic bedrock is typically very thin (less than $\frac{1}{2}$ m), except in fracture zones where it may be thicker. In humid areas, the depths to bedrock is typically 1 to 2 m. The residual soil texture in humid areas is typically loamy sand, sand loam, or sandy clay loam. Granitic rocks yield essentially no water, except in fracture zones. Limited water may be available from the sandy soil above the solid rock.

Areas of massive granitic rocks with residual soil cover are typically not well suited to urban development because of a combination of rugged topography, shallow depth to bedrock, and poor groundwater supply.

Image Identification of Granite Rocks

Topography: Massive, rounded, unbedded, domelike hills with variable summit elevations and steep side slopes. Often strongly jointed with an irregular and sometimes gently curving pattern. Joints may form topographic depressions in which soil and vegetation accumulate and along which water tends to flow. *Drainage and erosion:* Coarse-textured dendritic pattern with a tendency for streams to curve around the bases of domelike hills. Secondary drainage channels form along joints. Few gullies, except in areas of deeper residual soil. *Photographic image tone:* Light toned due to light rock color. Darker toned in depressions that form along joints. *Vegetation and land use:* Sparse vegetation in an arid climate. Often forested with some bare rock outcrops in a humid climate. Vegetation may be concentrated in depressions that form along some joints. *Other:* Granitic rocks are sometimes mistakenly identified as horizontally bedded sandstone. The principal difference in image identification of granitic rocks versus sandstone can be summarized as follows. (1) *Evidences of bedding:* Granitic rocks are unbedded; sandstone is bedded. (2) *Topography:* Granitic outcrops have variable summit elevations, sandstone caprocks form plateaus; granitic rocks have rounded cliffs, sandstone has vertical cliffs; granitic microfeatures are rounded, sandstone microfeatures are blocky. (3) *Joint pattern:* Granitic rocks have an irregular joint pattern with some distinct linear depressions; sandstone has a joint system consisting of two or three principal directions.

Figure 4.54 shows granitic rocks in an arid climate with very little soil or vegetative cover. Note the massive, unbedded formation with rounded cliffs. Note also that a number of joints are enlarged and form depressions with some soil and vegetative cover.

Extrusive Igneous Rocks

Extrusive igneous rocks consist principally of lava flows and pyroclastic materials. Lava flows are the rock bodies formed from the solidification of molten rock that issued from volcanic cones or fissures with little or no explosive activity. In contrast, pyroclastic materials, such as cinders and ash, were ejected from volcanic vents.

The form of lava flows depends principally on the viscosity of the flowing lava. The viscosity of lava increases with the proportion of silica (SiO_2) and alumina (Al_2O_3) in the lava. The least viscous (most fluid) lavas are the basaltic lavas, which contain about 65 percent silica and alumina. Andesitic lavas are intermediate in viscosity and contain about 75 percent silica and alumina. Rhyolitic lavas are very viscous and contain about 85 percent silica and alumina. Several basic volcanic forms are recognized.

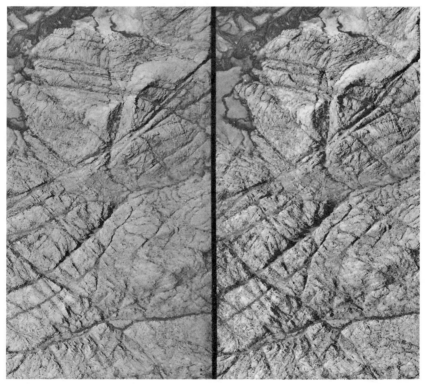

Figure 4.54 Granitic rock in an arid climate, Wyoming. Scale 1 : 37,300. (USGS panchromatic photos.)

Strato volcanoes (also called composite volcanoes) are steep-sided, cone-shaped volcanoes composed of alternating layers of lava and pyroclastic materials. The lava is typically andesitic or rhyolitic and side slopes can be 30° or more. Many strato volcanoes are graceful cones of striking beauty and grandeur. Each of the following mountains is a strato volcano: Shasta (California), Hood (Oregon), Ranier (Washington), St. Helens (Washington), Fuji (Japan), Vesuvius (Italy), and Kilimanjaro (Tanzania).

Shield volcanoes (also called Hawaiian-type volcanoes) are broad, gently sloping volcanic cones of flat domical shape built chiefly of overlapping basaltic lava flows. Side slopes generally range from about 4° to 10°. The Hawaiian volcanoes Haleakala, Mauna Kea, Mauna Loa, and Kilauea are shield volcanoes.

Flood basalt (also called plateau basalt) consists of large-scale eruptions of very fluid basalt that build broad, nearly level plains, some of which are at high elevation. Extensive flood basalt flows form the Columbia River and Snake River plains of the northwest United States.

Image Identification of Lava Flows

Topography: A series of tonguelike flows that may overlap and interbed, often with associated cinder and spatter cones. Viscous lavas (andesite and rhyolite) form thick flows with prominent, steep edges. Fluid lavas (basalt) form thin flows, seldom exceeding 15 m in thickness. *Drainage and erosion:* Lava is well drained internally and there is seldom a well-developed drainage pattern. *Photographic image tone and vegetation:* The color of unweathered, unvegetated, lava is dark toned in the case of basalt, medium toned for andesite, and light toned for rhyolite. In general, recent unvegetated flows are darker toned than weathered, vegetated flows. *Land use:* Recent flows are seldom farmed or developed.

Here, we illustrate only one example of a lava flow issuing from a volcano. Figure 4.55 shows a viscous lava flow that emanated from Mt. Shasta, California, a strato volcano. This flow is 60 m thick and has a 307 slope on its front face.

Metamorphic Rocks

Common metamorphic rocks are quartzite, slate, marble, gneiss, and schist. They are formed from preexisting sedimentary or igneous rocks due princi-

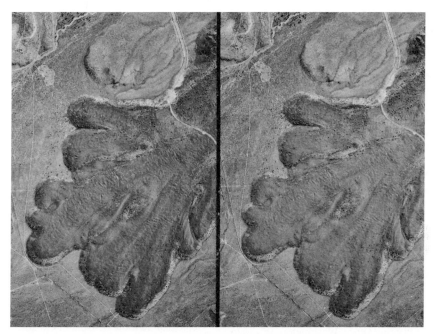

Figure 4.55 Viscous lava flow in an arid climate, Siskiyou County, CA. Scale 1:33,000. (USDA–ASCS panchromatic photos.)

pally to the action of heat and pressure. Occasionally, chemical action or shearing stresses are also involved.

Most metamorphic rocks have a distinct banding that can be seen via field observations and that sets them apart from sedimentary and igneous rocks.

Metamorphic rocks can be found throughout the world. However, since their extent is limited, the identification of metamorphic rocks is not covered here. In addition, the image identification of metamorphic rocks is more difficult than for sedimentary and igneous rocks, and interpretive techniques for metamorphic rocks are not well established.

SELECTED BIBLIOGRAPHY

Albert, D.P., et al., *Spatial Analysis, GIS, and Remote Sensing Applications in the Health Sciences*, Ann Arbor Press, Chelsea, MI, 2000.

Aldenderfer, M., and H. Maschner, eds., *Anthropology, Space, and Geographic Information Systems*, Oxford University Press, Oxford, UK, 1996.

Al-Rawi, K.R., et al., "Integrated Fire Evolution Monitoring System (IFEMS) for Monitoring Spatial-Temporal Behaviour of Multiple Fire Phenomena," *International Journal of Remote Sensing*, vol. 23, no. 10, 2002, pp. 1967–1983.

American Society of Photogrammetry (ASP), *Manual of Remote Sensing*, 2nd ed., ASP, Falls Church, VA, 1983.

American Society for Photogrammetry and Remote Sensing (ASPRS), *Air Spy*, ASPRS, Bethesda, MD, 1985.

American Society for Photogrammetry and Remote Sensing (ASPRS), *Our Secret Little War*, ASPRS, Bethesda, MD, 1991.

American Society for Photogrammetry and Remote Sensing (ASPRS), *Corona Between the Sun and the Earth: The First NRO Reconnaissance Eye in Space*, ASPRS, Bethesda, MD, 1997a.

American Society for Photogrammetry and Remote Sensing (ASPRS), *Manual of Photographic Interpretation*, 2nd ed., ASPRS, Bethesda, MD, 1997b.

American Society for Photogrammetry and Remote Sensing, *Remote Sensing for the Earth Sciences*, Manual of Remote Sensing, 3rd ed., vol. 3, Wiley, New York, 1999.

Anderson, J.E., R.L. Fischer, and S.R. Deloach, "Remote Sensing and Precision Agriculture: Ready for Harvest or Still Maturing?" *Photogrammetric Engineering and Remote Sensing*, vol. 65, no. 10, 1999, pp 1118–1123.

Anderson, J.R., et al., "A Land Use and Land Cover Classification System for Use with Remote Sensor Data," *Geological Survey Professional Paper 964*, U.S. Government Printing Office, Washington, DC, 1976.

Anderson, J., "Aerial Photo Interpretation of Archived Aerial Photography for Wetlands Resource Assessment Applications," *Technical Papers*, ASPRS-RTI 1998 Annual Conference, American Society for Photogrammetry and Remote Sensing, Bethesda, MD, 1998, pp. 1234–1240.

Arnold, R.H., *Interpretation of Airphotos and Remotely Sensed Imagery*, Prentice Hall, Upper Saddle River, NJ, 1997.

Arthus-Bertrand, Y., *Earth from Above*, Harry N. Abrams, New York, 1999.

Asner, G.P., et al., "Remote Sensing of Selective Logging in Amazonia—Assessing Limitations Based on Detailed Field Observations, Landsat ETM+, and Textural Analysis," *Remote Sensing of Environment*, vol. 80, no. 3, 2002, pp. 483–496.

Avery, T.E., and G.L. Berlin, *Fundamentals of Remote Sensing and Airphoto Interpretation*, 6th ed., Prentice Hall, Upper Saddle River, NJ, 2004.

Baber, J.J., Jr., and A.D. Flowerday, "Use of Low Altitude Aerial Biosensing with Color Infrared Photography as a Crop Management Service," *Technical Papers of the American Society of Photogrammetry*, ASP-ACSM Fall Technical Meeting, 1979, pp. 252–259.

Bailey, M.E., and H.C. Halls, "Use of Remote Sensing Data to Locate Groundwater Trapped by Dykes in Precambrian Basement Terrains," *Canadian Journal of Remote Sensing*, vol. 26, no. 2, 2002, pp. 111–120.

Blaszczynski, J.S., "Landform Characterization with Geographic Information Systems," *Photogrammetric Engineering and Remote Sensing*, vol. 63, no. 2, 1997, pp. 183–191.

Brisco, B., et al., "Precision Agriculture and the Role of Remote Sensing: A Review," *Canadian Journal of Remote Sensing*, vol. 24, no. 3, 1998, pp. 315–327.

Brooks, R.R., and D. Johannes, *Phytoarchaeology*, Dioscorides, Portland, OR, 1990.

Brown, D.G., and A.F. Arbogast, "Digital Photogrammetric Change Analysis as Applied to Active Coastal Dunes in Michigan," *Photogrammetric Engineering and Remote Sensing*, vol. 65, no. 4, 1999, pp. 467–474.

Bukata, R.P., et al., *Optical Properties and Remote Sensing of Inland and Coastal Waters*, Lewis Publishers, Boca Raton, FL, 1995.

Campbell, J.B., *Introduction to Remote Sensing*, 3rd ed., Guilford Press, New York, 2002.

Canadian Journal of Remote Sensing, Special Issue on Geological Remote Sensing in Canada, vol. 17, no. 2, 1991.

Canadian Journal of Remote Sensing, Special Issue on Agriculture, vol. 21, no. 2, 1995.

Chen, K., "An Approach to Linking Remotely Sensed Data and Areal Census Data," *International Journal of Remote Sensing*, vol. 23, no. 1, 2002, pp. 37–48.

Cowardin, L.M., et al., *Classification of Wetlands and Deepwater Habitats of the United States*, U.S. Fish and Wildlife Service Publication FWS/OBS-79/31, Washington, DC, 1979.

Cuevas-Jiménez, A., P.-L. Ardisson, and A.R. Condal, "Mapping of Shallow Coral Reefs by Colour Aerial Photography," *International Journal of Remote Sensing*, vol. 23, no. 18, 2002, pp. 3697–3712.

Curran, H.A., et al., *Atlas of Landforms*, 3rd ed., Wiley, New York, 1984.

Deuel, L., *Flights into Yesterday—The Story of Aerial Archaeology*, St. Martin's, New York, 1969.

Donnay, J-P., M.J. Barnsley, and P. Longley, *Remote Sensing and Urban Analysis*, Taylor & Francis, New York, 2001.

Drake, S., "Visual Interpretation of Vegetation Classes from Airborne Videography: An Evaluation of Observer Proficiency with Minimal Training," *Photogrammetric Engineering and Remote Sensing*, vol. 62, no. 8, 1996, pp. 969–978.

Drury, S.A., *Image Interpretation in Geology*, 3rd ed., Nelson Thornes, Cheltenham, UK, 2001.

Edwards, G., and K.E. Lowell, "Modeling Uncertainty in Photointerpreted Boundaries," *Photogrammetric Engineering and Remote Sensing*, vol. 62, no. 4, 1996, pp. 377–391.

Elvidge, C.D., et al., "Mapping City Lights with Nighttime Data from the DMSP Operational Linescan System," *Photogrammetric Engineering and Remote Sensing*, vol. 63, no. 6, 1997, pp. 727–734.

Epstein, J., K. Payne, and E. Kramer, "Techniques for Mapping Suburban Sprawl," *Photogrammetric Engineering and Remote Sensing*, vol. 68, no. 9, 2002, pp. 913–918.

Evans, B.M., "Aerial Photographic Analysis of Septic System Performance," *Photogrammetric Engineering and Remote Sensing*, vol. 48, no. 11, 1982, pp. 1709–1712.

Federal Interagency Committee for Wetland Delineation, *Federal Manual for Identifying and Delineating Jurisdictional Wetlands*, U.S. Army Corps of Engineers, U.S. Environmental Protection Agency, U.S. Fish and Wildlife Service, and USDA Soil Conservation Service, Washington, DC, Cooperative Technical Publication, 1989.

Fensham, R.J., et al., "Quantitative Assessment of Vegetation Structural Attributes from Aerial Photography," *International Journal of Remote Sensing*, vol. 23, no. 11, 2002, pp. 2293–2317.

Franklin, S.E., *Remote Sensing for Sustainable Forest Management*, Lewis, Boca Raton, FL, 2001.

Franklin, S.E., et al., "Aerial and Satellite Sensor Detection and Classification of Western Spruce Budworm Defoliation in a Subalpine Forest," *Canadian Journal of Remote Sensing*, vol. 21, no. 3, August 1995, pp. 299–308.

Friedman, R.A., and J.R. Stein, "Mapping Ancient Landscapes in the Four Corners Region: Integrating the Use of GIS, GPS, and Remote Sensing Technologies," *Proceedings: GIS/LIS '98*, American Society for Photogrammetry and Remote Sensing (and other organizations), 1998, pp. 193–201 (CD-ROM).

Gao, J., and S.M. O'Leary, "Estimation of Suspended Solids from Aerial Photographs in a GIS," *International Journal of Remote Sensing*, vol. 18, no. 10, 1997a, pp. 2073–2086.

Gao, J., and S.M. O'Leary, "The Role of Spatial Resolution in Quantifying SSC from Airborne Remotely Sensed Data," *Photogrammetric Engineering and Remote Sensing*, vol. 63, no. 3, 1997b, pp. 267–271.

Gauthier, R.P., M. Maloley, and K.B. Fung, "Land-Cover Anomaly Detection along Pipeline Rights-of-Way," *Photogrammetric Engineering and Remote Sensing*, vol. 67, no. 12, 2001, pp. 1377–1389.

Gerster, G., *Flights of Discovery—The Earth from Above*, Paddington, New York, 1978.

Gould, W.A., et al., "Canadian Arctic Vegetation Mapping," *International Journal of Remote Sensing*, vol. 23, no. 21, 2002, pp. 4597–4609.

Gupta, R.P. *Remote Sensing Geology*, 2nd ed., Springer, New York, 2003.

Haboudane, D., et al., "Land Degradation and Erosion Risk Mapping by Fusion of Spectrally-based Information and Digital Geomorphometric Attributes," *International Journal of Remote Sensing*, vol. 23, no. 18, 2002, pp. 3795–3820.

Hamblin, W.K., *Atlas of Stereoscopic Aerial Photographs and Remote Sensing Imagery of North America*, Crystal Productions, Glenview, IL, 1996.

Harvey, J.T., "Estimating Census District Populations from Satellite Imagery: Some Approaches and Limitations," *International Journal of Remote Sensing*, vol. 23, no. 10, 2002, pp. 2071–2095.

Heady, H.F., and R.D. Child, *Rangeland Ecology and Management*, Westview Press, Boulder, CO, 1994.

Hoffman, R.R., and A.B. Markman, *Interpreting Remote Sensing Imagery: Human Factors*, Lewis Publishers, Boca Raton, FL, 2001.

Huadong, G., *Applications of Radar Remote Sensing in China*, Taylor & Francis, New York, 2001.

International Journal of Remote Sensing, Special Issue on Global and Regional Land Cover Characterization from Remotely Sensed Data, vol. 21, nos. 6 & 7, 2000.

International Journal of Remote Sensing, Special Issue on Algal Blooms Detection, Monitoring, and Prediction, vol. 22, nos. 2 & 3, 2001.

Jensen, J.R., *Remote Sensing of the Environment: An Earth Resource Perspective*, Prentice Hall, Upper Saddle River, NJ, 2000.

Jensen, J.R., and D.C. Cowan, "Remote Sensing of Urban/Suburban Infrastructure and Socio-Economic Attributes," *Photogrammetric Engineering and Remote Sensing*, vol. 65, no. 5, 1999, pp. 611–622.

Jensen, M.E., and P.S. Bourgeron, eds., *A Guidebook for Integrated Ecological Assessments*, Springer, New York, 2001.

Jupp, D.L.B., et al., "The Application and Potential of Remote Sensing to Planning and Managing the Great Barrier Reef of Australia," *Proceedings: Eighteenth International Symposium on Remote Sensing of Environment*, Paris, France, 1984, pp. 121–137.

Kadmon, R., and R. Harari-Kremer, "Studying Long-Term Vegetation Dynamics Using Digital Processing of Historical Aerial Photographs," *Remote Sensing of Environment*, vol. 68, no. 2, 1999, pp. 164–176.

Kimura, H., and Y. Yamaguchi, "Detection of Landslide Areas Using Satellite Radar Interferometry," *Photogrammetric Engineering and Remote Sensing*, vol. 66, no. 3, 2000, pp. 337–344.

King, R.B., "Land Cover Mapping Principles: A Return to Interpretation Fundaments," *International Journal of Remote Sensing*, vol. 23, no. 18, 2002, pp. 3525–3545.

Kosok, P., *Life, Land and Water in Ancient Peru*, Long Island University Press, New York, 1965.

Kuehn, F., *Remote Sensing for Site Characterization*, Springer, New York, 2000.

Lark, C.D., S.M. Garod, and M.P. Pearson, "Landscape Archaeology and Remote Sensing in Southern Madagascar," *International Journal of Remote Sensing*, vol. 19, no. 8, 1998, pp. 1461–1477.

Leuder, D.R., *Aerial Photographic Interpretation*, McGraw-Hill, New York, 1959.

Light, D., "The National Aerial Photography Program as a Geographic Information System Resource," *Photogrammetric Engineering and Remote Sensing*, vol. 59, no. 1, 1993, pp. 61–65.

Liverman, D., et al., *People and Pixels: Linking Remote Sensing and Social Science*, National Academy Press, Washington, DC, 1998.

Loveland, T.R., et al., "Development of a Land-Cover Characteristics Database for the Conterminous U.S.," *Photogrammetric Engineering and Remote Sensing*, vol. 57, no. 11, 1991, pp. 1453–1463.

Loveland, T.R., et al., "A Strategy for Estimating the Rates of Recent United States Land-Cover Changes," *Photogrammetric Engineering and Remote Sensing*, vol. 68, no. 10, 2002, pp. 1091–1100.

Lyon, J.G., *Practical Handbook for Wetland Identification and Delineation*, Lewis Publishers, Boca Raton, FL, 1993.

Lyon, J.G., *Wetland Landscape Characterization GIS, Remote Sensing and Image Analysis*, Taylor & Francis, New York, 2001a.

Lyon, J.G., *Wetland Landscape Characterization*, Ann Arbor Press, Ann Arbor, MI, 2001b.

Lyon, J.G., and J. McCarthy, *Wetland and Environmental Applications of GIS*, Lewis Publishers, Boca Raton, FL, 1995.

Maeder, J., et al., "Classifying and Mapping General Coral-Reef Structure Using Ikonos Data," *Photogrammetric Engineering and Remote Sensing*, vol. 68, no. 12, 2002, pp. 1297–1306.

Maschner, H., ed., *New Methods, Old Problems: Geographic Information Systems in Archaeological Research*, Center for Archaeological Investigations Press, Southern Illinois University, Carbondale, IL, 1996.

Masek, J.G., et al., "Dynamics of Urban Growth in the Washington DC Metropolitan Area, 1973–1996, from Landsat Observations," *International Journal of Remote Sensing*, vol. 21, no. 18, 2000, pp. 3473–3486.

McRoberts, R.E., et al., "Using a Land Cover Classification Based on Satellite Imagery to Improve the Precision of Forest Inventory Area Estimates," *Remote Sensing of Environment*, vol. 81, no. 1, 2002, pp. 36–44.

Mesev, V., *Remotely-Sensed Cities*, Taylor & Francis, New York, 2003.

Meyer, M.P., "Place of Small-Format Aerial Photography in Resource Surveys," *Journal of Forestry*, vol. 80, no. 1, 1982, pp. 15–17.

Milfred, C.J., and R.W. Kiefer, "Analysis of Soil Variability with Repetitive Aerial Photography," *Soil Science Society of America Journal*, vol. 40, no. 4, 1976, pp. 553–557.

Miller, V.C., *Photogeology*, McGraw-Hill, New York, 1961.

Millington, A.C., et al., *GIS and Remote Sensing Applications in Biogeography and Ecology*, Kluwer Academic, Boston, 2001.

Mitchell, C., *Terrain Evaluation*, 2nd ed., Longman Scientific & Technical, copublished with Wiley, New York, 1991.

Mollard, J.D., "Techniques of Aerial-Photographic Study," *Reviews in Engineering Geology*, vol. 1, Geological Society of America, 1962, pp. 105–127.

Mollard, J.D., and J.R. Janes, *Airphoto Interpretation and the Canadian Landscape*, Canadian Government Publishing Centre, Hull, Quebec, Canada, 1984.

Moran, M.S., Y. Inoue, and E.M. Barnes, "Opportunities and Limitations for Image-Based Remote Sensing in Precision Crop Management," *Remote Sensing of Environment*, vol. 61, no. 3, 1997, pp. 319–346.

National Research Council, *Remote Sensing with Special Reference to Agriculture and Forestry*, National Research Council, National Academy of Science, Washington, DC, 1970.

Olson, Charles E., Jr., "Elements of Photographic Interpretation Common to Several Sensors," *Photogrammetric Engineering*, vol. 26, no. 4, 1960, pp. 651–656.

Olthof, I., and D.J. King, "Development of a Forest Health Index Using Multispectral Airborne Digital Camera Imagery," *Canadian Journal of Remote Sensing*, vol. 26, no. 3, 2002, pp. 166–186.

Paine, D.P., and J.D. Kiser, *Aerial Photography and Image Interpretation*, 2nd ed., Wiley, Hoboken, NJ, 2004.

Peters, A.J., et al., "Use of Remotely Sensed Data for Assessing Crop Hail Damage," *Photogrammetric Engineering and Remote Sensing*, vol. 66, no. 11, 2000, pp. 1349–1355.

Photogrammetric Engineering and Remote Sensing, Special Issue on Remote Sensing and GIS for Hazards, vol. 64, no. 10, 1998.

Photogrammetric Engineering and Remote Sensing, Special Issue on Global Land Cover Data Set Validation, vol. 65, no. 9, 1999.

Photogrammetric Engineering and Remote Sensing, Special Issue on Remote Sensing and Human Health, vol. 68, no. 2, 2002.

Pope, P., E. Van Eeckhout, and C. Rofer, "Waste Site Characterization through Digital Analysis of Historical Aerial Photographs," *Photogrammetric Engineering and Remote Sensing*, vol. 62, no. 12, 1996, pp. 1387–1394.

Prol-Ledesma, R.M., et al., "Use of Cartographic Data and Landsat TM Images to Determine Land Use Change in the Vicinity of Mexico City," *International Journal of Remote Sensing*, vol. 23, no. 9, 2002, pp. 1927–1933.

Prost, G.L., *Remote Sensing for Geologists: A Guide to Image Interpretation*, Taylor & Francis, New York, 2001.

Ray, R.G., *Aerial Photographs in Geologic Interpretation and Mapping*, Geological Survey Professional Paper No. 373, U.S. Government Printing Office, Washington, DC, 1960.

Robert, P.C., "Remote Sensing: A Potentially Powerful Technique for Precision Agriculture," *Proceedings: Land Satellite Information in the Next Decade II: Sources and Applications*, American Society for Photogrammetry and Remote Sensing, Bethesda MD, 1997, pp 19–25. (CD-ROM).

Robinson, J.A., et al., "Astronaut-Acquired Orbital Photographs as Digital Data for Remote Sensing: Spatial Resolution," *International Journal of Remote Sensing*, vol. 23, no. 20, 2002, pp. 4403–4438.

Rosas, U., "Vertical Exaggeration in Stereo-Vision: Theories and Facts," *Photogrammetric Engineering and Remote Sensing*, vol. 52, no. 11, 1986, pp. 1747–1751.

Rowe, N.C., and L.L. Grewe, "Change Detection for Linear Features in Aerial Photographs Using Edge-Finding," *IEEE Transactions on Geoscience and Remote Sensing*, vol. 39, no. 7, 2001, pp. 1608–1612.

Rundquist, D.C., et al., "Remote Measurement of Algal Chlorophyll in Surface Waters: The Case for the First Derivative of Reflectance Near 690 nm," *Photogrammetric Engineering and Remote Sensing*, vol. 62, no. 2, 1996, pp. 195–200.

Sabins, F.F., Jr., *Remote Sensing Laboratory Manual*, rev. 3rd ed., Kendall/Hunt, Dubuque, IA 1997a.

Sabins, F.F., Jr., *Remote Sensing—Principles and Interpretation*, 3rd ed., W.H. Freeman, New York, 1997b.

Saura, S., "Effects of Minimum Mapping Unit on Land Cover Data Spatial Configuration and Composition," *International Journal of Remote Sensing*, vol. 23, no. 22, 2002, pp. 4853–4880.

Sayn-Wittgenstein, L., "Recognition of Tree Species on Air Photographs by Crown Characteristics," *Photogrammetric Engineering*, vol. 27, no. 5, 1961, pp. 792–809.

Schultz, G.A., and E.T. Engman, *Remote Sensing in Hydrology and Water Management*, Springer, New York, 2000.

Scollar, I., et al., *Archaeological Prospecting and Remote Sensing*, Cambridge University Press, Cambridge, 1990.

Senay, G.B., and R.L. Elliott, "Capability of AVHRR Data in Discriminating Rangeland Cover Mixtures," *International Journal of Remote Sensing*, vol. 23, no. 2, 2002, pp. 299–312.

Sharkov, E.A., *Remote Sensing of Tropical Regions*, Wiley, New York, 1998.

Short, N.M., *The Remote Sensing Tutorial Online Handbook*, on the Internet at http://rst.gsfc.nasa.gov.

Siegal, B.S., and A.R. Gillespie, eds., *Remote Sensing in Geology*, Wiley, New York, 1980.

Singh, R.B., et al., *Land Use and Cover Change*, Science Publishers, Enfield, NH, 2001.

Singhroy, V., K. Mattar, and A.L. Gray, "Landslide Characterization in Canada Using Interferometric SAR and Combined SAR and TM Images," *Advances in Space Research*, 1998, vol. 21, no. 3, pp. 465–476.

Smith, A.M., M.S. Bullock, and C.A. Ivie, "Estimating Wheat Residue Cover Using Broad and Narrow Band Visible-Infrared Reflectance," *Canadian Journal of Remote Sensing*, vol. 26, no. 3 2002, pp. 241–252.

Smith, F.G.F., and J.R. Jensen, "The Multispectral Mapping of Seagrass: Application of Band Transformations for Minimization of Water Attenuation Using Landsat TM," *Technical Papers*, ASPRS-RTI 1998 Annual Conference, American Society for Photogrammetry and Remote Sensing, Bethesda, MD, 1998, pp. 592–603.

Sokhi, B.S., and S.M. Rashid, *Remote Sensing of Urban Environment*, Manak Publications, New Delhi, 1999.

Stolz, W., "Analysis of Large Format Camera Photographs of the Po Delta, Italy, for Topographic and Thematic Mapping," *International Journal of Remote Sensing*, vol. 9, nos. 10/11, 1988, pp. 1705–1714.

Strahler, A.H., and A.N. Strahler, *Physical Geography: Science and Systems of the Human Environment*, 2nd ed., Wiley, New York, 2002.

Strahler, A.H., and A.N. Strahler, *Introducing Physical Geography*, 3rd ed., Wiley, New York, 2003.

Strandberg, C.H., *Aerial Discovery Manual*, Wiley, New York, 1967.

Thierry, B., and K. Lowell, "An Uncertainty-Based Method of Photointerpretation," *Photogrammetric Engineering and Remote Sensing*, vol. 67, no. 1, 2001, pp. 65–72.

Thornbury, W.D., *Principles of Geomorphology*, 2nd ed., Wiley, New York, 1969.

Tomer, M.D., J.L. Anderson, and J.A. Lamb, "Assessing Corn Yield and Nitrogen Uptake Variability with Digitized Aerial Infrared Photographs," *Photogrammetric Engineering and Remote Sensing*, vol. 63, no. 3, 1997, pp. 299–306.

Tueller, P.T., "Rangeland Management," *The Remote Sensing Core Curriculum, Applications in Remote Sensing*, Vol. 4, on the Internet at http://www.asprs.org, 1996.

Twidale, C.R., *Analysis of Landforms*, Wiley, New York, 1976.

Twiss, S.D., et al., "Remote Estimation of Grey Seal Length, Width, and Body Mass from Aerial Photography," *Photogrammetric Engineering and Remote Sensing*, vol. 66, no. 7, 2000, pp. 859–866.

U.S. Department of Agriculture, *Aerial-Photo Interpretation in Classifying and Mapping Soils*, Agriculture Handbook 294, U.S. Government Printing Office, Washington, DC, 1969.

U.S. Department of Agriculture, *Soil Survey of Dane County, Wisconsin*, U.S. Government Printing Office, Washington, DC, 1977.

U.S. Department of Agriculture, *Forester's Guide to Aerial Photo Interpretation*, Agriculture Handbook 308, U.S. Government Printing Office, Washington, DC, 1978.

U.S. Department of Agriculture, Soil Survey Staff, Natural Resources Conservation Service, *National Soil Survey Handbook*, Title 430-VI, U.S. Government Printing Office, Washington, DC, 1997. (Available over the Internet at http://www.soils.usda.gov.)

Urai, M., "Volcano Monitoring with Landsat TM Short-Wave Infrared Bands: The 1990–1994 Eruption of Unzen Volcano, Japan," *International Journal of Remote Sensing*, vol. 21, no. 5, 2000, pp. 861–872.

Vaidyanathan, N.S., et al., "Mapping of Erosion Intensity in the Garhwal Himalaya," *International Journal of Remote Sensing*, vol. 23, no. 20, 2002, pp. 4125–4129.

Vincent, R.K., *Fundamentals of Geological and Environmental Remote Sensing*, Prentice Hall, Upper Saddle River, NJ, 1997.

Vogelmann, J.E., T. Sohl, and S.M. Howard, "Regional Characterization of Land Cover Using Multiple Sources of Data," *Photogrammetric Engineering and Remote Sensing*, vol. 64, no. 1, 1998, pp. 45–57.

Vogelmann, J.E., et al., "Completion of the 1990s National Land Cover Data Set for the Conterminous United States from Landsat Thematic Mapper Data and Ancillary Data Sources," *Photogrammetric Engineering and Remote Sensing*, vol. 67, no. 6, 2001, p. 650.

von Bandat, H.F., *Aerogeology*, Gulf, Houston, TX, 1962.

Wang, Yu, J.D. Colby, and K.A. Mulcahy, "An Efficient Method for Mapping Flood Extent in a Coastal

Floodplain Using Landsat TM and DEM Data," *International Journal of Remote Sensing*, vol. 23, no. 18, 2002, pp. 3681–3696.

Wheatley, D., and M. Gillings, *Spatial Technology and Archaeology the Archaeological Applications of GIS*, Taylor & Francis, New York, 2002.

Williams, R.S., and W.D. Carter, eds., *ERTS-1, A New Window on Our Planet*, USGS Professional Paper 929, Washington, DC, 1976.

Wilson, D.R., ed., *Aerial Reconnaissance for Archaeology*, Research Report No. 12, The Council for British Archaeology, London, 1975.

Wilson, D.R., *Air Photo Interpretation for Archaeologists*, St. Martin's, New York, 1982.

Wolf, P.R., and B.A. Dewitt, *Elements of Photogrammetry: With Applications in GIS*, 3rd ed., McGraw-Hill, New York, 2000.

Yang, X.J., "Satellite Monitoring of Urban Spatial Growth in the Atlanta Metropolitan Area," *Photogrammetric Engineering and Remote Sensing*, vol. 68, no. 7, 2002, pp. 725–734.

Zsilinszky, V.G., *Photographic Interpretation of Tree Species in Ontario*, Ontario Department of Lands and Forests, Ottawa, 1966.

5 MULTISPECTRAL, THERMAL, AND HYPERSPECTRAL SENSING

5.1 INTRODUCTION

In Section 2.14 we described multiband imaging using film-based, digital, and video camera systems. With these camera types, generally only three or four relatively wide wavelength bands ranging from 0.3 to 0.9 μm were sensed. A *multispectral scanner* operates on the same principle of selective sensing in multiple spectral bands, but such instruments can sense in many more bands over a greater range of the electromagnetic spectrum. Utilizing electronic detectors, multispectral scanners can extend the range of sensing from 0.3 to approximately 14 μm. (This includes the UV, visible, near-IR, mid-IR, and thermal IR spectral regions.) Furthermore, multispectral scanner systems can sense in very narrow bands.

We begin our discussion in this chapter with a discussion of how multispectral scanner images are acquired physically. We treat the basic processes of across-track and along-track scanning, followed by a description of the basic operating principles of multispectral scanners. After our treatment of multispectral scanning, we discuss thermal scanning. A thermal scanner can be thought of as merely a particular kind of multispectral scanner—one that senses only in the thermal portion of the spectrum (in one or more bands). However, thermal images must be interpreted with due regard for the basic thermal radiation principles involved. We discuss these principles as we describe how thermal scanner images can be interpreted visually, calibrated radiometrically, and processed digitally. We conclude the chapter with an introduction to *hyperspectral sensing*, the acquisition of images in many (often hundreds) very narrow, contiguous spectral bands throughout the visible, near-IR, and mid-IR portions of the spectrum.

In this chapter we stress *airborne* scanning systems. However, as we see in Chapter 6, the operating principles of multispectral, thermal, and hyperspectral scanners operated from space platforms are essentially identical to the airborne systems described in this chapter.

5.2   ACROSS-TRACK SCANNING

Airborne multispectral scanner systems build up two-dimensional images of the terrain for a swath beneath the aircraft. There are two different ways in which this can be done—using *across-track* (*whiskbroom*) scanning or *along-track* (*pushbroom*) scanning.

Figure 5.1 illustrates the operation of an across-track, or whiskbroom, scanner. Using a rotating or oscillating mirror, such systems scan the terrain along *scan lines* that are at right angles to the flight line. This allows the scanner to repeatedly measure the energy from one side of the aircraft to the other. Data are collected within an arc below the aircraft typically of 90° to 120°. Successive scan lines are covered as the aircraft moves forward, yielding a series of contiguous, or just touching, narrow strips of observation comprising a two-dimensional image of rows (scan lines) and columns.

At any instant in time, the scanner "sees" the energy within the system's *instantaneous field of view (IFOV)*. The IFOV is normally expressed as the cone angle within which incident energy is focused on the detector. (See β in Figure 5.1.) The angle β is determined by the instrument's optical system and size of its detectors. All energy propagating toward the instrument within the IFOV contributes to the detector response at any instant. Hence, more than one land cover type or feature may be included in the IFOV at any given instant and the composite signal response will be recorded. Thus, an image typically contains a combination of "pure" and "mixed" pixels, depending upon the IFOV and the spatial complexity of the ground features.

Figure 5.2 illustrates the segment of the ground surface observed when the IFOV of a scanner is oriented directly beneath the aircraft. This area can be expressed as a circle of diameter D given by

$$D = H'\beta \tag{5.1}$$

where

D = diameter of circular ground area viewed
H' = flying height above terrain
β = IFOV of system (expressed in radians)

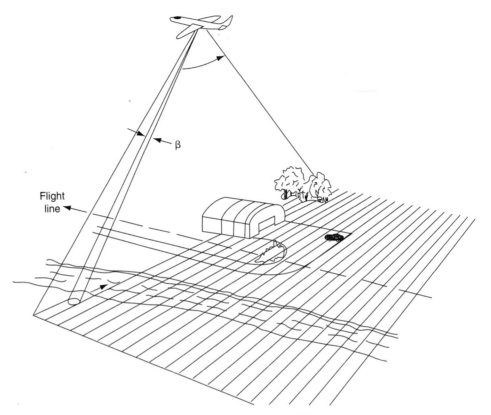

Figure 5.1 Across-track, or whiskbroom, scanner system operation.

The ground segment sensed at any instant is called the *ground resolution element* or *ground resolution cell*. The diameter D of the ground area sensed at any instant in time is loosely referred to as the system's *spatial resolution*. For example, the spatial resolution of a scanner having a 2.5-milliradian (mrad) IFOV and being operated from 1000 m above the terrain can be found from Eq. 5.1 as $D = 1000 \text{ m} \times (2.5 \times 10^{-3} \text{ rad}) = 2.5 \text{ m}$. That is, the ground resolution cell would be 2.5 m in diameter directly under the aircraft. (Depending on the optical properties of the system used, ground resolution cells directly beneath the aircraft can be either circular or square.) The size of the ground resolution cell increases symmetrically on each side of the nadir as the distance between the scanner and the ground resolution cell increases. Hence, the ground resolution cells are larger toward the edge of the image than near the middle. This causes a scale distortion that often must be accounted for in image interpretation or mathematically compensated for in image generation. (We discuss such distortions in Section 5.9.)

The IFOV for airborne multispectral scanner systems typically ranges from about 0.5 to 5 mrad. A small IFOV is desirable to record fine spatial de-

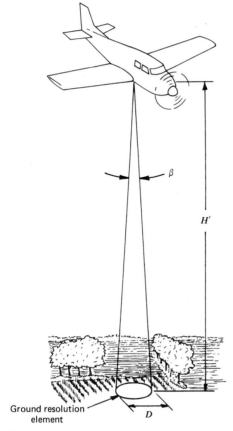

Ground resolution
element

Figure 5.2 Instantaneous field of view and resulting ground area sensed directly beneath an aircraft by a multispectral scanner.

tail. On the other hand, a larger IFOV means a greater quantity of total energy is focused on a detector as the scanner's mirror sweeps across a ground resolution cell. This permits more sensitive scene radiance measurements due to higher signal levels. The result is an improvement in the *radiometric resolution*, or the ability to discriminate very slight energy differences. Thus, there is a trade-off between high spatial resolution and high radiometric resolution in the design of multispectral scanner systems. A large IFOV yields a signal that is much greater than the background electronic *noise* (extraneous, unwanted responses) associated with any given system. Thus, other things being equal, a system with a large IFOV will have a higher *signal-to-noise ratio* than will one with a small IFOV. Again, a large IFOV results in a longer *dwell time,* or residence time of measurement, over any given ground area. What is sacrificed for these higher signal levels is spatial resolution. In a similar vein, the signal-to-noise ratio can be increased by broadening the wavelength band over which a given detector operates. What is sacrificed in this case is *spectral resolution*, that is, the ability to discriminate fine spectral differences.

Again, what we have described above as a scanner's ground resolution cell (the scanner's IFOV projected onto the ground) is often simply called a system's "resolution" (we use this term often in Chapters 5 and 6, especially in tables describing system characteristics).

Most scanners use square detectors, and across-track scan lines are sampled such that they are represented by a series of just-touching pixels projected to ground level. The ground track of the aircraft ideally advances a distance just equal to the size of the resolution cell between rotations of the scanning mirror. This results in sampling the ground without gaps or overlaps.

The ground distance between adjacent sampling points in a digital scanner image need not necessarily exactly equal the dimensions of the IFOV projected onto the ground. As illustrated in Figure 5.3, the ground pixel size, or *ground sampled distance* (GSD), is determined by the sampling time interval (ΔT) used during the A-to-D signal conversion process (Section 1.5). In the case of Figure 5.3, the sampling time interval results in a ground pixel width that is smaller than the width of the scanner's IFOV projected onto the ground. The ground pixel height in the case shown in Figure 5.3 has been sampled to be equal to the scanner's IFOV. This illustrates that there can often be a difference between the projection of a scanner's IFOV onto the ground and the GSD. Often the term *resolution* is loosely used to refer to either of these dimensions.

While on the topic of resolution, we must also point out that the meaning of the term with respect to images acquired by film cameras is not the same as that with reference to electro-optical systems.

The concept of photographic film resolution (described in Section 2.11) is based on being able to distinguish two objects from each other. When lines and spaces of equal width are wide enough that they can just be distinguished on the film, the center-to-center line spacing is called the film resolution. When these lines are projected onto the ground, the resulting distance is called the ground resolution distance (GRD), as expressed by Eq. 2.13 and shown in Figure 5.4a. Applying this concept to a scanner's ground IFOV, we can see that it

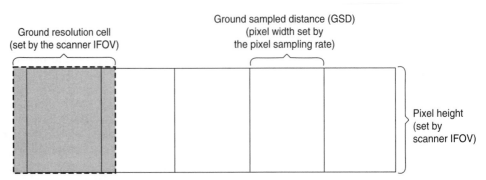

Figure 5.3 Ground sampled distance concept.

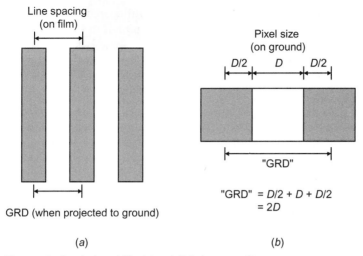

Figure 5.4 Resolution of film (a) and digital systems (b).

gives a misleading value of the ability of the scanner to resolve ground objects in that such objects must be separated from each other by at least the GSD to be distinguished individually (Figure 5.4b). Consequently, the nominal resolution of an electro-optical system must be approximately doubled to compare it with the GRD of a film-based camera system. For example, a 1-m-resolution scanner system would have the approximate detection capability of a 2-m-resolution film camera system. Looking at this concept another way, the optimal GRD that can be achieved with a digital imaging system is approximately twice its GSD.

As explained in Section 2.11, the image analyst is interested not only in object detection but also in object recognition and identification. In the case of scanner images, the larger the number of pixels that make up the image of an object on the ground, the more information that can be determined about that object. For example, one pixel on a vehicle might just be identified as an object on the ground (and this assumes adequate contrast between the vehicle and its background). Two pixels on the same object might be adequate to determine the object's orientation. Three pixels might permit identification of the object as a vehicle. And it might be possible to identify the type of vehicle when it includes five or more pixels. Thus, the effective spatial resolution of a digital imaging system depends on a number of factors in addition to its GSD (e.g., the nature of the ground scene, optical distortions, image motion, illumination and viewing geometry, and atmospheric effects).

5.3 ALONG-TRACK SCANNING

As with across-track systems, along-track, or pushbroom, scanners record multispectral image data along a swath beneath an aircraft. Also similar is the use of the forward motion of the aircraft to build up a two-dimensional image by recording successive scan lines that are oriented at right angles to the flight direction. However, there is a distinct difference between along-track and across-track systems in the manner in which each scan line is recorded. In an along-track system there is no scanning mirror. Instead, a *linear array* of detectors is used (Figure 5.5). Linear arrays typically consist of numerous *charge-coupled devices (CCDs)* positioned end to end. As illustrated in Figure 5.5, each detector element is dedicated to sensing the energy in a single column of data. The size of the ground resolution cell is determined by the IFOV of a single detector projected onto the ground. The GSD in the across-track direction is set by the detector IFOV. The GSD in the along-track direction is again set by the sampling interval (ΔT) used for the A-to-D signal conversion. Normally, the sampling results in just-touching square pixels comprising the image.

Linear array CCDs are designed to be very small, and a single array may contain over 10,000 individual detectors. Each spectral band of sensing requires its own linear array. Normally, the arrays are located in the focal plane of the scanner such that each scan line is viewed by all arrays simultaneously.

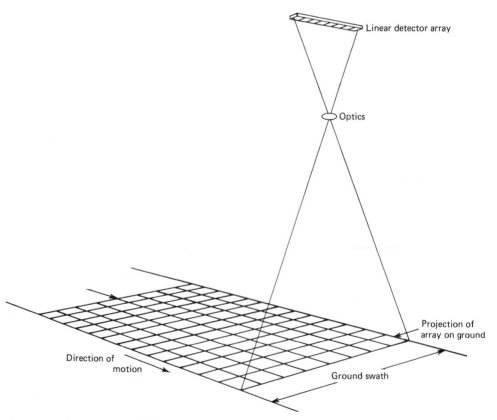

Figure 5.5   Along-track, or pushbroom, scanner system operation.

Linear array systems afford a number of advantages over across-track mirror scanning systems. First, linear arrays provide the opportunity for each detector to have a longer dwell time over which to measure the energy from each ground resolution cell. This enables a stronger signal to be recorded (and, thus, a higher signal-to-noise ratio) and a greater range in the signal levels that can be sensed, which leads to better radiometric resolution. In addition, the geometric integrity of linear array systems is greater because of the fixed relationship among detector elements recording each scan line. The geometry along each row of data (scan line) is similar to an individual photo taken by an aerial mapping camera. The geometric errors introduced into the sensing process by variations in the scan mirror velocity of across-track scanners are not present in along-track scanners. Because linear arrays are solid-state microelectronic devices, along-track scanners are generally smaller in size and weight and require less power for their operation than across-track scanners. Also, having no moving parts, a linear array system has higher reliability and longer life expectancy.

One disadvantage of linear array systems is the need to calibrate many more detectors. Another current limitation to commercially available solid-state arrays is their relatively limited range of spectral sensitivity. Linear array detectors that are sensitive to wavelengths longer than the mid-IR are not readily available.

5.4 OPERATING PRINCIPLES OF ACROSS-TRACK MULTISPECTRAL SCANNERS

Figure 5.6 illustrates the operation of a typical across-track, or whiskbroom, scanner. As previously mentioned, by using a rotating or oscillating mirror (Figure 5.6*b*), such systems scan the terrain along *scan lines* that are at right angles to the flight line. Successive scan lines are covered as the aircraft moves forward, yielding a series of contiguous, or just-touching, narrow strips of observation composing a two-dimensional image. The incoming energy is separated into several spectral components that are sensed independently. In Figure 5.6 we illustrate an example of a system that records both thermal and nonthermal wavelengths. A *dichroic grating* is used to separate these two forms of energy. The nonthermal wavelength component is directed from the grating through a prism (or diffraction grating) that splits the energy into a continuum of UV, visible, and near-IR wavelengths. At the same time, the dichroic grating disperses the thermal component of the incoming signal into its constituent wavelengths. By placing an array of electro-optical detectors at the proper geometric positions behind the grating and the prism, the incoming beam is essentially "pulled apart" into multiple narrow bands, each of which is measured independently. Each detector is designed to have its peak spectral sensitivity in a specific wavelength band. Figure 5.6 illustrates a five-band scanner. As we see later (Section 5.14), scanners with hundreds of bands are available.

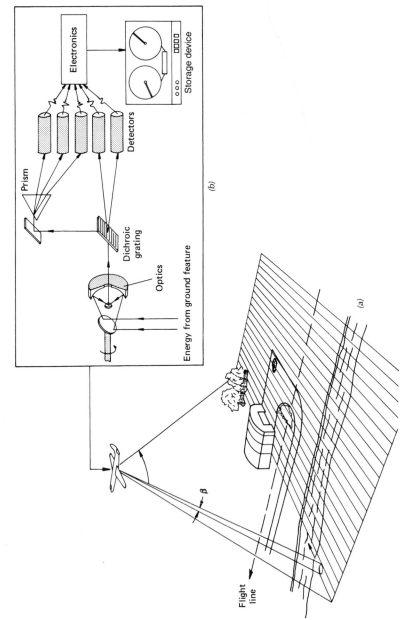

Figure 5.6 Across-track, or whiskbroom, multispectral scanner system operation: (a) scanning procedure during flight; (b) scanner schematic.

As shown in Figure 5.6, the electrical signals generated by each of the detectors of the multispectral scanner are amplified by the system electronics and recorded by a multichannel storage device. Usually, onboard A-to-D signal conversion (Section 1.5) is used to record the data digitally.

Figure 5.7 shows an across-track multispectral scanner system that has an IFOV of 2.5 mrad (1.25 mrad optional) and a total scan angle of 86°. The scanner acquires 8- or 12-bit (operator-selectable) digital data simultaneously in 6 bands (selected from a total of 10 available bands) within the wavelength range 0.32 to 12.5 μm and records these data on 8-mm digital tape. This system provides continuous video monitoring of the scene (see monitor at upper left in Figure 5.7b). An optional printer can provide continuous hardcopy images, and a VHS video recording can be made from the video monitor output.

Figure 5.8 shows six bands of multispectral scanner data acquired by the above scanner along a portion of the Yakima River in the state of Washington,

Figure 5.7 Sensys Technologies (formerly Daedalus Enterprises) Airborne Multispectral Scanner. At left is the scan head that is mounted in the belly of an aircraft (or in a separate pod). At right is the hardware used in the aircraft to control and monitor data acquisition and record the data. (Courtesy Sensys Technologies, Inc.)

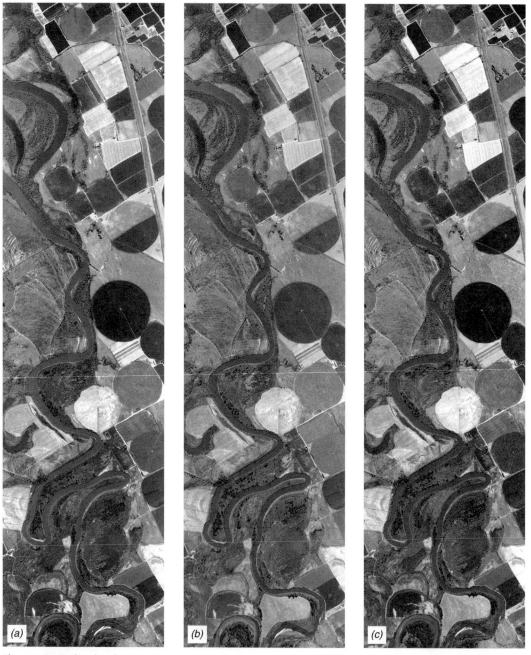

Figure 5.8 Six-band multispectral scanner data, Yakima River valley, Washington, mid-August 1997: (a) band 1, 0.42 to 0.52 μm (blue); (b) band 2, 0.52 to 0.60 μm (green); (c) band 3, 0.63 to 0.69 μm (red); (d) band 4, 0.76 to 0.90 μm (near IR); (e) band 5, 0.91 to 1.05 μm (near IR); (f) band 6, 8.0 to 12.5 μm (thermal IR). Scale 1 : 50,000. (Courtesy Sensys Technologies, Inc.)

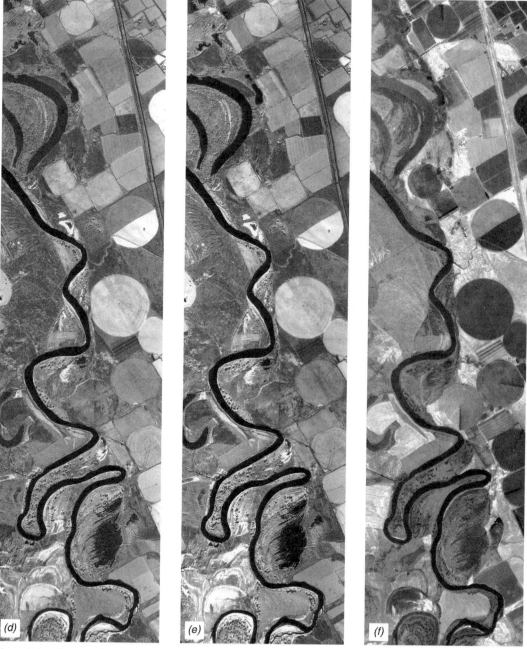

Figure 5.8   *(Continued)*

and Plate 14 shows normal color and color IR composites of three of these bands. (The circular areas appearing in these images result from the practice of center-pivot irrigation for agricultural production.) As is typical with multispectral data, the three bands in the visible part of the spectrum (bands 1, 2, and 3) show a great deal of correlation; that is, objects (fields) that are light toned in one visible band are light toned in the other two visible bands. There are, of course, some variations in reflectance, which is why different colors are seen (Plate 14*a*). In the near-IR bands, healthy vegetation is much lighter toned than in the visible bands, resulting in red tones where such vegetation is present (Plate 14*b*). As was seen in the photographs in Figure 2.17, water is extremely dark in the near-IR bands—compare the image tone of the Yakima River in (*a*) with its tone in (*d*) and (*e*). In the thermal band (*f*), healthy green vegetation (e.g., the darker toned large circular field near the middle of the images) is much cooler than less-vegetated areas in this daytime thermal image (additional thermal imagery is shown later in this chapter).

In the normal color and color IR combinations of multispectral scanner images shown in Plate 14, different crops are seen as different colors. In these mid-August images, the largest circular field (just above the center of the images) with the bright red color (in the color IR image) is seed grass. The bright red half-circle above the seed grass is alfalfa. The alfalfa and seed grass appear similar in color in the normal color and color IR images, but a close inspection of all bands shows that the alfalfa is brighter toned in the near-IR bands (bands 4 and 5), and thus the two crops can be distinguished from each other using these bands. The two circular fields below (and slightly to the right) of the seed grass field are also alfalfa, but at different growth stages. The upper half of the circular field near the top of the image that has alfalfa at the bottom contains asparagus in its upper half. In the normal color image, the tan field at the top of the image and the tan circular field just below the center of the image are winter wheat. Sweet corn is planted in the two nearly rectangular fields below the winter wheat at the top of the image. The bright pink color (in the color IR image) at the extreme left edge about a third of the way down from the top is an area of lily pads and algae in a lake. Note how the reflectance of this feature varies from that of the other cover types in the image among the various bands. Again, it is this difference in spectral reflectance that aids in the discrimination of cover types visually and in the process of automated image classification (discussed in Chapter 7).

5.5 EXAMPLE ALONG-TRACK MULTISPECTRAL SCANNER AND DATA

An example of an airborne along-track scanner is the *ADS40 Airborne Digital Sensor*, manufactured by LH Systems shown in Figure 5.9. It incorporates multiple linear arrays of CCDs for panchromatic and multispectral image ac-

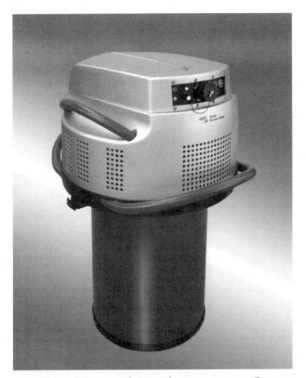

Figure 5.9 The ADS40 along-track scanner system. (Courtesy Leica Geosystems.)

quisition, all sharing a single lens system, with a focal length of 62.77 mm. The ADS40 panchromatic arrays are sensitive over the range of wavelengths from 0.465 to 0.680 μm. Three visible-wavelength bands are present, with sensitivities of 0.430 to 0.490 μm, 0.535 to 0.585 μm, and 0.610 to 0.660 μm, as is one near-IR band (0.835 to 0.885 μm). The sensor has a wide dynamic range, and a total field of view of 64° across-track.

Three linear arrays can be used to collect panchromatic data, with one forward-viewing, one nadir-viewing, and one rear-viewing. Data from these arrays can be used for stereoscopic viewing and for photogrammetric analyses. The forward-viewing array is positioned at an angle of 28.4° off nadir, while the rear-viewing array is 14.2° off nadir. There are thus three possible stereoscopic configurations of the panchromatic images, with angles of 14.2° (using the nadir and rear-viewing arrays), 28.4° (using the nadir and forward-viewing arrays), and 42.6° (using the rear-viewing and forward-viewing arrays). These three configurations provide stereoscopic base-height ratios of 0.25, 0.54, and 0.79, respectively. The blue, green, and red multispectral arrays are inclined 14.2° off nadir in the forward direction, and the near-IR array is oriented 2° to the rear of nadir.

Each of the three panchromatic arrays consists of two lines of 12,000 CCDs, with the two lines offset by 0.325 μm, or half the width of a single CCD. In high resolution mode, both of the lines are used, yielding an effective IFOV of 0.05 mrad; if only one line is used, the IFOV is 0.1 mrad. At a typical flying height of 2880 m above ground level, this results in a ground sampling distance of 15 cm in high resolution mode or 30 cm otherwise.

Figure 5.10 shows an example of ADS40 panchromatic imagery acquired over the Texas state capitol in Austin. Data from the forward-viewing, nadir-

Figure 5.10 ADS40 panchromatic images of the Texas state capital, Austin, TX. Flight direction is left to right. (a) Forward-viewing array, 28.4° off nadir; (b) nadir-viewing array; (c) backward-viewing array, 14.2° off-nadir. (Courtesy Leica Geosystems.)

(c)

Figure 5.10 *(Continued)*

viewing, and backward-viewing linear arrays show the effects of relief displacement. The magnitude of displacement is greater in the forward-viewing image than in the backward-viewing image because the forward-viewing array is tilted further off-nadir. The data shown in Figure 5.10 represent only a small portion of the full image dimensions. For this data acquisition the sensor was operated from a flying height of 1920 m above ground level. The original ground sampling distance for this data set was 20 cm.

5.6 ACROSS-TRACK THERMAL SCANNING

As mentioned earlier, a *thermal scanner* is merely a particular kind of across-track multispectral scanner, namely, one whose detector(s) only senses in the thermal portion of the spectrum. Due to atmospheric effects, these systems are restricted to operating in either (or both) the 3- to 5-μm or 8- to 14-μm range of wavelengths (Section 1.3). *Quantum* or *photon* detectors are typically used for this purpose. These detectors are capable of very rapid (less than 1-μsec) response. They operate on the principle of direct interaction between photons of radiation incident on them and the energy levels of electrical charge carriers within the detector material. For maximum sensitivity, the detectors must be cooled to temperatures approaching absolute zero to minimize their own thermal emissions. Normally, the detector is surrounded by a *dewar* containing liquid nitrogen at 77 K. A dewar is a double-walled insulated vessel that acts like a thermos bottle to prevent the liquid coolant from boiling away at a rapid rate.

TABLE 5.1 Characteristics of Photon Detectors in Common Use

Type	Abbreviation	Useful Spectral Range (μm)
Mercury-doped germanium	Ge:Hg	3–14
Indium antimonide	InSb	3–5
Mercury cadmium telluride	HgCdTe (MCT), or "trimetal"	8–14

The spectral sensitivity range and the operating temperatures of three photon detectors in common use today are included in Table 5.1.

Thermal scanners became commercially available during the late 1960s. Earlier models used only direct film recording for image generation. Newer systems record data digitally. In addition, scan line output signals are generally monitored in flight on an oscilloscope or some other real-time monitor. Present-day systems are capable of temperature resolution on the order of 0.1°C.

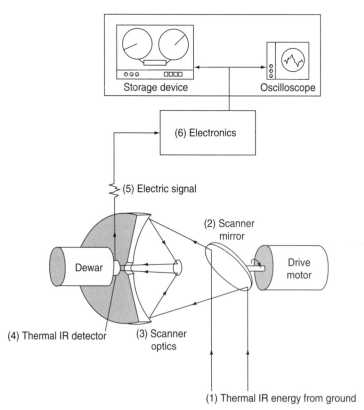

Figure 5.11 Across-track thermal scanner schematic.

Figure 5.11 illustrates schematically the basic operation of a thermal scanner system. The system works as follows. Thermal IR radiation from the ground (1) is received at the rotating scanner mirror (2). Additional optics (3) focus the incoming energy on the thermal IR radiation detector (4), which is encased by a dewar filled with a liquid nitrogen coolant. The detector converts the incoming radiation level to an electric signal (5) that is amplified by the system electronics (6). The signal (or an image) is displayed on the monitor and recorded digitally on a line-by-line basis after an A-to-D conversion.

A thermal scanner image is a pictorial representation of the detector response on a line-by-line basis. The usual convention when looking at the earth's surface is to have higher radiant temperature areas displayed as lighter toned image areas. For meteorological purposes, this convention is typically reversed so that clouds (cooler than the earth's surface) appear light toned.

5.7 THERMAL RADIATION PRINCIPLES

Implicit in the proper interpretation of a thermal scanner image, or *thermogram*, is at least a basic understanding of the nature of thermal radiation. In this section we review and extend some of the principles of blackbody radiation introduced in Section 1.2. We also treat how thermal radiation interacts with the atmosphere and various earth surface features.

Radiant Versus Kinetic Temperature

One normally thinks of temperature measurements as involving some measuring instrument being placed in contact with, or being immersed in, the body whose temperature is to be measured. When this is done, *kinetic temperature* is measured. Kinetic temperature is an "internal" manifestation of the average translational energy of the molecules constituting a body. In addition to this internal manifestation, objects radiate energy as a function of their temperature. This emitted energy is an "external" manifestation of an object's energy state. It is this external manifestation of an object's energy state that is remotely sensed using thermal scanning. The emitted energy is used to determine the *radiant temperature* of earth surface features. Later we describe how to relate kinetic and radiant temperature.

Blackbody Radiation

We have previously described the physics of electromagnetic radiation in accordance with the concepts of blackbody radiation (see Section 1.2). Recall that any object having a temperature greater than absolute zero (0 K, or −273°C) emits radiation whose intensity and spectral composition are a function of the material type involved and the temperature of the object under consideration. Figure 5.12 shows the spectral distribution of the energy radiated from the surface of a blackbody at various temperatures. All such blackbody curves have similar form, and their energy peaks shift toward shorter wavelengths with increases in temperature. As indicated earlier (Section 1.2), the relationship between the wavelength of peak spectral exitance and temperature for a blackbody is given by *Wien's displacement law*:

$$\lambda_m = \frac{A}{T} \tag{5.2}$$

where

λ_m = wavelength of maximum spectral radiant exitance, μm

A = 2898 μm K

T = temperature, K

The total radiant exitance coming from the surface of a blackbody at any given temperature is given by the area under its spectral radiant exitance curve. That is, if a sensor were able to measure the radiant exitance from a blackbody at all wavelengths, the signal recorded would be proportional to

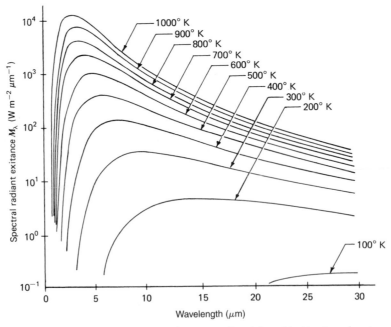

Figure 5.12   Spectral distribution of energy radiated from blackbodies of various temperatures.

the area under the blackbody radiation curve for the given temperature. This area is described mathematically by the *Stefan–Boltzmann law*, given in Eq. 1.4, and in expanded form here as:

$$M = \int_{0}^{\infty} M(\lambda)\, d\lambda = \sigma T^{4} \tag{5.3}$$

where

M = total radiant exitance, W m^{-2}
$M(\lambda)$ = spectral radiant exitance, W m^{-2} μm^{-1}
σ = Stefan–Boltzmann constant, 5.6697×10^{-8} W m^{-2} K^{-4}
T = temperature of blackbody, K

Equation 5.3 indicates that the total radiant exitance from the surface of a blackbody varies as the fourth power of absolute temperature. The remote measurement of radiant exitance M from a surface can therefore be used to infer the temperature T of the surface. In essence, it is this indirect approach to temperature measurement that is used in thermal sensing. Radiant exitance M is measured over a discrete wavelength range and used to find the radiant temperature of the radiating surface.

Radiation from Real Materials

While the concept of a blackbody is a convenient theoretical vehicle to describe radiation principles, real materials do not behave as blackbodies. Instead, all real materials emit only a fraction of the energy emitted from a blackbody at the equivalent temperature. The "emitting ability" of a real material, compared to that of a blackbody, is referred to as a material's *emissivity ε*.

Emissivity ε is a factor that describes how efficiently an object radiates energy compared to a blackbody. By definition,

$$\varepsilon(\lambda) = \frac{\text{radiant exitance of an object at a given temperature}}{\text{radiant exitance of a blackbody at the same temperature}} \tag{5.4}$$

Note that ε can have values between 0 and 1. As with reflectance, emissivity can vary with wavelength and viewing angle. Depending on the material, emissivity can also vary somewhat with temperature.

A *graybody* has an emissivity that is less than 1 but is constant at all wavelengths. At any given wavelength the radiant exitance from a graybody is a constant fraction of that of a blackbody. If the emissivity of an object varies with wavelength, the object is said to be a *selective radiator*. Figure 5.13 illustrates the comparative emissivities and spectral radiant exitances for a blackbody, a graybody (having an emissivity of 0.5), and a selective radiator.

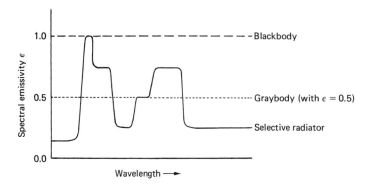

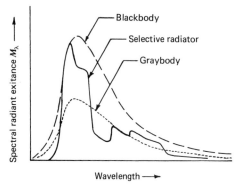

Figure 5.13 Spectral emissivities and radiant exitances for a blackbody, a graybody, and a selective radiator. (Adapted from Hudson, 1969.)

Many materials radiate like blackbodies over certain wavelength intervals. For example, as shown in Figure 5.14a, water is very close (ε is 0.98 to 0.99) to behaving as a blackbody radiator in the 6- to 14-μm range. Other materials, such as quartz, act as a selective radiator, with considerable variation in emissivity at different wavelengths in the 6- to 14-μm range (Figure 5.14b).

The 8- to 14-μm region of spectral radiant exitance is of particular interest since it not only includes an atmospheric window but also contains the peak energy emissions for most surface features. That is, the ambient temperature of earth surface features is normally about 300 K, at which temperature the peak emissions will occur at approximately 9.7 μm. For these reasons, most thermal sensing is performed in the 8- to 14-μm region of the spectrum. The emissivities of different objects vary greatly with material type in this range. However, for any given material type, emissivity is often considered constant in the 8- to 14-μm range when broadband sensors are being used. This means that within this spectral region materials are often treated as graybodies. However, close examination of emissivity versus wavelength for

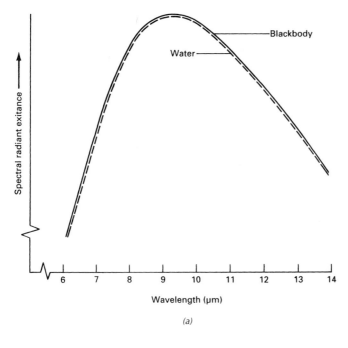

(a)

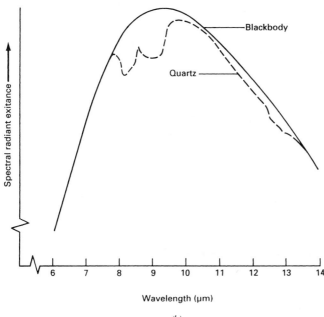

(b)

Figure 5.14   Comparison of spectral radiant exitances for (*a*) water versus a blackbody and (*b*) quartz versus a blackbody. (For additional curves, see Salisbury and D'Aria, 1992.)

materials in this wavelength range shows that values can vary considerably with wavelength. Therefore, the within-band emissivities of materials sensed in the 10.5- to 11.5-μm range [National Oceanic and Atmospheric Administration (NOAA) AVHRR band 4] would not necessarily be the same as the within-band emissivities of materials sensed in the 10.4- to 12.5-μm range (Landsat TM band 6). Furthermore, emissivities of materials vary with material condition. A soil that has an emissivity of 0.92 when dry could have an emissivity of 0.95 when wet (water-coated soil particles have an emissivity approaching that of water). Objects such as deciduous tree leaves can have a dif-

TABLE 5.2   **Typical Emissivities of Various Common Materials over the Range of 8–14 μm**

Material	Typical Average Emissivity ε over 8–14 μm[a]
Clear water	0.98–0.99
Wet snow	0.98–0.99
Human skin	0.97–0.99
Rough ice	0.97–0.98
Healthy green vegetation	0.96–0.99
Wet soil	0.95–0.98
Asphaltic concrete	0.94–0.97
Brick	0.93–0.94
Wood	0.93–0.94
Basaltic rock	0.92–0.96
Dry mineral soil	0.92–0.94
Portland cement concrete	0.92–0.94
Paint	0.90–0.96
Dry vegetation	0.88–0.94
Dry snow	0.85–0.90
Granitic rock	0.83–0.87
Glass	0.77–0.81
Sheet iron (rusted)	0.63–0.70
Polished metals	0.16–0.21
Aluminum foil	0.03–0.07
Highly polished gold	0.02–0.03

[a]Emissivity values (ordered from high to low) are typical average values for the materials listed over the range of 8–14 μm. Emissivities associated with only a portion of the 8–14-μm range can vary significantly from these values. Furthermore, emissivities for the materials listed can vary significantly, depending on the condition and arrangement of the materials (e.g., loose soil vs. compacted soil, individual tree leaves vs. tree crowns).

ferent emissivity when a single leaf is sensed (0.96) than when an entire tree crown is sensed (0.98). Table 5.2 indicates some typical values of emissivity over the 8- to 14-μm-wavelength range for various common materials.

It should be noted that as objects are heated above ambient temperature, their emissive radiation peaks shift to shorter wavelengths. In special-purpose applications, such as forest fire mapping, systems operating in the 3- to 5-μm atmospheric window may be used. These systems offer improved definition of hot objects at the expense of the surrounding terrain at ambient temperature.

Atmospheric Effects

As is the case with all passive remote sensing systems, the atmosphere has a significant effect on the intensity and spectral composition of the energy recorded by a thermal system. As mentioned, atmospheric windows (Figure 5.15) influence the selection of the optimum spectral bands within which to measure thermal energy signals. Within a given window, the atmosphere intervening between a thermal sensor and the ground can increase or decrease the apparent level of radiation coming from the ground. The effect that the atmosphere has on a ground signal will depend on the degree of atmospheric absorption, scatter, and emission at the time and place of sensing.

Gases and suspended particles in the atmosphere may absorb radiation emitted from ground objects, resulting in a decrease in the energy reaching a

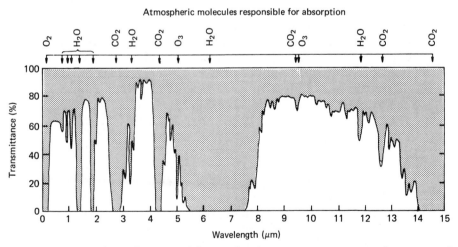

Figure 5.15 Atmospheric absorption of the wavelength range 0 to 15 μm. Note the presence of atmospheric windows in the thermal wavelength regions 3 to 5 μm and 8 to 14 μm. (Adapted from Hudson, 1969.)

thermal sensor. Ground signals can also be attenuated by scattering in the presence of suspended particles. On the other hand, gases and suspended particles in the atmosphere may emit radiation of their own, adding to the radiation sensed. Hence, atmospheric absorption and scattering tend to make the signals from ground objects appear colder than they are, and atmospheric emission tends to make ground objects appear warmer than they are. Depending on atmospheric conditions during imaging, one of these effects will outweigh the other. This will result in a biased sensor output. Both effects are directly related to the atmospheric path length, or distance, through which the radiation is sensed.

Thermal sensor measurements of temperature can be biased by as much as 2°C or more when acquired at altitudes as low as 300 m. Of course, meteorological conditions have a strong influence on the form and magnitude of the thermal atmospheric effects. Fog and clouds are essentially opaque to thermal radiation. Even on a clear day, aerosols can cause major modifications of signals sensed. Dust, carbon particles, smoke, and water droplets can all modify thermal measurements. These atmospheric constituents vary with site, altitude, time, and local weather conditions.

Atmospheric effects on radiant temperature measurements usually may not be ignored. The various strategies commonly employed to compensate for atmospheric effects are described later in this chapter. We now consider how thermal radiation interacts with ground objects.

Interaction of Thermal Radiation with Terrain Elements

In thermal sensing we are interested in the radiation emitted from terrain features. However, the energy radiated *from* an object usually is the result of energy incident *on* the feature. In Section 1.4 we introduced the basic notion that energy incident on the surface of a terrain element can be absorbed, reflected, or transmitted. In accordance with the principle of conservation of energy, we can state the relationship between incident energy and its disposition upon interaction with a terrain element as

$$E_I = E_A + E_R + E_T \tag{5.5}$$

where
E_I = energy incident on surface of terrain element
E_A = component of incident energy absorbed by terrain element
E_R = component of incident energy reflected by terrain element
E_T = component of incident energy transmitted by terrain element

If Eq. 5.5 is divided by the quantity E_I, we obtain the relationship

$$\frac{E_I}{E_I} = \frac{E_A}{E_I} + \frac{E_R}{E_I} + \frac{E_T}{E_I} \tag{5.6}$$

The terms on the right side of Eq. 5.6 comprise ratios that are convenient in further describing the nature of thermal energy interactions. We define

$$\alpha(\lambda) = \frac{E_A}{E_I} \quad \rho(\lambda) = \frac{E_R}{E_I} \quad \tau(\lambda) = \frac{E_T}{E_I} \tag{5.7}$$

where
   $\alpha(\lambda)$ = *absorptance* of terrain element
   $\rho(\lambda)$ = *reflectance* of terrain element
   $\tau(\lambda)$ = *transmittance* of terrain element

We can now restate Eq. 5.5 in the form

$$\alpha(\lambda) + \rho(\lambda) + \tau(\lambda) = 1 \tag{5.8}$$

which defines the interrelationship among a terrain element's absorbing, reflecting, and transmitting properties.

Another ingredient necessary is the *Kirchhoff radiation law*. It states that the spectral emissivity of an object equals its spectral absorptance:

$$\varepsilon(\lambda) = \alpha(\lambda) \tag{5.9}$$

Paraphrased, "good absorbers are good emitters." While Kirchhoff's law is based on conditions of thermal equilibrium, the relationship holds true for most sensing conditions. Hence, if we apply it in Eq. 5.8, we may replace $\alpha(\lambda)$ with $\varepsilon(\lambda)$, resulting in

$$\varepsilon(\lambda) + \rho(\lambda) + \tau(\lambda) = 1 \tag{5.10}$$

Finally, in most remote sensing applications the objects we deal with are assumed to be opaque to thermal radiation. That is, $\tau(\lambda) = 0$, and it is therefore dropped from Eq. 5.10 such that

$$\varepsilon(\lambda) + \rho(\lambda) = 1 \tag{5.11}$$

Equation 5.11 demonstrates the direct relationship between an object's emissivity and its reflectance in the thermal region of the spectrum. The lower an object's reflectance, the higher its emissivity. The higher an object's reflectance, the lower its emissivity. For example, water has nearly negligible reflectance in the thermal spectrum. Therefore, its emissivity is essentially 1. In contrast, a material such as sheet metal is highly reflective of thermal energy, so it has an emissivity much less than 1.

The emissivity of an object has an important implication when measuring radiant temperatures. Recall that the Stefan–Boltzmann law, as stated in Eq. 5.3 ($M = \sigma T^4$), applied to blackbody radiators. We can extend the blackbody radiation principles to real materials by reducing the radiant exitance M by the emissivity factor ε such that

$$M = \varepsilon \sigma T^4 \tag{5.12}$$

TABLE 5.3   Kinetic versus Radiant Temperature for Four Typical Material Types

Object	Emissivity ε	Kinetic Temperature T_{kin}		Radiant Temperature $T_{rad} = \varepsilon^{1/4}T_{kin}$	
		K	°C	K	°C
Blackbody	1.00	300	27	300.0	27.0
Vegetation	0.98	300	27	298.5	25.5
Wet soil	0.95	300	27	296.2	23.2
Dry soil	0.92	300	27	293.8	20.8

Equation 5.12 describes the interrelationship between the measured signal a thermal sensor "sees," M, and the parameters of temperature and emissivity. Note that because of emissivity differences, earth surface features can have the same temperature and yet have completely different radiant exitances.

The output from a thermal sensor is a measurement of the radiant temperature of an object, T_{rad}. Often, the user is interested in relating the radiant temperature of an object to its kinetic temperature, T_{kin}. If a sensor were to view a blackbody, T_{rad} would equal T_{kin}. For all real objects, however, we must account for the emissivity factor. Hence, the kinetic temperature of an object is related to its radiant temperature by

$$T_{rad} = \varepsilon^{1/4}T_{kin} \tag{5.13}$$

Equation 5.13 expresses the fact that for any given object the radiant temperature recorded by a remote sensor will always be less than the kinetic temperature of the object. This effect is illustrated in Table 5.3, which shows the kinetic versus the radiant temperatures for four objects having the same kinetic temperature but different emissivities. Note how kinetic temperatures are always underestimated if emissivity effects are not accounted for in analyzing thermal sensing data.

A final point to be made here is that *thermal sensors detect radiation from the surface (approximately the first 50 µm) of ground objects*. This radiation may or may not be indicative of the internal bulk temperature of an object. For example, on a day of low humidity, a water body having a high temperature will manifest evaporative cooling effects at its surface. Although the bulk temperature of the water body could be substantially warmer than that of its surface temperature, a thermal sensor would record only the surface temperature.

5.8   INTERPRETING THERMAL SCANNER IMAGERY

Successful interpretations of thermal imagery have been made in many fields of application. These include such diverse tasks as determining rock type and structure, locating geologic faults, mapping soil type and soil moisture, locating irrigation canal leaks, determining the thermal characteristics of volcanoes, studying evapotranspiration from vegetation, locating cold-water springs, locating hot springs and geysers, determining the extent and characteristics of thermal plumes in lakes and rivers, studying the natural circulation patterns in water bodies, determining the extent of active forest fires, and locating subsurface fires in landfills or coal refuse piles.

Most thermal scanning operations, such as geologic and soil mapping, are qualitative in nature. In these cases, it is not usually necessary to know absolute ground temperatures and emissivities, but simply to study relative differences in the radiant temperatures within a scene. However, some thermal scanning operations require quantitative data analysis in order to determine absolute temperatures. An example would be the use of thermal scanning as an enforcement tool by a state department of natural resources to monitor surface water temperatures of the effluent from a nuclear power plant.

Various times of day can be utilized in thermal scanning studies. Many factors influence the selection of an optimum time or times for acquiring thermal data. Mission planning and image interpretation must take into consideration the effects of diurnal temperature variation. The importance of diurnal effects is shown in Figure 5.16, which illustrates the relative radiant temperatures of soils and rocks versus water during a typical 24-hr period.

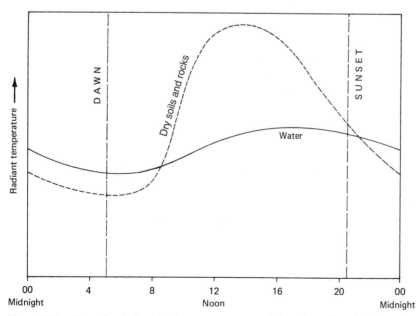

Figure 5.16 Generalized diurnal radiant temperature variations for soils and rocks versus water.

Note that just before dawn a quasi-equilibrium condition is reached where the slopes of the temperature curves for these materials are very small. After dawn, this equilibrium is upset and the materials warm up to a peak that is reached sometime after noon. Maximum scene contrast normally occurs at about this time and cooling takes place thereafter.

Temperature extremes and heating and cooling rates can often furnish significant information about the type and condition of an object. Note, for example, the temperature curve for water. It is distinctive for two reasons. First, its range of temperature is quite small compared to that of soils and rocks. Second, it reaches its maximum temperature an hour or two after the other materials. As a result, terrain temperatures are normally higher than water temperatures during the day and lower during the night. Shortly after dawn and near sunset, the curves for water and the other features intersect. These points are called thermal *crossovers* and indicate times at which no radiant temperature difference exists between two materials.

The extremes and rates of temperature variation of any earth surface material are determined, among other things, by the material's thermal conductivity, capacity, and inertia. *Thermal conductivity* is a measure of the rate at which heat passes through a material. For example, heat passes through metals much faster than through rocks. *Thermal capacity* determines how well a material stores heat. Water has a very high thermal capacity compared to other material types. *Thermal inertia* is a measure of the response of a material to temperature changes. It increases with an increase in material conductivity, capacity, and density. In general, materials with high thermal inertia have more uniform surface temperatures throughout the day and night than materials of low thermal inertia.

During the daytime, direct sunlight differentially heats objects according to their thermal characteristics and their sunlight absorption, principally in the visible and near-IR portion of the spectrum. Reflected sunlight can be significant in imagery utilizing the 3- to 5-μm band. Although reflected sunlight has virtually no direct effect on imagery utilizing the 8- to 14-μm band, daytime imagery contains thermal "shadows" in cool areas shaded from direct sunlight by objects such as trees, buildings, and some topographic features. Also, slopes receive differential heating according to their orientation. In the Northern Hemisphere, south-facing slopes receive more solar heating than north-facing slopes. Many geologists prefer "predawn" imagery for their work as this time of day provides the longest period of reasonably stable temperatures, and "shadow" effects and slope orientation effects are minimized. However, aircraft navigation over areas selected for thermal image acquisition is more difficult during periods of darkness, when ground features cannot be readily seen by the pilot. Other logistics also enter into the timing of thermal scanning missions. For example, scanning of effluents from power plant operations normally must be conducted during periods of peak power generation.

A number of thermal images are illustrated in the remainder of this chapter. In all cases, darker image tones represent cooler radiant temperatures and lighter image tones represent warmer radiant temperatures. This is the representation most commonly used in thermal images of earth surface features. In meteorological applications, the reverse representation is used to preserve the light-toned appearance of clouds.

Figure 5.17 illustrates the contrast between daytime (*a*) and nighttime (*b*) thermal images. The water in this scene (note the large lake at right and the small, lobed pond in lower center) appears cooler (darker) than its surroundings during the daytime and warmer (lighter) at night. The kinetic water temperature has changed little during the few hours of elapsed time between these images. However, the surrounding land areas have cooled considerably during the evening hours. Again, water normally appears cooler than its surroundings on daytime thermal images and warmer on nighttime thermal images, except for the case of open water surrounded by frozen or snow-covered ground where the water would appear warmer day and night. Trees can be seen many places in these images (note the area above and to the right of the small pond). Trees generally appear cooler than their surroundings during the daytime and warmer at night. Tree shadows appear in many places in the daytime image (note the residential area at upper left) but are not noticeable in the nighttime image. Paved areas (streets and parking lots) appear relatively warm both day and night. The pavement surfaces heat up to temperatures higher than their surroundings during the daytime and lose heat relatively slowly at night, thus retaining a temperature higher than their surroundings.

Figure 5.18 is a daytime thermal image showing the former shoreline of glacial Lake Middleton, an ephemeral glacial lake that is now primarily agricultural fields. This was a small lake, about 800 ha in extent at its maximum. At its lowest level, the lake was only about 80 ha in size. The beach ridge associated with this lowest lake level is shown at B. The ridge is most evident at the lower right because the prevailing winds at the time of its formation were from the upper left. The ridge is a small feature, only 60 m wide and $\frac{1}{2}$ to 1 m higher than the surrounding lakebed material. The beach ridge has a fine sandy loam surface soil 0.3 to 0.45 m thick underlain by deep sandy materials. The lakebed soils (A) are silt loam to a depth of at least 1.5 m and are seasonally wet with a groundwater table within 0.6 m of the ground surface in the early spring. At the time of this thermal image, most of the area shown here was covered with agricultural crops. The scanner sensed the radiant temperature of the vegetation over the soils rather than the bare soils themselves. Based on field radiometric measurements, the radiant temperature of the vegetation on the dry, sandy beach ridge soil is 16°C, whereas that over the wetter, siltier lakebed soil is 13°C. Although prominent on this thermal image, the beach ridge is often overlooked on panchromatic aerial photographs and is only partially mapped on a soil map of the area. Also seen on

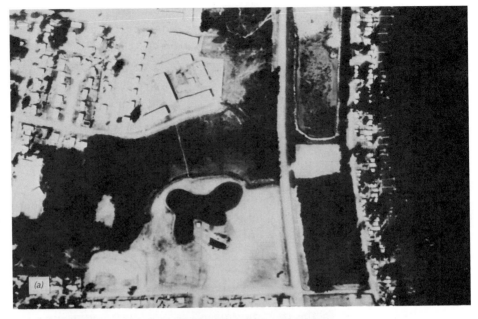

Figure 5.17 Daytime and nighttime thermal images, Middleton, WI: (*a*) 2:40 P.M., (*b*) 9:50 P.M.; 600 m flying height, 5 mrad IFOV. (Courtesy National Center for Atmospheric Research.)

Figure 5.18 Daytime thermal image. Middleton, WI, 9:40 A.M., 600 m flying height, 5 mrad IFOV. (Courtesy National Center for Atmospheric Research.)

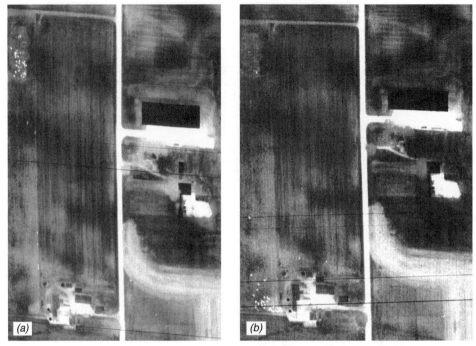

Figure 5.19 Nighttime thermal images, Middleton, WI: (*a*) 9:50 P.M., (*b*) 1:45 A.M.; 600 m flying height, 5 mrad IFOV. (Courtesy National Center for Atmospheric Research.)

this thermal image are trees at C, bare soil at D, and closely mowed grass (a sod farm) at E.

Figure 5.19 contains two nighttime thermal images illustrating the detectability of relatively small features on large-scale imagery. In Figure 5.19*a* (9:50 P.M.), a group of 28 cows can be seen as white spots near the upper left. In Figure 5.19*b* (1:45 A.M.), they have moved near the bottom of the image. (Deer located in flat areas relatively free of obstructing vegetation have also been detected, with mixed success, on thermal imagery.) The large, rectangular, very dark-toned object near the upper right of Figure 5.19*a* is a storage building with a sheet metal roof having a low emissivity. Although the kinetic temperature of the roof may be as warm as or warmer than the surrounding ground, its radiant temperature is low due to its low emissivity.

Figure 5.20 illustrates thermal imagery acquired with a high resolution scanner system. The image was acquired using a 0.25-mrad IFOV. The corresponding ground resolution cell size is approximately 0.3 m. This enlargement illustrates the spatial detail observable in the imagery. For example, several helicopters parked near hangers can be clearly seen in the enlarged image. Note also the "thermal shadow" left by two helicopters that are not in

Figure 5.20 Daytime thermal image, Quantico, VA, 0.25 mrad IFOV. (Courtesy RECON/OPTICAL, Inc.)

Figure 5.21 Daytime thermal image, Oak Creek Power Plant, WI, 1:50 P.M., 800 m flying height, 2.5 mrad IFOV. (Courtesy Wisconsin Department of Natural Resources.)

their original parked positions. Two helicopters can also be seen with their blades running while sitting on the helipad in the lower right.

Figure 5.21 illustrates the heated cooling water from a coal-fired power plant discharging into Lake Michigan. This daytime thermal image shows that the plant's cooling system is recirculating its heated discharge water. Heated water initially flows to the right. Winds from the upper right at 5 m/sec cause the plume to double back on itself, eventually flowing into the intake channel. The ambient lake temperature is about 4°C. The surface water temperatures in the plume are 11°C near the submerged discharge and 6°C in the intake channel. On a late winter day with cold lake temperatures, such as shown here, the recirculating plume does not cause problems with the power plant operation. However, such an event could cause operational problems during the summer because the intake water would be warmer than acceptable for the cooling system.

The use of aerial thermal scanning to study heat loss from buildings has been investigated in many cities. Figure 5.22 illustrates the typical form of imagery acquired in such studies. Note the striking differences among the radiant temperatures of various roofs as well as the temperature differences between various house roofs and garage/carport roofs of the same house. Such images are often useful in assessing inadequate or damaged insulation and roofing materials. The dark-toned streaks aligned from upper right to lower

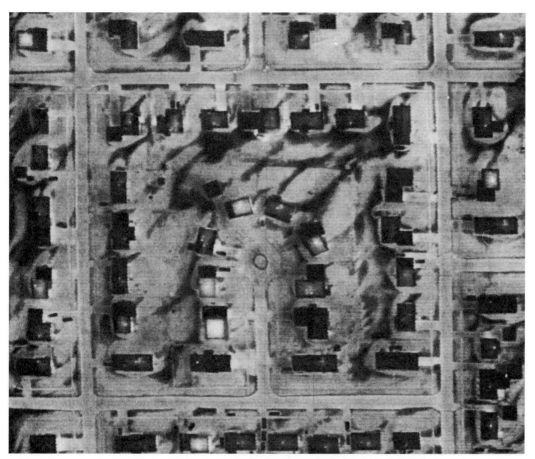

Figure 5.22 Nighttime thermal image depicting building heat loss in an Iowa City, approximately 2:00 A.M., snow-covered ground, air temperature approximately −4°C, 460 m flying height, 1 mrad IFOV. (Courtesy Iowa Utility Association.)

left on the ground areas between houses result from the effects of wind on the snow-covered ground.

Although aerial thermal scanning can be used to estimate the amount of energy radiated from the roofs of buildings, *the emissivity of the roof surfaces must be known to determine the kinetic temperature of the roof surfaces*. With the exception of unpainted sheet metal roofs (Figure 5.19), which have a very low emissivity, roof emissivities vary from 0.88 to 0.94. (Painted metal surfaces assume the emissivity characteristics of the particular paint used.)

Thermal scanning of roofs to estimate heat loss is best accomplished on cold winter nights at least 6 to 8 hr after sunset, in order to minimize the effects of solar heating. Alternatively, surveys can be conducted on cold overcast winter days. In any case, roofs should be neither snow covered nor wet. Because of the side-looking characteristics of scanner images, the scanner vertically views the roof tops of buildings only directly beneath the plane. At the edge of the scan, it views the rooftops plus a portion of building sides. Roof pitch affects the temperature of roofs. A flat roof directly faces a night sky that is 20 to 30°C cooler than ambient air temperature and will therefore lose heat by radiation. Sloping roofs often receive radiation from surrounding buildings and trees, keeping their surfaces warmer than flat roofs. Attic ventilation characteristics must also be taken into account when analyzing roof heat loss.

When mapping heat loss by aerial thermal scanning, it must also be realized that heat loss from roofs constitutes only a portion of the heat lost from buildings, as heat is also lost through walls, doors, windows, and foundations. It is estimated that a house with comparable insulation levels in all areas loses about 10 to 15 percent of its heat through the roof. Homes with well-insulated walls and poorly insulated ceilings may lose more heat through the roof.

An alternative and supplement to aerial thermal scanning is the use of ground-based thermal scanning systems. Applications of these imaging systems range from detecting faulty electrical equipment (such as overheated transformers) to monitoring industrial processes to medical diagnosis. Depending on the detector used, the systems operate in the 3- to 5-μm or 8- to 14-μm wavelength bands. When operating in the range 3 to 5 μm, care must be taken not to include reflected sunlight in the scene.

Figure 5.23 illustrates the use of aerial thermal scanning to detect heat loss from a steam line located on an old stone arch bridge that crosses a deep gorge on the Cornell University campus. The surface of the bridge is constructed of steel-reinforced concrete. Maintenance personnel were aware of the steam leak before thermal scanning of the area, but one entire lane of the bridge was apparently hot and they were reluctant to cut through the reinforced concrete in an attempt to locate the defective section of steam line. Using various enhancements of the thermal image shown as Figure 5.23a, they were able to locate the heat source, which turned out to be a 6-mm-diameter hole in a 10-cm-diameter steam line. The line is insulated with asbestos and encased in a 30-cm-diameter clay tile located 60 cm below the road surface. The clay tile casing compounded the problem of precisely locating the steam leak using conventional means. That is, the casing ducted the steam a substantial distance from the leak. This caused the illusion of a general steam line failure, but close inspection of the thermal images indicated the exact site of the leak. Figure 5.23b shows the same bridge after correction of the steam leak.

NASA's *Thermal Infrared Multispectral Scanner (TIMS)* is a multispectral scanner that operates exclusively in the thermal infrared portion of the spectrum. It utilizes six narrow bands located between 8.2 and 12.2 μm and has an IFOV of 2.5 mrad and a total field of view of 80°. This instrument has great potential for geologic mapping because the spectral exitance of various minerals, especially silicates, is distinct in the 8- to 14-μm-wavelength region.

Figure 5.23 Thermal images showing steam line heat loss, Cornell University, Ithaca, NY: (a) prior to repair; (b) following repair. (Courtesy Cornell University and Daedalus Enterprises, Inc.)

5.9 GEOMETRIC CHARACTERISTICS OF ACROSS-TRACK SCANNER IMAGERY

The airborne thermal images illustrated in this discussion were all collected using across-track, or whiskbroom, scanning procedures. Not only are across-track scanners (multispectral and thermal) subject to altitude and attitude variations due to the continuous and dynamic nature of scanning, but also their images contain systematic geometric variations due to the geometry of across-track scanning. We discuss these sources of systematic and random geometric variations under separate headings below, although they occur simultaneously. Also, although we use thermal scanner images to illustrate the various geometric characteristics treated here, it should be emphasized that these characteristics hold for *all* across-track multispectral scanner images, not simply across-track thermal scanner images.

Spatial Resolution and Ground Coverage

Airborne across-track scanning systems are generally operated at altitudes in the range 300 to 12,000 m. Table 5.4 summarizes the spatial resolution and ground coverage that would result at various operating altitudes when using a system having a 90° total field of view and a 2.5-mrad IFOV. The ground resolution at nadir is calculated from Eq. 5.1 ($D = H'\beta$). The swath width W can be calculated from

$$W = 2H' \tan \theta \qquad (5.14)$$

where

W = swath width
H' = flying height above terrain
θ = one-half total field of view of scanner

TABLE 5.4 Ground Resolution at Nadir and Swath Width for Various Flying Heights of an Across-Track Scanner Having a 90° Total Field of View and a 2.5-mrad IFOV

Altitude	Flying Height above Ground (m)	Ground Resolution at Nadir (m)	Swath Width (m)
Low	300	0.75	600
Medium	6,000	15	12,000
High	12,000	30	24,000

Many of the geometric distortions characterizing across-track scanner imagery can be minimized by constraining one's analysis to the near-center portion of the imaging swath. Also, as we will discuss, several of these distortions can be compensated for mathematically. However, their effect is difficult to negate completely. As a consequence, across-track images are rarely used as a tool for precision mapping. Instead, data extracted from the images are normally registered to some base map when positional accuracy is required in the interpretation process.

Tangential-Scale Distortion

Unless it is geometrically rectified, across-track imagery manifests severe scale distortions in a direction perpendicular to the flight direction. The problem arises because a scanner mirror rotating at constant angular velocity does not result in a constant speed of the scanner's IFOV over the terrain. As shown in Figure 5.24, for any increment of time, the mirror sweeps through a constant incremental arc, $\Delta\theta$. Because the mirror rotates at a constant angular velocity, $\Delta\theta$ is the same at any scan angle θ. However, as the distance between the nadir and the ground resolution cell increases, the linear ground velocity of the resolution cell increases. Hence, the ground element, ΔX, covered per unit time increases with increasing distance from the nadir. This results in image scale compression at points away from the nadir, as the ground spot covers a greater distance at its increasing ground speed. The resulting distortion is known as *tangential-scale distortion*. Note that it occurs only in the along-scan direction, perpendicular to the direction of flight. Image scale in the direction of flight is essentially constant.

Figure 5.25 schematically illustrates the effect of tangential distortion. Shown in Figure 5.25a is a hypothetical vertical aerial photograph taken over flat terrain containing patterns of various forms. An unrectified across-track

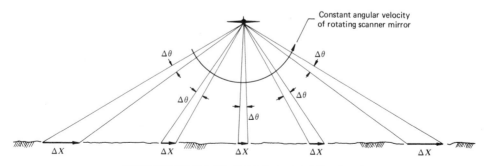

Figure 5.24 Source of tangential-scale distortion.

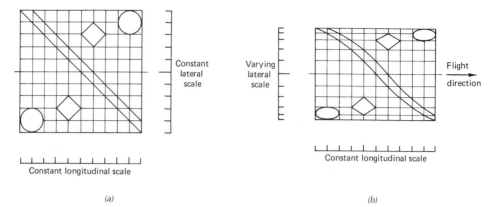

Figure 5.25 Tangential-scale distortion in unrectified across-track scanner imagery: (a) vertical aerial photograph; (b) across-track scanner imagery.

scanner image of the same area is shown in Figure 5.25*b*. Note that because of the constant longitudinal scale and varying lateral scale of the scanner imagery, objects do not maintain their proper shapes. Linear features—other than those parallel or normal to the scan lines—take on an S-shaped *sigmoid curvature*. Extreme compression of ground features characterizes the image near its edges. These effects are illustrated in Figure 5.26, which shows an aerial photograph and thermal image of the same area. The flight line for the thermal image is vertical on the page. Note that the photograph and thermal image are reproduced with the same scale along the flight line but that the scale of the thermal image is compressed in a direction perpendicular to the flight line. Two diagonal roads that are straight on the aerial photograph take on a sigmoid curvature on the thermal image. Note also the light-toned water and trees on this nighttime thermal image (similar to Figure 5.17*b*).

Figure 5.27 further illustrates tangential-scale distortion. This scanner image shows a group of cylindrical oil storage tanks. The flight line was from left to right. Note how the scale becomes compressed at the top and bottom of the image, distorting the circular shape of the tank tops. Note also that the scanner views the sides as well as the tops of features located away from the flight line.

Tangential-scale distortion normally precludes useful interpretation near the edges of unrectified across-track scanner imagery. Likewise, geometric measurements made on unrectified scanner imagery must be corrected for this distortion. Figure 5.28 shows the elements involved in computing true ground positions from measurements made on a distorted image. On unrectified imagery, *y* coordinates will relate directly to *angular* dimensions, not to lineal dimensions. This results in the geometric relationship depicted in the figure, where the effective focal plane of image formation is shown as a curved surface below the aircraft. In order to determine the

Figure 5.26 Comparison of aerial photograph and across-track thermal scanner image illustrating tangential distortion, Iowa County, WI: (*a*) panchromatic aerial photograph, 3000-m flying height; (*b*) nonrectified thermal image, 6:00 A.M., 300-m flying height. [(*a*) Courtesy USDA–ASCS. (*b*) Courtesy National Center for Atmospheric Research.]

Figure 5.27 Across-track thermal scanner image illustrating tangential distortion, 100-m flying height. (Courtesy Texas Instruments, Inc.)

ground position Y_p corresponding to image point p, we must first compute θ_p from the relationship

$$\frac{y_p}{y_{max}} = \frac{\theta_p}{\theta_{max}}$$

Rearranging yields

$$\theta_p = \frac{y_p \theta_{max}}{y_{max}} \tag{5.15}$$

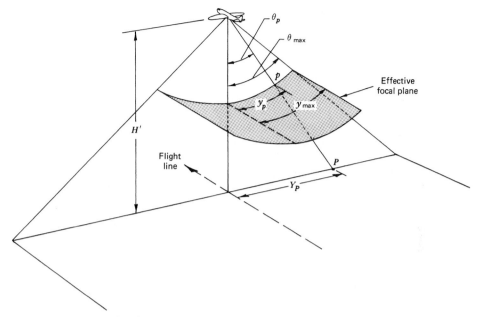

Figure 5.28   Tangential-scale distortion correction.

where

y_p = distance measured on image from nadir line to point p

y_{max} = distance from nadir line to edge of image

θ_{max} = one-half total field of view of scanner

Once θ_p has been computed, it may be trigonometrically related to ground distance Y_p by

$$Y_p = H' \tan \theta_p \qquad (5.16)$$

When determining ground positions on unrectified imagery, the above process must be applied to each y coordinate measurement. Alternatively, the correction can be implemented electronically, or digitally, in the image recording process, resulting in *rectilinearized* images. In addition to permitting direct measurement of positions, rectilinearized imagery improves the ability to obtain useful interpretations in areas near the edge of images.

Resolution Cell Size Variations

Across-track scanners sense energy over ground resolution cells of continuously varying size. An increased cell size is obtained as the IFOV of a scanner moves outward from the flight nadir.

The geometric elements defining the size of the ground resolution cell are shown in Figure 5.29. At the nadir line, the ground resolution cell has a dimension of $H'\beta$. At a scan angle θ, the distance from the aircraft to the cell becomes $H'_\theta = H' \sec \theta$. Hence the size of the resolution cell increases. The cell has dimensions of $(H' \sec \theta)\beta$ in the direction of flight and $(H' \sec^2 \theta)\beta$ in the direction of scanning. These are actually the *nominal* dimensions of the measurement cell. The *true* size and shape of a ground resolution cell are a function not only of β, H', and θ but also of the *response time* of a particular scanner's electronics. The response time is a measure of the time that a scanner takes to respond electronically to a change in ground reflected or emitted energy. With this added restriction, we see that optics control the resolution cell size in the direction of flight, while both optics and electronics can influence the cell size in the direction of scan. Because of system response time limitations, the resolution cell size along a scan line can be as much as three to four times that in the direction of flight.

Although it is rarely critical to know the precise degree of resolution cell size variation, it is important to realize the effect this variation has on the interpretability of the imagery at various scan angles. The scanner output at any point represents the integrated radiance from all features within the ground resolution cell. Because the cell increases in size near the edge of the image, only larger terrain features will completely fill the IFOV in these

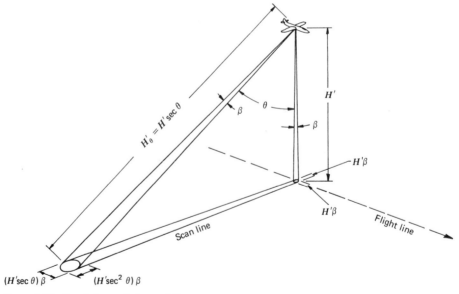

Figure 5.29 Resolution cell size variation.

areas. When objects smaller than the area viewed by the IFOV are imaged, background features also contribute to the recorded signal. This is particularly important in thermal scanning applications where accurate temperature information is required. *For an object to be registered with the proper radiant temperature, its size must be larger than the ground resolution cell.* This effect may again limit the image analysis to the central portion of the image, even after rectilinearization is performed. However, an advantage of the changing size of the ground resolution cell is that it compensates for off-nadir radiometric falloff. If the ground resolution cell area were constant, the irradiance received by the scanner would decrease as $1/H_\theta'^2$. But since the ground resolution cell area increases as $H_\theta'^2$, the irradiance falloff is precisely compensated and a consistent radiance is recorded over uniform surfaces.

One-Dimensional Relief Displacement

Figure 5.30 illustrates the nature of relief displacement characterizing across-track scanner images. Since all objects are viewed by the scanner only along "side-looking" scan lines, relief displacement occurs only in this single direction. (See also Figure 5.27.) An advantage to relief displacement is that it affords an opportunity to see a side view of objects. On the other hand, it can obscure the view of objects of interest. For example, a thermal mission might be planned to detect heat losses in steam lines in an urban setting. Tall buildings proximate to the objects of interest may completely obscure their view.

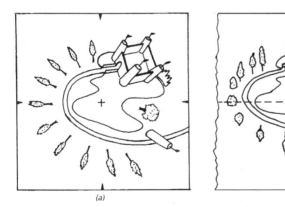

(a)   (b)

Figure 5.30   Relief displacement on a photograph versus an across-track scanner image. (*a*) In a vertical aerial photograph vertical features are displaced radially from the principal point. (*b*) In an across-track scanner image vertical features are displaced at right angles from the nadir line.

In such cases it is often necessary to cover the study area twice, in perpendicular flight directions.

Figure 5.31 is a thermal image illustrating one-dimensional relief displacement. Note that the displacement of the tops of the tall buildings is greater with increasing distance from the nadir.

Figure 5.31 Across-track thermal scanner image illustrating one-dimensional relief displacement, San Francisco, CA, predawn, 1500-m flying height, 1 mrad IFOV. (Courtesy NASA.)

Flight Parameter Distortions

Because across-track scanner imagery is collected in a continuous fashion, it lacks the consistent relative orientation of image points found on instantaneously imaged aerial photographs. That is, across-track scanning is a dynamic continuous process rather than an intermittent sampling of discrete perspective projections as in photography. Because of this, any variations in the aircraft flight trajectory during scanning affect the relative positions of points recorded on the resulting imagery.

A variety of distortions associated with aircraft attitude (angular orientation) deviations are shown in Figure 5.32. The figure shows the effect that each type of distortion would have on the image of a square grid on the ground. This grid is shown in Figure 5.32*a*.

Figure 5.32*b* gives a sketch of an across-track scanner image acquired under constant aircraft altitude and attitude conditions. Only the tangential-scale distortion is present in this case. In Figure 5.32*c* the effect of aircraft *roll* about its flight axis is shown. Roll causes the ground grid lines to be imaged at varying times in the mirror rotation cycle. Consequently, the image takes on a wavy appearance. This effect may be negated by the process of *roll compensation*. This involves using a gyroscope to monitor aircraft roll on a line-by-line basis and appropriately advancing or retarding the start time of each scan line on the final image.

When extreme crosswind is encountered during data acquisition, the axis of the aircraft must be oriented away from the flight axis slightly to counteract the wind. This condition is called *crab* and causes a skewed image

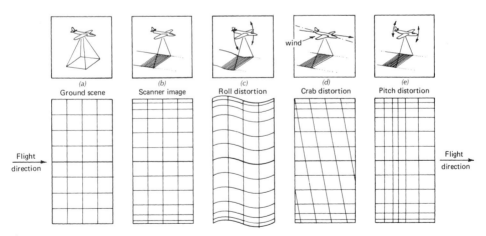

Figure 5.32 Across-track scanner imagery distortions induced by aircraft attitude deviations: (*a*) ground scene; (*b*) scanner image; (*c*) roll distortion; (*d*) crab distortion; (*e*) pitch distortion.

(Figure 5.32*d*). Crab distortion can be corrected by rotating the scanner in its ring mount during flight or by computer processing the distorted data. Most often, crab distortion is avoided by not acquiring data under high crosswind conditions.

Finally, as illustrated in Figure 5.32*e*, variations in aircraft *pitch* might distort scanner imagery. The resulting local-scale changes in the flight direction due to pitch are generally slight enough that they may be ignored in most analyses.

5.10 GEOMETRIC CHARACTERISTICS OF ALONG-TRACK SCANNER IMAGERY

The geometric characteristics of along-track scanner images are quite different from those of across-track scanner images. Along-track scanners have no scanning mirror, and there is a fixed geometric relationship among the solid-state detector elements recording each scan line. In essence, the geometry along each scan line of an along-track scanner image is similar to that of an aerial photograph. Line-to-line variation in imaging geometry is caused purely by any random variation in the altitude or attitude (angular orientation) of the aircraft along a flight line. Often, on-board inertial measurement units and GPS systems are used to measure these variations and geometrically correct the data from along-track scanners.

5.11 RADIOMETRIC CALIBRATION OF THERMAL SCANNERS

As mentioned previously, the general lack of geometric integrity of thermal scanner imagery precludes its use as a precision mapping tool. Because of this, photographic imagery is often acquired simultaneously with thermal imagery. Naturally, when nighttime thermal missions are flown, simultaneous photography is usually not feasible. In such cases, photography can be taken on the day before or after a mission. Sometimes, new photography is not needed and existing photography is used. In any case, the photography expedites object identification and study of spatial detail and affords positional accuracy. The thermal imagery is then used solely for its radiometric information content. In order to obtain accurate radiometric information from the scanner data, the scanner must be radiometrically calibrated.

There are numerous approaches to scanner calibration, each with its own degree of accuracy and efficiency. What form of calibration is used in any given circumstance is a function of not only the equipment available for data acquisition and processing but also the requirements of the application at hand. We limit our discussion here to a general description of the two most commonly used calibration methods:

1. Internal blackbody source referencing
2. Air-to-ground correlation

As will become apparent in the following discussion, a major distinction between the two methods is that the first does not account for atmospheric effects but the second does. Other calibration methods are described by Schott (1997).

Internal Blackbody Source Referencing

Current generations of thermal scanners normally incorporate internal temperature references. These take the form of two "blackbody" radiation sources positioned so that they are viewed by the scanner mirror during each scan. The temperatures of these sources can be precisely controlled and are generally set at the "cold" and "hot" extremes of the ground scene to be monitored. Along each scan line, the scanner optics sequentially view one of the radiation standards, scan across the ground scene below the aircraft, and then view the other radiation standard. This cycle is repeated for each scan line.

Figure 5.33 illustrates the configuration of an internally calibrated scanner. The arrangement of the reference sources, or "plates," relative to the field of view of the scanner is shown in Figure 5.33a. The detector signal typical of one scan line is illustrated in Figure 5.33b. The scanner mirror sequentially looks at the cold-temperature reference plate (T_1), then sweeps the ground, and finally looks at the hot reference plate (T_2). The scanner output at the two temperature plates is recorded along with the image data. This provides a continuously updated reference by which the other scanner output values can be related to absolute radiant temperature.

The internal source referencing procedure permits acceptable calibration accuracy for many applications. Actual versus predicted temperature discrepancies of less than 0.3°C are typical for missions flown at altitudes up to 600 m under clear, dry weather conditions. However, internal calibration still does not account for atmospheric effects. As indicated earlier (Section 5.7), under many prevailing mission conditions the atmosphere can bias scanner temperature measurements by as much as 2°C.

Air-to-Ground Correlation

Atmospheric effects can be accounted for in thermal scanner calibration by using empirical or theoretical atmospheric models. Theoretical atmospheric

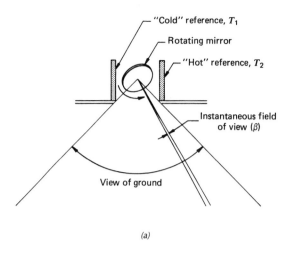

(a)

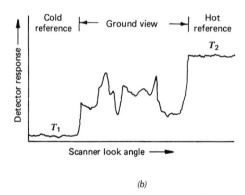

(b)

Figure 5.33 Internal blackbody source calibration: (a) reference plate arrangement; (b) typical detector output for one scan line.

models use observations of various environmental parameters (such as temperature, pressure, and CO_2 concentration) in mathematical relationships that predict the effect the atmosphere will have on the signal sensed. Because of the complexity of measuring and modeling the factors that influence atmospheric effects, these effects are normally eliminated by correlating scanner data with actual surface measurements on an empirical basis.

Air-to-ground correlation is frequently employed in calibration of thermal water quality data, such as those acquired over heated water discharges. Surface temperature measurements are taken on the ground simultaneously with the passage of the aircraft. Thermometers, thermistors, or thermal radiome-

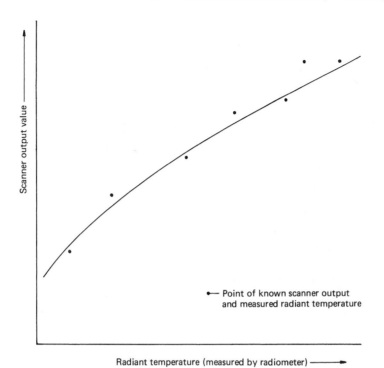

Figure 5.34   Sample calibration curve used to correlate scanner output with radiant temperature measured by radiometer.

ters operated from boats are commonly used for this purpose. Observations are typically made at points where temperatures are assumed to be constant over large areas. The corresponding scanner output value is then determined for each point of ground-based surface temperature measurement. A calibration curve is then constructed relating the scanned output values to the corresponding ground-based radiant temperature (Figure 5.34). Once the calibration relationship is defined (typically using linear regression procedures), it is used to estimate the temperature at all points on the scanner images where no ground data exist.

Figure 5.35 illustrates the type of thermal radiometer that can be used for air-to-ground correlation measurements. This particular instrument is an Everest Interscience hand-held "infrared thermometer" that operates in the 8- to 14-μm range and displays radiant temperatures on a liquid-crystal display (LCD) panel. It reads temperatures in the range from -40 to $+100°C$, with a radiometric resolution of 0.1°C and an accuracy of 0.5°C. The instrument was designed as a plant stress monitor, and its LCD panel also displays dry-bulb air temperature and differential temperature (the difference between radiant temperature and air temperature).

Figure 5.35 Infrared thermometer (thermal radiometer). Spot radiant temperatures are read from LCD display panel. (Courtesy Everest Interscience, Inc.)

5.12 TEMPERATURE MAPPING WITH THERMAL SCANNER DATA

In many applications of thermal scanning techniques, it is of interest to prepare "maps" of surface temperature distributions. The digital data recorded by a thermal scanner can be processed, analyzed, and displayed in a variety of ways. For example, consider scanner data for which a correlation has been developed to relate scanner output values to absolute ground temperatures. This calibration relationship can be applied to each point in the digital data set, producing a matrix of absolute temperature values.

The precise form of a calibration relationship will vary with the temperature range in question, but for the sake of the example, we assume that a linear fit of the digital data to radiant exitance is appropriate. Under this assumption, a digital number, DN, recorded by a scanner, can be expressed by

$$\mathrm{DN} = A + B\varepsilon T^4 \qquad (5.17)$$

where

A, B = system response parameters determined from sensor calibration procedures described earlier

ε = emissivity at point of measurement

T = kinetic temperature at point of measurement

Once *A* and *B* are determined, kinetic temperature *T* for any observed digital number DN is given by

$$T = \left(\frac{\text{DN} - A}{B\varepsilon}\right)^{1/4} \tag{5.18}$$

The parameters *A* and *B* can be obtained from internal blackbody calibration, air-to-ground correlation, or any of the other calibration procedures. At a minimum, two corresponding temperature (*T*) and digital number (DN) values are needed to solve for the two unknowns *A* and *B*. Once parameters *A* and *B* are known, Eq. 5.18 can be used to determine the kinetic temperature for any ground point for which DN is observed *and* the emissivity is known. The calibrated data may be further processed and displayed in a number of different forms (e.g., isotherm maps, color-coded images, GIS layers).

5.13 FLIR SYSTEMS

To this point in our discussion of aerial thermography, we have emphasized scanners that view the terrain directly beneath an aircraft. *Forward-looking infrared (FLIR)* systems can be used to acquire oblique views of the terrain ahead of an aircraft. Figure 5.36 shows two images acquired by FLIR systems. Figure 5.36*a* was acquired during the daytime over a storage tank facility. Conceptually, the FLIR system used to produce this image operates on the same basic principles as an across-track line scanning system. However, the mirror for the system points forward and optically sweeps the field of view of a linear array of thermal detectors across the scene of interest. Figure 5.36*b* was acquired using a helicopter-borne FLIR system. Thermal shadows can be seen next to the parked vehicles at the left side of the image, while the

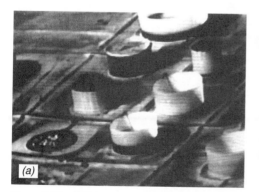

Figure 5.36 FLIR images. (*a*) Storage tank facility. Note level of liquid in each tank. (Courtesy Texas Instruments, Inc.) (*b*) City street. Note thermal shadows adjacent to parked vehicles and the image of a person on the right side of the street. (Courtesy FLIR Systems, Inc.)

moving vehicle at the center of the image does not have a thermal shadow associated with it. Note also the heat emitted from the hood of those cars that have been operated most recently and the image of a person near the right side of the image.

Modern FLIR systems are extremely portable (typically weighing less than 30 kg) and can be operated on a wide variety of fixed-wing aircraft and helicopters as well as from ground-based mobile platforms. Forward-looking imagery has been used extensively in military applications. Civilian use of FLIR is increasing in applications such as fire fighting, electrical transmission line maintenance, law enforcement activities, and nighttime vision systems for automobiles.

5.14   HYPERSPECTRAL SENSING

Hyperspectral sensors (sometimes referred to as *imaging spectrometers*) are instruments that acquire images in many, very narrow, contiguous spectral bands throughout the visible, near-IR, mid-IR, and thermal IR portions of the spectrum. (They can employ across-track or along-track scanning or two-dimensional framing arrays.) These systems typically collect 200 or more bands of data, which enables the construction of an effectively continuous reflectance (emittance in the case of thermal IR energy) spectrum for every pixel in the scene (Figure 5.37). These systems can discriminate among earth

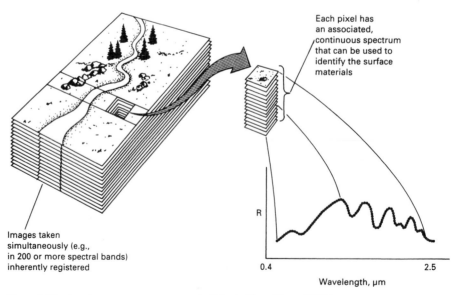

Figure 5.37   Imaging spectrometry concept. (Adapted from Vane, 1985.)

surface features that have diagnostic absorption and reflection characteristics over narrow wavelength intervals that are "lost" within the relatively coarse bandwidths of the various bands of conventional multispectral scanners. This concept is illustrated in Figure 5.38, which shows the laboratory-measured reflectance spectra for a number of common minerals over the wavelength range 2.0 to 2.5 μm. Note the diagnostic absorption features for these various material types over this spectral range. Also shown in this figure is the bandwidth of band 7 of the Landsat TM and ETM+ (Chapter 6). Whereas these sensors obtain only one data point corresponding to the integrated response

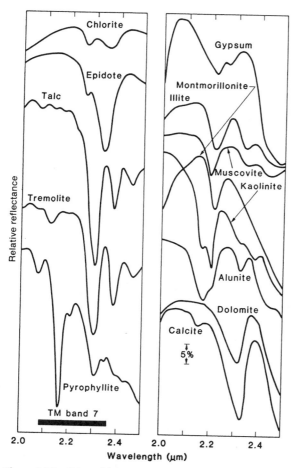

Figure 5.38 Selected laboratory spectra of minerals showing diagnostic absorptance and reflectance characteristics. The spectra are displaced vertically to avoid overlap. The bandwidth of band 7 of the Landsat TM (Chapter 6) is also shown. (From Goetz et al., 1985. Copyright 1985 by the AAAS. Courtesy NASA Jet Propulsion Laboratory.)

over a spectral band 0.27 μm wide, a hyperspectral sensor is capable of obtaining many data points over this range using bands on the order of 0.01 μm wide. Thus, hyperspectral sensors can produce data of sufficient spectral resolution for direct identification of the materials, whereas the broader band TM cannot resolve these diagnostic spectral differences. Hence, while a broadband system can only discriminate general differences among material types, a hyperspectral sensor affords the potential for detailed identification of materials and better estimates of their abundance.

Because of the large number of very narrow bands sampled, hyperspectral data enable the use of remote sensing data collection to replace data collection that was formerly limited to laboratory testing or expensive ground site surveys. Some application areas of hyperspectral sensing include determinations of: surface mineralogy; water quality; bathymetry; soil type and erosion; vegetation type, plant stress, leaf water content, and canopy chemistry; crop type and condition; and snow and ice properties.

Extensive initial hyperspectral sensing research was conducted with data acquired by the *Airborne Imaging Spectrometer* (AIS). This system collected 128 bands of data, approximately 9.3 nm wide. In its "tree mode," the AIS collected data in contiguous bands between 0.4 and 1.2 μm; in its "rock mode," between 1.2 and 2.4 μm. The IFOV of the AIS was 1.9 mrad, and the system was typically operated from an altitude of 4200 m above the terrain. This yielded a narrow swath 32 pixels wide beneath the flight path for AIS-1 (64 pixels wide for AIS-2) with a ground pixel size of approximately 8 $\times$ 8 m.

Figure 5.39 shows an AIS image acquired over Van Nuys, California, on the first engineering test flight of the AIS. This image covers the area outlined in black lines on the corresponding photographic mosaic. Notable features include a field in the lower portion of the image, a condominium complex in the center, and a school in the upper half of the image. On this test flight, images were acquired in only 32 contiguous spectral bands in the region from 1.50 to 1.21 μm. A composite of the 32 AIS images, each in a different 9.3-nm-wide spectral band and each 32 pixels wide, is shown below the photographic mosaic. The most obvious feature in the AIS images is the loss of detail in the atmospheric water absorption band centered around 1.4 μm. However, there is some detail visible in the spectral images in this band. Details associated with reflectance variations are identified with arrows. For instance, the reflectance of a well-watered courtyard lawn inside the school grounds (location *a*) drops significantly beyond 1.4 μm in comparison with the reflectance of the unwatered field (location *b*).

Figure 5.39 also illustrates the pronounced effect water vapor can have on hyperspectral sensor data acquired near the 1.40-μm absorption band. Other atmospheric water vapor absorption bands occur at approximately 0.94, 1.14, and 1.88 μm. In addition to being a strong absorber, atmospheric water vapor can vary dramatically in its temporal and spatial distribution. Within a single scene, the distribution of water vapor can be very patchy and change on the

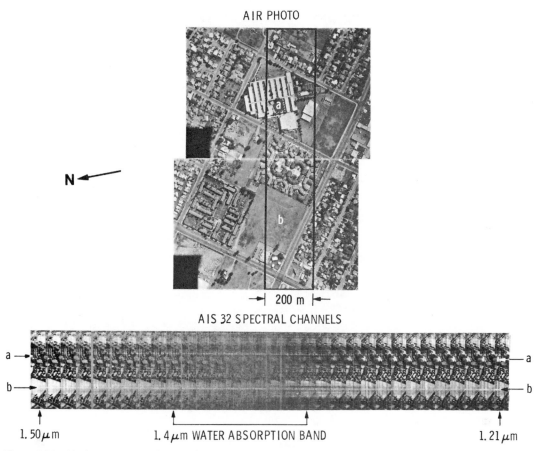

Figure 5.39 Airphoto mosaic and 32 spectral bands of AIS images covering a portion of the same area, Van Nuys, CA: (a) A school courtyard and (b) an open field are shown on the mosaic. The vertical black lines outline the coverage of the corresponding 32-pixel-wide AIS images taken in 32 spectral bands between 1.50 and 1.21 μm. The individual AIS images taken in 9.3-nm-wide, contiguous, spectral intervals are shown at the bottom. The spectral reflectance behavior of the well-watered courtyard and the open field are quite different and are primarily associated with differences in irrigation practices. (From Goetz et al., 1985. Copyright 1985 by the AAAS. Courtesy NASA Jet Propulsion Laboratory.)

time scale of minutes. Also, large variations in water vapor can result simply from changes in the atmospheric path length between the sensor and the ground due to changes in surface elevation. Given the strength and variability of water vapor effects, accounting for their influence on hyperspectral data collection and analysis is the subject of continuing research.

Figure 5.40 illustrates the appearance of 20 discrete hyperspectral sensor bands ranging from 1.98 to 2.36 μm used for the identification of minerals in a hydrothermally altered volcanic area at Cuprite, Nevada. These 20 bands represent a sequence of all odd-numbered bands between bands 171 and 209 of the AVIRIS airborne hyperspectral sensor. Broadband sensors such as the

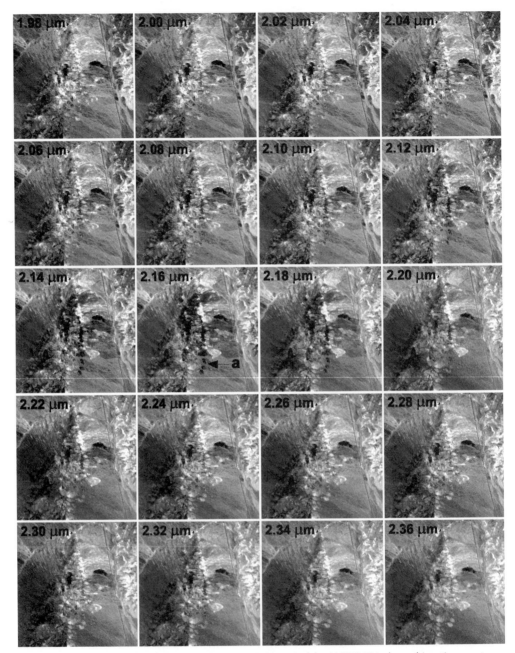

Figure 5.40 Images from bands 171 to 209 (1.98 to 2.36 μm) of the AVIRIS 224-channel imaging spectrometer, showing hydrothermically altered volcanic rocks at Cuprite, NV. Only odd-numbered channels shown. (Courtesy NASA-JPL.)

Landsat TM and ETM+ (Chapter 6) have been unable to discriminate clearly among various mineral types in this spectral region. Figure 5.40 displays the 20 bands in sequence, from band 171 at the upper left to band 209 at the lower right. By choosing one area within the image and viewing that area across the images of all 20 bands, one can obtain a visual estimation of the relative spectral behavior of the materials within the image. For example, labeled as *a* in band 189 (2.16 μm) are several dark spots near the center of the image that are occurrences of the mineral alunite. These alunite areas are darkest in bands 187 to 191 because of their absorption characteristics in wavelengths from 2.14 through 2.18 μm. In other bands (e.g., band 209, 2.36 μm), alunite areas are very light toned, meaning that they reflect strongly in the corresponding wavelengths.

Color composites of hyperspectral data are limited to displays of three bands at a time (with one band displayed as blue, one band as green, and one band as red). In an attempt to convey the spectral nature and complexity of hyperspectral images, hyperspectral data are commonly displayed in the manner shown in Plate 15. In this isometric view, the data can be thought of as a "cube" of the dimensions lines times columns times bands. The front of the cube is a color composite, with the bands centered at 0.570, 0.672, and 0.773 μm displayed as blue, green, and red, respectively. The top and right side of the cube represent color-coded radiance values corresponding to 187 spectral bands for each of the edge pixels along the top and right of the front of the cube. These spectral bands range from 0.468 μm, just behind the front of the cube, to 2.345 μm, at the back. In this color coding scheme, "cool" colors such as blue correspond to low reflectance values and "hot" colors such as red correspond to high reflectance values. In the visible spectrum, both vegetation and water have relatively low radiance values, as indicated by the predominantly blue hues along the edges of the cube near the front. The abrupt transition from visible to near-IR wavelengths can be seen along the top and side of the cube, where the high radiance values from vegetation are indicated by green, yellow, and red hues, while the very low radiance from water is shown as dark blue and violet. Bare soil, at upper left, shows relatively higher radiance values in the visible than does vegetation, but lower radiance in the near IR.

Several types of atmospheric effects also can be seen in the hyperspectral image cube in Plate 15. The dark bands from left to right along the top and from top to bottom along the side of the cube represent atmospheric absorption regions, where there is little signal reaching the sensor (irrespective of the surface relectance involved). The light blue and cyan hues just "behind" the front of the cube are caused by Rayleigh scattering of blue light within the atmosphere, resulting in increased radiance reaching the sensor at short wavelengths.

Substantial research has been undertaken to automate the analysis of the detailed spectral reflectance curves associated with individual pixels in hyperspectral data sets. Among the most fruitful techniques to date has been the use of expert-system-based procedures for matching the remotely sensed spectrum for each pixel to libraries of laboratory spectra for various materials (Kruse and Lefkoff, 1992). This matching is based primarily on the unique absorption characteristics of various material types (the "valleys" in the spectral reflectance curve for each material type). Features such as the number of low points in a curve, their wavelength location, width, depth, and symmetry are used for this purpose. Additional hyperspectral data analysis techniques are described in Chapter 7.

A number of government agencies and mapping/consulting companies maintain data sets of spectral reflectance curves. For example, the USGS maintains a "digital spectral library" that contains the reflectance spectra of more than 500 minerals, plants, and other materials. NASA's Jet Propulsion Laboratory maintains the Advanced Spaceborne Thermal Emission and Reflection Radiometer (ASTER) spectral library, a compilation of nearly 2000 spectra of natural and manufactured materials. The U.S. Army Topographic Engineering Center maintains "hyperspectral signatures" in the range of 0.400 to 2.500 μm for vegetation, minerals, soils, rock, and cultural features, based on laboratory and field measurements. (See Appendix B for URLs.) Note that atmospheric corrections to hyperspectral data must be made in order to make valid comparisons with library spectra.

Figure 5.41 shows the effect of atmospheric scattering on the spectral response pattern of a lake. The figure shows at-sensor radiance in bands 12 to

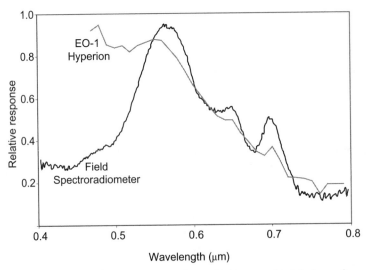

Figure 5.41 Spectral response pattern from a highly eutrophic lake in southern Wisconsin, late summer. Black line shows spectral reflectance measured *in situ* with a hand-held spectroradiometer. Gray line shows spectral radiance recorded by the EO-1 Hyperion hyperspectral sensor.

44 of the Hyperion hyperspectral sensor on the EO-1 satellite (described in Chapter 6). The image from which this spectrum was extracted was acquired over a highly eutrophic lake in midsummer. Note the elevated radiance values recorded by the satellite sensor at short wavelengths, particularly below 0.5 μm. These represent the effects of Rayleigh scattering along the path between the lake surface and the sensor. For comparison, Figure 5.41 also shows a spectral reflectance curve collected with a hand-held field spectroradiometer from a boat on the same lake during the Hyperion image acquisition. The pattern in this curve is typical of the spectral reflectance of highly eutrophic lakes, with chlorophyll-induced absorption of blue wavelengths, peak reflectance between 0.5 and 0.6 μm, and lower peaks in the red wavelengths.

Many airborne hyperspectral scanning instruments have been developed. Here, we will describe some of the more widely used instruments.

The first commercially developed, programmable airborne hyperspectral scanner was the *Compact Airborne Spectrographic Imager (CASI)*, available since 1989. A redesigned version (CASI 2) has improved upon the original CASI. This linear array system collects data in up to 288 bands between 0.4 and 1.0 μm at 1.8-nm spectral intervals. The precise number of bands, their locations, and bandwidths can all be programmed to suit particular application requirements. The IFOV of the system's detectors is 1.3 mrad, and the total field of view is 37.8°. The system can also be used in combination with downwelling irradiance measurements for atmospheric corrections, and with a GPS, an altimeter, and inertial sensors to correct the recorded data for variations in aircraft altitude and attitude during flight.

The *Airborne Visible–Infrared Imaging Spectrometer (AVIRIS)* typically collects data in 224 bands that are approximately 9.6 nm wide in contiguous bands between 0.40 and 2.45 μm. When flown on NASA's ER-2 research aircraft at an altitude of 20 km, the swath width of this across-track scanner is approximately 10 km with a ground pixel resolution of approximately 20 m.

The *Hyperspectral Digital Imagery Collection Experiment (HYDICE)* is a hyperspectral sensor developed by the Remote Sensing Division of the U.S. Naval Research Laboratory. It is an along-track sensor that operates over the range of 0.4 to 2.5 μm.

The Geophysical and Environmental Research Corporation (GER) has produced several hyperspectral scanners, including the *EPS-H*, with 136 bands from 0.3 to 2.5 μm, and 12 additional bands from 8 to 12 μm.

Now in its third generation, the *Advanced Airborne Hyperspectral Imaging Spectrometer (AAHIS)* is a commercially produced along-track hyperspectral scanner that senses in 384 channels in the 0.39- to 0.84-μm range.

AISA+, AISA Eagle, and *AISA Birdie* are commercially available along-track hyperspectral sensors manufactured by Specim. They feature 240 to 273 spectral bands, in the visible and near IR (AISA+ and AISA Eagle) and the mid IR (AISA Birdie).

The *HyMap (Hyperspectral Mapping)* system is an across-track airborne hyperspectral scanner built by Integrated Spectronics that senses in up to 200 bands. A variation of this scanner is the *Probe-1* built by Integrated Spectronics for ESSI (Earth Search Sciences Incorporated), which collects 128 bands of data in the 0.40- to 2.5-μm range.

The HyMap data shown in Plate 16 illustrate how it is possible to use airborne hyperspectral data to construct image-derived spectral reflectance curves for various minerals that are similar to laboratory-based spectral reflectance curves and to use these data to identify and plot the location of various minerals in an image. The area shown in (*a*) and (*b*) of Plate 16 is approximately 2.6 × 4.0 km in size and is located in the Mt. Fitton area of South Australia. This is an area of strongly folded sedimentary rocks (with some intrusive rocks). The HyMap data were collected in 128 narrow spectral bands over the range of 0.40 to 2.5 μm, with a ground resolution cell size of 5 m per pixel. Part (*a*) is a color IR composite image using bands centered at 0.557, 0.665, and 0.863 μm, displayed as blue, green, and red, respectively. Part (*c*) shows laboratory-derived spectral reflectance curves for selected minerals occurring in the image. Part (*d*) shows spectral reflectance curves for these same minerals, based directly on the HyMap hyperspectral data. These curves were tentatively identified using visual comparison with the laboratory curves. Part (*b*) shows color overlays of these selected minerals on a grayscale image of the HyMap data. The colors in (*b*) correspond with those used in the spectral plots (*c*) and (*d*). Six different minerals are shown here. A preliminary analysis of the HyMap data without a priori information, and using only the 2.0- to 2.5-μm range, indicated the presence of over 15 distinct minerals.

5.15   CONCLUSION

In composite, hyperspectral sensing holds the potential to provide a quantum jump in the quality of spectral data obtained about earth surface features. Hyperspectral sensing also provides a quantum jump in the quantity of data acquired. Hence, research is continuing on how to optimize the analysis of the high volumes of data acquired by these systems.

Research is also underway on the development of "ultraspectral" sensors that will provide data from thousands of very narrow spectral bands. This level of spectral sensitivity is seen as necessary for such applications as identifying specific materials and components of aerosols, gas plumes, and other effluents (Logicon, 1997b).

In this chapter, we have treated the basic theory and operation of multispectral, thermal, and hyperspectral systems. We have illustrated selected examples of image interpretation, processing, and display, with an emphasis on data gathered using airborne systems. In Chapter 6, we discuss multispectral, thermal, and hyperspectral sensing from space platforms. In Chapter 7, we further describe how the data from these systems are processed digitally.

SELECTED BIBLIOGRAPHY

Abrams, M., R. Bianchi, and D. Pieri, "Revised Mapping of Lava Flows on Mount Etna, Sicily," *Photogrammetric Engineering and Remote Sensing*, vol. 62, no. 12, 1996, pp. 1353–1359.

Adams, M.L., W.D. Philpot, and W.A. Norvell, "Yellowness Index: An Application of Spectral Second Derivatives to Estimate Chlorosis of Leaves in Stressed Vegetation," *International Journal of Remote Sensing*, vol. 20, no. 18, 1999, pp. 3663–3675.

Alberotanza, L., et al., "Hyperspectral Aerial Images: A Valuable Tool for Submerged Vegetation Recognition in the Orbetello Lagoons, Italy," *International Journal of Remote Sensing*, vol. 20, no. 3, 1999, pp. 523–534.

American Society of Photogrammetry (ASP), *Manual of Remote Sensing*, 2nd ed., ASP, Falls Church, VA, 1983.

Avery, T.E., and G.L. Berlin, *Fundamentals of Remote Sensing and Airphoto Interpretation*, 6th ed., Prentice Hall, Upper Saddle River, NJ, 2004.

Ben-Dor, E., et al., "Mapping of Several Soil Properties Using DAIS-7915 Hyperspectral Scanner Data—A Case Study over Clayey Soils in Israel," *International Journal of Remote Sensing*, vol. 23, no. 6, 2002, pp. 1043–1062.

Bielski, C.M., et al., "S-Space: A New Concept for Information Extraction from Imaging Spectrometer Data," *International Journal of Remote Sensing*, vol. 23, no. 10, 2002, pp. 2005–2022.

Blackburn, G.A., "Quantifying Chlorophylls and Carotenoids at Leaf and Canopy Scales: An Evaluation of Some Hyperspectral Approaches," *Remote Sensing of Environment*, vol. 66, no. 3, 1998, pp. 273–285.

Boochs, F., et al., "Shape of the Red Edge as Vitality Indicator for Plants," *International Journal of Remote Sensing*, vol. 11, no. 10, 1990, pp. 1741–1753.

Bowers, T.L., and L.C. Rowan, "Remote Mineralogic and Lithologic Mapping of the Ice River Alkaline Complex, British Columbia, Canada, Using AVIRIS Data," *Photogrammetric Engineering and Remote Sensing*, vol. 62, no. 12, 1996, pp. 1379–1385.

Bruce, L.M., and J. Li, " Wavelets for Computationally Efficient Hyperspectral Derivative Analysis," *IEEE Transactions on Geoscience and Remote Sensing*, vol. 39, no. 7, 2001, pp. 1540–1546.

Bruce, L.M., J. Li, and Y. Huang, "Automated Detection of Subpixel Hyperspectral Targets with Adaptive Multichannel Discrete Wavelet Transform," *IEEE Transactions on Geoscience and Remote Sensing*, vol. 40, no. 4, 2002, pp. 977–980.

Campbell, J.B., *Introduction to Remote Sensing*, 3rd ed., Guilford, New York, 2002.

Canadian Journal of Remote Sensing, Special Issue on Hyperspectral Remote Sensing, vol. 24, no. 2, 1998.

Chang, C.I., H. Ren, and S.S. Chiang, "Real-Time Processing Algorithms for Target Detection and Classification in Hyperspectral Imagery," *IEEE Transactions on Geoscience and Remote Sensing*, vol. 39, no. 4, 2001, pp. 760–768.

Cialella, A.T., et al., "Predicting Soil Drainage Class Using Remotely Sensed and Digital Elevation Data," *Photogrammetric Engineering and Remote Sensing*, vol. 63, no. 2, 1997, pp. 171–178.

Cloutis, E.A., "Hyperspectral Geological Remote Sensing: Evaluation of Analytical Techniques," *International Journal of Remote Sensing*, vol. 17, no. 12, 1996, pp. 2215–2242.

Cochrane, M.A., "Using Vegetation Reflectance Variability for Species Level Classification of Hyperspectral Data," *International Journal of Remote Sensing*, vol. 21, no. 10, 2000, pp. 2075–2087.

Courault, D., et al., "Airborne Thermal Data for Evaluating the Spatial Distribution of Actual Evapotranspiration over a Watershed in Oceanic-Climatic Conditions—Application of Semi-Empirical Models," *International Journal of Remote Sensing*, vol. 17, no. 12, 1996, pp. 2281–2302.

Cracknell, A.P., and Y. Xue, "Thermal Inertia Determination from Space—A Tutorial Review," *International Journal of Remote Sensing*, vol. 17, no. 3, 1996, pp. 431–461.

Curtis, J.O., "Moisture Effects on the Dielectric Properties of Soils," *IEEE Transactions on Geoscience and Remote Sensing*, vol. 39, no. 1, 2001, pp. 125–128.

Dash, P., et al., "Land Surface Temperature and Emissivity Estimation from Passive Sensor Data: Theory and Practice—Current Trends," *International Journal of Remote Sensing*, vol. 23, no. 13, 2002, pp. 2563–2594.

Elvidge, C.D., and F.P. Portigal, "Reflectance Spectra of Green and Dry Vegetation Derived from 1989 AVIRIS Data," *Proceedings of the Fifth Australasian Remote Sensing Conference*, Perth, Australia, October 1990, pp. 185–195.

Fraser, R.N., "Hyperspectral Remote Sensing of Turbidity and Chlorophyll *a* Among Nebraska Sand Hills Lakes," *International Journal of Remote Sensing*, vol. 19, no. 8, 1998, pp. 1579–1589.

Gao, B., and A.F.H. Goetz, "Column Atmospheric Water Vapor and Vegetation Liquid Water Retrievals from Airborne Imaging Spectrometer Data," *Journal of Geophysical Research*, vol. 95, no. D4, 1990, pp. 3549–3564.

Gao, B., K.B. Heidebrecht, and A.F.H. Goetz, "Derivation of Scaled Surface Reflectances from AVIRIS Data," *Remote Sensing of Environment*, vol. 44, nos. 2–3, 1993, pp. 165–178.

Goetz, A.H., et al., "Imaging Spectrometry for Earth Remote Sensing," *Science*, vol. 228, no. 4704, June 7, 1985, pp. 1147–1153.

Heinz, D.C., and C.I. Chang, "Fully Constrained Least Squares Linear Spectral Mixture Analysis Method for Material Quantification in Hyperspectral Imagery," *IEEE Transactions on Geoscience and Remote Sensing*, vol. 39, no. 3, 2001, pp. 529–545.

Hepner, G. F., et al., "Investigation of the Integration of AVIRIS and IFSAR for Urban Analysis," *Photogrammetric Engineering and Remote Sensing*, vol. 64, no. 8, 1998, pp. 813–820.

Hill, R.A., et al., "Landscape Modelling Using Integrated Airborne Multi-Spectral and Laser Scanning Data," *International Journal of Remote Sensing*, vol. 23, no. 11, 2002, pp. 2327–2334.

Hoffbeck, J.P., and D.A. Landgrebe, "Classification of Remote Sensing Images Having High Spectral Resolution," *Remote Sensing of Environment*, vol. 57, no. 3, 1996, pp. 119–126.

Hudson, R.D., Jr., *Infrared System Engineering*, Wiley, New York, 1969.

IEEE Transactions on Geoscience and Remote Sensing, Special Issue on Analysis of Hyperspectral Image Data, vol. 39, no. 7, 2001.

International Journal of Remote Sensing, Special Issue on Environmental Research and Systems Development Using Airborne Remote Sensing, vol. 18, no. 9, 1997.

International Journal of Remote Sensing, Special Issue on Imaging Spectroscopy, vol. 65, no. 3, 1998.

Irvine, J.M., et al., "The Detection and Mapping of Buried Waste," *International Journal of Remote Sensing*, vol. 18, no. 7, 1997, pp. 1583–1595.

Kahle, A.B., and A.F.H. Goetz, "Mineralogic Information from a New Airborne Thermal Infrared Multispectral Scanner," *Science*, vol. 222, no. 4619, October 7, 1983, pp. 24–27.

Kant, Y., and K.V.S. Badarinath, "Ground-Based Method for Measuring Thermal Infrared Effective Emissivities: Implications and Perspectives on the Measurement of Land Surface Temperature from Satellite Data," *International Journal of Remote Sensing*, vol. 23, no. 11, 2002, pp. 2179–2191.

Kirkland, L., et al., "First Use of an Airborne Thermal Infrared Hyperspectral Scanner for Compositional Mapping," *Remote Sensing of Environment*, vol. 80, no. 3, 2002, pp. 447–459.

Kruse, F.A., and A.B. Lefkoff, "Hyperspectral Imaging of the Earth's Surface—An Expert System-Based Analysis Approach, *Proceedings of the International Symposium on Spectral Sensing Research*, Maui, Hawaii, November 1992.

Kruse, F.A., K.S. Kierein-Young, and J.W. Boardman, "Mineral Mapping at Cuprite, Nevada with a 63-Channel Imaging Spectrometer," *Photogrammetric Engineering and Remote Sensing*, vol. 56, no. 1, 1990, pp. 83–92.

Lee, C., and J. Bethel, "Georegistration of Airborne Hyperspectral Image Data," *IEEE Transactions on Geoscience and Remote Sensing*, vol. 39, no. 7, 2001, pp. 1347–1351.

Lee, C., et al., "Rigorous Mathematical Modeling of Airborne Pushbroom Imaging Systems," *Photogrammetric Engineering and Remote Sensing*, vol. 66, no. 4, 2000, pp. 385–392.

Lelong, C.D., P.C. Pinet, and H. Poilve, "Hyperspectral Imaging and Stress Mapping in Agriculture: A Case Study on Wheat in Beauce (France)," *Remote Sensing of Environment*, vol. 66, 1998, pp. 179–191.

Lewis, M., V. Jooste, and A.A. de Gasparis, "Discrimination of Arid Vegetation with Airborne Multispectral Scanner Hyperspectral Imagery," *IEEE Transactions on Geoscience and Remote Sensing*, vol. 39, no. 7, 2001, pp. 1471–1479.

Li, Z.L., et al., "Evaluation of Six Methods for Extracting Relative Emissivity Spectra from Thermal Infrared Images," *Remote Sensing of Environment*, vol. 69, no. 3, 1999, pp. 197–214.

Liang, S.L., "An Optimization Algorithm for Separating Land Surface Temperature and Emissivity from Multispectral Thermal Infrared Imagery," *IEEE Transactions on Geoscience and Remote Sensing*, vol. 39, no. 2, 2001, pp. 264–274.

Lo, C.P., D.A. Quattrochi, and J.C. Luvall, "Application of High-Resolution Thermal Infrared Remote Sensing and GIS to Assess the Urban Heat Island Effect," *International Journal of Remote Sensing*, vol. 18, no. 2, 1997, pp. 287–304.

Logicon, *Multispectral Imagery Reference Guide*, Logicon Geodynamics, Fairfax, VA, 1997a.

Logicon, *Multispectral Users Guide, Logicon Geodynamics*, Fairfax, VA, 1997b.

McGwire, K.C., "Mosaicking Airborne Scanner Data with the Multiquadric Rectification Techniques," *Photogrammetric Engineering and Remote Sensing*, vol. 64, no. 6, 1998, pp. 601–606.

Naesset, E., and T. Okland, "Estimating Tree Height and Tree Crown Properties Using Airborne Scanning Laser in a Boreal Nature Reserve," *Remote Sensing of Environment*, vol. 79, no. 1, 2002, pp. 105–115.

Nolin, A.W., and J. Dozier, "A Hyperspectral Method for Remotely Sensing the Grain Size of Snow," *Remote Sensing of Environment*, vol. 74, no. 2, 2000, pp. 207–216.

Puniard, D.J., "The ARIES Project—A Window of Opportunity for Australia," *Proceedings: Land Satellite Information in the Next Decade II: Sources and Applications*, American Society for Photogrammetry and Remote Sensing, Bethesda, MD, 1997, pp. 273–282 (CD-ROM).

Rapier, C.B., and K.J. Michael, "The Calibration of a Small, Low-Cost Thermal Infrared Radiometer," *Remote Sensing of Environment*, vol. 56, no. 2, 1996, pp. 97–103.

Remote Sensing of Environment, Special Issue on Spectral Signatures of Objects, vol. 68, no. 3, 1999.

Resmini, R.G., et al., "Mineral Mapping with Hyperspectral Digital Imagery Collection Experiment (HYDICE) Sensor-Data at Cuprite, Nevada, USA," *International Journal of Remote Sensing*, vol. 18, no. 7, 1997, pp. 1553–1570.

Riano, D., et al., "Assessment of Vegetation Regeneration after Fire through Multitemporal Analysis of AVIRIS Images in the Santa Monica Mountains," *Remote Sensing of Environment*, vol. 79, no.1, 2002, pp. 60–71.

Richter, R., and D. Schlapfer, "Geo-Atmospheric Processing of Airborne Imaging Spectrometry Data. Part 2: Atmospheric/Topographic Correction," *International Journal of Remote Sensing*, vol. 23, no. 13, 2002, pp. 2631–2649.

Richter, R., A. Müller, and V. Heiden, "Aspects of Operational Atmospheric Correction of Hyperspectral Imagery," *International Journal of Remote Sensing*, vol. 23, no. 1, 2002, pp. 145–157.

Roberts, D.A., R.O. Green, and J.B. Adams, "Temporal and Spatial Patterns in Vegetation and Atmospheric Properties from AVIRIS," *Remote Sensing of Environment*, vol. 62, no. 3, 1997, pp. 223–240.

Ryan, M.J., and J.F. Arnold, "Lossy Compression of Hyperspectral Data Using Vector Quantization," *Remote Sensing of Environment*, vol. 61, no. 3, 1997, pp. 419–436.

Salisbury, J.W., and D.M. D'Aria, "Emissivity of Terrestrial Materials in the 8–14 μm Atmospheric Window," *Remote Sensing of Environment*, vol. 42, 1992, pp. 83–106.

Sandmeier, S., et al., "Physical Mechanisms in Hyperspectral BRDF Data of Grass and Watercress," *Remote Sensing of Environment*, vol. 66, 1998, pp. 222–233.

Sathyendranath, S., et al., "Aircraft Remote Sensing of Toxic Phytoplankton Blooms: A Case Study from Cardigan River, Prince Edward Island," *Canadian Journal of Remote Sensing*, vol. 23, no. 1, 1997, pp. 15–23.

Schlapfer, D., and R. Richter, "Geo-Atmospheric Processing of Airborne Imaging Spectrometry Data. Part 1: Parametric Orthorectification," *International Journal of Remote Sensing*, vol. 23, no. 13, 2002, pp. 2609–2630.

Schmugge, T., S.J. Hook, and C. Coll, "Recovering Surface Temperature and Emissivity from Thermal Infrared Multispectral Data," *Remote Sensing of Environment*, vol. 65, no. 2, 1998, pp. 121–131.

Schott, J.R., *Remote Sensing: The Image Chain Approach*, Oxford University Press, New York, 1997.

Secker, J., et al., "Vicarious Calibration of Airborne Hyperspectral Sensors in Operational Environments," *Remote Sensing of Environment*, vol. 76, no. 1, 2001, pp. 81–92.

Simon, I., *Infrared Radiation*, Litton Educational, New York, 1966.

Staenz, K., "Classification of a Hyperspectral Agricultural Data Set Using Band Moments for Reduction of the Spectral Dimensionality," *Canadian Journal of Remote Sensing*, vol. 22, no. 3 1996, pp. 248–257.

Strachan, I.B., E. Pattey, and J.B. Boisvert, "Impact of Nitrogen and Environmental Conditions on Corn as Detected by Hyperspectral Reflectance," *Remote Sensing of Environment*, vol. 80, no. 2, 2002, pp. 213–224.

Thenkabail, P.S., R.B. Smith, and E. De Pauw, "Hyperspectral Vegetation Indices and Their Relationships with Agricultural Crop Characteristics," *Remote Sensing of Environment*, vol. 71, no. 2, 2000, pp. 158–182.

Thenkabail, P.S., R.B. Smith, and E. De Pauw, "Evaluation of Narrowband and Broadband Vegetation Indices for Determining Optimal Hyperspectral Wavebands for Agricultural Crop Characterization," *Photogrammetric Engineering and Remote Sensing*, vol. 68, no. 6, 2002, pp. 607–621.

Theobold, E.G., "When Is a Meter a Meter?" *Imaging Notes*, Space Imaging, Thornton, CO, September/October 1998.

Torgersen, C.E., et al., "Airborne Thermal Remote Sensing for Water Temperature Assessment in Rivers and Streams," *Remote Sensing of Environment*, vol. 76, no. 3, 2001, pp. 386–398.

Tsai, F., and W. Philpot, "Derivative Analysis of Hyperspectral Data," *Remote Sensing of Environment*, vol. 66, 1998, pp. 41–51.

Tsai, F., and W.D. Philpot, "A Derivative-Aided Hyperspectral Image Analysis System for Land-Cover Classification," *IEEE Transactions on Geoscience and Remote Sensing*, vol. 40, no. 2, 2002, pp. 416–425.

Van der Meer, F.D., and S.M. de Jong, eds., *Imaging Spectrometry: Basic Principles and Prospective Applications*, Kluwer Academic Publishers, Boston, 2001.

Vane, G., "High Spectral Resolution Remote Sensing of the Earth," *Sensors*, 1985, no. 2, pp. 11–20.

Vane, G., and A.F.H. Goetz, "Terrestrial Imaging Spectrometry," *Remote Sensing of Environment*, vol. 24, 1988, pp. 1–29.

Wolfe, W.L., and G.J. Zissis, eds., *The Infrared Handbook*, U.S. Government Printing Office, Washington, DC, 1978.

6 EARTH RESOURCE SATELLITES OPERATING IN THE OPTICAL SPECTRUM

6.1 INTRODUCTION

Probably no combination of two technologies has generated more interest and application over a wider range of disciplines than the merger of remote sensing and space exploration. Although many aspects of the process are still in the developmental stage, studying the earth from space has evolved from the realm of pure research to that of worldwide, day-to-day application. Currently we depend on spaceborne sensors to assist in tasks ranging from weather prediction, crop forecasting, and mineral exploration to applications as diverse as pollution detection, rangeland monitoring, and commercial fishing. All this has happened in a very short period of time, and the status of remote sensing from space continues to change as new and/or improved spacecraft are placed into earth orbit.

Without question, the most important outcome of development in space exploration and remote sensing has been the role these technologies have played in conceiving of the earth as a system. Space remote sensing has brought a new dimension to understanding not only the natural wonders and processes operative on our planet but also the impacts of humankind on earth's fragile and interconnected resource base.

397

This chapter emphasizes satellite systems that operate within the *optical spectrum*, which extends from approximately 0.3 to 14 μm. This range includes UV, visible, near-, mid-, and thermal IR wavelengths. (It is termed the optical spectrum because lenses and mirrors can be used to refract and reflect such energy.) Substantial remote sensing from space is also performed using systems that operate in the *microwave* portion of the spectrum (approximately 1 mm to 1 m wavelength). Microwave remote sensing is the subject of Chapter 8.

The subject of remote sensing from space is a rapidly changing one, with numerous countries and commercial firms developing and launching new systems on a regular basis. In this dynamic environment, it is more important than ever that potential users of data from spaceborne remote sensing systems understand the basic design characteristics of these systems and the various trade-offs that determine whether a particular sensor will be suitable for a given application. Hence, we begin this discussion with an overview of the general characteristics common to all satellite systems. We then examine a wide range of satellite systems, with reference to the ways their designs exemplify the fundamental characteristics discussed in this section.

Although we aim to be current in our coverage of spaceborne remote sensing systems, the rapid pace of change in this field suggests that some of the material in this chapter about certain specific systems will likely be dated rapidly. At the same time, as researchers and practitioners focus on regional to global change issues, knowledge about the characteristics of many of the early systems that are no longer in operation is essential. Thus, after a brief historical overview of the early development of remote sensing from space, we treat the U.S. Landsat and French SPOT satellite systems in some degree of detail. These were the first and most robust global monitoring systems to acquire moderate resolution data on a systematic basis. Also, many of their principles of operation apply to the numerous other systems available today and planned for the future.

After our discussion of the Landsat and SPOT satellite systems and their future counterparts, we describe briefly numerous other earth resources satellites, including high resolution systems and hyperspectral sensors. We then discuss meteorological satellites, ocean monitoring satellites, the Earth Observing System, and space station remote sensing. Depending upon one's academic or work setting, one may wish to "skip over" certain of the sections in this chapter. We have written this discussion with this possibility in mind.

There are many characteristics that describe any given satellite remote sensing system and determine whether or not it will be suitable for a particular application. Among the most fundamental of these characteristics is the satellite's orbit. A satellite in orbit about a planet moves in an elliptical path with the planet at one of the foci of the ellipse. Important elements of the orbit include its *altitude, period, inclination,* and *equatorial crossing time*. For most earth observation satellites, the orbit is approximately circular, with altitudes more than 400 km above the earth's surface. The period of a satellite, or time to complete one orbit, is related to its altitude according to the formula (Elachi, 1987)

$$T_o = 2\pi(R_p + H') \sqrt{\frac{R_p + H'}{g_s R_p^2}} \tag{6.1}$$

where

T_o = orbital period, sec

R_p = planet radius, km (about 6380 km for earth)

H' = orbit altitude (above planet's surface), km

g_s = gravitational acceleration at planet's surface (0.00981 km/sec^2 for earth)

The inclination of a satellite's orbit refers to the angle at which it crosses the equator. An orbit with an inclination close to 90° is referred to as near polar because the satellite will pass near the north and south poles on each orbit. An equatorial orbit, in which the spacecraft's ground track follows the line of the equator, has an inclination of 0°. Two special cases are *sun-synchronous* orbits and *geostationary* orbits. A sun-synchronous orbit results from a combination of orbital period and inclination such that the satellite keeps pace with the sun's westward progress as the earth rotates. Thus, the satellite always crosses the equator at precisely the same local sun time (the local clock time varies with position within each time zone). A geostationary orbit is an equatorial orbit at the altitude (approximately 36,000 km) that will produce an orbital period of exactly 24 hr. A geostationary satellite thus completes one orbit around the earth in the same amount of time needed for the earth to rotate once about its axis and remains in a constant relative position over the equator.

The above elements of a satellite's orbit are illustrated in Figure 6.1, for the Landsat-4 and -5 satellites. As discussed in Section 6.5, the Landsat-4 and -5 satellites were launched into circular, near-polar orbits at an altitude of 705 km above the earth's surface. Solving Eq. 6.1 for this altitude yields an orbital period of about 98.9 min, which corresponds to approximately 14.5 orbits per day. The orbital inclination chosen for Landsat-4 and -5 was 98.2°, resulting in a sun-synchronous orbit. (The orbit for Landsat-7 is nearly identical to that of Landsat-4 and -5).

Another important characteristic of a satellite remote sensing system is its spatial resolution. The principles discussed in Chapter 5 that determine the spatial resolution for airborne sensors also apply to spaceborne systems. Thus, at nadir the spatial resolution is determined by the orbital altitude and the instantaneous field of view, while off nadir the ground resolution cell size increases in both the along-track and across-track dimensions. However, for spaceborne systems the curvature of the earth further degrades the spatial resolution for off-nadir viewing. This becomes particularly important for pointable systems and for systems with a wide total field of view.

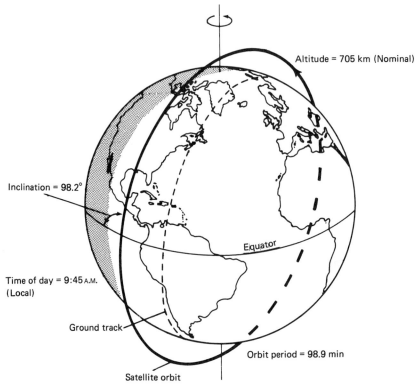

Figure 6.1   Sun-synchronous orbit of Landsat-4 and -5. (Adapted from NASA diagram.)

The spectral characteristics of a satellite remote sensing system include the number, width, and position of the sensor's spectral bands. The extreme cases are panchromatic sensors, with a single broad spectral band, and hyperspectral sensors, with a hundred or more narrow, contiguous spectral bands. During sensor calibration, the relative sensitivity of each band as a function of wavelength may be determined.

The radiometric properties of a remote sensing system include the radiometric resolution, typically expressed as the number of data bits used to record the observed radiance. They also include the gain setting (Section 7.2) that determines the range of variation in brightness over which the system will be sensitive, and the signal-to-noise ratio of the sensor. In some cases, the gain setting on one or more bands may be programmable from the ground, to permit the acquisition of data under different conditions, such as over the dark ocean and the bright polar ice caps. It should be noted that increasing the spatial and spectral resolution both result in a decrease in the energy available to be sensed. A high spatial resolution means that each detector is receiving energy from a smaller area, while a high spectral resolution means that each detector is receiving energy in a narrower range of wavelengths. Thus, there are trade-offs among the spatial, spectral, and radiometric properties of a spaceborne remote sensing system.

Some other factors to consider in evaluating a satellite system include its coverage area, revisit period (time between successive coverages), and off-nadir imaging capabilities. For sensors that cannot be steered to view off nadir, the swath width is determined by the orbital altitude and the sensor's total field of view, while the revisit period is determined by the orbital period and the swath width. For example, the Landsat-7 ETM+ and Terra/Aqua MODIS instruments are on satellite platforms that share the same orbital altitude and period, but MODIS has a much wider field of view. This gives it a much greater swath width, which in turn means that any given point on the earth will appear in MODIS imagery more frequently. If across-track pointing is used, both the swath width and revisit period will be affected. For certain time-sensitive applications, such as monitoring the effects of flooding, fires, and other natural disasters, a frequent revisit cycle is particularly important. Either across-track or along-track pointing can be used to collect stereoscopic imagery, which can assist in image interpretation and topographic analysis. The primary advantage of along-track pointing for stereoscopic image acquisition is that both images of the stereopair will be acquired within a very brief period of time, typically no more than seconds or minutes apart. Across-track stereoscopic images are often acquired several days apart (with the attendant changes in atmospheric and surface conditions between dates).

Many other practical considerations determine the potential utility of any given satellite remote sensing system in any given application context. These include such factors as the procedures for tasking the satellite, data delivery options, data archiving, license/copyright restrictions, and data cost. These characteristics, many of which are often overlooked by prospective users of satellite imagery, have the potential to radically alter the usability of the imagery from any given system.

6.2 EARLY HISTORY OF SPACE IMAGING

Remote sensing from space received its first impetus through remote sensing from rockets. As early as 1891, a patent was granted to Ludwig Rahrmann of Germany for a "New or Improved Apparatus for Obtaining Bird's Eye Photographic Views." The apparatus was a rocket-propelled camera system that was recovered by parachute. By 1907, another German, Alfred Maul, had added the concept of gyrostabilization to rocket–camera systems. In 1912, he successfully boosted a 41-kg payload containing a 200 × 250-mm format camera to a height of 790 m (ASP, 1983).

Space remote sensing began in earnest during the period 1946 to 1950 when small cameras were carried aboard captured V-2 rockets that were fired from the White Sands Proving Ground in New Mexico. Over the succeeding years, numerous flights involving photography were made by rockets, ballistic missiles, satellites, and manned spacecraft. However, the photographs produced during early space flights were generally of inferior quality because early missions were made primarily for purposes other than photography. But crude as they were by today's standards, the early photographs demonstrated the potential value of remote sensing from space.

In many respects, the initial efforts aimed at imaging the earth's surface from space were rather incidental outgrowths of the development of meteorological satellites. Beginning with the first Television and Infrared Observation Satellite (TIROS-1) in 1960, early weather satellites returned rather coarse views of cloud patterns and virtually indistinct images of the earth's surface. With refinements in the imaging sensors aboard the meteorological satellites, images of both atmospheric and terrestrial features became more distinct. Eventually, meteorologists began intensive study of surface areas to collect data on water, snow, and ice features. Looking *through*, not just *at*, the earth's atmosphere was becoming possible.

Also beginning in 1960 was an early U.S. military space imaging reconnaissance program, called Corona. This program went through many developments during its lifetime (its final mission was flown in 1972), but the entire effort was classified until 1995 (ASPRS, 1997a).[1] Consequently, the exciting future for remote sensing from space only became apparent to the civilian community as part of the manned space programs of the 1960s: Mercury, Gemini, and Apollo. On May 5, 1961, Alan B. Shepard, Jr., made a 15-min suborbital Mercury flight on which 150 excellent photographs were taken. These pictures were shot with an automatic Mauer 70-mm camera. Because of the trajectory of Shepard's flight, the photographs showed only sky, clouds, and ocean, but the images did indeed substantiate Shepard's statement, "What a beautiful view." On February 20, 1962, John Glenn, Jr., made three historic orbits around the earth and took 48 color photographs during Mercury mission MA-6. The photographs were taken on color negative film with a 35-mm camera and showed mostly clouds and water, although several pictured the deserts of northwest Africa. On later Mercury missions, color pho-

---

[1]Since 1995, the U.S. government has released two sets of formerly classified imagery taken from photo-reconnaissance satellites. *Declass-1* imagery includes approximately 880,000 frames of photography from the years 1959–1972. *Declass-2* imagery includes approximately 50,000 images taken from 1963–1980. Much of this declassified imagery was taken with various panoramic camera systems. The geographic areas covered by photographs in the Declass-1 and -2 archives are quite extensive, though the images do not provide systematic or complete coverage of the earth. One of the most valuable contributions of these data today is to provide a record of space observations prior to the launch of Landsat-1 in 1972. Historical trends in coastal erosion, glacier movement, urbanization, land use and land cover change, and numerous other time-sensitive phenomena can be studied using these images. Additional information about these data sources may be obtained by consulting the relevant URLs listed in Appendix B.

tographs were taken with 70-mm Hasselblad cameras. A specially modified Hasselblad camera, with an 80-mm lens, soon became the standard for the photographic experiments conducted in the Gemini program. Mission GT-4 of this program included the first formal photographic experiment from space specifically directed at geology. Coverage included nearly vertical overlapping photographs of the southwestern United States, northern Mexico, and other areas of North America, Africa, and Asia. These images soon led to new and exciting discoveries in tectonics, volcanology, and geomorphology.

With the success of the Gemini GT-4 photographic experiments in geology, subsequent missions included a host of similar experiments aimed at investigating various geographic and oceanographic phenomena. Photography comparable to that of the GT-4 experiments was acquired over areas extending between approximately 32° north and south latitudes. Each image had a nominal scale of about 1 : 2,400,000 and included about 140 km on a side. By the end of the Gemini program, some 1100 high quality color photographs had been taken for earth resource applications, and the value of remote sensing from space had become well recognized. Serious thinking began about systematic, repetitive image coverage of the globe.

The scientific community's knowledge and experience with space photography was further extended with the Apollo program. One of the Apollo earth orbit flights (Apollo 9) made prior to the lunar landings included the first controlled experiment involving the acquisition of *multispectral* orbital photography for earth resource studies. A four-camera array of electrically driven and triggered 70-mm Hasselblad cameras was used in the experiment. Photographs were produced using panchromatic film with green and red filters, black and white IR film, and color IR film. Some 140 sets of imagery were thus obtained over the course of 4 days. The imagery covered parts of the southwestern, south central, and southeastern United States as well as parts of Mexico and the Caribbean–Atlantic area.

In 1973, Skylab, the first American space workshop, was launched and its astronauts took over 35,000 images of the earth with the *Earth Resources Experiment Package* (*EREP*) on board. The EREP included a six-camera multispectral array, a long focal length "earth terrain" camera, a 13-channel multispectral scanner, a pointable spectroradiometer, and two microwave systems. The EREP experiments were the first to demonstrate the complementary nature of photography and electronic imaging from space.

Another early (1975) space station experiment having a remote sensing component was the joint *U.S.–USSR Apollo–Soyuz Test Project* (*ASTP*). Regrettably—because earth resource imaging was not a primary goal of this venture—hand-held 35- and 70-mm cameras were again used. For various reasons, the overall quality of most of the images from the ASTP was disappointing. However, like Skylab, the ASTP mission demonstrated that trained crew members could obtain useful, and sometimes unique, earth resource data from visual observation and discretionary imaging. The results of training crew members to look for specific earth resource phenomena and selectively record important events crystallized the complementary nature of manned and unmanned observation systems.

6.3 LANDSAT SATELLITE PROGRAM OVERVIEW

With the exciting glimpses of earth resources being provided by the early meteorological satellites and the manned spacecraft missions, NASA, with the cooperation of the U.S. Department of the Interior, began a conceptual study of the feasibility of a series of *Earth Resources Technology Satellites* (*ERTSs*). Initiated in 1967, the program resulted in a planned sequence of six satellites that were given before-launch designations ERTS-A, -B, -C, -D, -E, and -F. (After a successful launch into prescribed orbits, they were to become ERTS-1, -2, -3, -4, -5, and -6.)

ERTS-1 was launched by a Thor-Delta rocket on July 23, 1972, and it operated until January 6, 1978. The platform used for the ERTS-1 sensors was a Nimbus weather satellite, modified for the ERTS mission objectives. It represented the first unmanned satellite specifically designed to acquire data about earth resources on a systematic, repetitive, medium resolution, multispectral basis. It was primarily designed as an *experimental* system to test the *feasibility* of collecting earth resource data from unmanned satellites. All data would be collected in accordance with an *"open skies"* principle, meaning there would be nondiscriminatory access to data collected anywhere in the world. All nations of the world were invited to take part in evaluating ERTS-1 data, and the results of this worldwide experimentation with the system were overwhelmingly favorable. In fact, these results probably exceeded most of the expectations of the scientific community. About 300 individual ERTS-1 experiments were conducted in 43 U.S. states and 36 nations.

Just prior to the launch of ERTS-B on January 22, 1975, NASA officially renamed the ERTS program the "Landsat" program (to distinguish it from the planned Seasat oceanographic satellite program). Hence, ERTS-1 was retroactively named Landsat-1 and all subsequent satellites in the series carried the Landsat designation. As of the time of this writing (2002), six Landsat satellites have been launched successfully, namely Landsat-1 to -5 and Landsat-7. Landsat-6 suffered a launch failure.

Table 6.1 highlights the characteristics of the Landsat-1 through -7 missions. It should be noted that five different types of sensors have been included in various combinations on these missions. These are the *Return Beam Vidicon* (*RBV*), the *Multispectral Scanner* (*MSS*), the *Thematic Mapper* (*TM*), the *Enhanced Thematic Mapper* (*ETM*), and the *Enhanced Thematic Mapper Plus* (*ETM+*). Table 6.2 summarizes the spectral sensitivity and spatial resolution of each of these systems as included on the various missions.

TABLE 6.1 Characteristics of Landsat-1 to -7 Missions

Satellite	Launched	Decommissioned	RBV Bands	MSS Bands	TM Bands	Orbit
Landsat-1	July 23, 1972	January 6, 1978	1–3 (simultaneous images)	4–7	None	18 days/900 km
Landsat-2	January 22, 1975	February 25, 1982	1–3 (simultaneous images)	4–7	None	18 days/900 km
Landsat-3	March 5, 1978	March 31, 1983	A–D (one-band side-by-side images)	4–8[a]	None	18 days/900 km
Landsat-4	July 16, 1982[b]	—	None	1–4	1–7	16 days/705 km
Landsat-5	March 1, 1984	—	None	1–4	1–7	16 days/705 km
Landsat-6	October 5, 1993	Failure upon launch	None	None	1–7 plus panchromatic band (ETM)	16 days/705 km
Landsat-7	April 15, 1999	—	None	None	1–7 plus panchromatic band (ETM+)	16 days/705 km

[a]Band 8 (10.4–12.6 μm) failed shortly after launch.
[b]TM data transmission failed in August 1993.

TABLE 6.2   Sensors Used on Landsat-1 to -7 Missions

Sensor	Mission	Sensitivity (μm)	Resolution (m)
RBV	1, 2	0.475–0.575	80
		0.580–0.680	80
		0.690–0.830	80
	3	0.505–0.750	30
MSS	1–5	0.5–0.6	79/82[a]
		0.6–0.7	79/82[a]
		0.7–0.8	79/82[a]
		0.8–1.1	79/82[a]
	3	10.4–12.6[b]	240
TM	4, 5	0.45–0.52	30
		0.52–0.60	30
		0.63–0.69	30
		0.76–0.90	30
		1.55–1.75	30
		10.4–12.5	120
		2.08–2.35	30
ETM[c]	6	Above TM bands plus 0.50–0.90	30 (120 m thermal band) 15
ETM+	7	Above TM bands plus 0.50–0.90	30 (60 m thermal band) 15

[a]79 m for Landsat-1 to -3 and 82 m for Landsat-4 and -5.
[b]Failed shortly after launch (band 8 of Landsat-3).
[c]Landsat-6 launch failure.

Because Landsat-1, -2, and -3 were so similar in their operation, as were Landsat-4 and -5, it is convenient to discuss these systems as two distinct groups, followed by discussion of the Landsat-6 planned mission and Landsat-7.

6.4   LANDSAT-1, -2, AND -3

Figure 6.2 illustrates the basic configuration of Landsat-1, -2, and -3. These butterfly-shaped systems were about 3 m tall and 1.5 m in diameter, with solar panels extending to about 4 m. The satellites weighed about 815 kg and were launched into circular orbits at a nominal altitude of 900 km (the altitude varied between 880 and 940 km) that passed within 9° of the North and South Poles.

Figure 6.3 shows the north-to-south ground traces of the satellite orbits for a single day. Note that they cross the equator at an angle of about 9° from normal, and successive orbits are about 2760 km apart at the equator. Because the sensors aboard the satellite imaged only a 185-km swath, there are

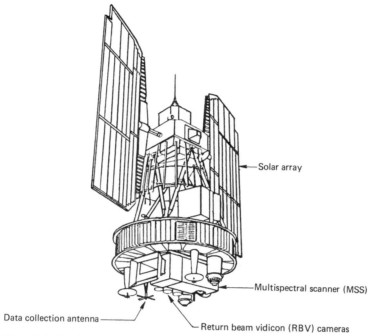

Solar array

Multispectral scanner (MSS)

Data collection antenna

Return beam vidicon (RBV) cameras

Figure 6.2 Landsat-1, -2, and -3 observatory configuration. (Adapted from NASA diagram.)

large gaps in image coverage between successive orbits on a given day. However, with each new day the satellite orbit progressed slightly westward, just overshooting the orbit pattern of the previous day. (See orbit 15 in Figure 6.3.) This satellite orbit–earth rotation relationship thus yielded images that overlap those of the previous day. The overlap is a maximum at 817 north and south latitudes (about 85 percent) and a minimum (about 14 percent) at the equator. Figure 6.4 shows the set of orbital paths covering the conterminous United States. It took 18 days for the orbit pattern to progress westward to the point of coverage repetition. Thus, the satellites had the capability of covering the globe (except the 82° to 90° polar latitudes) once every 18 days, or about 20 times per year. The satellite orbits were corrected occasionally to compensate for orbital precession caused by atmospheric drag. This ensured that repetitive image centers were maintained to within about 37 km.

At the 103-min orbital period, the 2760-km equatorial spacing between successive orbits caused the satellites to keep precise pace with the sun's westward progress as the earth rotated. As a result, the satellite always crossed the equator at precisely the same local *sun* time (the local clock time varied with location within a time zone). Again, such orbits are referred to as *sun synchronous*.

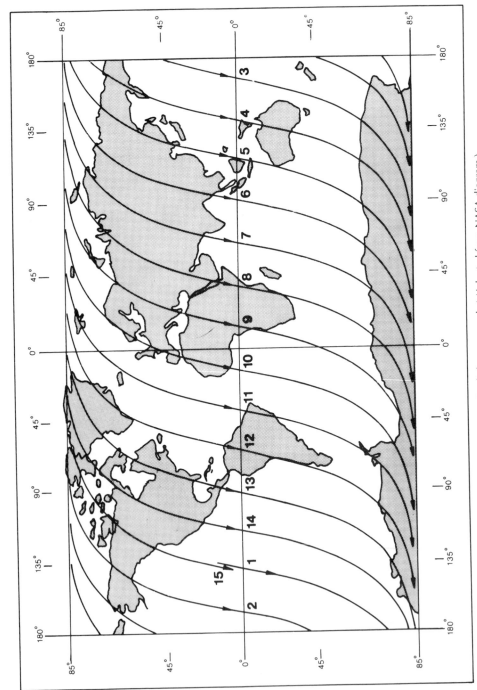

Figure 6.3 Typical Landsat-1, -2, and -3 daily orbit pattern. (Daylight passes only.) (Adapted from NASA diagram.)

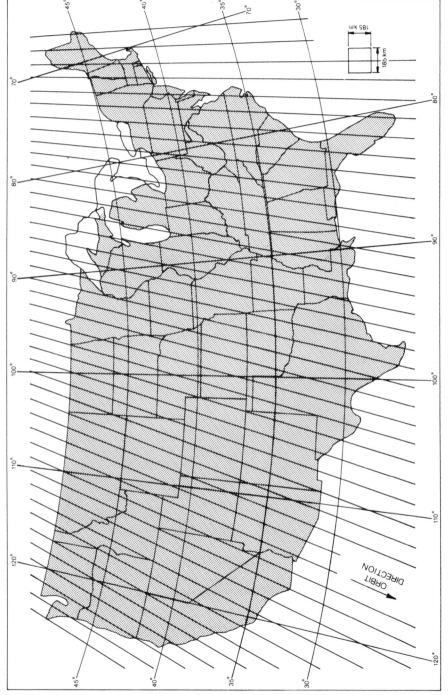

Figure 6.4 Landsat-1, -2, and -3 orbital passes over the conterminous United States. (Adapted from NASA diagram.)

409

Landsat-1, -2, and -3 were launched into orbits that crossed the equator at 9:42 A.M. local sun time on each pass; however, orbital perturbations caused the crossing times to vary somewhat. This time was selected to take advantage of early morning skies that are generally clearer than later in the day. Because the system's orbital velocity was constant, all other points in its orbit were also passed at a relatively constant local sun time, either slightly after 9:42 A.M. in the Northern Hemisphere or slightly before in the Southern Hemisphere. The important implication of the sun-synchronous orbit is that it ensures repeatable sun illumination conditions during the specific seasons. Repeatable illumination conditions are desirable when mosaicking adjacent tracks of imagery and comparing annual changes in land cover.

Although sun-synchronous orbits ensure repeatable illumination conditions, these conditions vary with location and season. That is, the sun's rays strike the earth at varying solar elevation angles as a function of both latitude and time. For example, the sun's rays strike Sioux Falls, South Dakota, at approximately 20° in December and at 60° in July. Along a single January orbit, the solar elevation changes from 4° in Alaska to 45° near the equator. Likewise, the azimuth direction of solar illumination changes with season and latitude. In short, a sun-synchronous orbit does not compensate for changes in solar altitude, azimuth, or intensity. These factors are always changing and are compounded by variations in atmospheric conditions between scenes.

Landsat-1 and -2 were launched with two identical remote sensing systems onboard: (1) a three-channel RBV system and (2) a four-channel MSS system. The RBV system consisted of three television-like cameras aimed to view the same 185 × 185-km ground area simultaneously. The nominal ground resolution of the cameras was about 80 m, and the spectral sensitivity of each camera was essentially akin to that of a single layer of color IR film: 0.475 to 0.575 μm (green), 0.580 to 0.680 μm (red), and 0.690 to 0.830 μm (near IR). These bands were designated as bands 1, 2, and 3. The RBVs did not contain film, but instead their images were exposed by a shutter device and stored on a photosensitive surface within each camera. This surface was then scanned in raster form by an internal electron beam to produce a video signal just as in a conventional television camera.

Because RBVs image an entire scene instantaneously, in camera fashion, their images have greater inherent cartographic fidelity than those acquired by the Landsat MSS. Also, the RBVs contained a *reseau grid* in their image plane to facilitate geometric correction of the imagery. This resulted in an array of tick marks being precisely placed in each image. By knowing the observed image position versus the theoretical calibration position of these marks, almost all image distortion can be compensated for in the image recording process.

The RBV on Landsat-1 produced only 1690 scenes between July 23 and August 5, 1972, when a tape recorder switching problem (malfunctioning relay switch) forced a system shutdown. The RBV on Landsat-2 was operated primarily for engineering evaluation purposes and only occasional RBV imagery was obtained, primarily for cartographic uses in remote areas. Two major changes were introduced in the design of the RBV system onboard Landsat-3: The system sensed in a single broad band rather than multispectrally, and the spatial resolution of the system was improved by a factor of about 2.6 compared with the previous RBVs. The spectral sensitivity range of the system was 0.505 to 0.750 μm (green to near IR). The change to 30 m nominal ground resolution was achieved by doubling the focal length of the camera lens system, decreasing the exposure time to reduce image motion during exposure, and maintaining adequate exposure by removing the spectral filters of the previous RBVs. To compensate for the decrease in the ground area covered by doubling the focal length, a two-camera side-by-side configuration was employed. The two cameras were aligned to view adjacent 98-km-square ground scenes with a 13-km sidelap and a 17-km endlap, yielding a 183 × 98-km scene pair (Figure 6.5). Two successive scene pairs coincided nominally with one MSS scene. The four RBV scenes that filled each MSS scene were designated A, B, C, and D. Figure 6.6 shows one frame of RBV imagery from Landsat-3.

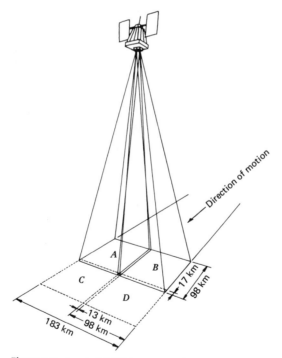

Figure 6.5   Landsat-3 RBV system configuration. (Adapted from NASA diagram.)

Figure 6.6 Landsat-3 RBV image, Cape Canaveral, FL. Scale 1 : 500,000.

While not intended, the RBV systems onboard Landsat-1, -2, and -3 became secondary sources of data in comparison to the MSS systems flown on these satellites. Two factors contributed to this situation. First, RBV operations were plagued with various technical malfunctions. More importantly, the MSS systems were the first global monitoring systems capable of producing multispectral data in a digital format. The advantages in being able to process the MSS data by computer led to their widespread application during the Landsat-1, -2, and -3 era. Several tens of billions of square kilometers of the earth's surface (unique in time but repetitive in area) were imaged by the MSS systems onboard these satellites.

The MSS onboard Landsat-1, -2, and -3 covered a 185-km swath width in four wavelength bands: two in the visible spectrum at 0.5 to 0.6 μm (green) and 0.6 to 0.7 μm (red) and two in the near IR at 0.7 to 0.8 μm and 0.8 to 1.1 μm. These bands were designated as bands 4, 5, 6, and 7. The MSS onboard Landsat-3 also incorporated a thermal band (band 8) operating in the region 10.4 to 12.6 μm. However, operating problems caused this channel to fail shortly after launch. Thus, all three MSS systems effectively produced data in the same four bands. In fact, the identical bands were also used in the MSS systems flown on Landsat-4 and -5 but were designated as bands 1, 2, 3, and 4. In this discussion, we use only the Landsat-1 to -3 MSS band numbers (4, 5, 6, and 7). In Figure 6.7 these bands of operation are compared with the spectral bands associated with color and color IR film.

The MSS operating configuration is shown in Figure 6.8. The instantaneous field of view (IFOV) of the scanner is square and results in a ground resolution cell of approximately 79 m on a side. The total field of view scanned is approximately 11.56°. Because this angle is so small (compared to 90 to 120° in airborne scanners), an oscillating, instead of spinning, scan mirror is employed. The mirror oscillates once every 33 msec. Six contiguous lines are scanned

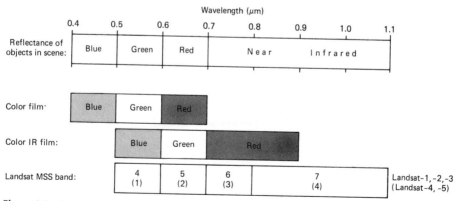

Figure 6.7 Spectral sensitivity of the four Landsat MSS bands compared with the spectral sensitivity of the three emulsion layers used in color and color IR film.

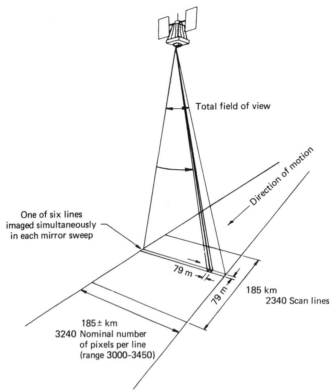

Total field of view

Direction of motion

One of six lines
imaged simultaneously
in each mirror sweep

79 m

79 m

185 km
2340 Scan lines

185± km
3240 Nominal number
of pixels per line
(range 3000–3450)

Figure 6.8 Landsat MSS operating configuration. (Adapted from NASA diagram.)

simultaneously with each mirror oscillation. This permits the ground coverage rate to be achieved at one-sixth the single-line scan rate, resulting in improved system response characteristics. This arrangement requires four arrays (one for each band) of six detectors each (one for each line). When not viewing the earth, the detectors are exposed to internal light and sun calibration sources.

The analog signal from each detector is converted to digital form by an onboard A-to-D converter. A digital number range of 0 to 63 (6 bits) is used for this purpose. These data are then scaled to other ranges during subsequent ground-based processing. (Normally, bands 4 to 6 are scaled to a range of 0 to 127 and band 7 is scaled to 0 to 63.)

The A-to-D converter samples the output of the detectors about 100,000 times a second, resulting in a ground sampled distance of 56 m between readings. Because of this spacing, the image values form a matrix of 56 × 79-m cells (as shown in Figure 6.9). Note, however, that the brightness value for each pixel is actually derived from the full 79 × 79-m-ground-resolution cell (shaded area in Figure 6.9).

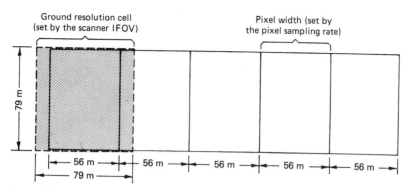

Figure 6.9 Ground resolution cell size versus MSS pixel size. (Adapted from Taranik, 1978.)

The MSS scans each line from west to east with the southward motion of the spacecraft providing the along-track progression of the scan lines. Each Landsat MSS scene is "framed" from the continuous MSS data swath so that it covers approximately a 185 × 185-km area with 10 percent endlap between successive scenes. A nominal scene consists of some 2340 scan lines, with about 3240 pixels per line, or about 7,581,600 pixels per channel. With four spectral observations per pixel, each image data set contains over 30 million observations.

Figure 6.10 is a full-frame, band 5 Landsat MSS scene covering a portion of central New York. Note that the image area is a parallelogram, not a square, because of the earth's rotation during the 25 sec it takes the satellite to travel from the top of the scene to the bottom. The tick marks and numbers around the margins of this image refer to an approximate latitude-and-longitude (degrees and minutes) grid for the image. At the bottom of the image is a step wedge containing 15 steps corresponding to the full potential range of brightness values detected by the MSS. Not all steps are visible on this image because only a limited portion of the full scale was used in printing this scene.

Above the step wedge is an annotation block giving specific information about the acquisition of this image. For Figure 6.10, the block shows, from left to right, the date (10JUN75); the latitude and longitude of the center of the image in degrees and minutes (N43–11/W075–36); the latitude and longitude of the ground point directly beneath the satellite (the nadir) (N43–08/W075–29)—the discrepancy between this location and the center indicates a slight degree of tilt in the image; the sensor and band (MSS 5); the reception mode (D), specifying direct versus recorded; sun elevation and azimuth to the nearest degree (SUN EL57 AZ117); various orbital and processing parameters (191–4668-N-I-N-D-2L); identification as Landsat satellite

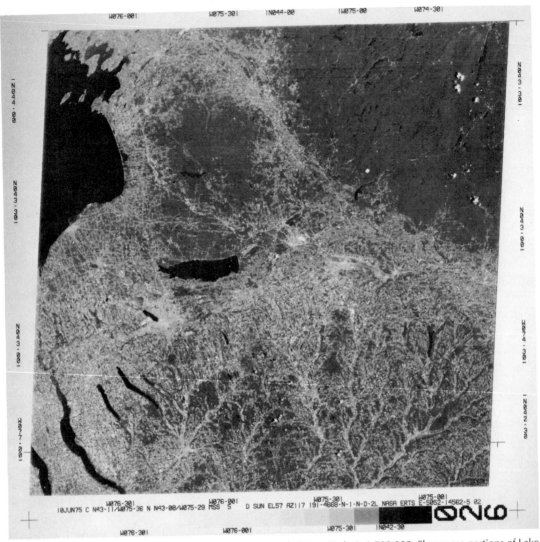

Figure 6.10 Full-frame, band 5 (red), Landsat MSS scene, central New York. 1 : 1,700,000. Shown are portions of Lake Ontario (upper left), Adirondack Mountains (upper right), and Finger Lakes region (lower left).

(NASA ERTS); and a unique scene identification number (E-5052–14562–5). The exact format of the annotation block has changed periodically through the course of the Landsat program.

In addition to black and white images of single bands, *color composites* can be generated for any MSS data set by printing three MSS bands in registration onto color film or by displaying the three-band data digitally. Generally, band 4 is displayed in blue, band 5 in green, and band 7 in red. This combination simulates the color rendition of color IR film (Figure 6.7).

The distribution of Landsat data in the United States has gone through four distinct phases: experimental, transitional, commercial, and governmental [as part of NASA's Earth Science Enterprise Program (Section 6.18)]. During the experimental phase of Landsat-1, -2, and -3 all imagery and computer-compatible tapes were disseminated by the Earth Resources Observation System (EROS) Data Center at Sioux Falls, South Dakota. The satellites were operated by NASA and the data distribution process was operated by the USGS within the Department of the Interior. Gradually, all operations were assumed by the National Oceanic and Atmospheric Administration (NOAA) within the U.S. Department of Commerce during the transitional period. During this period, the operation of the Landsat program was transferred from the federal government to a commercial firm—the Earth Observation Satellite Company (EOSAT). This transfer was provided for within the Land Remote Sensing Commercialization Act of 1984. With the launch of Landsat-7 on April 15, 1999, operation of the Landsat program reverted to the government (Section 6.7) and the EROS Data Center assumed the role of processing and distributing the data.

As the above changes in the dissemination of Landsat data were being made, several technical improvements were also being made in the manner in which Landsat data were processed prior to distribution. Hence, the precise form of Landsat-1, -2, and -3 data products varied considerably over the course of time. For example, digital MSS data supplied in computer-compatible tape (CCT) format after 1979 were resampled into pixels having a nominal dimension of 57×57 m (compared to the 56×79-m size used previously).

During the Landsat-1, -2, and -3 era, several countries throughout the world established data-receiving stations. The precise form of data products produced at these facilities likewise varied considerably. Accordingly, prospective users of data from these early missions are advised to closely investigate the exact form of data processing employed to produce the products they may wish to analyze. Also, prospective users of Landsat-3 data obtained between 1979 and 1983 should be aware that the MSS developed a scanning line-start synchronization problem in early 1979. This resulted in loss of all (or portions) of the data from the western 30 percent of each scene. The remaining 70 percent of each scene was normal.

Landsat-1, -2, and -3 images are cataloged according to their location within the *Worldwide Reference System* (*WRS*). In this system each orbit within a cycle is designated as a path. Along these paths, the individual nominal sensor frame centers are designated as rows. Thus, a scene can be uniquely defined by specifying a path, a row, and a date. The WRS for Landsat-1, -2, and -3 has 251 paths corresponding to the number of orbits required to cover the earth in one 18-day cycle. Paths are numbered from 001 to 251, east to west. The rows are numbered so that row 60 coincides with the equator on the orbit's descending node. The U.S. archive of Landsat-1, -2, and -3 data contains 500,000 MSS and RBV scenes. The worldwide database for MSS and RBV data contains some 1.3 million scenes. These data represent an irreplaceable resource for long term global change studies.

6.5 LANDSAT-4 AND -5

Landsat-4 and -5, like their predecessors, were launched into repetitive, circular, sun-synchronous, near-polar orbits. However, these orbits were lowered from 900 to 705 km. These lower orbits were chosen to make the satellites potentially retrievable by the space shuttle and to aid in the improvement of the ground resolution of the sensors onboard.

As was shown in Figure 6.1, Landsat-4 and -5 orbits have an inclination angle of 98.2° (8.2° from normal) with respect to the equator. The satellite crosses the equator on the north-to-south portion of each orbit at 9:45 A.M. local sun time. Each orbit takes approximately 99 min, with just over 14.5 orbits being completed in a day. Due to earth rotation, the distance between ground tracks for consecutive orbits is approximately 2752 km at the equator (Figure 6.11).

The above orbit results in a 16-day repeat cycle for each satellite. The orbits of Landsat-4 and -5 were established 8 days out of phase, such that when both satellites were operational, an 8-day repeat coverage cycle could be maintained with alternating coverage by each satellite. As shown in Figure 6.12, the time interval between adjacent coverage tracks of the same satellite is 7 days. This coverage pattern is quite different from that of the previous three satellites, which had 18-day orbital cycles and a 1-day interval between orbits over adjacent tracks. Consequently, Landsat-4 and -5 images are cataloged according to a set of WRS paths different from those used to reference data from Landsat-1, -2, and -3. The Landsat-4 and -5 WRS is made up of 233 paths numbered 001 to 233, east to west, with path 001 crossing the equator at longitude 64°36' W. The same number of rows is used as in the previous WRS system. That is, row 60 coincides with the equator at the orbit's descending node. Row 1 of each path starts at 80°47' N latitude.

Figure 6.13 shows the design of the Landsat-4 and -5 satellites, which include both the MSS and the TM. This spacecraft weighs approximately 2000 kg and includes four 1.5 × 2.3-m solar panels that are mounted to one side. The high gain antenna shown protruding above the spacecraft can be used to relay data through geosynchronous communication satellites included in the *Tracking and Data Relay Satellite System* (TDRSS). Direct transmission of MSS and TM data to ground receiving stations is made possible via the X-band and S-band antennas onboard the satellite. The data transmission rates involved are substantial; the MSS transmits 15 megabits per second (Mbps) and the TM transmits 85 Mbps. (The MSS has been flown on these missions primarily to ensure continuity of data for receiving stations unable to receive and process TM data.)

The MSS onboard Landsat-4 and -5 is essentially identical to the MSS sensors on the previous Landsat satellites. The across-track swath of 185 km has been maintained at the lower orbit altitude by increasing the total field of view to 14.92° (from 11.56° on previous systems). The optics of the MSS system have also been modified to yield an 82-m-ground-resolution cell to

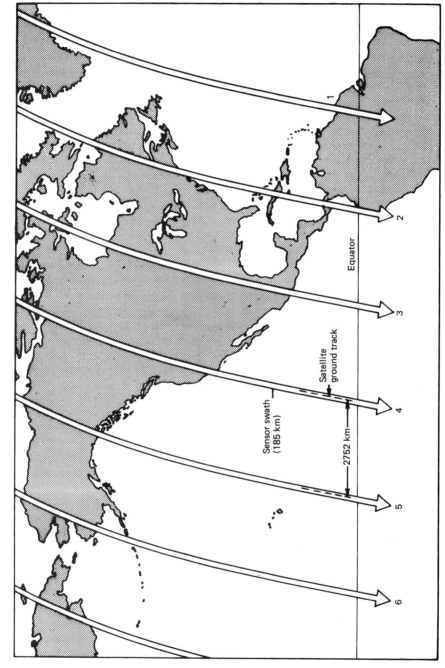

Figure 6.11 Spacing between adjacent Landsat-4 or -5 orbit tracks at the equator. The earth revolves 2752 km to the east at the equator between passes. (Adapted from NASA diagram.)

420

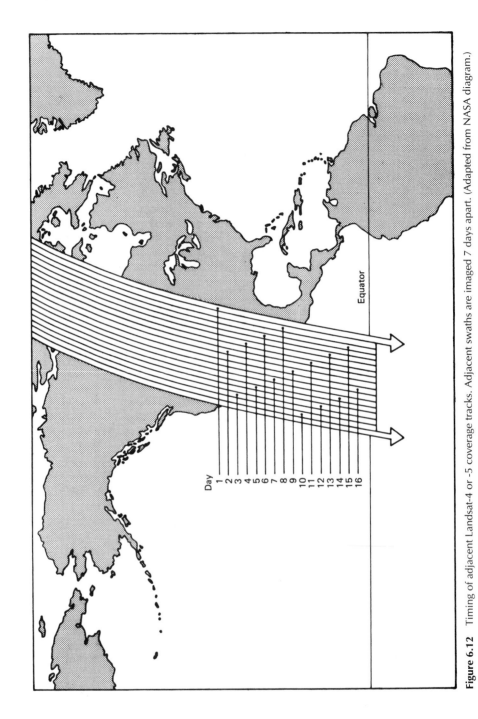

Figure 6.12 Timing of adjacent Landsat-4 or -5 coverage tracks. Adjacent swaths are imaged 7 days apart. (Adapted from NASA diagram.)

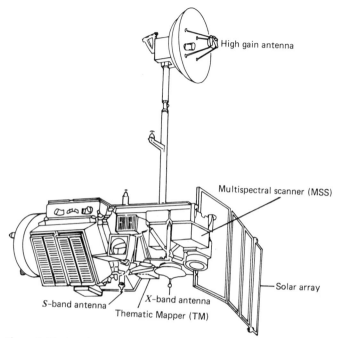

High gain antenna

Multispectral scanner (MSS)

Solar array

S–band antenna

X–band antenna

Thematic Mapper (TM)

Figure 6.13   Landsat-4 and -5 observatory configuration. (Adapted from NASA diagram.)

approximate the 79-m-ground-resolution cell of the previous systems. The same four spectral bands are used for data collection, but they have been renumbered. That is, bands 1 to 4 of the Landsat-4 and -5 MSS correspond directly to bands 4 to 7 of the previous MSS systems (Figure 6.7).

The TM is a highly advanced sensor incorporating a number of spectral, radiometric, and geometric design improvements relative to the MSS. Spectral improvements include the acquisition of data in seven bands instead of four, with new bands in the visible (blue), mid-IR, and thermal portions of the spectrum. Also, based on experience with MSS data and extensive field radiometer research results, the wavelength range and location of the TM bands have been chosen to improve the spectral differentiability of major earth surface features.

Radiometrically, the TM performs its onboard A-to-D signal conversion over a quantization range of 256 digital numbers (8 bits). This corresponds to a fourfold increase in the gray scale range relative to the 64 digital numbers (6 bits) used by the MSS. This finer radiometric precision permits observation of smaller changes in radiometric magnitudes in a given band and provides greater sensitivity to changes in relationships between bands. Thus, differences in radiometric values that are lost in one digital number in MSS data may now be distinguished.

Geometrically, TM data are collected using a 30-m-ground-resolution cell (for all but the thermal band, which has 120 m resolution). This represents a decrease in the lineal dimensions of the ground resolution cell of approximately 2.6 times, or a reduction in the area of the ground resolution cell of approximately 7 times. At the same time, several design changes have been incorporated within the TM to improve the accuracy of the geodetic positioning of the data. Most geometrically corrected TM data are supplied using a 28.5 × 28.5-m GSD registered to the Space Oblique Mercator (SOM) cartographic projection. The data may also be fit to the Universal Transverse Mercator (UTM) or polar stereographic projections.

Table 6.3 lists the seven spectral bands of the TM along with a brief summary of the intended principal applications of each. The TM bands are more finely tuned for vegetation discrimination than those of the MSS for several reasons. The green and red bands of the TM (bands 2 and 3) are narrower

TABLE 6.3 Thematic Mapper Spectral Bands

Band	Wavelength (μm)	Nominal Spectral Location	Principal Applications
1	0.45–0.52	Blue	Designed for water body penetration, making it useful for coastal water mapping. Also useful for soil/vegetation discrimination, forest-type mapping, and cultural feature identification.
2	0.52–0.60	Green	Designed to measure green reflectance peak of vegetation (Figure 1.10) for vegetation discrimination and vigor assessment. Also useful for cultural feature identification.
3	0.63–0.69	Red	Designed to sense in a chlorophyll absorption region (Figure 1.10) aiding in plant species differentiation. Also useful for cultural feature identification.
4	0.76–0.90	Near IR	Useful for determining vegetation types, vigor, and biomass content, for delineating water bodies, and for soil moisture discrimination.
5	1.55–1.75	Mid IR	Indicative of vegetation moisture content and soil moisture. Also useful for differentiation of snow from clouds.
6[a]	10.4–12.5	Thermal IR	Useful in vegetation stress analysis, soil moisture discrimination, and thermal mapping applications.
7[a]	2.08–2.35	Mid IR	Useful for discrimination of mineral and rock types. Also sensitive to vegetation moisture content.

[a]Bands 6 and 7 are out of wavelength sequence because band 7 was added to the TM late in the original system design process.

than their MSS counterparts. Also, the near-IR TM band (4) is narrower than the combined bands of the MSS in this region and centered in a region of maximum sensitivity to plant vigor. Sensitivity to plant water stress is obtained in both of the TM mid-IR bands (5 and 7). Plant stress discrimination is also aided by data from the TM blue band (1).

In addition to improved discrimination of vegetation, the TM has been designed to afford expanded or improved use of satellite data in a number of other application areas. Among these is the use of TM data (particularly from band 1) in the field of bathymetry. Likewise, the mid-IR bands (5 and 7) have proven to be extremely valuable in discrimination of rock types. Band 5 is also ideal for differentiating between snow- and cloud-covered areas (snow has very low reflectance in this band, while the reflectance of clouds is relatively high). Finally, band 6 makes TM data potentially useful in a range of thermal mapping applications. We treat the visual interpretation of TM data in Section 6.8 and the digital processing of TM data in Chapter 7. In the remainder of this section we discuss the basic differences in the design and operation of the TM and the MSS.

Whereas the MSS collects data only when its IFOV is traversing in the west-to-east direction along a scan line, the TM acquires data during both the forward (west-to-east) and reverse (east-to-west) sweeps of its scan mirror. This bidirectional scanning procedure is employed to reduce the rate of oscillation of the scan mirror and to increase the time an individual detector is able to dwell upon a given portion of the earth within its IFOV. The TM scans through a total field of view of 15.4° (±7.7° from nadir). It completes approximately seven combined forward and reverse scan cycles per second. This relatively slow rate limits the acceleration of the scan mirror, improving the geometric integrity of the data collection process and improving the signal-to-noise performance of the system.

Another major difference between the TM and the MSS is the number of detectors used for the various bands of sensing. Whereas the MSS employs six detectors to record data in each of its four bands of sensing (total of 24 detectors), the TM uses 16 detectors for all nonthermal bands and four detectors for the thermal band (total of 100 detectors). That is, 16 lines of each nonthermal band and four lines of thermal data are acquired with each sweep of the scan mirror. Silicon detectors are used for bands 1 to 4, and these are located within a primary focal plane assembly (Figure 6.14). The detectors for bands 5 to 7 are located in a second focal plane assembly incorporating passive radiation cooling to increase their radiometric sensitivity. Indium antimonide (InSb) detectors are employed for bands 5 and 7, and mercury cadmium telluride (HgCdTe) detectors are used for band 6.

At any instant in time, all 100 detectors view a different area on the ground due to the spatial separation of the individual detectors within the two TM focal planes. Figure 6.14 illustrates the projection of the detector IFOVs onto the ground. Accurate band-to-band data registration requires knowledge of the relative projection of the detectors in both focal planes as a function of time. This information is derived from data concerning the relative position

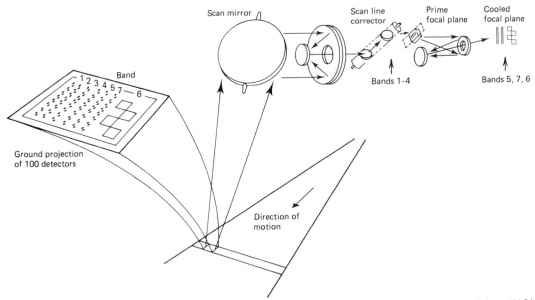

Figure 6.14 Thematic Mapper optical path and projection of detector IFOVs on earth surface. (Adapted from NASA diagram.)

of the individual detector arrays with respect to the optical axis, the spacecraft position and attitude, and the motion of the scan mirror during successive scan cycles. A *scan angle monitor* on the scan mirror generates signals indicating the mirror's angular position as a function of time. These signals are called scan mirror correction data and are transmitted to the ground for incorporation into the geometric processing of TM image data.

Signals from the scan angle monitor are also used to guide the motions of a *scan line corrector* located in front of the primary focal plane (Figure 6.14). The function of the scan line corrector is illustrated in Figure 6.15. During each scan mirror sweep, the scan line corrector rotates the TM line of sight backward along the satellite ground track to compensate for the forward motion of the spacecraft. This prevents the overlap and underlap of scan lines and produces straight scan lines that are perpendicular to the ground track.

The TM also employs an internal radiometric calibration source consisting of three tungsten filament lamps, a blackbody for the thermal band, and a pivot-mounted shutter. The shutter passes through the field of view of the instrument's detectors each time the scan mirror changes directions. The shutter permits light from the lamps to pass into the field of view of the nonthermal bands directly, and a mirror on the shutter directs energy from the thermal calibration source into the field of view of the thermal detectors. These calibration sources are used to monitor the radiometric response of the various detectors over the sensor's service life.

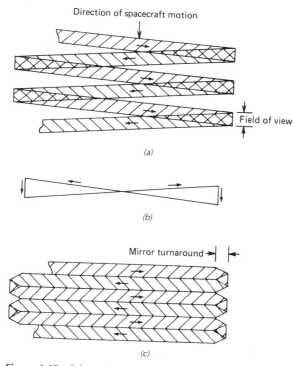

Figure 6.15 Schematic of TM scan line correction process: (a) uncompensated scan lines; (b) correction for satellite motion; (c) compensated scan lines. (Adapted from NASA diagram.)

6.6 LANDSAT-6 PLANNED MISSION

After more than two decades of success, the Landsat program realized its first unsuccessful mission with the October 5, 1993, launch failure of Landsat-6. This satellite was designed to have occupied an orbit identical to that of Landsat-4 and -5. The sensor included on board this ill-fated mission was the Enhanced Thematic Mapper (ETM). To provide data continuity with Landsat-4 and -5, the ETM incorporated the same seven spectral bands and the same spatial resolutions as the TM. The ETM's major improvement over the TM was the addition of an eighth, "panchromatic" band operating in the 0.50- to 0.90-μm range with a spatial resolution of 15 m. The 15-m panchromatic band data could have been merged with various combinations of the 30-m data from ETM bands 1 to 5 and 7 to produce color images with essentially 15 m resolution. (We discuss this "pan sharpening" procedure in Chapter 7). Changes in the design of the detector system for the ETM were to also permit the data for all bands to be automatically coregistered as they were acquired. This capability is referred to as a "monolithic" detector design.

Another improvement made in the design of the ETM was the ability to set the gains for the individual bands from the ground (Section 7.2). The ETM had a 9-bit A-to-D converter that provided either a high or low gain 8-bit setting. The high gain was to be used over areas of low reflectance (such as water) and the low gain was designed to provide reliable readings over very bright regions (such as deserts) without detector saturation.

6.7 LANDSAT-7

The Land Remote Sensing Policy Act of 1992 once again transferred the management responsibility for the Landsat program with the design and operation of Landsat-7. Originally, a joint NASA–U.S. Air Force program was envisioned. This concept was superseded by a 1994 presidential directive and subsequent deliberations that established a joint program between NASA and USGS. The system was also made a part of NASA's Earth Science Enterprise Program.

Landsat-7 was launched on April 15, 1999. The earth-observing instrument onboard this spacecraft is the Enhanced Thematic Mapper Plus (ETM+). The design of the ETM+ stresses the provision of data continuity with Landsat-4 and -5. Similar orbits and repeat patterns are used, as is the 185-km swath width for imaging. As with the ETM planned for Landsat-6, the system is designed to collect 15-m-resolution "panchromatic" data (which actually extend to 0.90 μm in the near IR, well outside the visible spectral range normally associated with panchromatic imagery) and six bands of data in the visible, near-IR, and mid-IR spectral regions at a resolution of 30 m. A seventh, thermal, band is incorporated with a resolution of 60 m (vs. 120 m for the ETM). As with the ETM, high and low gain settings for the individual channels may be controlled from the ground. The system also includes a "dual-mode solar calibrator," in addition to an internal lamp calibrator. This results in onboard absolute radiometric calibration to within an accuracy of 5 percent.

Landsat-7 data can be transmitted to ground either directly or by playback of data stored by an onboard, solid-state, recorder. The primary receiving station for the data is located at the EROS Data Center (EDC) in Sioux Falls, South Dakota. The ground system at the EDC is capable of processing a complete global view of earth's land areas seasonally, or approximately four times per year. In addition, a worldwide network of receiving stations is able to receive real-time direct downlink of image data when and where the satellite is in sight of the receiving station. Landsat-7 data supplied by the EDC can be obtained at the cost of fulfilling user requests.

Figure 6.16 illustrates a portion of the first image acquired by the Landsat-7 ETM+. This is a panchromatic band image depicting Sioux Falls, South Dakota, and vicinity. Features such as the airport (upper center), major roads, and new residential development (especially evident in lower right) are clearly discernible in this 15-m-resolution image. This band can be merged with the

Figure 6.16 The first image acquired by the Landsat-7 ETM+. Panchromatic band (15 m resolution), Sioux Falls, SD, April 18, 1999. Scale 1 : 88,000. (Courtesy EROS Data Center.)

30-m data from ETM+ bands 1 to 5 and 7 to produce "pan-sharpened" color images with essentially 15 m resolution (Section 7.6).

6.8 LANDSAT IMAGE INTERPRETATION

The utility of Landsat image interpretation has been demonstrated in many fields, such as agriculture, botany, cartography, civil engineering, environmental monitoring, forestry, geography, geology, geophysics, land resource analysis, land use planning, oceanography, and water resource analysis.

As shown in Table 6.4, the image scale and area covered per frame are very different for Landsat images than for conventional aerial photographs. For example, more than 1600 aerial photographs at a scale of 1:20,000 with no overlap are required to cover the area of a single Landsat image! Because of scale and resolution differences, Landsat images should be considered as a complementary interpretive tool instead of a replacement for low altitude aerial photographs. For example, the existence and/or significance of certain geologic features trending for tens or hundreds of kilometers, and clearly evident on a Landsat image, might escape notice on low altitude aerial photographs. On the other hand, housing quality studies from aerial imagery would certainly be more effective using low altitude aerial photographs rather than Landsat images, since individual houses cannot be studied on Landsat images. In addition, most Landsat images can only be viewed in two dimensions, whereas most aerial photographs are acquired and viewed in stereo.

The ground resolution cell size of Landsat data is about 79 m for the MSS and about 30 m for the TM. However, linear features as narrow as a few meters, having a reflectance that contrasts sharply with that of their surroundings, can often be seen on such images (e.g., two-lane roads, concrete bridges crossing water bodies). On the other hand, objects much larger than the ground resolution cell size may not be apparent if they have a very low reflectance contrast with their surroundings, and features visible in one band may not be visible in another.

TABLE 6.4 Comparison of Image Characteristics

Image Format	Image Scale	Area Covered per Frame (km^2)
Low altitude USDA–ASCS aerial photographs (230 × 230 mm)	1:20,000	21
High altitude NASA aerial photographs (RB-57 or ER-2) (230 × 230 mm)	1:120,000	760
Landsat scene (185 × 185 mm)	1:1,000,000	34,000

Because the Landsat MSS and TM are across-track scanning systems, they produce images having one-dimensional relief displacement. Because there is displacement only in the scan direction and not in the flight track direction, Landsat images can be viewed in stereo only in areas of sidelap on adjacent orbit passes. This sidelap varies from about 85 percent near the poles to about 14 percent at the equator. Consequently, only a limited area of the globe may be viewed in stereo. Also, the vertical exaggeration when viewing Landsat images in stereo is quite small compared to conventional airphotos. This stems from the high platform altitude (ranging from 705 to 900 km) of the Landsat satellites compared to the base distance between images. Whereas stereo airphotos may have a 4× vertical exaggeration, stereo Landsat vertical exaggeration ranges from about 1.3× at the equator to less than 0.4× at latitudes above about 70°. Subtle as this stereo effect is, geologists in particular have found stereoviewing in Landsat overlap areas quite valuable in studying topographic expression. However, most interpretations of Landsat imagery are made monoscopically, either because sidelapping imagery does not exist or because the relief displacement needed for stereoviewing is small.

Figure 6.17, a small portion of a Landsat scene, illustrates the comparative appearance of the seven Landsat TM bands. Here, the blue-green water of the lake, river, and ponds in the scene has moderate reflection in bands 1 and 2 (blue and green), a small amount of reflection in band 3 (red), and virtually no reflection in bands 4, 5, and 7 (near and mid IR). Reflection from roads and urban streets is highest in bands 1, 2, and 3 and least in band 4 (other cultural features such as new subdivisions, gravel pits, and quarries would have similar reflectances). Overall reflection from agricultural crops is highest in band 4 (near IR). Note also the high band 4 reflectance of the golf courses appearing to the right of the river and the lake. The distinct tonal lineations from upper right (northeast) to lower left (southwest) in these images are a legacy from the most recent glaciation of Wisconsin. Glacial ice movement from northeast to southwest left a terrain characterized by many drumlins and scoured bedrock hills. Present-day crop and soil moisture patterns reflect the alignment of this grooved terrain. The thermal band (band 6) has a less distinct appearance than the other bands because the ground resolution cell of this band is 120 m. It has an indistinct, rather than blocky, appearance because the data have been resampled into the 30-m format of the other bands. As would be expected on a summertime thermal image recorded during the daytime, the roads and urban areas have the highest radiant temperature, and the water bodies have the lowest radiant temperature.

Plate 17 shows six color composite images of the same area shown in Figure 6.17. Table 6.5 shows the color combinations used to generate each of these composites. Note that (*a*) is a "normal color" composite, (*b*) is a "color infrared" composite, and (*c*) to (*f*) are some of the many other "false color" combinations that can be produced. A study at the USGS EDC (NOAA, 1984) showed an interpreter preference for several specific band–color combinations

Figure 6.17 Individual Landsat TM bands, suburban Madison, WI, late August (scale 1:115,000): (a) band 1, 0.45 to 0.52 μm (blue); (b) band 2, 0.52 to 0.60 μm (green); (c) band 3, 0.63 to 0.69 μm (red); (d) band 4, 0.76 to 0.90 μm (near IR); (e) band 5, 1.55 to 1.75 μm (mid IR); (f) band 7, 2.08 to 2.35 μm (mid IR); (g) band 6, 10.4 to 12.5 μm (thermal IR).

Figure 6.17   *(Continued)*

Figure 6.17 *(Continued)*

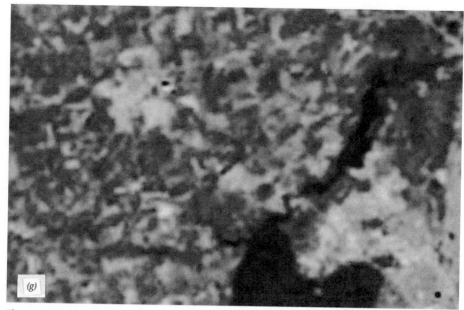

Figure 6.17 *(Continued)*

for various features. For the mapping of water sediment patterns, a normal color composite of bands 1, 2, and 3 (displayed as blue, green, and red) was preferred. For most other applications, such as mapping urban features and vegetation types, the combinations of (1) bands 2, 3, and 4 (color IR composite), (2) bands 3, 4, and 7, and (3) bands 3, 4, and 5 (all in the order blue, green, and red) were preferred. In general, vegetation discrimination is enhanced through the incorporation of data from one of the mid-IR bands (5 or 7). Combinations of any one visible (bands 1 to 3), the near-IR (band 4), and one mid-IR (band 5 or 7) band are also very useful. However, a great deal of

TABLE 6.5 TM Band–Color Combination Shown in Plate 17

Plate 17	TM Band-Color Assignment in Composite		
	Blue	Green	Red
(a)	1	2	3
(b)	2	3	4
(c)	3	4	5
(d)	3	4	7
(e)	3	5	7
(f)	4	5	7

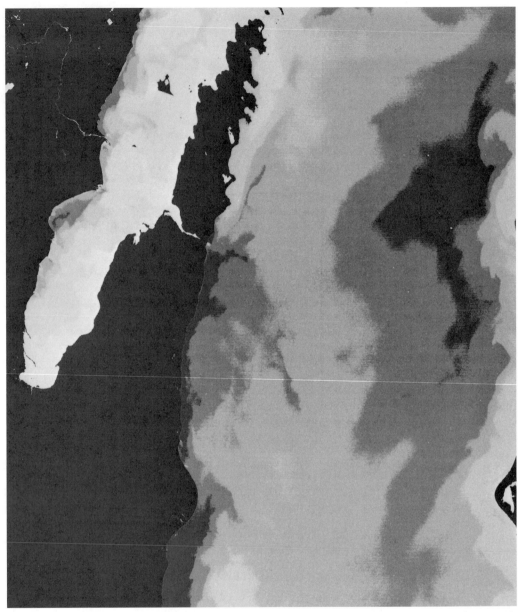

Figure 6.18 Landsat TM band 6 (thermal IR) image, Green Bay and Lake Michigan, Wisconsin–Michigan, mid-July. Scale 1 : 225,000.

personal preference is involved in band–color combinations for interpretive purposes, and for specific applications, other combinations could be optimum.

Figure 6.18 shows a Landsat TM band 6 (thermal) image of Green Bay and Lake Michigan (between the states of Wisconsin and Michigan). In this image, the land area has been masked out and is shown as black (using techniques described in Chapter 7). Based on a correlation with field observations of water surface temperature, the image data were "sliced" into six gray levels, with the darkest tones having a temperature less than 12°C, the brightest tones having a temperature greater than 20°C, and each of the four intermediate levels representing a 2°C range between 12 and 20°C.

Each Landsat satellite passes over the same area on the earth's surface during daylight hours about 20 times per year. The actual number of times per year a given ground area is imaged depends on amount of cloud cover, sun angle, and whether or not the satellite is in operation on any specific pass. This provides the opportunity for many areas to have Landsat images available for several dates per year. Because the appearance of the ground in many areas with climatic change is dramatically different in different seasons, the image interpretation process is often improved by utilizing images from two or more dates.

Figure 6.19 shows MSS band 5 (red) images of a portion of Wisconsin as imaged in September and December. The ground is snow covered (about 200 mm deep) in the December image and all water bodies are frozen, except for a small stretch of the Wisconsin River. The physiography of the area can be better appreciated by viewing the December image, due in part to the low solar elevation angle in winter that accentuates subtle relief. A series of stream valleys cuts into the horizontally bedded sedimentary rock in the upper left portion of this scene. The snow-covered upland areas and valley floors have a very light tone, whereas the steep, tree-covered valley sides have a darker tone. The identification of urban, agricultural, and water areas can better be accomplished using the September image. The identification of forested areas can be more positively done using the December image.

The synoptic view afforded by space platforms can be particularly useful for observing short-lived phenomena. However, the use of Landsat images to capture such ephemeral events as floods, forest fires, and volcanic activity is, to some degree, a hit-or-miss proposition. If a satellite passes over such an event on a clear day when the imaging system is in operation, excellent images of such events can be obtained. On the other hand, such events can easily be missed if there are no images obtained within the duration of the event or, as is often true during floods, extensive cloud cover obscures the earth's surface. However, some of these events do leave lingering traces. For example, soil is typically wet in a flooded area for at least several days after the flood waters have receded, and this condition may be imaged even if the flood waters are not. Thematic Mapper images of the Missouri River in preflood and flooding conditions were shown in Figure 4.25.

Figure 6.20 is a black and white copy of a band 3, 4, and 5 color composite Landsat TM image showing extensive deforestation near Rondônia, Brazil.

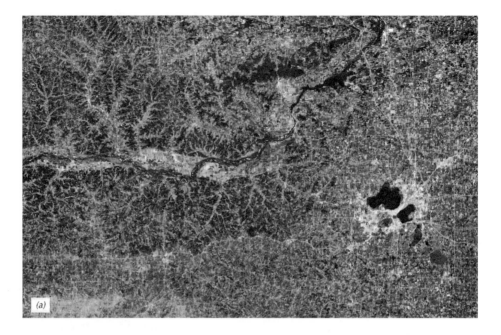

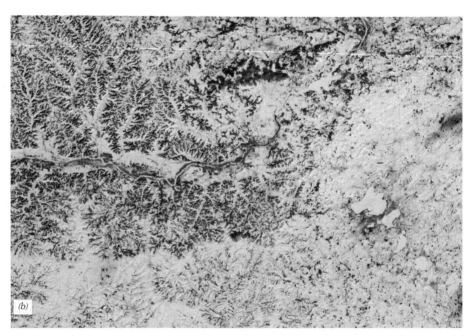

Figure 6.19   Landsat MSS band 5 (red) images, southwestern Wisconsin. Scale 1:1,000,000: (*a*) mid-September; (*b*) mid-December.

Figure 6.20   Landsat TM image, Rondônia, Brazil, and vicinity, mid-August. Black and white copy of band 3–4–5 color composite image. Scale 1:1,000,000. (Courtesy EOSAT.)

This band combination has been found to be effective for monitoring deforestation in the Amazon region during the dry season. The light-toned "fishbonelike" pattern in this image represents clearings for small farm settlements in the Brazilian jungle.

Thematic Mapper data have been used extensively to prepare image maps over a range of mapping scales. Such maps have proven to be useful tools for resource assessment in that they depict the terrain in actual detail, rather than in the line-and-symbol format of conventional maps. Image maps are often used as map supplements to augment conventional map coverage and to provide coverage of unmapped areas.

As we will see in Chapter 7, there are several digital image processing procedures that may be applied to the image mapping process. These include large-area digital mosaicking, image enhancement procedures, merging of image data with conventional cartographic information, and streamlining the map production and printing process using highly automated cartographic systems. Extensive research continues in the area of image mapping, with numerous sources of image data.

6.9 LANDSAT DATA CONTINUITY MISSION

The Land Remote Sensing Policy Act (1992) instructed the USGS and NASA to examine options for the Landsat program beyond the operational life of Landsat-7. The *Landsat Data Continuity Mission* (*LDCM*) is a joint effort of the USGS and NASA to plan for the continued collection of "Landsat-like" imagery in the future. Major elements of the LDCM program include maintaining data continuity, serving a variety of public interests, incorporating improved technology, providing data at low cost, and relying on private sector organizations for a significant portion of LDCM development and operations. The latter goal may be achieved by a privately operated system that would acquire more valuable (e.g., higher resolution) data for commercial distribution in combination with the acquisition of moderate resolution "Landsat continuity" data on behalf of the U.S. government.

The current (2002) plan for the LDCM calls for a privately developed and operated system that would provide an average of 250 images per day for the USGS archive, about half the number of scenes now provided by Landsat-7. In the words of the draft Request for Proposals, these data would be "sufficiently consistent in terms of acquisition geometry, coverage characteristics, spectral characteristics, output product quality, and data availability to ensure continuity of the Landsat mission." More specifically, the proposed system would have a set of nine spectral bands, including six that correspond to the 30-m bands on the Landsat-7 ETM+, an additional band in the blue portion of the spectrum (0.43 to 0.45 μm), one "Sharpening" band analogous to the panchromatic band on SPOT and Landsat-7, and a "Cirrus" band for high altitude cloud detection in the mid IR. (At the present time, the LDCM plans do not necessarily call for inclusion of a thermal-IR band.) The Sharpening band would have a spatial resolution of 15 m or better, the Cirrus band 120 m or better, and the remaining bands 30 m or better. The swath width would be at least as wide as that of the ETM+ (185 km), and the orbit would correspond to the World Wide Reference System used for Landsat-4 through -7.

It is expected that LDCM data would begin to be collected during 2006, slightly after the end of the minimum expected lifetime of Landsat-7. Whether Landsat-7 will continue operating long enough to ensure true "continuity" from the LDCM thus remains to be seen.

6.10   SPOT SATELLITE PROGRAM

In early 1978 the French government decided to undertake the development of the *Système Pour l'Observation de la Terre*, or *SPOT*, program. Shortly thereafter Sweden and Belgium agreed to participate in the program with the aim of launching the first of a series of SPOT earth observation satellites. From its inception, SPOT was designed as a commercially oriented program that was to be operational, rather than experimental, in character.

Conceived and designed by the French Centre National d'Etudes Spatiales (CNES), SPOT has developed into a large-scale international program with ground receiving stations and data distribution outlets located in more than 20 countries. The first satellite in the program, SPOT-1, was launched from the Kourou Launch Range in French Guiana on February 21, 1986, onboard an Ariane launch vehicle. This satellite began a new era in space remote sensing, for it is the first earth resource satellite system to include a linear array sensor and employ pushbroom scanning techniques. It is also the first system to have pointable optics. This enables side-to-side off-nadir viewing capabilities, and it affords full-scene stereoscopic imaging from two different satellite tracks permitting coverage of the same area.

SPOT-1 was retired from full-time service on December 31, 1990 (it was used in a backup mode thereafter). The SPOT-2 satellite was launched on January 21, 1990; SPOT-3 was launched on September 25, 1993; and SPOT-4 was launched on March 23, 1998. In that SPOT-1, -2, and -3 have identical orbits and sensor systems, we describe them together in the following section. Section 6.12 treats SPOT-4.

6.11   SPOT-1, -2, AND -3

Like the Landsat satellites, SPOT-1, -2, and -3 have a circular, near-polar, sun-synchronous orbit. The nominal orbit of SPOT-1, -2, and -3 has an altitude of 832 km and an inclination of 98.7°. They descend across the equator

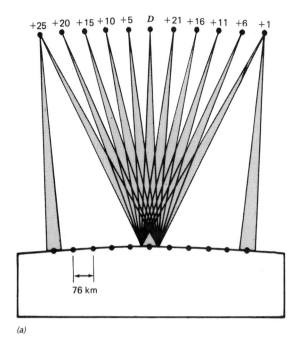

(a)

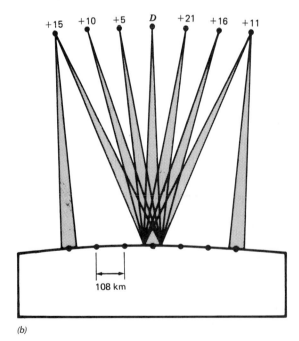

(b)

Figure 6.21 SPOT revisit pattern: (a) latitude 45°; (b) latitude 0°. (Adapted from CNES diagram.)

at 10:30 A.M. local solar time, with somewhat later crossings in northern latitudes and somewhat earlier crossings in southern latitudes. For example, SPOT crosses areas at a latitude of 40° N at approximately 11:00 A.M. and areas at a latitude of 40° S at 10:00 A.M.

The orbit pattern for SPOT-1, -2, and -3 repeats every 26 days. This means any given point on the earth can be imaged using the same viewing angle at this frequency. However, the pointable optics of the system enable off-nadir viewing during satellite passes separated alternatively by 1 and 4 (and occasionally 5) days, depending on the latitude of the area viewed (Figure 6.21). For example, during the 26-day period separating two successive satellite passes directly over a point located at the equator, seven viewing opportunities exist (day D and days $D + 5$, $+10$, $+11$, $+15$, $+16$, and $+21$). For a point located at a latitude of 45° a total of 11 viewing opportunities exist (day D and days $D + 1$, $+5$, $+6$, $+10$, $+11$, $+15$, $+16$, $+20$, $+21$, and $+25$). This "revisit" capability is important in two respects. First, it increases the potential frequency of coverage of areas where cloud cover is a problem. Second, it provides an opportunity for viewing a given area at frequencies ranging from successive days to several days to a few weeks. Several application areas, particularly within agriculture and forestry, require repeated observations over these types of time frames.

Figure 6.22 is a schematic of the SPOT-1, -2, and -3 satellites. The systems weigh approximately 1750 kg and the main body of the satellites is approximately $2 \times 2 \times 3.5$ m. The solar panel has an overall length of approximately 15.6 m. The SPOT platform is of modular design such that it is compatible with a variety of sensor payloads. Thus, in subsequent SPOT missions, changes in sensor design can be implemented without significant modification of the platform.

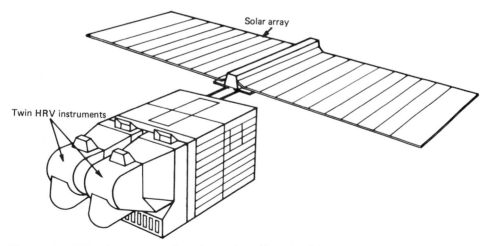

Figure 6.22   SPOT observatory configuration. (Adapted from CNES diagram.)

The sensor payload for SPOT-1, -2, and -3 consists of two identical *high resolution visible* (HRV) imaging systems and auxiliary magnetic tape recorders. Each HRV is designed to operate in either of two modes of sensing: (1) a 10-m-resolution 'panchromatic' (black and white) mode over the range 0.51 to 0.73 μm or (2) a 20-m-resolution multispectral (color IR) mode over the ranges 0.50 to 0.59 μm, 0.61 to 0.68 μm, and 0.79 to 0.89 μm.

The HRV employs along-track, or pushbroom, scanning, as described in Section 5.3. Each HRV actually contains four CCD subarrays. A 6000-element subarray is used in the panchromatic mode to record data at 10 m resolution. Three 3000-element subarrays are employed in the multispectral mode at 20 m resolution. Data are effectively encoded over a 256-digital-number range and are transmitted at a rate of 25 Mbps. Each instrument's field of view is 4.13°, such that the ground swath of each HRV scene is 60 km wide under nadir viewing conditions.

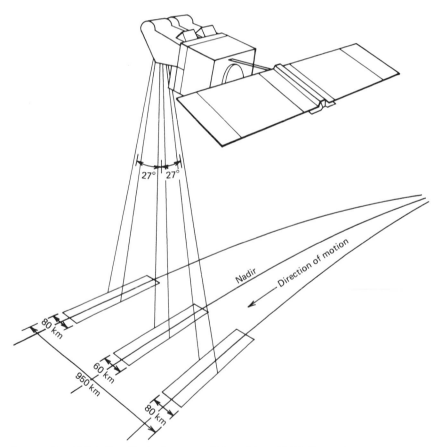

Figure 6.23 SPOT off-nadir viewing range. (Adapted from CNES diagram.)

The first element in the optical system for each HRV is a plane mirror that can be rotated to either side by ground command through an angle of ±27° (in 45 steps of 0.6° each). This allows each instrument to image any point within a strip extending 475 km to either side of the satellite ground track (Figure 6.23). The size of the actual ground swath covered naturally varies with the pointing angle employed. At the 27° maximum value, the swath width for each instrument is 80 km. When the two instruments are pointed so as to cover adjacent image fields at nadir, the total swath width is 117 km and the two fields overlap by 3 km (Figure 6.24). While each HRV instrument is capable of collecting panchromatic and multispectral data simultaneously, resulting in four data streams, only two data streams can be transmitted at one time. Thus, either panchromatic or multispectral data can be transmitted over a 117-km-wide swath, but not both simultaneously.

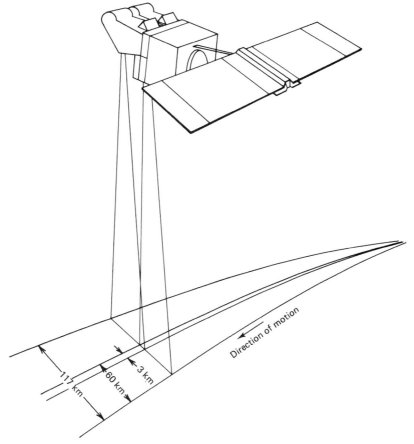

Figure 6.24 SPOT ground coverage with HRVs recording adjacent swaths. (Adapted from CNES diagram.)

Stereoscopic imaging is also possible due to the off-nadir viewing capability of the HRV. That is, images of an area recorded on different satellite tracks can be viewed in stereo (Figure 6.25). The frequency with which stereoscopic coverage can be obtained, being tied directly to the revisit schedule for the satellite, varies with latitude. At a latitude of 45° (Figure 6.21*a*), there are six possible occasions during the 26-day orbit cycle on which *successive-day* stereo coverage may be obtained (day D with $D + 1$, $D + 5$ with $D + 6$, $D + 10$ with $D + 11$, $D + 15$ with $D + 16$, $D + 20$ with $D + 21$, and $D + 25$ with day D of the next orbit cycle). At the equator (Figure 6.21*b*), only two stereoviewing opportunities on successive days are possible ($D + 10$ with $D + 11$ and $D + 15$ with $D + 16$). The base–height ratio also varies with latitude, from approximately 0.50 at 45° to approximately 0.75 at the equator. (If stereoscopic coverage need not be acquired on a successive-day basis, the range of possible viewing opportunities and viewing geometries greatly increases.)

It should now be clear that the SPOT-1, -2, and -3 system affords a broad range of viewing conditions and spectral modes of operation. The particular observation sequence for a given day of satellite operation is loaded into the system's onboard computer by the Toulouse, France, ground control station

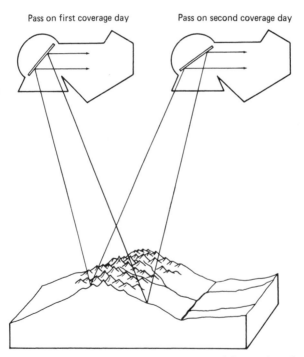

Pass on first coverage day   Pass on second coverage day

Figure 6.25 SPOT stereoscopic imaging capability. (Adapted from CNES diagram.)

while the satellite is within its range. The day's operating sequence for each HRV is controlled entirely independently. This includes the viewing angles of the two instruments, the spectral mode of operation (panchromatic or multi-spectral), timing of image acquisition, and modes of data transmission. Data are normally transmitted directly when the satellite is within range of a ground receiving station (approximately a 2600-km-radius around the station). The onboard tape recorders are used when images are acquired over areas outside the range of a ground receiving station. In such cases, the recorded image data are subsequently transmitted to the Toulouse or Kiruna, Sweden, station when the satellite reenters its range.

6.12   SPOT-4

Like its Landsat counterpart, the SPOT program has been designed to provide long term continuity of data collection but with successive improvements in the technical capability and performance of the sensing systems involved. SPOT-4, launched on March 23, 1998, reflects this continuity with improved performance design philosophy. The primary imaging systems on board this satellite are the *high resolution visible and infrared* (*HRVIR*) sensors and the *Vegetation* instrument.

HRVIR

Similar to the HRV system incorporated in SPOT-1 to -3, the *HRVIR* system includes two identical sensors capable of imaging a total nadir swath width of 120 km. Among the major improvements in the system is the addition of a 20-m-resolution band in the mid-IR portion of the spectrum (between 1.58 and 1.75 μm). This band improves the vegetation monitoring, mineral discriminating, and soil moisture mapping capability of the data. Furthermore, mixed 20- and 10-m data sets are coregistered onboard instead of during ground processing. This has been accomplished by replacing the panchromatic band of SPOT-1, -2, and -3 (0.49 to 0.73 μm) with the "red" band from these systems (0.61 to 0.68 μm). This band is used to produce both 10-m black and white images and 20-m multispectral data.

Vegetation

Another major addition to SPOT-4's payload is the Vegetation instrument. While designed primarily for vegetation monitoring, this system is useful in a range of applications where frequent, large-area coverage is important. (We illustrate several of these applications in our discussion of meteorological and

TABLE 6.6   Comparison of Spectral Bands Used by the SPOT-4 HRVIRs and the Vegetation Instrument

Spectral Band (μm)	Nominal Spectral Location	HRVIR	Vegetation Instrument
0.43–0.47	Blue	—	Yes
0.50–0.59	Green	Yes	—
0.61–0.68	Red	Yes	Yes
0.79–0.89	Near IR	Yes	Yes
1.58–1.75	Mid IR	Yes	Yes

ocean-monitoring satellites.) The instrument uses linear-array technology to provide a very wide angle image swath 2250 km wide with a spatial resolution at nadir of approximately 1 km and global coverage on a daily basis. (A few zones near the equator are covered every other day.)

As shown in Table 6.6, the Vegetation instrument uses three of the same spectral bands as the HRVIR system, namely the red, near-IR, and mid-IR bands, but it also incorporates a blue band (0.43 to 0.47 μm) for oceano-graphic applications.

Flown in combination, the Vegetation instrument and the HRVIR system afford the opportunity for coincident sampling of comparable large-area, coarse-resolution data and small-area, fine-resolution data. Intercomparison of the data from both systems is facilitated by the use of the same geometric reference system for both instruments and the spectral bands common to the two systems (Table 6.6). Given the flexibility of the combined system, data users can tailor sampling strategies to combine the benefits of Vegetation's revisit capability and the HRVIR system's high spatial resolution to validate the lower resolution data. Applications of the combined data sets range from regional forecasting of crop yield to monitoring forest cover and observing long term environmental change.

6.13   SPOT-5

On May 3, 2002, the SPOT program entered a new era with the successful launch of SPOT-5. The satellite carries two *high resolution geometric (HRG)* instruments, a single *high resolution stereoscopic (HRS)* instrument, and a Vegetation instrument similar to that on SPOT-4. The HRG systems are designed to provide high spatial resolution, with either 2.5- or 5-m-resolution panchromatic imagery, 10 m resolution in the green, red, and near-IR multispectral bands, and 20 m resolution in the mid-IR band. The panchromatic band has a spectral range similar to that of the HRV on SPOT-1, -2, and -3 (0.48 to 0.71 μm), rather than the high resolution red band that was adopted

TABLE 6.7   Sensors Carried on SPOT-5

Sensor	Spectral Band (μm)	Spatial Resolution (m)	Swath Width (km)	Stereoscopic Coverage
HRG	Pan: 0.48–0.71	2.5 or 5[a]	60–80	Cross-track pointing to $\pm 31.06°$
	B1: 0.50–0.59	10		
	B2: 0.61–0.68	10		
	B3: 0.78–0.89	10		
	B4: 1.58–1.75	20		
HRS	Pan: 0.49–0.69	5–10[b]	120	Along-track stereo coverage
Vegetation 2[c]	B0: 0.45–0.52	1000	2250	
	B2: 0.61–0.68	1000		
	B3: 0.78–0.89	1000		
	B4: 1.58–1.75	1000		

[a]The HRG panchromatic band has two linear arrays with 5 m spatial resolution that can be combined to yield 2.5 m resolution.

[b]The HRS panchromatic band has a spatial resolution of 10 m in the along-track direction and 5 m in the across-track direction.

[c]For consistency with the band numbering on other SPOT sensors, the Vegetation instrument's blue band is numbered 0 and the red band is numbered 2; there is no band 1.

for the SPOT-4 HRVIR. The high resolution mode in the panchromatic band is achieved by using two linear arrays of CCDs, offset horizontally within the focal plane by one-half the width of a CCD. Each linear array has a nominal ground sampling distance of 5 m. During processing at the ground station, data from the two linear arrays are interlaced and interpolated, resulting in a single panchromatic image with 2.5 m resolution.

The HRS instrument incorporates fore-and-aft stereo collection of panchromatic imagery, to facilitate the preparation of digital elevation models (DEMs) at a resolution of 10 m. It is also possible to acquire across-track stereoscopic imagery from the twin HRG instruments, which are pointable to $\pm 31°$ off nadir. Table 6.7 summarizes the spatial and spectral characteristics of the HRG and HRS instruments on SPOT-5.

6.14   SPOT IMAGE INTERPRETATION

The use of SPOT HRV and HRVIR data for various interpretive purposes is facilitated by the system's combination of multispectral sensing with moderate spatial resolution, high geometric fidelity, and the provision for multidate and stereo imaging.

The spatial resolution of SPOT-1, -2, and -3 panchromatic imagery is illustrated in Figure 6.26, which shows Baltimore's Inner Harbor. Many individual

Figure 6.26 SPOT-1 panchromatic image, Baltimore, MD, Inner Harbor, late August. Scale 1:70,000. (Copyright © 1986 CNES. Courtesy SPOT Image Corp.)

buildings, industrial and harbor facilities, and the transportation network are clearly visible.

The stereoscopic imaging capability of the SPOT satellites is illustrated in Figure 6.27, a stereopair acquired on the third and fourth days after the launch of SPOT-1. In these images of Libya, an eroded plateau is seen at the top and center, and alluvial fans are shown at the bottom. Some vertical streaking can be seen, especially in the left-hand image. This artifact is present in this early image from the system because the scene was processed using preflight calibration parameters to record this "engineering data set" and the individual detectors of the CCDs were not fully calibrated at that time.

Using the parallax resulting when SPOT data are acquired from two different orbit tracks, perspective views of a scene can be calculated and displayed. Figure 6.28 shows a perspective view of the Albertville area in the French Alps generated entirely from SPOT data acquired from two different

Figure 6.27 SPOT-1 panchromatic image, stereopair, February 24 and 25, 1986, Libya. (Copyright © 1986 CNES. Courtesy SPOT Image Corp.)

orbit tracks. Perspective views can also be produced by processing data from a single image with digital elevation data of the same scene (as discussed in Chapter 7).

Figure 6.29 shows SPOT panchromatic images of an area in the Mojave Desert located approximately 60 km southeast of Barstow, California. These images were acquired before and after a major earthquake (magnitude 7.5), the Landers earthquake of June 28, 1992. The Emerson fault, which runs

Figure 6.28 Perspective view of the Albertville area in the French Alps, generated entirely from SPOT data, on successive days in late July. (Produced by ISTAR. Copyright © 1993 CNES. Provided by SPOT Image Corp.)

from upper left to lower right in these images, cracked extensively during the earthquake with motion across the fault of about 4 m. The trace of ground cracks created during the earthquake is clearly evident in the postearthquake images (*b*) and (*d*). Although individual cracks are not visible, they are so numerous within the fault zone that collectively they appear darkened in the postearthquake images. The preearthquake images (*a*) and (*c*) do not show any fault cracks.

Figure 6.30 shows a portion of a SPOT-5 image collected over Stockholm, Sweden. This image was among the first acquired after SPOT-5's launch in May, 2002. The 10-m red band is shown in (*a*), and 2.5-m panchromatic imagery for the central city is shown in (*b*). As discussed in Section 6.13, this 2.5 m resolution is achieved through interpolation of the data collected by two separate 5 m resolution linear array sensors. Stockholm is situated along a waterway between Lake Mälaren, on the left side of (*a*), and an inlet of the

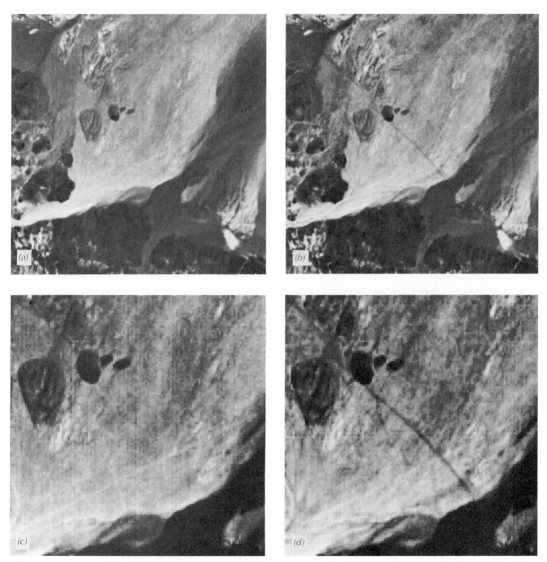

Figure 6.29 SPOT panchromatic images, Mohave Desert, 60 km southeast of Barstow, CA, before and after a major earthquake: (*a, c*) July 27, 1991; (*b, d*) July 25, 1992, 27 days after the earthquake. Scale 1 : 36,000 for (*a*) and (*b*); scale 1 : 18,000 for (*c*) and (*d*). (Copyright © 1992 CNES. Courtesy NASA Jet Propulsion Laboratory.)

Baltic Sea. Between the city and the sea lies an archipelago of some 24,000 islands, only some of which are visible at the right side of this image. The Old City area includes the royal palace, located near the north end of the island of Stadsholmen, in the center of image (*b*). At 2.5 m resolution, numerous bridges, docks, and boats can be seen along the waterways of the city.

Figure 6.30 SPOT-5 HRG imagery of Stockholm, Sweden, May, 2002: (*a*) Band 2, 10 m resolution (red); (*b*) panchromatic band, 2.5 m resolution. (Courtesy CNES and SPOT Image, Inc.)

452

Plate 18 shows four images obtained with the SPOT Vegetation instrument. In (a), the red tones associated with the healthy green vegetation in western Europe contrast sharply with the yellow-brown colors of the North African desert. Snow cover in the Alps is also readily apparent in this image. In (b), most of the area around the Red Sea does not support much vegetation, but the Nile River floodplain and delta stand out in red tones near the upper left of the image. In (c) much of the land area in this portion of Southeast Asia is a bright red color associated with lush tropical vegetation. Plate 18d is a combination of red, near-IR, and mid-IR bands, in contrast to the blue, red, and near-IR combination used for (a), (b), and (c). Here the resulting colors are quite different than seen in the previous color band combinations. This image covers portions of the U.S. Pacific Northwest and western Canada. Mountainous areas with significant snow cover and bare rock are a distinct pink color. The areas with the most lush vegetation are depicted in brown tones in areas of Washington and Oregon west of the Cascade Mountains and on Vancouver Island, British Columbia. (The brown colors result from high reflection in the near-IR and mid-IR bands.) The white streaks over the ocean area at upper left are aircraft contrails. The bright white area in the lower left part of the image is coastal fog.

6.15   OTHER EARTH RESOURCE SATELLITES

Due to their historical long term use and widespread availability, Landsat and SPOT data have been emphasized in this discussion. However, it should be pointed out that many other earth observation satellite systems have been operated or are planned for launch in the not too distant future. These can be generally classified into the categories of moderate resolution systems, high resolution systems, and hyperspectral systems. Representative examples of each of these categories are discussed in this section.

Moderate Resolution Systems

The Republic of India has successfully developed, launched, and operated several moderate resolution *Indian Remote Sensing (IRS)* satellite systems, beginning with IRS-1A in 1988 and continuing with three additional IRS-series satellites in the 1990s. IRS-1A and -B included the *Linear Imaging Self-scanning Sensor (LISS)-I and -II*, multispectral sensors roughly equivalent to the MSS and TM systems from the Landsat program. IRS-1C and IRS-1D carry three sensors: the *LISS-III*, a panchromatic sensor, and a *Wide Field Sensor (WiFS)*. Figure 6.31 is a portion of a LISS-III panchromatic image that includes the Denver International Airport and illustrates the level of detail

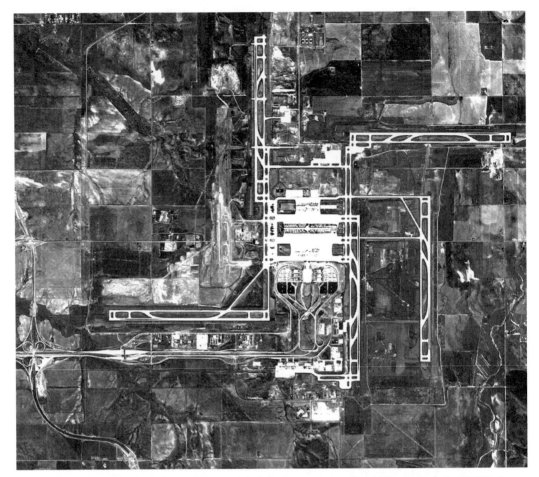

Figure 6.31 IRS panchromatic image (5.8 m resolution), Denver International Airport. Scale 1 : 85,000. (Image courtesy of Space Imaging, Thornton, CO.)

available with the 5.8-m data. Table 6.8 summarizes the characteristics of IRS-1A through IRS-1D.

The former Soviet Union launched two *RESURS-O1* satellites in 1985 and 1988. The RESURS program was then continued by Russia, with launches of RESURS-O1-3 in 1994 and RESURS-O1-4 in 1998. The latter two were the first in the series from which data were readily available outside the former Soviet Union; these data are now distributed by Eurimage. The orbital altitude, spatial resolution, and swath width for the RESURS-O1 satellites have changed significantly over the duration of the program. The characteristics of RESURS-O1-3 and RESURS-O1-4 are summarized in Table 6.9.

TABLE 6.8   Summary of IRS-1A through IRS-1D Satellite Characteristics

Satellite	Launch Year	Sensor	Spatial Resolution (m)	Spectral Band (μm)	Swath Width (km)
IRS-1A	1988	LISS-I	72.5	0.45–0.52	148
				0.52–0.59	
				0.62–0.68	
				0.77–0.86	
		LISS-II	36.25	0.45–0.52	146
				0.52–0.59	
				0.62–0.68	
				0.77–0.86	
IRS-1B	1991	Same as IRS-1A			
IRS-1C	1995	Pan	5.8	0.50–0.75	70
		LISS-III	23	0.52–0.59	142
				0.62–0.68	
				0.77–0.86	
			70	1.55–1.70	148
		WiFS	188	0.62–0.68	774
			188	0.77–0.86	774
IRS-1D	1997	Same as IRS-1C			

TABLE 6.9   Summary of RESURS-O1-3 and RESURS-O1-4 Satellite Characteristics

Satellite	Launch Year	Sensor	Spatial Resolution (m)		Spectral Band (μm)	Swath Width (km)
			Along Track	Across Track		
RESURS-O1-3	1994	MSU-E	35	45	0.5–0.6	45
					0.6–0.7	
					0.8–0.9	
		MSU-SK	140	185	0.5–0.6	600
					0.6–0.7	
					0.7–0.8	
					0.8–1.0	
			560	740	10.4–12.6	
RESURS-O1-4	1998	MSU-E	33	29	0.5–0.6	58
					0.6–0.7	
					0.8–0.9	
		MSU-SK	130	170	0.5–0.6	710
					0.6–0.7	
					0.7–0.8	
					0.8–1.0	
			520	680	10.4–12.6	

The RESURS-O1 satellites carry the *MSU-E* and *MSU-SK* multispectral scanners. The MSU-E is a linear array system that can be pointed up to 30° off nadir. In contrast, the MSU-SK employs a conical scanning procedure. In this approach to data acquisition, the IFOV of the system is aimed obliquely forward and the scan lines become a successive series of circular segments where the conical path of the field of view intersects the surface of the earth below. This procedure yields a constant resolution and viewing angle for all pixels, resulting in high radiometric accuracy. The unusual geometry of the MSU-SK image data is accounted for in the system correction of the image data during ground processing.

Launched by Japan on August 17, 1996, the *Advanced Earth Observing Satellite* (*ADEOS*) carried two primary sensing systems: the *Advanced Visible and Near Infrared Radiometer* (*AVNIR*) and the *Ocean Color and Temperature Sensor* (*OCTS*). The ADEOS acquired global earth observation data for approximately 7 months, prior to structural damage to its solar array. The AVNIR instrument is summarized in Table 6.10, while OCTS is discussed with other ocean-monitoring systems in Section 6.17.

While designed primarily as a radar remote sensing mission (see Chapter 8), Japan's *JERS-1* satellite (launched February 11, 1992) also carried the Optical Sensor (OPS), an along-track multispectral sensor. The OPS system included along-track stereoscopic coverage in one band, for topographic mapping purposes. Table 6.11 characterizes the JERS-1 OPS.

China and Brazil pursued joint development of *CBERS-1*, launched on October 14, 1999. CBERS-1 carried three sensors, including a CCD camera, a moderate resolution multispectral scanner, and a coarse resolution wide-field imaging system (Table 6.12). The CCD camera was designed to provide across-track stereoscopic coverage.

The *New Millennium Program* (*NMP*), managed by NASA's Jet Propulsion Laboratory, is aimed at developing and testing new technology in space missions. The NMP's first *Earth Orbiter* (*EO-1*) mission, launched on November 21, 2000, focused on validating a range of new developments in lightweight materials, detector arrays, spectrometers, communication, power generation, propulsion, and data storage. The EO-1 mission includes three land-imaging

TABLE 6.10 Summary of the AVNIR Sensor on ADEOS

Spectral Band (μm)	Spatial Resolution (m)	Swath Width (km)	Stereoscopic Coverage
0.52–0.69	8	80	Cross-track pointing to ±40°
0.42–0.50	16	80	Cross-track pointing to ±40°
0.52–0.60	16		
0.61–0.69	16		
0.76–0.89	16		

TABLE 6.11   Summary of the OPS Sensor on JERS-1

Spectral Band (μm)	Spatial Resolution (m)	Swath Width (km)	Stereoscopic Coverage
0.52–0.60	18 × 24	75	None
0.63–0.69			None
0.76–0.86			Along-track
1.60–1.71			None
2.01–2.12			None
2.13–2.25			None
2.27–2.40			None

instruments: the *Atmospheric Corrector* (*AC*) and *Hyperion*, both of which are discussed below in the context of hyperspectral satellite systems, and the *Advanced Land Imager* (*ALI*).

The ALI represents a technology verification project designed to demonstrate comparable or improved Landsat ETM+ spatial and spectral resolution with substantial reduction in sensor mass, volume, and cost. For example, the ALI consists of only 25 percent of the mass, requires only 20 percent of the electric power, and involves only approximately 40 percent of the overall mission cost of launching the ETM+. The ALI employs pushbroom scanning and features 10-m-resolution panchromatic and 30-m-resolution multispectral image acquisition. Table 6.13 provides a comparison of the spectral bands and resolutions for the ETM+ and the ALI. The ALI bands have been designed to mimic several of the ETM+ bands as well as to test the utility of bands covering 0.433 to 0.453 μm, 0.845 to 0.890 μm, and 1.20 to 1.30 μm. Finally, in order to facilitate the comparison of ALI and ETM+ imagery, the

TABLE 6.12   Sensors Carried on CBERS-1

Sensor	Spectral Band (μm)	Spatial Resolution (m)	Swath Width (km)	Stereoscopic Coverage
CCD camera	0.51–0.73	20	113	Cross-track pointing to ±32°
	0.45–0.52			
	0.52–0.59			
	0.63–0.69			
	0.77–0.89			
IR-MSS	0.50–1.10	80	120	None
	1.55–1.75			
	2.08–2.35			
	10.40–12.50	160	120	None
WFI	0.63–0.69	260	890	None
	0.76–0.90			

TABLE 6.13   Comparison of Landsat-7 ETM+ and ALI Spectral Bands and Spatial Resolutions

ETM+		ALI	
Band (μm)	Resolution (m)	Band (μm)	Resolution (m)
0.450–0.515	30	0.433–0.453	30
0.525–0.605	30	0.450–0.510	30
0.630–0.690	30	0.525–0.605	30
0.750–0.900	30	0.630–0.690	30
1.55–1.75	30	0.775–0.805	30
10.40–12.50	60	0.845–0.890	30
2.09–2.35	30	1.20–1.30	30
0.520–0.900 (pan)	15	1.55–1.75	30
		2.08–2.35	30
		0.480–0.680 (pan)	10

EO-1 satellite was maneuvered into an orbit that matched that of Landsat-7 to within 1 min.

A partial list of those countries (or agencies) operating, planning, or developing various remote sensing satellites has been provided by Glackin (1998). The list includes Argentina, Australia, Brazil, Canada, China, European Space Agency (ESA), France, Germany, India, Israel, Italy, Japan, Korea, Malaysia, Russia, South Africa, Spain, Taiwan, Thailand, United Kingdom, Ukraine, and the United States. Space limits our description of these satellites, and many of the details of the launch and operation of these satellites are subject to change.

High Resolution Systems

Numerous systems have been launched, or are in the development phase, that achieve much higher spatial resolution than those described earlier in this discussion. Following are descriptions of some of the most widely available sources of high resolution data. These systems are summarized in Table 6.14.

The first successful launch of a commercially developed high resolution earth observation satellite occurred on September 24, 1999, with the launch of Space Imaging's *IKONOS* system. IKONOS occupies a 682-km sun-synchronous orbit with an equatorial crossing time of 10:30 A.M. The ground track of the system repeats every 11 days, but the revisit time for

TABLE 6.14   High Resolution Earth Observation Satellite Systems

Satellite	Launch Date	Spatial Resolution (m)	Spectral Bands (μm)	Swath Width (km)	Altitude (km)
IKONOS	Sept. 24, 1999	1 4	Pan: 0.45–0.90 1: 0.45–0.52 2: 0.52–0.60 3: 0.63–0.69 4: 0.76–0.90	11	681
EROS-A	Dec. 5, 2000	1.8	Pan: 0.50–0.90	13.5	480
QuickBird	Oct. 18, 2001	0.61 2.40	Pan: 0.45–0.90 1: 0.45–0.52 2: 0.52–0.60 3: 0.63–0.69 4: 0.76–0.90	16.5	450
OrbView-3	2003 (planned)	1 4	Pan: 0.45–0.90 1: 0.45–0.52 2: 0.52–0.60 3: 0.63–0.69 4: 0.76–0.90	8	470
EROS-B1	2004 (planned)	0.82 3.48	Pan: 0.50–0.90 Four multispectral bands planned	13	600

imaging is less than 11 days, based on latitude and the angular orientation of the system selected to acquire any given image. The system includes the capability to collect data at angles of up to 45° from vertical both in the across-track and along-track directions. This not only provides the opportunity for frequently covering a given area but also enables the collection of both side-by-side (cross-track) and fore-and-aft (along-track) stereoscopic imagery. At nadir, the system has a swath width of 11 km. A typical image is 11 × 11 km in size, but user-specified image strips and mosaics can also be collected.

IKONOS employs linear array technology and collects data in four multispectral bands at a nominal ground resolution of 4 m, as well as a 1-m-resolution panchromatic band. The panchromatic band and the multispectral bands can be combined to produce "pan-sharpened" multispectral imagery (with an effective resolution of 1 m). As noted in Table 6.14, the spectral ranges for these bands are essentially identical to their counterparts on the Landsat-7 ETM+. Data are collected over 2048 gray levels (11 bits).

The IKONOS satellite is designed to be highly maneuverable. The system is capable of pointing to a new target and stabilizing itself within a few seconds. This enables the system to be programmed to follow meandering features or power lines and similar features. The entire spacecraft, not just the optical system, is pointed in the direction of data collection.

The first IKONOS image, collected on September 30, 1999, is shown in Figure 6.32*a*. Enlargements of selected portions of the image, including a runway at National Airport (*b*) and the Washington Monument (*c*), illustrate the level of detail that can be seen in 1-m-resolution panchromatic imagery. At the time this image was acquired, the Washington Monument was undergoing renovation, and scaffolding can be seen on the side facing the sensor. The monument also shows the effects of relief displacement in this image.

Plate 19 shows a small portion of an IKONOS image acquired during early summer over an agricultural area in eastern Wisconsin. Included are a false-color composite of IKONOS multispectral bands 2, 3, and 4 at original 4 m resolution; the panchromatic band at 1 m resolution; and a merged image

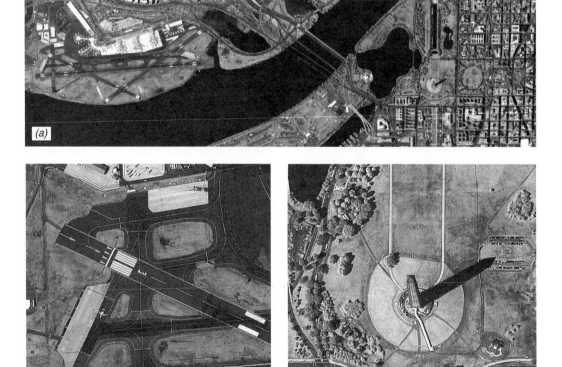

Figure 6.32 First IKONOS image, acquired on September 30, 1999, over Washington, DC: (*a*) Panchromatic image; (*b*) enlargement showing runway at National Airport; (*c*) enlargement showing Washington Monument (with scaffolding present during renovation activity). (Courtesy Space Imaging, Inc.)

showing the results of the pan-sharpening process (Section 7.6). The sharpened image has the spectral information present in the original multispectral data, but with an effective spatial resolution of 1 m.

ImageSat International launched *EROS-A* on December 5, 2000. The first of a planned series of several high resolution satellites, EROS-A has a single pushbroom panchromatic sensor with 1.8 m resolution. Off-nadir pointing (up to 45°) in any direction is possible, by changing the angular orientation of the satellite. Current plans call for a second system, *EROS-B1*, to be launched in 2004, with a 0.82-m-resolution panchromatic band and four 3.48-m-resolution multispectral bands. If the proposed constellation of six EROS satellites is successfully brought into operation, it will permit at least daily, if not more frequent, imaging of any point within a ground station's coverage area. Currently, EROS-A data are distributed through ImageNet, a service of Core Software Technology.

The highest resolution satellite imagery currently available to the public is provided by *QuickBird*, launched on October 18, 2001, and operated by DigitalGlobe, Inc. The system is in a sun-synchronous, relatively low orbit, at an altitude of 450 km. The average revisit time is 1 to 3.5 days, depending on latitude and image collection angle. QuickBird features a 0.61-m-resolution panchromatic sensor and a four-band multispectral sensor having a resolution of 2.40 m. As shown in Table 6.14, the spectral ranges of these linear array sensors are essentially equivalent to those of the IKONOS system. Similarly, 11-bit data recording is employed.

Figure 6.33 shows a panchromatic image (61 cm resolution) of the Taj Mahal of India, collected by QuickBird on February 15, 2002. The complex was built by the Muslim Emperor Shah Jahan as a mausoleum for his wife, queen Mumtaz Mahal. The height of the minarets at the four corners of the main structure is indicated by the length of their shadows. In addition, the peaks of the domes appear to be offset from the centers of the domes, due to relief displacement in this image.

Plate 20 illustrates the use of pan-sharpening techniques to produce color QuickBird images with an effective 61 cm resolution. The image shows a true-color composite derived from the merger of data from the QuickBird panchromatic band and multispectral bands 1, 2, and 3 shown as blue, green, and red, respectively. It was acquired on May 10, 2002, over the Burchardkai Container Terminal at the Port of Hamburg. The outlines and colors of individual cargo containers can be readily differentiated at the resolution of this image.

ORBIMAGE's *OrbView-3* system is currently scheduled for launch in 2003. The satellite is to be operated at an altitude of 470 km in a sun-synchronous orbit having a 10:30 A.M. equatorial crossing time. The system orbit does not repeat exactly, but approximate repetition is accomplished in less than 3 days depending on latitude. The sensing system on board will include a panchromatic band (with 1 m resolution) and four multispectral bands (with 4 m resolution).

Figure 6.33 QuickBird panchromatic image of the Taj Mahal, India, 61 cm resolution. (Courtesy DigitalGlobe.)

The swath width of the instrument is 8 km at nadir, and the sensor can be tilted up to 45° cross-track. The spectral bands of the system are functionally equivalent to those of the IKONOS and QuickBird instruments.

While this discussion has focused on electro-optical systems, high resolution photographic satellite imagery has been available for several years from a series of formerly classified Russian military satellite systems, known as *Space Information-2 Meter* (*SPIN-2*). SPIN-2 data are produced by digitizing panchromatic (0.51 to 0.76 μm) photographs taken by the KUR-1000 camera, a panoramic camera system employing a 1-m-focal-length lens. Operated from an altitude of 220 km, the camera acquires photographs at an average scale of 1:220,000 with individual scenes covering an area of 40 × 160 km, with a GRD of approximately 1 m at nadir. The average ground cell size for the digitized data is 1.56 m per pixel. Both photographic and digital data derived from the system are available through TerraServer on the Internet (www.terraserver.com).

SPIN-2 image data are rectified by operating the KUR-1000 camera in conjunction with a 10-m-resolution "topographic camera," the TK-350. The TK-350 has a 350-mm focal length and produces photographs at a scale of 1:660,000 with ground coverage of 200 × 300 km. These photographs are acquired with 80 percent overlap and are used to prepare DEMs to rectify the KUR-1000 images. The exterior orientation parameters for the camera are determined by two onboard systems that base their calculations on a combination of celestial readings and laser altimeter data. The quoted geometric accuracy of the 2-m data is 10 m without the use of supplemental ground control and 3 to 4 m with supplemental control. The quoted vertical accuracy of the DEMs produced from the 10-m data is 10 m without ground control and 5 m when ground control is used.

Hyperspectral Satellite Systems

In Chapter 5, we discussed the concept of airborne hyperspectral scanning. This involves collecting data in numerous, narrow, contiguous spectral bands. At the time of this writing (2002), there was only one operating spaceborne hyperspectral scanning system, the Hyperion instrument carried on the previously mentioned EO-1 spacecraft. Several other attempts to launch hyperspectral satellite systems have ended in failure. With other such systems currently in development, it is expected that spaceborne hyperspectral imaging will experience a period of significant growth over the coming decade.

The EO-1 Hyperion instrument nominally provides 242 spectral bands of data over the 0.36- to 2.6-μm range, each of which has a width of 0.010 to 0.011 μm. Some of the bands, particularly those at the lower and upper ends of the range, exhibit a poor signal-to-noise ratio. As a result, during Level 1 processing, only 198 of the 242 bands are calibrated; radiometric values in the remaining bands are set to 0 for most data products. Hyperion consists of two distinct linear array sensors, one with 70 bands in the UV, visible, and near IR, and the other with 172 bands in the near IR and mid IR (with some overlap in the spectral sensitivity of the two arrays). The spatial resolution of this experimental sensor is 30 m, and the swath width is 7.5 km. Data from the Hyperion system are distributed by the USGS.

The EO-1 AC (also referred to as the *LEISA AC or LAC*) is a hyperspectral imager of coarse spatial resolution covering the 0.85- to 1.5-μm-wavelength range. It was designed to correct imagery from other sensors for atmospheric variability due primarily to water vapor and aerosols. The AC has a spatial resolution of 250 m at nadir. Acquisition of AC data was ended after the first year of the EO-1 mission, though Hyperion and ALI data continue to be collected as of the time of this writing (2002).

Several additional hyperspectral systems have been proposed or are in development. The U.S. Navy's Office of Naval Research has sponsored the development of a hyperspectral imaging system to be launched on the proposed *Naval EarthMap Observer* (*NEMO*) satellite. This hyperspectral instrument on NEMO would acquire data in 200 spectral bands over the range of 0.4 to 2.5 μm. It would permit data acquisition at resolutions of either 30 or 60 m over a 30-km-wide swath, with cross-track pointing capability. The satellite would be launched into a 605-km, sun-synchronous orbit with a 10:30 A.M. ascending equatorial crossing time. A 5-m-resolution *Panchromatic Imaging Camera* (*PIC*) would be flown in combination with the hyperspectral instrument, covering the same 30-km swath.

The NEMO program features many design characteristics that are directly responsive to the Navy's needs for data collection in the coastal zone of the world's oceans. Among these needs are information on shallow water bathymetry, bottom-type composition, currents, oil slicks, underwater hazards, water clarity, atmospheric visibility, beach characterization, and near-shore soil and vegetation. Several of these requirements dictate that the system have a very high signal-to-noise ratio for maximum water penetration over the very low reflectance range typical of ocean water. Because of this, along-track pointing is used to increase the integration time over which data are collected in the 30-m mode. This technique is called *ground motion compensation* (*GMC*). At the beginning of a scene the spacecraft is pointed forward and continues to stare at the same ground position as the satellite moves forward. This mode of operation results in a fivefold increase in the dwell time over which the system views the scene elements below and a corresponding increase in signal-to-noise performance.

Another unique aspect of the NEMO system is onboard feature recognition and data compression based on the *Optical Real-Time Adaptive Spectral Identification System* (*ORASIS*) data processing algorithm. This computationally intensive process is implemented using an *Imagery On-Board Processor* (*IOBP*). The basic approach employed by the ORASIS system is to analyze the spectrum from each pixel in a scene sequentially, discarding duplicate spectra, and subsequently analyzing only the unique spectra using such procedures as subpixel spectral demixing and feature extraction (Chapter 7).

Another proposed hyperspectral satellite system is the *Australian Resource Information and Environmental Satellite* (*ARIES*). Its nominal specifications include a hyperspectral imager with up to 105 bands located between 0.4 and 2.5 μm, 30 m resolution, a 15-km swath, up to 30° cross-track pointing, and a 7-day revisit period. The system also would include a 10-m panchromatic imager. While designed primarily for mineral exploration, ARIES would have substantial utility in applications ranging from agriculture and forestry to wetland mapping and environmental monitoring. Indeed, the future for this and similar hyperspectral satellite systems is an extremely bright one.

6.16 METEOROLOGICAL SATELLITES FREQUENTLY APPLIED TO EARTH SURFACE FEATURE OBSERVATION

Designed specifically to assist in weather prediction and monitoring, *meteorological satellites*, or *metsats*, generally incorporate sensors that have very coarse spatial resolution compared to land-oriented systems. On the other hand, metsats afford the advantages of global coverage at very high temporal resolution. Accordingly, metsat data have been shown to be useful in natural resource applications where frequent, large-area mapping is required and fine detail is not. Apart from the advantage of depicting large areas at high temporal resolution, the coarse spatial resolution of metsats also greatly reduces the volume of data to be processed for a particular application.

Numerous countries have launched various types of metsats with a range of orbit and sensing system designs. In the remainder of this section we treat only three representative series of metsats that are operated by the United States. The first is the *NOAA* series named after the *National Oceanic and Atmospheric Administration*. These satellites are in near-polar, sun-synchronous orbits similar to those of Landsat and SPOT. In contrast, the *GOES* series of satellites are geostationary, remaining in a constant relative position over the equator. Solving Eq. 6.1 for an orbital period of 24 hr yields an orbital altitude of about 36,000 km above the earth's surface for geostationary satellites. GOES is an acronym for *Geostationary Operational Environmental Satellite*. The final systems we discuss are part of the *Defense Meteorological Satellite Program* (*DMSP*). All three of these satellite series carry a range of meteorological sensors. We treat only the salient characteristics of those sensors used most often in land remote sensing applications.

NOAA Satellites

Several generations of satellites have been flown in the NOAA series. Germane to this discussion are the NOAA-6 through NOAA-17 missions that contained the *Advanced Very High Resolution Radiometer* (*AVHRR*). NOAA-6, -8, -10, -12, -15, and -17 have daytime (7:30 A.M., or 10:00 A.M. for NOAA-17) north-to-south equatorial crossing times while NOAA-7, -9, -11, -14, and -16 have nighttime (1:30 to 2:30 A.M.) north-to-south equatorial crossing times. Table 6.15 lists the basic characteristics of these missions and the AVHRR instrument. Figure 6.34 shows the 2400-km swath width characteristic of the system. Coverage is acquired at a ground resolution of 1.1 km at nadir. This

TABLE 6.15 Characteristics of NOAA-6 to -17 Missions

Parameter	NOAA-6, -8, -10, -12, -15, and -17	NOAA-7, -9, -11, -14, and -16[a]
Launch	6/27/79, 3/28/83, 9/17/86, 5/14/91, 5/13/98, 6/24/02	6/23/81, 12/12/84, 9/24/88, 12/30/94, 9/21/00
Altitude, km	833	870
Period of orbit, min	101	102
Orbit inclination	98.7°	98.9°
Orbits per day	14.2	14.1
Distance between orbits	25.6°	25.6°
Day-to-day orbital shift[b]	5.5° E	3.0° E
Orbit repeat period (days)[c]	4–5	8–9
Scan angle from nadir	±55.4°	±55.4°
Optical field of view, mrad	1.3	1.3
IFOV at nadir, km	1.1	1.1
IFOV off-nadir maximum, km		
Along track	2.4	2.4
Across track	6.9	6.9
Swath width, km	2400	2400
Coverage	Every 12 hr	Every 12 hr
Northbound equatorial crossing[d]	7:30 P.M.	1:30–2:30 P.M.
Southbound equatorial crossing[d]	7:30 A.M.	1:30–2:30 A.M.
AVHRR spectral channels, μm		
1	0.58–0.68	0.58–0.68
2	0.72–1.10	0.72–1.10
3	3.55–3.93[e]	3.55–3.93
4	10.5–11.50	10.3–11.30
5	Channel 4 repeat[f]	11.5–12.50

[a]NOAA-13 failed due to a short circuit in its solar array.

[b]Satellite differences due to differing orbital alignments.

[c]Caused by orbits per day not being integers.

[d]NOAA-17 has equatorial crossing times of 10:00 P.M. (northbound) and 10:00 A.M. (southbound).

[e]NOAA-15, -16, and -17 include two separate channels: 3A (1.58–1.64 μm) and 3B (3.55–3.93 μm).

[f]NOAA-12, -15, and -17 include a separate channel 5.

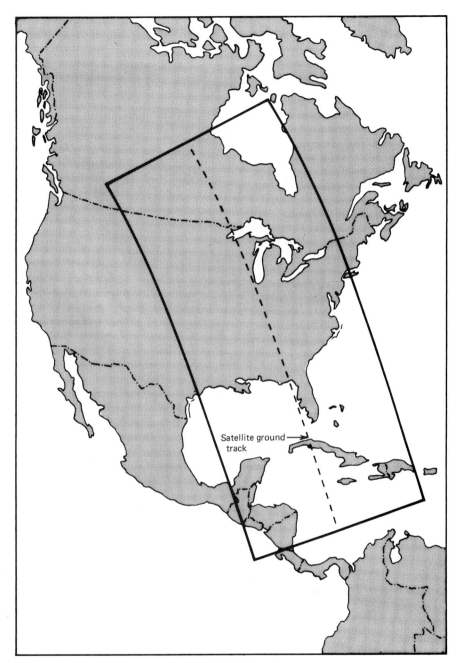

Figure 6.34 Example coverage of the NOAA AVHRR.

resolution naturally becomes coarser with increases in the viewing angle off nadir. NOAA receives AVHRR data at full resolution and archives them in two different forms. Selected data are recorded at full resolution, referred to as local area coverage (LAC) data. All of the data are sampled down to a nominal resolution of 4 km, referred to as global area coverage (GAC) data. Figure 6.35 summarizes the spectral sensitivity of the AVHRR relative to the Landsat MSS, the Landsat TM, the Landsat ETM+, the SPOT HRV, SPOT HRVIR, and the SPOT Vegetation sensor.

NOAA satellites provide daily (visible) and twice-daily (thermal IR) coverage. Images and digital tapes are used operationally in a host of applications requiring timely data. For example, Plate 21 illustrates the use of NOAA AVHRR thermal data for water temperature mapping. Here, the AVHRR data provide a synoptic view of water surface temperatures in all five of the Laurentian Great Lakes simultaneously.

In addition to surface water temperature mapping, AVHRR data have been used extensively in applications as varied as snow cover mapping, flood monitoring, vegetation mapping, regional soil moisture analysis, wildfire fuel mapping, fire detection, dust and sandstorm monitoring, and various geologic applications including observation of volcanic eruptions and mapping of regional drainage and physiographic features. Both AVHRR and GOES data (described below) have also been used in regional climate change studies. For example, Wynne et al. (1998) have used 14 years of such data to monitor the phenology of lake ice formation and breakup as a potential climate change indicator in the U.S. Upper Midwest and portions of Canada.

AVHRR data have been used extensively for large-area vegetation monitoring. Typically, the spectral bands used for this purpose have been the channel 1 visible band (0.58 to 0.68 μm) and the channel 2 near-IR band (0.73 to 1.10 μm). Various mathematical combinations of the AVHRR channel 1 and 2 data have been found to be sensitive indicators of the presence and condition of green vegetation. These mathematical quantities are thus referred to as *vegetation indices*. Two such indices have been routinely calculated from AVHRR data—a simple vegetation index (VI) and a normalized difference vegetation index (NDVI). These indices are computed from the equations

$$VI = Ch_2 - Ch_1 \tag{6.2}$$

and

$$NDVI = \frac{Ch_2 - Ch_1}{Ch_2 + Ch_1} \tag{6.3}$$

where Ch_1 and Ch_2 represent data from AVHRR channels 1 and 2, preferably expressed in terms of radiance or reflectance (see Chapter 7).

Vegetated areas will generally yield high values for either index because of their relatively high near-IR reflectance and low visible reflectance. In contrast, clouds, water, and snow have larger visible reflectance than near-IR reflectance.

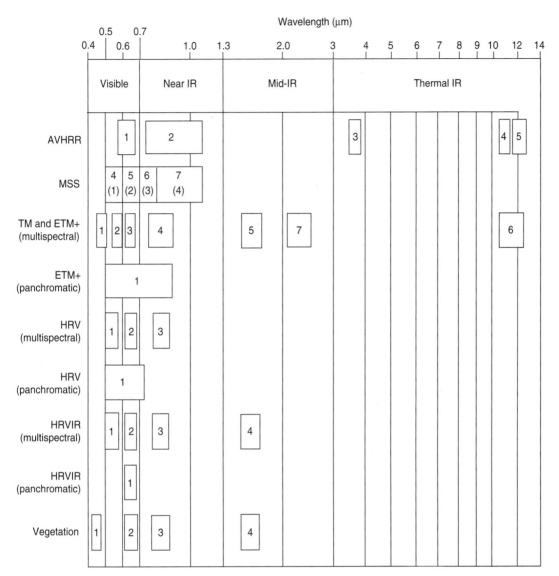

Figure 6.35 Summary of spectral sensitivity of NOAA AVHRR, Landsat MSS, TM, ETM+, and SPOT HRV, HRVIR, and Vegetation Instrument. Shown are the band numbers and associated bandwidths for each sensing system.

Thus, these features yield negative index values. Rock and bare soil areas have similar reflectances in the two bands and result in vegetation indices near zero.

The normalized difference vegetation index is preferred to the simple index for global vegetation monitoring because the NDVI helps compensate for changing illumination conditions, surface slope, aspect, and other extraneous factors.

Plate 22 shows NDVI maps of the conterminous United States for six 2-week periods during 1998 prepared from AVHRR data. The NDVI values for vegetation range from a low of 0.05 to a high of 0.66+ in these maps and are displayed in various colors (see key at bottom of plate). Clouds, snow, and bright nonvegetated surfaces have NDVI values of less than zero. The NDVI selected for each pixel is the greatest value on any day during the 14-day period (the highest NDVI value is assumed to represent the maximum vegetation "greenness" during the period). This process eliminates clouds from the composite (except in areas that are cloudy for all 14 days).

Scientists associated with the EDC have also produced numerous global NDVI composites to meet the needs of the international community. In these efforts, data collected at 29 international receiving stations, in addition to that recorded by NOAA, were acquired (starting April 1, 1992) on a daily global basis. Several tens of thousands of AVHRR images have been archived for this program and numerous 10-day maximum NDVI composites have been produced (Eidenshink and Faundeen, 1994).

Numerous investigators have related the NDVI to several vegetation phenomena. These phenomena have ranged from vegetation seasonal dynamics at global and continental scales to tropical forest clearance, leaf area index measurement, biomass estimation, percentage ground cover determination, and photosynthetically active radiation estimation. In turn, these vegetation attributes are used in various models to study photosynthesis, carbon budgets, water balance, and related processes.

Notwithstanding the widespread use of the AVHRR NDVI, it should be pointed out that a number of factors can influence NDVI observations that are unrelated to vegetation conditions. Among these factors are variability in incident solar radiation, radiometric response characteristics of the sensor, atmospheric effects (including the influence of aerosols), and off-nadir viewing effects. Recall that the AVHRR scans over ±55° off nadir (compared to ±7.7° for Landsat and ±2.06° within a SPOT scene). This causes a substantial change not only in the size of the ground resolution cell along an AVHRR scan line but also in the angles and distances over which the atmosphere and earth surface features are viewed. Normalizing for such effects is the subject of continuing research. (We further discuss vegetation indices in Chapter 7.)

GOES Satellites

The Geostationary Operational Environmental Satellites (GOES) are part of a network of meteorological satellites located in geostationary orbit around the globe. The United States normally operates two such systems, typically located at 75° W longitude (and the equator) and at 135° W longitude (and the equator). These systems are referred to as GOES-EAST and GOES-WEST, respectively. Similar systems have been placed in operation by several other countries as part of a corporate venture within the World Meteorological Organization.

From its vantage point, GOES can see an entire hemispherical disk (Figure 6.36). The repeat frequency is therefore limited only by the time it takes to scan and relay an image. The current generation of GOES imagers (GOES-8 and -10) are five-band systems having one visible band (0.55 to 0.75 μm) and four thermal bands (3.80 to 4.00 μm, 6.50 to 7.00 μm, 10.20 to 11.20 μm, and 11.50 to 12.50 μm). The spatial resolution at nadir of the visible band is 1 km, and the thermal bands have resolutions of 4, 8, 4, and 4 km, respectively. The IFOVs for all bands are simultaneously swept east

Figure 6.36 GOES-2 visible band (0.55 to 0.7 μm) image of hemispherical disk including North and South America, early September. A hurricane is clearly discernible in the Gulf of Mexico. (Courtesy NOAA/National Environmental Satellite, Data, and Information Service.)

to west and west to east along a north-to-south path by mirrors of a two-axis mirror scan system. The instrument can produce scans of the full hemispherical disks of the earth, "sector" images that contain the edges of the earth, or images of local regions. (Prior to GOES-8, only hemispherical disk images were produced in a visible and a single thermal band.) GOES-1 was launched in 1975; GOES-8 and -10 were launched in 1994 and 1997, respectively.

GOES images are by now very familiar to us all. They are distributed in near real time for use in local weather forecasting. They have also been used in numerous other large-area analyses, such as regional snow and ice cover mapping, when higher resolution data are not required.

Defense Meteorological Satellite Program

NOAA also coordinates the Defense Meteorological Satellite Program (DMSP), which was originally administered by the U.S. Air Force. Some of the data produced from the satellites in this program have been available to civilian users on an unclassified basis since April 1973. The DMSP satellites carry a range of meteorological sensors. Scanners onboard the satellites produce images in the 0.4- to 1.1-μm (visible and near-IR) band and the 8- to 13-μm (thermal IR) band. System resolution is on the order of 3 km and sun-synchronous polar orbits permit day and night global coverage. A unique capability of the DMSP scanner is nighttime visible band imaging. This comes about through the ability to "tune" the amplifiers of the system to obtain images under low illumination conditions. The system produces vivid images of phenomena such as the urban light patterns that were shown in Figures 4.28 and 4.29. Auroral displays, volcanoes, oil and gas fields, and forest fires have also been detected with the low light sensor. Both the thermal and daytime visible images of the DMSP have been used for such civilian applications as snow extent mapping.

6.17   OCEAN MONITORING SATELLITES

The oceans, which cover more than two-thirds of the earth's surface, have important influences on global weather and climate; yet they represent a natural resource about which comparatively little is known. Satellite imaging can provide synoptic views of the oceans over large areas and extended time periods. This task is virtually impossible to accomplish with traditional oceanographic measurement techniques.

Seasat (Section 8.11) carried several instruments dedicated to ocean monitoring that operated in the microwave portion of the spectrum. Another satellite carrying ocean-monitoring sensors was the Nimbus-7 satellite, launched in Oc-

**TABLE 6.16   Spectral Bands of the Coastal Zone
Color Scanner**

Channel	Wavelength (μm)	Principal Parameter Measured
1	0.43–0.45	Chlorophyll absorption
2	0.51–0.53	Chlorophyll absorption
3	0.54–0.56	*Gelbstoffe* (yellow substance)
4	0.66–0.68	Chlorophyll concentration
5	0.70–0.80	Surface vegetation
6	10.50–12.50	Surface temperature

tober 1978. This satellite carried the *Coastal Zone Color Scanner* (*CZCS*), which was employed on a limited-coverage "proof-of-concept" mission. The CZCS was designed specifically to measure the color and temperature of the coastal zones of the oceans. Table 6.16 lists the six bands in which the CZCS operated. These included bands in the visible, near-IR, and thermal regions of the spectrum. The system had a 1566-km swath width and an 825-m-resolution cell at nadir.

The first four (visible) bands of the CZCS were very narrow (0.02 μm wide) and centered to enhance the discrimination of very subtle water reflectance differences. Data from these bands were used to successfully map phytoplankton concentrations and inorganic suspended matter such as silt. The near-IR channel was used to map surface vegetation and to aid in separating water from land areas prior to processing data in the other bands. The thermal IR channel was used to measure sea surface temperatures. In short, the CZCS was used to successfully detect chlorophyll, temperature, suspended solids, and *gelbstoffe* (yellow substance) in the combinations and concentrations typical of near-shore and coastal waters. The CZCS ceased operation in mid-1986.

Seasat and the CZCS vividly demonstrated the capability to measure important ocean properties from space. Based on this experience, various other ocean-monitoring systems have been launched and several others are in the design phase. For example, Japan launched its first *Marine Observations Satellite* (*MOS-1*) on February 19, 1987, and a successor to this system (MOS-1b) on February 7, 1990. These satellites employed the three instruments included in Table 6.17, namely a four-channel *Multispectral Electronic Self-Scanning Radiometer* (*MESSR*), a four-channel *Visible and Thermal Infrared Radiometer* (*VTIR*), and a two-channel *Microwave Scanning Radiometer* (*MSR*). (We describe passive microwave radiometers in Section 8.20.)

MOS-1 and MOS-1b operated from an orbit height of 909 km and had a revisit period of 17 days. The design life for each system was 2 years. Note that the MESSR's spectral bands were very similar to those of the Landsat MSS, making MESSR data applicable to many land as well as marine applications.

TABLE 6.17   Instruments Included in MOS-1 and MOS-1b Missions

	Instrument		
	MESSR	VTIR	MSR
Spectral bands	0.51–0.59 μm	0.50–0.70 μm	1.26 cm
	0.61–0.69 μm	6.0–7.0 μm	0.96 cm
	0.72–0.80 μm	10.5–11.5 μm	
	0.80–1.10 μm	11.5–12.5 μm	
Ground resolution	50 m	900 m (visible)	32 km (λ = 1.26 cm)
		2700 m (thermal)	23 km (λ = 0.96 cm)
Swath width	100 km	1500 km	317 km

Another major source of ocean-oriented remote sensing data has been the *Sea-viewing Wide-Field-of-View Sensor* (*SeaWiFS*). SeaWiFS incorporates an eight-channel across-track scanner operating over the range of 0.402 to 0.885 μm (See Table 6.18). This system is designed primarily for the study of ocean biogeochemistry and is a joint venture between NASA and private industry.

Under a procurement arrangement known as a "data buy," NASA contracted with Orbital Sciences Corporation (OSC) to build, launch, and operate SeaWiFS on the OSC OrbView-2 satellite to meet NASA's science requirements for ocean-monitoring data. Orbital Sciences retains the rights to market the resulting data for commercial applications. This is the first time that industry has led the development of an entire mission. The system was launched on August 1, 1997, and has produced data since September 18, 1997.

From a scientific perspective, SeaWiFS data have been used for the study of such phenomena as ocean primary production and phytoplankton processes; cycles of carbon, sulfur, and nitrogen; and ocean influences on

TABLE 6.18   Major Characteristics of the SeaWiFS System

Spectral bands	402–422 nm
	433–453 nm
	480–500 nm
	500–520 nm
	545–565 nm
	660–680 nm
	745–785 nm
	845–885 nm
Ground resolution	1.1 km
Swath width	2800 km

physical climate, including heat storage in the upper ocean and marine aerosol formation. Commercial applications have ranged from fishing to navigation, weather prediction, and agriculture.

SeaWiFS produces two major types of data. *Local Area Coverage* (*LAC*) data with 1.13 km nadir resolution are broadcast directly to receiving stations and *Global Area Coverage* (*GAC*) data (subsampled onboard every fourth line, every fourth pixel) are recorded on board the OrbView-2 spacecraft for subsequent transmission. The system was designed to obtain full global coverage (GAC) data every 2 days. The LAC data are collected less frequently on a research priority basis. A 705-km orbit, with a southbound equatorial crossing time of 12:00 noon and a ±58.3° scan angle, characterize the system. This provides a scan swath width of approximately 2800 km.

Although originally designed for ocean observation, SeaWiFS has provided a unique opportunity to study land and atmospheric processes as well. In its first year of operation alone, SeaWiFS provided new scientific insights into a number of phenomena. Among others, these include the El Niño and La Niña climate processes; a range of natural disasters, including fires in Florida, Canada, Indonesia, Mexico, and Russia; floods in China; dust storms in the Sahara and Gobi Deserts; hurricanes in numerous locations; and unprecedented phytoplankton blooms. More than 800 scientists representing 35 countries accessed the data in the first year of system operation and more than 50 ground stations throughout the world began receiving the data. In a sense, with its ability to observe ocean, land, and atmospheric processes simultaneously, SeaWIFS has acted as a precursor to the series of instruments comprising the EOS.

Plate 23 is a composite of SeaWiFS GAC data for an 11-month time period showing chlorophyll *a* concentrations in the ocean areas and normalized difference vegetation index values over the land areas.

As part of its Envisat-1 payload, the European Space Agency will operate the *Medium Resolution Imaging Spectrometer* (*MERIS*). This system is also primarily aimed at addressing the needs of oceanographic observations (with additional utility in atmospheric and land observations on a secondary basis).

The MERIS is a pushbroom instrument having a 1150-km-wide swath that is divided into five segments for data acquisition. Five identical sensor arrays collect data side by side with a slight overlap between adjacent swaths. Data are collected at 300 m nadir resolution over coastal zones and land areas. Open-ocean data can also be collected at a reduced resolution of 1200 m through onboard combinations of 4 × 4 adjacent pixels across and along track.

The MERIS has the ability to record data in as many as 15 bands. This system is programmable by ground command such that the number, location, and width of each spectral band can be varied within the 0.4- to 1.05-μm range.

6.18   EARTH OBSERVING SYSTEM

The *Earth Observing System* (*EOS*) is one of the primary components of a NASA-initiated concept originally referred to as *Mission to Planet Earth* (*MTPE*), which was renamed the *Earth Science Enterprise* (*ESE*) during 1998. The ESE is an international earth science program aimed at providing the observations, understanding, and modeling capabilities needed to assess the impacts of natural events and human-induced activities on the earth's environment. The program incorporates both space- and ground-based measurement systems to provide the basis for documenting and understanding global change with an initial emphasis on climate change. The program also focuses on the necessary data and information systems to acquire, archive, and distribute the data and information collected about the earth. The intent is to further international understanding of the earth as a system.

The EOS component of the ESE includes the currently operational observing systems (beginning with Landsat-7), new programs under development, and planned programs for the future. Clearly, programs of this magnitude, cost, and complexity are subject to change. Also, the EOS program includes numerous platforms and sensors that are outside the realm of land-oriented remote sensing. We make no attempt to describe the overall program. Rather, we limit our attention to the first two EOS-dedicated platforms, the *Terra* and *Aqua* spacecrafts. Sometimes described as the "flagship" of EOS, Terra was launched on December 18, 1999. It was followed on May 4, 2002, by Aqua. Both of these platforms are complex systems with multiple remote sensing instruments. Five sensors are included on Terra:

ASTER:	Advanced Spaceborne Thermal Emission and Reflection Radiometer
CERES:	Clouds and the Earth's Radiant Energy System
MISR:	Multi-Angle Imaging Spectro-Radiometer
MODIS:	Moderate Resolution Imaging Spectro-Radiometer
MOPITT:	Measurements of Pollution in the Troposphere

Aqua carries six instruments, two of which (MODIS and CERES) are also present on Terra. The remaining four instruments on Aqua include:

AMSR/E:	Advanced Microwave Scanning Radiometer-EOS
AMSU:	Advanced Microwave Sounding Unit
AIRS:	Atmospheric Infrared Sounder
HSB:	Humidity Sounder for Brazil

Table 6.19 summarizes the salient characteristics and applications associated with each of the instruments on Terra and Aqua.

TABLE 6.19   Sensors Carried on Terra and Aqua

Sensor	Terra/ Aqua	General Characteristics	Primary Applications
ASTER	Terra	Three scanners operating in the visible and near, mid, and thermal IR, 15–90 m resolution, along-track stereo	Study vegetation, rock types, clouds, volcanoes; produce DEMs; provide high resolution data for overall mission requirements
MISR	Terra	Four-channel CCD arrays providing nine separate view angles	Provide multiangle views of earth surface features, data on clouds and atmospheric aerosols, and correction for atmospheric effects to data from ASTER and MODIS
MOPITT	Terra	Three channel near-IR scanner	Measure carbon monoxide and methane in the atmospheric column
CERES	Both	Two broadband scanners	Measure radiant flux at top of the atmosphere to monitor earth's total radiation energy balance
MODIS	Both	Thirty-six-channel imaging spectrometer; 250 m to 1 km resolution	Useful for multiple land and ocean applications, cloud cover, cloud properties
AIRS	Aqua	Hyperspectral sensor with 2378 channels, 2 to 14 km spatial resolution	Measure atmospheric temperature and humidity, cloud properties, and radiative energy flux
AMSR/E	Aqua	Twelve-channel microwave radiometer, 6.9 to 89 GHz	Measure precipitation, land surface wetness and snow cover, sea surface characteristics, cloud properties
AMSU	Aqua	Fifteen-channel microwave radiometer, 50 to 89 GHz	Measure atmospheric temperature and humidity
HSB	Aqua	Five-channel microwave radiometer, 150 to 183 MHz	Measure atmospheric humidity

The intent in the design of Terra and Aqua is to provide a suite of highly synergistic instruments on each platform. For example, four of the five instruments on Terra are used in a complementary manner to obtain data about cloud properties. Similarly, measurements made by one instrument (e.g., MISR) can be used to atmospherically correct the data for another (e.g., MODIS). In composite, each spacecraft provides detailed measurements contributing to a number of interrelated scientific objectives. Figure 6.37 illustrates the basic relationships among the science objectives, measurements, and instruments included in the Terra mission.

Both Terra and Aqua are quite massive for earth-observing satellites. Terra is approximately 6.8 m long and 3.5 m in diameter and weighs 5190 kg, while Aqua is 16.7 × 8.0 × 4.8 m and weighs 2934 kg. The satellites are in near-polar, sun-synchronous orbits at 705 km altitude, and their orbits fit those of the WRS-2 numbering system developed for Landsat-4, -5, and -7.

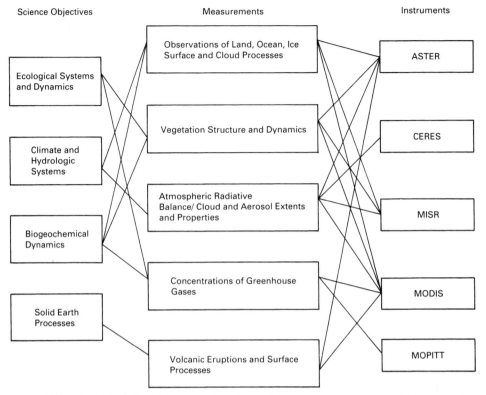

Figure 6.37 Principal relationships among the science objectives, measurements, and instruments involved in the Terra mission. (After NASA diagram.)

Terra has a 10:30 A.M. equatorial crossing time (descending), a time chosen to minimize cloud cover. Aqua has a 1:30 P.M. equatorial crossing time (ascending). Among the instruments carried on Terra and Aqua, the three of greatest interest to the readership of this book are the MODIS, ASTER, and MISR instruments.

MODIS

Flown on both the Terra and Aqua satellites, MODIS is a sensor that is intended to provide comprehensive data about land, ocean, and atmospheric processes simultaneously. Its design is rooted in various earlier sensors, or "heritage instruments," such as the AVHRR and CZCS. However, MODIS is a highly improved successor to these earlier systems. MODIS not only provides 2-day repeat global coverage with greater spatial resolution (250, 500, or 1000 m, depending on wavelength) than the AVHRR, but it also collects data in 36 carefully chosen spectral bands (Table 6.20), with 12-bit radiometric sensitivity. In addition,

TABLE 6.20   MODIS Spectral Bands

Primary Use	Band	Bandwidth	Resolution (m)
Land/cloud	1	620–670 nm	250
boundaries	2	841–876 nm	250
Land/cloud	3	459–479 nm	500
properties	4	545–565 nm	500
	5	1230–1250 nm	500
	6	1628–1652 nm	500
	7	2105–2155 nm	500
Ocean color/	8	405–420 nm	1000
phytoplankton/	9	438–448 nm	1000
biogeochemistry	10	483–493 nm	1000
	11	526–536 nm	1000
	12	546–556 nm	1000
	13	662–672 nm	1000
	14	673–683 nm	1000
	15	743–753 nm	1000
	16	862–877 nm	1000
Atmospheric	17	890–920 nm	1000
water vapor	18	931–941 nm	1000
	19	915–965 nm	1000
Surface/cloud	20	3.660–3.840 μm	1000
temperature	21[a]	3.929–3.989 μm	1000
	22	3.929–3.989 μm	1000
	23	4.020–4.080 μm	1000
Atmospheric	24	4.433–4.498 μm	1000
temperature	25	4.482–4.549 μm	1000
Cirrus clouds	26[b]	1.360–1.390 μm	1000
Water vapor	27	6.538–6.895 μm	1000
	28	7.175–7.475 μm	1000
	29	8.400–8.700 μm	1000
Ozone	30	9.580–9.880 μm	1000
Surface/cloud	31	10.780–11.280 μm	1000
temperature	32	11.770–12.270 μm	1000
Cloud top	33	13.185–13.485 μm	1000
altitude	34	13.485–13.758 μm	1000
	35	13.785–14.085 μm	1000
	36	14.085–14.385 μm	1000

[a]Band 21 and 22 are similar, but band 21 saturates at 500 K versus 328 K.

[b]Wavelength out of sequence due to change in sensor design.

MODIS data are characterized by improved geometric rectification and radiometric calibration. Band-to-band registration for all 36 MODIS channels is specified to be 0.1 pixel or better. The 20 reflected solar bands are absolutely calibrated radiometrically with an accuracy of 5 percent or better. The calibrated accuracy of the 16 thermal bands is specified to be 1 percent or better.

These stringent calibration standards are a consequence of the EOS/ESE requirement for a long term continuous series of observations aimed at documenting subtle changes in global climate. This data set must not be sensor dependent; hence, the emphasis on sensor calibration.

The total field of view of MODIS is ±55°, providing a swath width of 2330 km. A large variety of data products can be derived from MODIS data. Among the principal data products available are the following:

Cloud mask at 250 m and 1000 m resolution during the day and 1000 m resolution at night.

Aerosol concentration and optical properties at 5 km resolution over oceans and 10 km over land during the day.

Cloud properties (optical thickness, effective particle radius, thermodynamic phase, cloud top altitude, cloud top temperature) at 1 to 5 km resolution during the day and 5 km resolution at night.

Vegetation and land surface cover, conditions, and productivity, defined as vegetation indices corrected for atmospheric effects, soil, polarization, and directional effects; surface reflectance; land cover type; and net primary productivity, leaf area index, and intercepted photosynthetically active radiation.

Snow and sea-ice cover and reflectance.

Surface temperature with 1 km resolution, day and night, with absolute accuracy goals of 0.3 to 0.5°C over oceans and 1°C over land.

Ocean color (ocean-leaving spectral radiance measured to 5 percent), based on data acquired from the MODIS visible and near-IR channels.

Concentration of chlorophyll *a* (within 35 percent) from 0.05 to 50 mg/m^3 for case 1 waters.

Chlorophyll fluorescence (within 50 percent) at surface water concentration of 0.5 mg/m^3 of chlorophyll *a*.

Plates 24 and 25 show examples of high-level products derived from MODIS imagery. Plate 24 shows a global rendition of land surface reflectance and sea surface temperature. Both data sets were developed from composites of multiple dates of MODIS imagery in March and April 2000, with 7 dates being used for the sea surface temperature data and 20 dates for the land surface reflectance. Plate 25 shows a MODIS-based sea surface temperature map of the western Atlantic, including a portion of the Gulf Stream. Land areas and clouds have been masked out. Cold water, approximately 7°C, is shown in purple, with blue, green, yellow, and red representing increasingly warm water, up to approximately 22°C. The warm current of the Gulf Stream stands out clearly in deep red colors.

ASTER

The ASTER is an imaging instrument that is a cooperative effort between NASA and Japan's Ministry of International Trade and Industry. In a sense, ASTER serves as a "zoom lens" for the other instruments aboard Terra in that it has the highest spatial resolution of any of them. ASTER consists of three separate instrument subsystems, each operating in a different spectral region, using a separate optical system, and built by a different Japanese company. These subsystems are the *Visible and Near Infrared* (*VNIR*), the *Short Wave Infrared* (*SWIR*), and the *Thermal Infrared* (*TIR*), respectively. Table 6.21 indicates the basic characteristics of each of these subsystems.

The VNIR subsystem incorporates three spectral bands that have 15 m ground resolution. The instrument consists of two telescopes—one nadir looking with a three-band CCD detector and the other backward looking (27.7° off nadir) with a single-band (band 3) detector. This configuration enables

TABLE 6.21   ASTER Instrument Characteristics

Characteristic	VNIR	SWIR	TIR
Spectral range	Band 1: 0.52–0.60 μm, nadir looking	Band 4: 1.600–1.700 μm	Band 10: 8.125–8.475 μm
	Band 2: 0.63–0.69 μm, nadir looking	Band 5: 2.145–2.185 μm	Band 11: 8.475–8.825 μm
	Band 3: 0.76–0.86[a] μm, nadir looking	Band 6: 2.185–2.225 μm	Band 12: 8.925–9.275 μm
	Band 3: 0.76–0.86[a] μm, backward looking	Band 7: 2.235–2.285 μm	Band 13: 10.25–10.95 μm
		Band 8: 2.295–2.365 μm	Band 14: 10.95–11.65 μm
		Band 9: 2.360–2.430 μm	
Ground resolution (m)	15	30	90
Cross-track pointing (deg)	±24	±8.55	±8.55
Cross-track Pointing (km)	±318	±116	±116
Swath width (km)	60	60	60
Quantization (bits)	8	8	12

[a]Stereoscopic imaging subsystem.

along-track stereoscopic image acquisition in band 3 with a base-height ratio of 0.6. This permits the construction of DEMs from the stereo data with a vertical accuracy of approximately 7 to 50 m. Cross-track pointing to 24° on either side of the orbit path is accomplished by rotating the entire telescope assembly.

The SWIR subsystem operates in six spectral bands through a single, nadir-pointing telescope that provides 30 m resolution. Cross-track pointing to 8.55° is accomplished through the use of a pointing mirror.

The TIR subsystem operates in 5 bands in the thermal IR region with a resolution of 90 m. Unlike the other instrument subsystems, it incorporates a whiskbroom scanning mirror. Each band uses 10 detectors, and the scanning mirror functions both for scanning and cross-track pointing to 8.55°.

All ASTER bands cover the same 60-km imaging swath with a pointing capability in the cross-track direction to cover ±116 km from nadir. This means that any point on the globe is accessible at least once every 16 days with full 14-band spectral coverage from the VNIR, SWIR, and TIR. In that the VNIR subsystem has a larger pointing capability, it can collect data up to ±318 km from nadir, with an average revisit period of 4 days at the equator.

Plate 26 shows examples of ASTER imagery for the San Francisco Bay region in California, acquired on March 3, 2000. The color infrared composite (a) uses ASTER bands 1, 2, and 3, from the visible and near IR. Vegetation appears red, urban areas are gray, and sediment in the bays shows up as lighter shades of blue. At the spatial resolution of 15 m, shadows of the towers along the Bay Bridge can be seen (but not at the scale shown in this plate). In Plate 26b, a composite of bands in the mid IR displays differences in soils and rocks in the mountainous areas. Even though these regions appear entirely vegetated in the visible and near IR, there are enough openings in the vegetation to allow the ground to be imaged. Plate 26c shows a composite of three ASTER thermal bands, revealing differences in urban materials in varying colors. Separation of materials is due to differences in thermal emission properties. Where the thermal emissivity is known (e.g., for the water of San Francisco Bay), the ASTER thermal bands can be used to calculate kinetic temperature, as shown in Plate 26d. This is a color-coded temperature image of water temperature, with warm waters in white and yellow and colder waters in blue. Suisun Bay in the upper right is fed directly from the cold Sacramento River. As the water flows through San Pablo and San Francisco Bays on the way to the Pacific, the waters warm up.

Three individual bands of an ASTER data set for a 13.3 × 54.8-km area in Death Valley, California, are shown as Figure 6.38. Note that this area represents only a portion of the full ASTER swath width. Plate 27 shows three color composites of ASTER data for the same area as shown in Figure 6.38. Plate 27a is a color composite of three VNIR (visible and near-IR) bands, part (b) shows three SWIR (mid-IR) bands, and part (c) shows three TIR (thermal IR) bands. Various geological and vegetation differences are displayed in the three parts of this plate. For example, the bright red areas in (*a*) are patches of vegetation on the Furnace Creek alluvial fan that reflect more highly in the near-IR than in the green or red parts of the spectrum. These are seen as blue in (*b*), but are difficult to discern in (*c*), which shows thermal bands with 90 m resolution as contrasted with the 15 m resolution of (*a*) and the 30 m resolution of (*b*). A striking difference in geological materials is seen in (*c*), where surfaces containing materials high in the mineral quartz have a red color.

Because the nadir-looking and backward-looking band 3 sensors can obtain data of the same area from two different angles, stereo ASTER data can be used to produce DEMs. Various bands of ASTER image data can then be "draped" over the DEMs to produce perspective views of an area, as illustrated in Figure 6.39. (The area shown in Figure 6.39 roughly corresponds with that in Figure 6.38 and Plate 27.) Such perspective views can help image analysts to obtain an overview of the area under study. For example, in this figure, bedrock mountains are located at left (west) and right (east). The broad, very light toned area through the center of the image is the floor of Death Valley. Between the mountains and the valley floor, many alluvial fans (essentially continuous on the west side of the valley floor) can be seen.

MISR

The MISR instrument differs from most other remote sensing systems in that it has nine sensors viewing the earth simultaneously at different angles. Each sensor consists of a set of linear arrays providing coverage in four spectral bands (blue, green, red, and near IR). One sensor is oriented toward nadir (0°); four sensors are viewing forward at angles of 26.1°, 45.6°, 60.0°, and 70.5°; and four are viewing backward at similar angles. The system has a swath width of 360 km, and a spatial resolution of 250 m in the nadir-viewing sensor and 275 m in the off-nadir sensors.

The MISR viewing angles were selected to address several different objectives. The nadir-viewing sensor provides imagery with the minimum interference from the atmosphere. The two sensors at 26.1° off nadir were selected to provide stereoscopic coverage for measurement of cloud-top altitudes, among other purposes. The 45.6° sensors provide information about aerosols in the atmosphere, which are of great importance for studies of global change and the earth's radiation budget. The 60.0° sensors minimize the effects of directional differences in the scattering of light from clouds and can be used to estimate the hemispherical reflectance of land surface features. Finally, the 70.5° sensors were chosen to provide the most oblique angle that could be achieved within practical limitations, in order to maximize the impact of off-nadir scattering phenomena.

Analysis of MISR imagery contributes to studies of the earth's climate and helps monitor the global distribution of particulate pollution and different

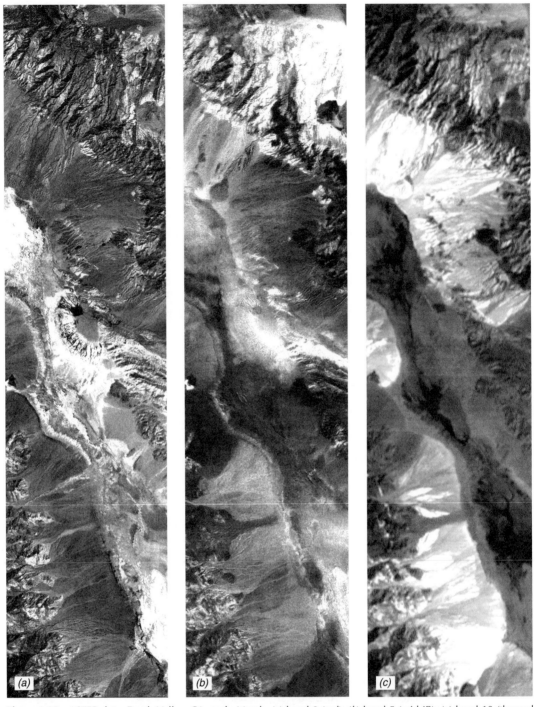

Figure 6.38 ASTER data, Death Valley, CA, early March: (*a*) band 2 (red); (*b*) band 5 (mid IR); (*c*) band 13 (thermal IR). Scale 1:300,000.

484

Figure 6.39   ASTER data draped over a digital elevation model, Death Valley, CA (view to the north). Black and white copy of a color composite derived from thermal IR bands. (Courtesy NASA/JPL.)

types of haze. For example, Plate 28 shows a series of MISR images of the eastern United States, acquired during a single orbit (on March 6, 2000) at four different angles. The area shown ranges from Lakes Ontario and Erie to northern Georgia, and covers a portion of the Appalachian Mountains. The eastern end of Lake Erie was covered by ice on this date, and thus appears bright in these images. Plate 28*a* was acquired at an angle of 0° (nadir-viewing), while (*b*), (*c*), and (*d*) were acquired at forward-viewing angles of 45.6°, 60.0°, and 70.5°, respectively. Areas of extensive haze over the Appalachian Mountains are almost invisible in the 0° image but have a major effect on the 70.5° image, due to the increased atmospheric path length. The images in (*c*) and (*d*) appear brighter due to increased path radiance, but also have a distinct color shift toward the blue end of the spectrum. This is a result of the fact that Rayleigh scattering in the atmosphere is inversely proportional to the fourth power of the wavelength, as discussed in Chapter 1.

6.19   SPACE STATION REMOTE SENSING

Remote sensing has been a part of previous space station missions and will likely continue in this vein in the *International Space Station* (*ISS*) program. As mentioned previously (Section 6.2), the first American space workshop, Skylab, was used to acquire numerous forms of remote sensing data in 1973.

The Russian MIR space station was fitted with a dedicated remote sensing module called **PRIRODA** that was launched in 1996. With contributions from 12 nations, the **PRIRODA** mission consisted of a broad variety of different sensors, including numerous optical systems as well as both active and passive microwave equipment.

The ISS provides an excellent platform for remote sensing from space, although its potential has barely begun to be realized. The orbital coverage of the ISS extends from 52° N to 52° S latitude. It orbits the earth every 90 min at an altitude of 400 km, and the revisit period over any given point at the equator (within 9° off nadir) is 32 hr. Hence, approximately 75 percent of the land surface of the globe, 100 percent of the rapidly changing tropics, and 95 percent of the world's population can be observed from this platform. To facilitate the collection of earth observation imagery, a 50-cm-diameter, high-optical-quality window has been installed in one of the laboratory modules on the ISS. As of the date of this writing (2002), no organized effort to collect remote sensing data from the ISS has been undertaken. However, ISS crewmembers have taken thousands of frames of imagery using a variety of film and digital cameras, sometimes in combination with telescopes. These images are being contributed to an ever-growing database of earth observation photographs at the NASA Johnson Space Center that dates back to the early years of the U.S. space program. In the future, data collected from the ISS could be used for applications ranging from validating other global terrestrial data sets to real-time monitoring of ephemeral events and human-induced global change. Real-time human oversight of the acquisition of such data complements greatly the capabilities of the various satellite systems discussed in this chapter.

SELECTED BIBLIOGRAPHY

Al-Rawi, K.R., et al., "Integrated Fire Evolution Monitoring System (IFEMS) for Monitoring Spatial-Temporal Behaviour of Multiple Fire Phenomena," *International Journal of Remote Sensing*, vol. 23, no. 10, 2002, pp. 1967–1983.

American Society of Photogrammetry (ASP), *Manual of Remote Sensing*, 2nd ed., ASP, Falls Church, VA, 1983.

American Society for Photogrammetry and Remote Sensing (ASPRS), *GAP Analysis Program Proceedings*, ASPRS, Bethesda, MD, 1996.

American Society for Photogrammetry and Remote Sensing (ASPRS), *Corona Between the Sun and the Earth: The First NRO Reconnaissance Eye in Space*, ASPRS, Bethesda, MD, 1997a.

American Society for Photogrammetry and Remote Sensing (ASPRS), *Earth Observing Platforms and Sensors*, Manual of Remote Sensing, 3rd ed. (A Series), Version 1.1, ASPRS, Bethesda, MD, 1997b (CD-ROM).

American Society for Photogrammetry and Remote Sensing (ASPRS), *Proceedings: Land Satellite Information in the Next Decade II: Sources and Applications*, ASPRS, Bethesda, MD, 1997c (CD-ROM).

American Society for Photogrammetry and Remote Sensing (ASPRS), *Remote Sensing for the Earth Sciences*, Manual of Remote Sensing, 3rd. ed., Vol. 3, Wiley, New York, 1999.

American Society for Photogrammetry and Remote Sensing (ASPRS) and Rand Corp.; *Commercial Observation Satellites at the Leading Edge of Global Trans-*

parency, ASPRS and RAND Corp., Bethesda, MD, 2001.

Asrar, G., S.G. Tilford, and D.M. Butler, "Mission to Planet Earth: Earth Observing System," *Palaeogeography, Palaeoclimatology, Palaeoecology (Global and Planetary Change Section)*, vol. 98, 1992, pp. 3–8.

Boles, S.H., and D.L. Verbyla, "Comparison of Three AVHRR-Based Fire Detection Algorithms for Interior Alaska," *Remote Sensing of Environment*, vol. 72, no.1, 2000, pp. 1–16.

Borzelli, G., et al., "A New Perspective on Oil Slick Detection from Space by NOAA Satellites," *International Journal of Remote Sensing*, vol. 17, no. 7, 1996, pp. 1279–1292.

Breaker, L.C., V.M. Krasnopolsky, and E.M. Maturi, "GOES-8 Imagery as New Source of Data to Conduct Ocean Feature Tracking," *Remote Sensing of Environment*, vol. 73, no. 2, 2000, pp. 198–206.

Brondizio, E., et al., "Land Cover in the Amazon Estuary: Linking of the Thematic Mapper with Botanical and Historical Data," *Photogrammetric Engineering and Remote Sensing*, vol. 62, no. 8, 1996, pp. 921–929.

Bukata, R.P., et al., *Optical Properties and Remote Sensing of Inland and Coastal Waters*, CRC Press, Boca Raton, FL, 1995.

Bussieres, N., D. Verseghy, and J.I. MacPherson, "The Evolution of AVHRR-Derived Water Temperatures over Boreal Lakes," *Remote Sensing of Environment*, vol. 80, no. 3, 2002, pp. 373–384.

Cohen, W.B., et al., "An Efficient and Accurate Method for Mapping Forest Clearcuts in the Pacific Northwest Using Landsat Imagery," *Photogrammetric Engineering and Remote Sensing*, vol. 64, no. 4, 1998, pp. 293–300.

Derenyi, E.E., "Skylab in Retrospect," *Photogrammetric Engineering and Remote Sensing*, vol. 47, no. 4, 1981, pp. 495–499.

Dwivedi, R.S., et al., "The Utility of IRS-1C LISS-III and PAN-Merged Data for Mapping Salt-Affected Soils," *Photogrammetric Engineering and Remote Sensing*, vol. 67, no. 10, 2001, pp. 1167–1175.

Eidenshink, J.C., "The 1990 Conterminous U.S. AVHRR Data Set," *Photogrammetric Engineering and Remote Sensing*, vol. 58, no. 6, 1992, pp. 809–813.

Eidenshink, J.C., and J.L. Faundeen, "The 1-km AVHRR Global Land Data Set: First Stages in Implementation," *International Journal of Remote Sensing*, vol. 15, no. 17, 1994, pp. 3443–3462.

Ekstrand, S., "Landsat TM-Based Forest Damage Assessment: Correction for Topographic Effects," *Photogrammetric Engineering and Remote Sensing*, vol. 62, no. 2, 1996, pp. 151–161.

Elachi, C., *Introduction to the Physics and Techniques of Remote Sensing*, Wiley, New York, 1987.

El-Manadili, Y., and K. Novak, "Precision Rectification of SPOT Imagery Using the Direct Linear Transformation Model," *Photogrammetric Engineering and Remote Sensing*, vol. 62, no. 1, 1996, pp. 67–72.

Elvidge, C.D., et al., "Mapping City Lights with Nighttime Data from the DMSP Operational Linescan System," *Photogrammetric Engineering and Remote Sensing*, vol. 63, no. 6, 1997, pp. 727–734.

Elvidge, C.D., et al., "Radiance Calibration of DMSP-OLS Low-Light Imaging Data of Human Settlement," *Remote Sensing of Environment*, vol. 68, no. 1, 1999, pp. 77–88.

Elvidge, C.D., et al., "Night-Time Lights of the World: 1994–1995," *ISPRS Journal of Photogrammetry and Remote Sensing*, vol. 56, no. 2, 2001, pp. 81–99.

Fraser, C.S., et al., "Processing of Ikonos Imagery for Submetre 3D Positioning and Building Extraction," *ISPRS Journal of Photogrammetry and Remote Sensing*, vol. 56, no. 3, 2002, pp. 177–194.

Froidefond, J.M., et al., "SeaWiFS Data Interpretation in A Coastal Area in the Bay of Biscay," *International Journal of Remote Sensing*, vol. 23, no. 5, 2002, pp. 881–904.

Gao, J., and M.B. Lythe, "Effectiveness of the MCC Methods in Detecting Oceanic Circulation Patterns at a Local Scale from Sequential AVHRR Images," *Photogrammetric Engineering and Remote Sensing*, vol. 64, no. 4, 1998, pp. 301–308.

Gao, J., and D. Skillcorn, "Capability of SPOT XS Data in Producing Detailed Land Cover Maps at the Urban-Rural Periphery," *International Journal of Remote Sensing*, vol. 19, no. 15, 1998, pp. 2877–2891.

Garguet-Duport, B., et al., "The Use of Multiresolution Analysis and Wavelets Transform for Merging SPOT Panchromatic and Multispectral Image Data," *Photogrammetric Engineering and Remote Sensing*, vol. 62, no. 9, 1996, pp. 1057–1066.

Giles, P.T., and S.E. Franklin, "Comparison of Derivative Topographic Surfaces of a DEM Generated from Stereoscopic SPOT Images with Field Measurements," *Photogrammetric Engineering and Remote Sensing*, vol. 62, no. 10, 1996, pp. 1165–1171.

Glackin, D.L., "International Space-Based Remote Sensing Overview: 1980–2007," *Canadian Journal of Remote Sensing*, vol. 24, no. 3, 1998, pp. 307–314.

Glasser, M.E., and K.P. Lulla, "NASA Astronaut Photography Depiction of the Spatial and Temporal Characteristics of Biomass Burning," *International Journal of Remote Sensing*, vol. 21, no. 10, 2000, pp. 1971–1986.

Goward, S.N., et al., "Normalized Difference Vegetation Index Measurements from the Advanced Very High Resolution Radiometer," *Remote Sensing of Environment*, vol. 35, 1991, pp. 257–277.

Guindon, B., "Computer-Based Aerial Image Understanding: A Review and Assessment of Its Application to Planimetric Information Extraction from Very High Resolution Satellite Images," *Canadian Journal of Remote Sensing*, vol. 23, no. 1, 1997, pp. 38–47.

Hu, B.X., et al., "Validation of Kernel-Driven Semiempirical Models for the Surface Bidirectional Reflectance Distribution Function of Land Surfaces," *Remote Sensing of Environment*, vol. 62, no. 3, 1997, pp. 201–214.

IEEE Transactions on Geoscience and Remote Sensing, Special Issue on EOS AM-1 Platform, Instruments, and Scientific Data, vol. 36, no. 4, 1998.

Ikeda, M., and F. Dobson, *Oceanographic Applications of Remote Sensing*, Lewis, Boca Raton, FL, 1995.

Imhoff, M.L., et al., "A Technique for Using Composite DMSP/OLS 'City Lights' Satellite Data to Map Urban Area," *Remote Sensing of Environment*, vol. 61, no. 3, 1997, pp. 361–370.

International Journal of Remote Sensing, Special Issue: European Achievements in Remote Sensing, vol. 13, nos. 6 and 7, 1992.

International Journal of Remote Sensing, Special Issue on ESA Medium Resolution Imaging Spectrometer, vol. 20, no. 9, 1999a.

International Journal of Remote Sensing, Special Issue on Remote Sensing of Polar Regions, vol. 20, nos. 15 & 16, 1999b.

International Journal of Remote Sensing, Special Issue on the OCEAN Project, vol. 20, no. 7, 1999c.

International Journal of Remote Sensing, Special Issue on Global and Regional Land Cover Characterization from Remotely Sensing Data, vol. 21, nos. 6 & 7, 2000.

International Journal of Remote Sensing, Special Issue on Algal Blooms Detection, Monitoring, and Prediction, vol. 22, nos. 2 & 3, 2001.

Jakubauskas, M.E., D.R. Legates, and J.H. Kastens, "Harmonic Analysis of Time-Series AVHRR NDVI Data," *Photogrammetric Engineering and Remote Sensing*, vol. 67, no. 4, 2001, pp. 461–470.

Kidwell, K.B., *NOAA Polar Orbiter Data Users Guide*, NOAA, Satellite Data Services Division, World Weather Building, Washington, DC, 1991.

Kramer, H.J., *Observation of the Earth and Its Environment: Survey of Missions and Sensors*, Springer, New York, 2002.

Krishnamurthy, J., "The Evaluation of Digitally Enhanced Indian Remote Sensing Satellite (IRS) Data for Lithological and Structural Mapping," *International Journal of Remote Sensing*, vol. 18, no. 16, 1997, pp. 3409–3437.

Lambin, E.F., and D. Ehrlich, "Land-Cover Changes in Sub-Saharan Africa (1982–1991): Application of a Change Index Based on Remotely Sensed Surface Temperature and Vegetation Indices at a Continental Scale," *Remote Sensing of Environment*, vol. 61, no. 2, 1997, pp. 181–200.

Lathrop, R.G., Jr., and T.M. Lillesand, "Calibration of Thematic Mapper Thermal Data for Water Surface Temperature Mapping: Case Study on the Great Lakes," *Remote Sensing of Environment*, vol. 22, no. 2, 1987, pp. 297–307.

Li, R., "Potential of High-Resolution Satellite Imagery for National Mapping Products," *Photogrammetric Engineering and Remote Sensing*, vol. 64, no. 12, 1998, pp. 1165–1169.

Light, D.L., "Characteristics of Remote Sensors for Mapping and Earth Science Applications," *Photogrammetric Engineering and Remote Sensing*, vol. 56, no. 12, 1990, pp. 1613–1623.

Lo, C.P., "Modeling the Population of China Using DMSP Operational Linescan System Nighttime Data," *Photogrammetric Engineering and Remote Sensing*, vol. 67, no. 9, 2001, pp. 1037–1047.

Lo, C.P., and B.J. Faber, "Integration of Landsat Thematic Mapper and Census Data for Quality of Life Assessment," *Remote Sensing of Environment*, vol. 62, no. 2, 1997, pp. 143–157.

Lopez, A. S., J. San-Miguel-Ayanz, and R.E. Burgan, "Integration of Satellite Sensor Data, Fuel Type Maps and Meteorological Observations for Evaluation of Forest Fire Risk at the Pan-European Scale," *Interna-*

tional *Journal of Remote Sensing*, vol. 23, no. 13, 2002, pp. 2713–2719.

Loveland, T.R., et al., "Development of a Land-Cover Characteristics Database for the Conterminous U.S.," *Photogrammetric Engineering and Remote Sensing*, vol. 57, no. 11, 1991, pp. 1453–1463.

Lowman, P.D., Jr., "Space Photography—A Review," *Photogrammetric Engineering*, vol. 31, no. 1, 1965, pp. 76–86.

Lowman, P. D., "Landsat and Apollo: The Forgotten Legacy," *Photogrammetric Engineering and Remote Sensing*, vol. 65, no. 10, 1999, pp. 1143–1147.

Lu, D., et al., "Assessment of Atmospheric Correction Methods for Landsat TM Data Applicable to Amazon Basin LBA Research," *International Journal of Remote Sensing*, vol. 23, no. 13, 2002, pp. 2651–2671.

Mantovani, A.C.D., and A.W. Setzer, "Deforestation Detection in the Amazon with an AVHRR-Based System," *International Journal of Remote Sensing*, vol. 18, no. 2, 1997, pp. 273–286.

Maselli, F., L. Petkov, and G. Maracchi, "Extension of Climate Parameters over the Land Surface by the Use of NOAA-AVHRR and Ancillary Data," *Photogrammetric Engineering and Remote Sensing*, vol. 64, no. 3, 1998, pp. 199–206.

NASA, *Skylab Earth Resources Data Catalog*, Document No. 3300–00586, U.S. Government Printing Office, Washington, DC, 1974.

NASA, *Skylab Explores The Earth*, SP-380, NASA Scientific and Technical Information Office, Washington, DC, 1977.

National Geographic, *Satellite Atlas of the World*, The National Geographic Society, Washington, DC, 1998.

NOAA, "Visual Interpretation of TM Band Combinations Being Studied," *Landsat Data Users Notes*, no. 30, March 1984.

Pal, S.R., and P.K. Mohanty, "Use of IRS-1B Data for Change Detection in Water Quality and Vegetation of Chilka Lagoon, East Coast of India," *International Journal of Remote Sensing*, vol. 23, no. 6, 2002, pp. 1027–1042.

Peters, A.J., et al., "Drought Monitoring with NDVI-Based Standardized Vegetation Index," *Photogrammetric Engineering and Remote Sensing*, vol. 68, no. 1, 2002, pp. 71–75.

Photogrammetric Engineering and Remote Sensing, Special Issue: Remote Sensing for Monitoring Tropical Moist Forests, vol. 56, no. 10, 1990.

Photogrammetric Engineering and Remote Sensing, Special Issue on the Twenty-Fifth Anniversary of Landsat, vol. 63, no. 7, 1997.

Ramsey, E.W., D.E. Chapel, and D.E. Baldwin, "AVHRR Imagery Used to Identify Hurricane Damage in a Forested Wetland of Louisiana," *Photogrammetric Engineering and Remote Sensing*, vol. 63, no. 3, 1997, pp. 293–297.

Ramsey, E.W., et al., "Resource Management of Forested Wetlands: Hurricane Impact and Recovery Mapped by Combining Landsat TM and NOAA AVHRR Data," *Photogrammetric Engineering and Remote Sensing*, vol. 64, no. 7, 1998, pp. 733–738.

Rand Corp., National Security Research Division, *U.S. Commercial Remote Sensing Satellite Industry: An Analysis of Risks*, Rand, Santa Monica, 2001.

Reddy, M.A., and K.M. Reddy, "Performance Analysis of IRS-Bands for Land Use Land Cover Classification System Using Maximum Likelihood Classifier," *International Journal of Remote Sensing*, vol. 17, no. 13, 1996, pp. 2505–2515.

Remote Sensing of Environment, Special Issue on Validating MODIS Terrestrial Ecology Products, vol. 70, no. 1, 1999.

Remote Sensing of Environment, Special Issue on Landsat 7, vol. 78, nos. 1/2, 2001.

Sannier, C.A.D., et al., "Real-Time Vegetation Monitoring with NOAA-AVHRR in Southern Africa for Wildlife Management and Food Security Assessment," *International Journal of Remote Sensing*, vol. 19, no. 4, 1998, pp. 621–639.

Sannier, C.A.D., J.C. Taylor, and W. DuPlessis, "Real-Time Monitoring of Vegetation Biomass with NOAA-AVHRR in Etosha National Park, Namibia, for Fire Risk Assessment," *International Journal of Remote Sensing*, vol. 23, no. 1, 2002, pp. 71–89.

Segl, K., and H. Kaufmann, "Detection of Small Objects from High-Resolution Panchromatic Satellite Imagery Based on Supervised Image Segmentation," *IEEE Transactions on Geoscience and Remote Sensing*, vol. 39, no. 9, 2001, pp. 2080–2083.

Senay, G.B., and R.L. Elliott, "Capability of AVHRR Data in Discriminating Rangeland Cover Mixtures," *International Journal of Remote Sensing*, vol. 23, no. 2, 2002, pp. 299–312.

Sheng, Y.W., Y. Su, and Q. Xiao, "Challenging the Cloud-Contamination Problem in Flood Monitoring with NOAA/AVHRR Imagery," *Photogrammetric*

Engineering and Remote Sensing, vol. 64, no. 3, 1998, pp. 191–198.

Short, N.M., *The Remote Sensing Tutorial Online Handbook*, on the Internet at http://rst.gsfc.nasa.gov (CD-ROM, 1998, available from NASA Goddard Space Flight Center).

Slonecker, E.T., D.M. Shaw, and T.M. Lillesand, "Emerging Legal and Ethical Issues in Advanced Remote Sensing Technology," *Photogrammetric Engineering and Remote Sensing*, vol. 64, no. 6, 1998, pp. 589–595.

Stoms, D.M., M.J. Bueno, and F.W. Davis, "Viewing Geometry of AVHRR Image Composites Derived Using Multiple Criteria," *Photogrammetric Engineering and Remote Sensing*, vol. 63, no. 6, 1997, pp. 681–689.

Stow, D.A., and D.M. Chen, "Sensitivity of Multitemporal NOAA AVHRR Data of an Urbanizing Region to Land-Use/Land-Cover Changes and Misregistration," *Remote Sensing of Environment*, vol. 80, no. 2, 2002, pp. 297–307.

Sutton, P., et al., "A Comparison of Nighttime Satellite Imagery and Population Density for the Continental United States," *Photogrammetric Engineering and Remote Sensing*, vol. 63, no. 11, 1997, pp. 1303–1313.

Sutton, P., et al., "Census from Heaven: An Estimate of the Global Human Population Using Night-Time Satellite Imagery," *International Journal of Remote Sensing*, vol. 22, no. 16, 2001, pp. 3061–3076.

Tappan, G.G., et al., "Use of Argon, Corona, and Landsat Imagery to Assess 30 Years of Land Resource Changes in West-Central Senegal," *Photogrammetric Engineering and Remote Sensing*, vol. 66, no. 6, 2000, pp. 727–735.

Taranik, J.V., *Characteristics of the Landsat Multispectral Data System*, Open File Report 78–187, USGS, Sioux Falls, SD, 1978.

Teillet, P.M., et al., "An Evaluation of the Global 1-km AVHRR Land Dataset," *International Journal of Remote Sensing*, vol. 21, no. 10, 2000, pp. 1987–2021.

Thornbury, W.D., *Regional Geomorphology of the United States*, Wiley, New York, 1965.

Valadan Zoej, M.J., and G. Petrie, "Mathematical Modelling and Accuracy Testing of SPOT Level 1B Stereopairs," *Photogrammetric Record*, vol. 16, no. 91, 1998, pp. 67–82.

Vazquez, D.P., F.J.O. Reyes, and L.A. Arboledas, "A Comparative Study of Algorithms for Estimating Land Surface Temperature from AVHRR Data," *Remote Sensing of Environment*, vol. 62, no. 3, 1997, pp. 215–222.

Verbyla, D.L., *Satellite Remote Sensing of Natural Resources*, Lewis, Boca Raton, FL, 1995.

Verstraete, M.M., et al., *Observing Land from Space: Science, Customers, and Technology*, Kluwer Academic, Boston, 2000.

Wald, L., R. Ranchin, and M. Mangolini, "Fusion of Satellite Images of Different Spatial Resolutions: Assessing the Quality of Resulting Images," *Photogrammetric Engineering and Remote Sensing*, vol. 63, no. 6, 1997, pp. 691–699.

Wan, Z.M., et al., "Preliminary Estimate of Calibration of the Moderate Resolution Imaging Spectroradiometer Thermal Infrared Data Using Lake Titicaca," *Remote Sensing of Environment*, vol. 80, no. 3, 2002, pp. 497–515.

Westin, T., "Inflight Calibration of SPOT CCD Detector Geometry," *Photogrammetric Engineering and Remote Sensing*, vol. 58, no. 9, 1992, pp. 1313–1319.

Wynne, R.H., et al., "Satellite Monitoring of Lake Ice Breakup on the Laurentian Shield (1980–1994)," *Photogrammetric Engineering and Remote Sensing*, vol. 64, no. 6, 1998, pp. 607–617.

Xiao, X.M., Z. Shen, and X. Qin, "Assessing the Potential of VEGETATION Sensor Data for Mapping Snow and Ice Cover: A Normalized Difference Snow and Ice Index," *International Journal of Remote Sensing*, vol. 22, no. 13, 2001, pp. 2479–2487.

Zhou, C.H., et al., "Flood Monitoring Using Multi-Temporal AVHRR and RADARSAT Imagery," *Photogrammetric Engineering and Remote Sensing*, vol. 66, no. 5, 2000, pp. 633–638.

Zhou, G.Q., and R. Li, "Accuracy Evaluation of Ground Points from IKONOS High-Resolution Satellite Imagery," *Photogrammetric Engineering and Remote Sensing*, vol. 66, no. 9, 2000, pp. 1103–1112.

Zhu, G.B., and D.G. Blumberg, "Classification using ASTER Data and SVM Algorithms; The Case Study of Beer Sheva, Israel," *Remote Sensing of Environment*, vol. 80, no. 2, 2002, pp. 233–240.

7 DIGITAL IMAGE PROCESSING

7.1 INTRODUCTION

Digital image processing involves the manipulation and interpretation of digital images (Section 1.5) with the aid of a computer. This form of remote sensing actually began in the 1960s with a limited number of researchers analyzing airborne multispectral scanner data and digitized aerial photographs. However, it was not until the launch of Landsat-1, in 1972, that digital image data became widely available for land remote sensing applications. At that time, not only was the theory and practice of digital image processing in its infancy, but also the cost of digital computers was very high and their computational efficiency was very low by modern standards. Today, access to low cost, efficient computer hardware and software is commonplace, and the sources of digital image data are many and varied. These sources range from commercial and governmental earth resource satellite systems, to the meteorological satellites, to airborne scanner data, to airborne digital camera data, to image data generated by photogrammetric scanners and other high resolution digitizing systems. All of these forms of data can be processed and analyzed using the techniques described in this chapter.

Digital image processing is an extremely broad subject, and it often involves procedures that can be mathematically complex. Our objective in this

491

chapter is to introduce the basic principles of digital image processing without delving into the detailed mathematics involved. Also, we avoid extensive treatment of the rapidly changing hardware associated with digital image processing. The references at the end of this chapter are provided for those wishing to pursue such additional detail.

The central idea behind digital image processing is quite simple. The digital image is fed into a computer one pixel at a time. The computer is programmed to insert these data into an equation, or series of equations, and then store the results of the computation for each pixel. These results form a new digital image that may be displayed or recorded in pictorial format or may itself be further manipulated by additional programs. The possible forms of digital image manipulation are literally infinite. However, virtually all these procedures may be categorized into one (or more) of the following seven broad types of computer-assisted operations:

1. **Image rectification and restoration.** These operations aim to correct distorted or degraded image data to create a more faithful representation of the original scene. This typically involves the initial processing of raw image data to correct for geometric distortions, to calibrate the data radiometrically, and to eliminate noise present in the data. Thus, the nature of any particular image restoration process is highly dependent upon the characteristics of the sensor used to acquire the image data. Image rectification and restoration procedures are often termed *preprocessing* operations because they normally precede further manipulation and analysis of the image data to extract specific information. We briefly discuss these procedures in Section 7.2 with treatment of various geometric corrections, radiometric corrections, and noise corrections.

2. **Image enhancement.** These procedures are applied to image data in order to more effectively display or record the data for subsequent visual interpretation. Normally, image enhancement involves techniques for increasing the visual distinctions between features in a scene. The objective is to create "new" images from the original image data in order to increase the amount of information that can be visually interpreted from the data. The enhanced images can be displayed interactively on a monitor or they can be recorded in a hardcopy format, either in black and white or in color. There are no simple rules for producing the single "best" image for a particular application. Often several enhancements made from the same "raw" image are necessary. We summarize the various broad approaches to enhancement in Section 7.3. In Section 7.4, we treat specific procedures that manipulate the contrast of an image (level slicing and contrast stretching). In Section 7.5, we discuss spatial feature manipulation (spatial filtering, convolution, edge enhancement, and Fourier analysis). In Section 7.6, we consider en-

hancements involving multiple spectral bands of imagery (spectral ratioing, principal and canonical components, vegetation components, and intensity–hue–saturation color space transformations).

3. **Image classification.** The objective of these operations is to replace visual analysis of the image data with quantitative techniques for automating the identification of features in a scene. This normally involves the analysis of multispectral image data and the application of statistically based decision rules for determining the land cover identity of each pixel in an image. When these decision rules are based solely on the spectral radiances observed in the data, we refer to the classification process as *spectral pattern recognition*. In contrast, the decision rules may be based on the geometric shapes, sizes, and patterns present in the image data. These procedures fall into the domain of *spatial pattern recognition*. In either case, the intent of the classification process is to categorize all pixels in a digital image into one of several land cover classes, or "themes." These categorized data may then be used to produce *thematic maps* of the land cover present in an image and/or produce summary statistics on the areas covered by each land cover type. Due to their importance, image classification procedures comprise the subject of more than one-third of the material in this chapter (Sections 7.7 to 7.16). We emphasize *spectral* pattern recognition procedures because the current state-of-the-art for these procedures is more advanced than for *spatial* pattern recognition approaches. (Substantial research is ongoing in the development of spatial and combined spectral/spatial image classification.) We emphasize "supervised," "unsupervised," and "hybrid" approaches to spectrally based image classification. We also describe various procedures for assessing the accuracy of image classification results.

4. **Data merging and GIS integration.** These procedures are used to combine image data for a given geographic area with other geographically referenced data sets for the same area. These other data sets might simply consist of image data generated on other dates by the same sensor or by other remote sensing systems. Frequently, the intent of data merging is to combine remotely sensed data with other sources of information in the context of a GIS. For example, image data are often combined with soil, topographic, ownership, zoning, and assessment information. We discuss data merging in Section 7.17. In this section, we highlight multitemporal data merging, change detection procedures, and multisensor image merging. We also illustrate the incorporation of GIS data in land cover classification.

5. **Hyperspectral image analysis.** Virtually all of the image processing principles introduced in this chapter in the context of multispectral image analysis may be extended directly to the analysis of

hyperspectral data. However, the basic nature and sheer volume of hyperspectral data sets is such that various image processing procedures have been developed to analyze such data specifically. We introduce these procedures in Section 7.18.

6. **Biophysical modeling.** The objective of biophysical modeling is to relate quantitatively the digital data recorded by a remote sensing system to biophysical features and phenomena measured on the ground. For example, remotely sensed data might be used to estimate such varied parameters as crop yield, pollution concentration, or water depth. Likewise, remotely sensed data are often used in concert with GIS techniques to facilitate environmental modeling. The intent of these operations is to simulate the functioning of environmental systems in a spatially explicit manner and to predict their behavior under altered ("what if") conditions, such as global climate change. We discuss the overall concept of biophysical modeling in Section 7.19. In Section 7.20 we highlight the important role spatial resolution plays in the incorporation of remotely sensed data in the environmental modeling process.

7. **Image transmission and compression.** Given the increasingly high volume of data available from remote sensing systems and the distribution of image data over the Internet, image compression techniques are the subject of continuing image processing research. We briefly introduce the use of wavelet transforms in this context in Section 7.21.

We have made the above subdivisions of the topic of digital image processing to provide the reader with a conceptual roadmap for studying this chapter. Although we treat each of these procedures as distinct operations, they all interrelate. For example, the restoration process of noise removal can often be considered an enhancement procedure. Likewise, certain enhancement procedures (such as principal components analysis) can be used not only to enhance the data but also to improve the efficiency of classification operations. In a similar vein, data merging can be used in image classification in order to improve classification accuracy. Hence, the boundaries between the various operations we discuss separately here are not well defined in practice.

7.2 IMAGE RECTIFICATION AND RESTORATION

As previously mentioned, the intent of image rectification and restoration is to correct image data for distortions or degradations that stem from the image acquisition process. Obviously, the nature of such procedures varies considerably with such factors as the digital image acquisition type (digital camera, along-track scanner, across-track scanner), platform (airborne versus satellite), and total field of view. We make no attempt to describe the entire range of image rectification and restoration procedures applied to each of these various types of factors. Rather, we treat these operations under the generic headings of geometric correction, radiometric correction, and noise removal.

Geometric Correction

Raw digital images usually contain geometric distortions so significant that they cannot be used directly as a map base without subsequent processing. The sources of these distortions range from variations in the altitude, attitude, and velocity of the sensor platform to factors such as panoramic distortion, earth curvature, atmospheric refraction, relief displacement, and nonlinearities in the sweep of a sensor's IFOV. The intent of geometric correction is to compensate for the distortions introduced by these factors so that the corrected image will have the highest practical geometric integrity.

The geometric correction process is normally implemented as a two-step procedure. First, those distortions that are *systematic*, or predictable, are considered. Second, those distortions that are essentially *random*, or unpredictable, are considered.

Systematic distortions are well understood and easily corrected by applying formulas derived by modeling the sources of the distortions mathematically. For example, a highly systematic source of distortion involved in multispectral scanning from satellite altitudes is the eastward rotation of the earth beneath the satellite during imaging. This causes each optical sweep of the scanner to cover an area slightly to the west of the previous sweep. This is known as *skew distortion*. The process of *deskewing* the resulting imagery involves offsetting each successive scan line slightly to the west. The skewed-parallelogram appearance of satellite multispectral scanner data is a result of this correction.

Random distortions and residual unknown systematic distortions are corrected by analyzing well-distributed ground control points (GCPs) occurring in an image. As with their counterparts on aerial photographs, GCPs are features of known ground location that can be accurately located on the digital imagery. Some features that might make good control points are highway intersections and distinct shoreline features. In the correction process numerous GCPs are located both in terms of their two image coordinates (column, row numbers) on the distorted image and in terms of their ground coordinates (typically measured from a map, or GPS located in the field, in terms of UTM coordinates or latitude and longitude). These values are then submitted to a least squares regression analysis to determine coefficients for two *coordinate transformation equations* that can be used to interrelate the geometrically correct (map) coordinates and the distorted-image coordinates. (Appendix C describes one of the more common forms of coordinate transformation, the *affine transformation.*) Once the coefficients for these equations are determined, the distorted-image coordinates for any map position can be precisely estimated. Expressing this in mathematic notation,

$$x = f_1(X, Y) \qquad y = f_2(X, Y) \qquad (7.1)$$

where

(x, y) = distorted-image coordinates (column, row)
(X, Y) = correct (map) coordinates
f_1, f_2 = transformation functions

Intuitively, it might seem as though the above equations are stated backward! That is, they specify how to determine the distorted-image positions corresponding to correct, or undistorted, map positions. But that is exactly what is done during the geometric correction process. We first define an undistorted output matrix of "empty" map cells and then fill in each cell with the gray level of the corresponding pixel, or pixels, in the distorted image. This process is illustrated in Figure 7.1. This diagram shows the geometrically

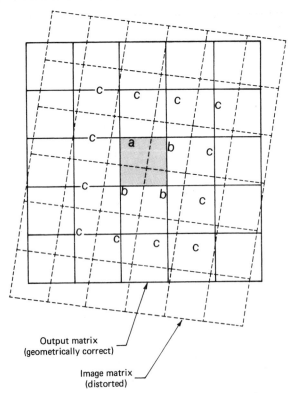

Output matrix
(geometrically correct)

Image matrix
(distorted)

Figure 7.1 Matrix of geometrically correct output pixels superimposed on matrix of original, distorted input pixels.

correct output matrix of cells (solid lines) superimposed over the original, distorted matrix of image pixels (dashed lines). After producing the transformation function, a process called *resampling* is used to determine the pixel values to fill into the output matrix from the original image matrix. This process is performed using the following operations:

1. The coordinates of each element in the undistorted output matrix are transformed to determine their corresponding location in the original input (distorted-image) matrix.

2. In general, a cell in the output matrix will not directly overlay a pixel in the input matrix. Accordingly, the intensity value or digital number (DN) eventually assigned to a cell in the output matrix is determined on the basis of the pixel values that surround its transformed position in the original input matrix.

A number of different resampling schemes can be used to assign the appropriate DN to an output cell or pixel. To illustrate this, consider the shaded output pixel shown in Figure 7.1. The DN for this pixel could be assigned simply on the basis of the DN of the closest pixel in the input matrix, disregarding the slight offset. In our example, the DN of the input pixel labeled *a* would be transferred to the shaded output pixel. This approach is called *nearest neighbor* resampling. It offers the advantage of computational simplicity and avoids having to alter the original input pixel values. However, features in the output matrix may be offset spatially by up to one-half pixel. This can cause a disjointed appearance in the output image product. Figure 7.2*b* is an example of a nearest neighbor resampled Landsat TM image. Figure 7.2*a* shows the original, distorted image.

More sophisticated methods of resampling evaluate the values of several pixels surrounding a given pixel in the input image to establish a "synthetic" DN to be assigned to its corresponding pixel in the output image. The *bilinear interpolation* technique takes a distance-weighted average of the DNs of the four nearest pixels (labeled *a* and *b* in the distorted-image matrix in Figure 7.1). This process is simply the two-dimensional equivalent to linear interpolation. As shown in Figure 7.2*c*, this technique generates a smoother appearing resampled image. However, because the process alters the gray levels of the original image, problems may be encountered in subsequent spectral pattern recognition analyses of the data. (Because of this, resampling is often performed after, rather than prior to, image classification procedures.)

An improved restoration of the image is provided by the *bicubic interpolation* or *cubic convolution* method of resampling. In this approach, the transferred synthetic pixel values are determined by evaluating the block of 16 pixels in the input matrix that surrounds each output pixel (labeled *a*, *b*, and *c* in Figure 7.1). Cubic convolution resampling (Figure 7.2*d*) avoids the disjointed appearance of the nearest neighbor method and provides a slightly

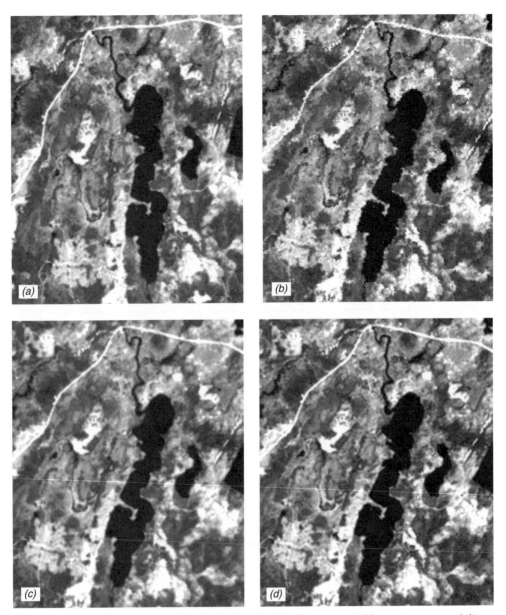

Figure 7.2 Resampling results: (a) original Landsat TM data; (b) nearest neighbor assignment; (c) bilinear interpolation; (d) cubic convolution. Scale 1:100,000.

sharper image than the bilinear interpolation method. (Again, this method alters the original image gray levels to some extent and other types of resampling can be employed to minimize this effect.)

As we discuss later, resampling techniques are important in several digital processing operations besides the geometric correction of raw images. For example, resampling is used to overlay or register multiple dates of imagery. It is also used to register images of differing resolution. Also, resampling procedures are used extensively to register image data and other sources of data in GISs. (Recall that resampling was discussed as an important part of digital orthophoto production in Section 3.9.) Appendix C contains additional details about the implementation of the various resampling procedures discussed in this section.

Radiometric Correction

As with geometric correction, the type of radiometric correction applied to any given digital image data set varies widely among sensors. Other things being equal, the radiance measured by any given system over a given object is influenced by such factors as changes in scene illumination, atmospheric conditions, viewing geometry, and instrument response characteristics. Some of these effects, such as viewing geometry variations, are greater in the case of airborne data collection than in satellite image acquisition. Also, the need to perform correction for any or all of these influences depends directly upon the particular application at hand.

In the case of satellite sensing in the visible and near-infrared portion of the spectrum, it is often desirable to generate mosaics of images taken at different times or to study the changes in the reflectance of ground features at different times or locations. In such applications, it is usually necessary to apply a *sun elevation correction* and an *earth–sun distance correction*. The sun elevation correction accounts for the seasonal position of the sun relative to the earth (Figure 7.3). Through this process, image data acquired under different solar illumination angles are normalized by calculating pixel brightness values assuming the sun was at the zenith on each date of sensing. The correction is usually applied by dividing each pixel value in a scene by the sine of the solar elevation angle for the particular time and location of imaging. Alternatively, the correction is applied in terms of the sun's angle from the zenith, which is simply 90° minus the solar elevation angle. (In this case each pixel value is divided by the cosine of the sun's angle from the zenith, resulting in the identical correction.) In either case, the correction ignores topographic and atmospheric effects.

The earth–sun distance correction is applied to normalize for the seasonal changes in the distance between the earth and the sun. The earth–sun distance

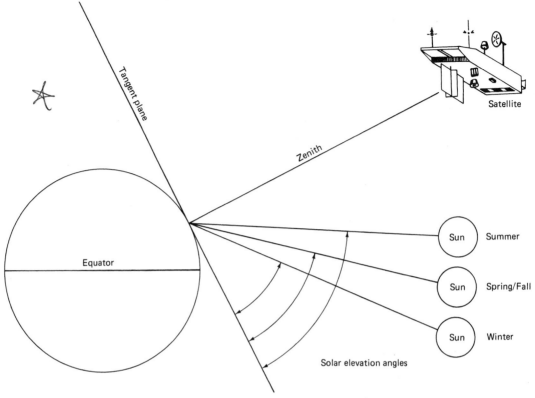

Figure 7.3   Effects of seasonal change on solar elevation angle. (The solar zenith angle is equal to 90° minus the solar elevation angle.)

is usually expressed in astronomical units. (An astronomical unit is equivalent to the mean distance between the earth and the sun, approximately 149.6 × 10⁶ km.) The irradiance from the sun decreases as the square of the earth–sun distance.

Ignoring atmospheric effects, the combined influence of solar zenith angle and earth–sun distance on the irradiance incident on the earth's surface can be expressed as

$$E = \frac{E_0 \cos \theta_0}{d^2} \tag{7.2}$$

where

E = normalized solar irradiance
E_0 = solar irradiance at mean earth–sun distance
θ_0 = sun's angle from the zenith
d = earth–sun distance, in astronomical units

(Information on the solar elevation angle and earth–sun distance for a given scene is normally part of the ancillary data supplied with the digital data.)

As initially discussed in Section 1.4, the influence of solar illumination variation is compounded by atmospheric effects. The atmosphere affects the radiance measured at any point in the scene in two contradictory ways. First, it attenuates (reduces) the energy illuminating a ground object. Second, it acts as a reflector itself, adding a scattered, extraneous "path radiance" to the signal detected by a sensor. Thus, the composite signal observed at any given pixel location can be expressed by

$$L_{\text{tot}} = \frac{\rho ET}{\pi} + L_{\text{p}} \tag{7.3}$$

where

L_{tot} = total spectral radiance measured by sensor
ρ = reflectance of object
E = irradiance on object
T = transmission of atmosphere
L_{p} = path radiance

(All of the above quantities depend on wavelength.)

Only the first term in the above equation contains valid information about ground reflectance. The second term represents the scattered path radiance, which introduces "haze" in the imagery and reduces image contrast. (Recall that scattering is wavelength dependent, with shorter wavelengths normally manifesting greater scattering effects.) *Haze compensation* procedures are designed to minimize the influence of path radiance effects. One means of haze compensation in multispectral data is to observe the radiance recorded over target areas of essentially zero reflectance. For example, the reflectance of deep clear water is essentially zero in the near-infrared region of the spectrum. Therefore, any signal observed over such an area represents the path radiance, and this value can be subtracted from all pixels in that band.

For convenience, haze correction routines are often applied uniformly throughout a scene. This may or may not be valid, depending on the uniformity of the atmosphere over a scene. When extreme viewing angles are involved in image acquisition, it is often necessary to compensate for the influence of varying the atmospheric path length through which the scene is recording. In such cases, off-nadir pixel values are usually normalized to their nadir equivalents.

Another radiometric data processing activity involved in many quantitative applications of digital image data is *conversion of DNs to absolute radiance values*. This operation accounts for the exact form of the A-to-D response functions for a given sensor and is essential in applications where measurement of absolute radiances is required. For example, such conversions are necessary when changes in the absolute reflectance of objects are to be measured over time using different sensors (e.g., the multispectral scanner on Landsat-3 versus that on Landsat-5). Likewise, such conversions are important in the development of mathematical models that physically relate image data to quantitative ground measurements (e.g., water quality data).

Normally, detectors and data systems are designed to produce a linear response to incident spectral radiance. For example, Figure 7.4 shows the linear radiometric response function typical of an individual TM channel. Each spectral band of the sensor has its own response function, and its characteristics are monitored using onboard calibration lamps (and temperature references for the thermal channel). The absolute spectral radiance output of the calibration sources is known from prelaunch calibration and is assumed to be stable over the life of the sensor. Thus, the onboard calibration sources form the basis for constructing the radiometric response function by relating known radiance values incident on the detectors to the resulting DNs.

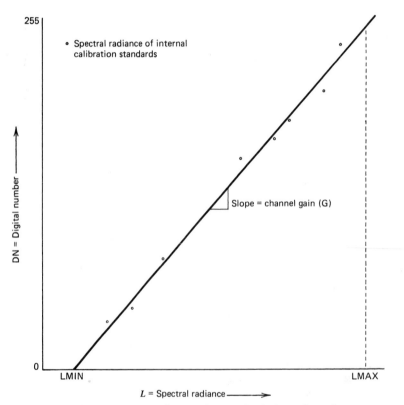

Figure 7.4 Radiometric response function for an individual TM channel.

It can be seen from Figure 7.4 that a linear fit to the calibration data results in the following relationship between radiance and DN values for any given channel:

$$DN = GL + B \qquad (7.4)$$

where

DN = digital number value recorded
G = slope of response function (channel gain)
L = spectral radiance measured (over the spectral bandwidth of the channel)
B = intercept of response function (channel offset)

Note that the slope and intercept of the above function are referred to as the *gain* and *offset* of the response function, respectively. In Figure 7.4 LMIN is the spectral radiance corresponding to a DN response of 0 and LMAX is the minimum radiance required to generate the maximum DN (here 255). That is, LMAX represents the radiance at which the channel saturates. The range from LMIN to LMAX is the dynamic range for the channel.

Figure 7.5 is a plot of the inverse of the radiometric response. Here we have simply interchanged the axes from Figure 7.4. The equation for this line is

$$L = \left(\frac{LMAX - LMIN}{255}\right) DN + LMIN \qquad (7.5)$$

Equation 7.5 can be used to convert any DN in a particular band to absolute units of spectral radiance in that band if LMAX and LMIN are known from the sensor calibration.

Often the LMAX and LMIN values published for a given sensor are expressed in units of mW cm^{-2} sr^{-1} μm^{-1}. That is, the values are often specified in terms of radiance per unit wavelength. To estimate the total within-band radiance in such cases, the value obtained from Eq. 7.5 must be multiplied by the width of the spectral band under consideration. Hence, a precise estimate of within-band radiance requires detailed knowledge of the spectral response curves for each band.

Noise Removal

Image noise is any unwanted disturbance in image data that is due to limitations in the sensing, signal digitization, or data recording process. The potential sources of noise range from periodic drift or malfunction of a detector, to electronic interference between sensor components, to intermittent "hiccups" in the data transmission and recording sequence. Noise can either degrade or totally mask the true radiometric information content of a digital image.

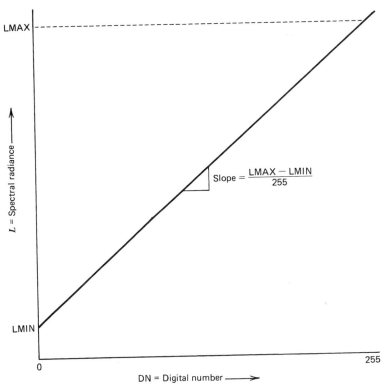

Figure 7.5 Inverse of radiometric response function for an individual TM channel.

Hence, noise removal usually precedes any subsequent enhancement or classification of the image data. The objective is to restore an image to as close an approximation of the original scene as possible.

As with geometric restoration procedures, the nature of noise correction required in any given situation depends upon whether the noise is systematic (periodic), random, or some combination of the two. For example, multispectral scanners that sweep multiple scan lines simultaneously often produce data containing systematic *striping* or *banding*. This stems from variations in the response of the individual detectors used within each band. Such problems were particularly prevalent in the collection of early Landsat MSS data. While the six detectors used for each band were carefully calibrated and matched prior to launch, the radiometric response of one or more tended to drift over time, resulting in relatively higher or lower values along every sixth line in the image data. In this case valid data are present in the defective lines, but they must be normalized with respect to their neighboring observations.

Several *destriping* procedures have been developed to deal with the type of problem described above. One method is to compile a set of histograms for

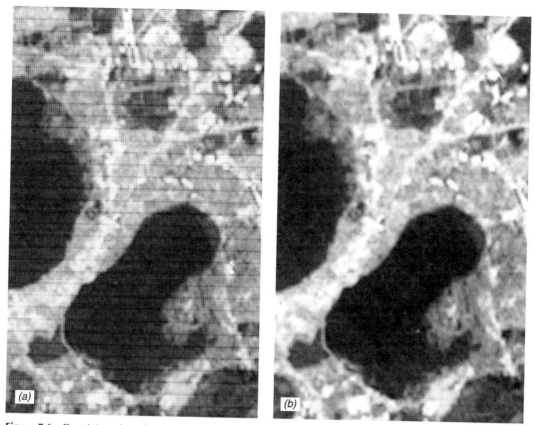

Figure 7.6 Destriping algorithm illustration: (*a*) original image manifesting striping with a six-line frequency; (*b*) restored image resulting from applying histogram algorithm.

the image—one for each detector involved in a given band. For MSS data this means that for a given band one histogram is generated for scan lines 1, 7, 13, and so on; a second is generated for lines 2, 8, 14, and so on; and so forth. These histograms are then compared in terms of their mean and median values to identify the problem detector(s). A gray-scale adjustment factor(s) can then be determined to adjust the histogram(s) for the problem lines to resemble those for the normal data lines. This adjustment factor is applied to each pixel in the problem lines and the others are not altered (Figure 7.6).

Another line-oriented noise problem sometimes encountered in digital data is *line drop*. In this situation, a number of adjacent pixels along a line (or an entire line) may contain spurious DNs. This problem is normally addressed by replacing the defective DNs with the average of the values for the pixels occurring in the lines just above and below (Figure 7.7). Alternatively, the DNs from the preceding line can simply be inserted in the defective pixels.

Figure 7.7 Line drop correction: (*a*) original image containing two line drops; (*b*) restored image resulting from averaging pixel values above and below defective line.

Random noise problems in digital data are handled quite differently than those we have discussed to this point. This type of noise is characterized by nonsystematic variations in gray levels from pixel to pixel called *bit errors*. Such noise is often referred to as being "spikey" in character, and it causes images to have a "salt and pepper" or "snowy" appearance.

Bit errors are handled by recognizing that noise values normally change much more abruptly than true image values. Thus, noise can be identified by comparing each pixel in an image with its neighbors. If the difference between a given pixel value and its surrounding values exceeds an analyst-specified threshold, the pixel is assumed to contain noise. The noisy pixel value can then be replaced by the average of its neighboring values. Moving neighborhoods or windows of 3 × 3 or 5 × 5 pixels are typically used in such procedures. Figure 7.8 illustrates the concept of a moving window comprising a 3 × 3-pixel neighborhood, and Figure 7.9 illustrates just one of many noise suppression algorithms using such a neighborhood. Finally, Figure 7.10 illustrates the results of applying the algorithm included in Figure 7.9 to an actual image.

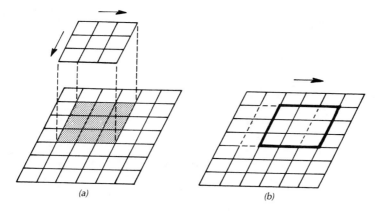

(a)   (b)

(c)

Figure 7.8 The moving window concept: (a) projection of 3 × 3-pixel window in image being processed; (b) movement of window along a line from pixel to pixel; (c) movement of window from line to line.

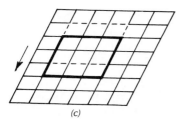

DN_1	DN_2	DN_3
DN_4	DN_5	DN_6
DN_7	DN_8	DN_9

$AVE_A = (DN_1 + DN_3 + DN_7 + DN_9)/4$

$AVE_B = (DN_2 + DN_4 + DN_6 + DN_8)/4$

$DIFF = |AVE_A - AVE_B|$

$THRESH = DIFF \times WEIGHT$

IF: $|DN_5 - AVE_A|$ or $|DN_5 - AVE_B| > THRESH$

THEN: $DN_5' = AVE_B$ OTHERWISE $DN_5' = DN_5$

Figure 7.9 Typical noise correction algorithm employing a 3 × 3-pixel neighborhood. *Note:* "WEIGHT" is an analyst-specified weighting factor. The lower the weight, the greater the number of pixels considered to be noise in an image.

Figure 7.10 Result of applying noise reduction algorithm: (*a*) original image data with noise-induced "salt-and-pepper" appearance; (*b*) image resulting from application of algorithm shown in Figure 7.9.

7.3 IMAGE ENHANCEMENT

As previously mentioned, the goal of image enhancement is to improve the visual interpretability of an image by increasing the apparent distinction between the features in the scene. The process of visually interpreting digitally enhanced imagery attempts to optimize the complementary abilities of the human mind and the computer. The mind is excellent at interpreting spatial attributes on an image and is capable of selectively identifying obscure or subtle features. However, the eye is poor at discriminating the slight radiometric or spectral differences that may characterize such features. Computer enhancement aims to visually amplify these slight differences to make them readily observable.

The range of possible image enhancement and display options available to the image analyst is virtually limitless. Most enhancement techniques may be categorized as either point or local operations. *Point operations* modify the brightness value of each pixel in an image data set independently. *Local operations* modify the value of each pixel based on neighboring brightness values.

Either form of enhancement can be performed on single-band (monochrome) images or on the individual components of multi-image composites. The resulting images may also be recorded or displayed in black and white or in color. Choosing the appropriate enhancement(s) for any particular application is an art and often a matter of personal preference.

Enhancement operations are normally applied to image data after the appropriate restoration procedures have been performed. Noise removal, in particular, is an important precursor to most enhancements. Without it, the image interpreter is left with the prospect of analyzing enhanced noise!

Below, we discuss the most commonly applied digital enhancement techniques. Three techniques can be categorized as *contrast manipulation, spatial feature manipulation,* or *multi-image manipulation.* Within these broad categories, we treat the following:

1. **Contrast manipulation.** Gray-level thresholding, level slicing, and contrast stretching.

2. **Spatial feature manipulation.** Spatial filtering, edge enhancement, and Fourier analysis.

3. **Multi-image manipulation.** Multispectral band ratioing and differencing, principal components, canonical components, vegetation components, intensity–hue–saturation (IHS) color space transformations, and decorrelation stretching.

7.4 CONTRAST MANIPULATION

Gray-Level Thresholding

Gray-level thresholding is used to *segment* an input image into two classes—one for those pixels having values below an analyst-defined gray level and one for those above this value. Below, we illustrate the use of thresholding to prepare a *binary mask* for an image. Such masks are used to segment an image into two classes so that additional processing can then be applied to each class independently.

Shown in Figure 7.11*a* is a TM1 image that displays a broad range of gray levels over both land and water. Let us assume that we wish to show the brightness variations in this band in the water areas only. Because many of the gray levels for land and water overlap in this band, it would be impossible to separate these two classes using a threshold set in this band. This is not the case in the TM4 band (Figure 7.11*b*). The histogram of DNs for the TM4 image (Figure 7.11*c*) shows that water strongly absorbs the incident energy in this near-infrared band (low DNs), while the land areas are highly reflective (high DNs). A threshold set at DN = 40 permits separation of these two classes in the TM4 data. This binary classification can then be applied to the

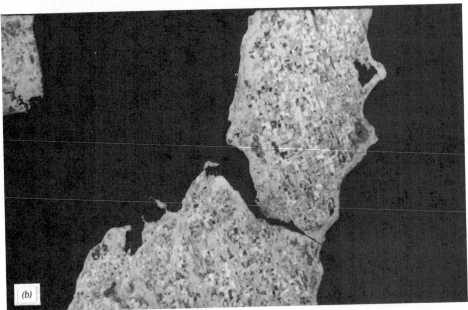

Figure 7.11 Gray-level thresholding for binary image segmentation: (*a*) original TM1 image containing continuous distribution of gray tones; (*b*) TM4 image; (*c*) TM4 histogram; (*d*) TM1 brightness variation in water areas only.

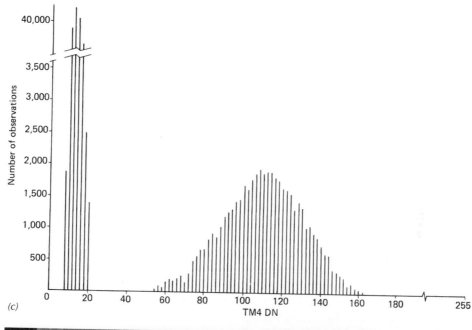

(c)

(d)

Figure 7.11   *(Continued)*

TM1 data to enable display of brightness variations in only the water areas. This is illustrated in Figure 7.11d. In this image, the TM1 land pixel values have all been set to 0 (black) based on their classification in the TM4 binary mask. The TM1 water pixel values have been preserved for display.

Level Slicing

Level slicing is an enhancement technique whereby the DNs distributed along the x axis of an image histogram are divided into a series of analyst-specified intervals or "slices." All of the DNs falling within a given interval in the input image are then displayed at a single DN in the output image. Consequently, if six different slices are established, the output image contains only six different gray levels. The result looks something like a contour map, except that the areas between boundaries are occupied by pixels displayed at the same DN. Each level can also be shown as a single color.

Figure 7.12 illustrates the application of level slicing to the "water" portion of the scene illustrated in Figure 7.11. Here, TM1 data have been level sliced into multiple levels in those areas previously determined to be water from the TM4 binary mask.

Figure 7.12   Level slicing operation applied to TM1 data in areas determined to be water in Figure 7.11.

Level slicing is used extensively in the display of thermal infrared images in order to show discrete temperature ranges coded by gray level or color. (See Plate 21 and Figure 6.18.)

Contrast Stretching

Image display and recording devices often operate over a range of 256 gray levels (the maximum number represented in 8-bit computer encoding). Sensor data in a single image rarely extend over this entire range. Hence, the intent of contrast stretching is to expand the narrow range of brightness values typically present in an input image over a wider range of gray values. The result is an output image that is designed to accentuate the contrast between features of interest to the image analyst.

To illustrate the contrast stretch process, consider a hypothetical sensing system whose image output levels can vary from 0 to 255. Figure 7.13*a* illustrates a histogram of brightness levels recorded in one spectral band over a scene. Assume that our hypothetical output device (e.g., computer monitor) is also capable of displaying 256 gray levels (0 to 255). Note that the histogram shows scene brightness values occurring only in the limited range of 60 to 158. If we were to use these image values directly in our display device (Figure 7.13*b*), we would be using only a small portion of the full range of possible display levels. Display levels 0 to 59 and 159 to 255 would not be utilized. Consequently, the tonal information in the scene would be compressed into a small range of display values, reducing the interpreter's ability to discriminate radiometric detail.

A more expressive display would result if we were to expand the range of image levels present in the scene (60 to 158) to fill the range of display values (0 to 255). In Figure 7.13*c*, the range of image values has been uniformly expanded to fill the total range of the output device. This uniform expansion is called a *linear stretch*. Subtle variations in input image data values would now be displayed in output tones that would be more readily distinguished by the interpreter. Light tonal areas would appear lighter and dark areas would appear darker.

In our example, the linear stretch would be applied to each pixel in the image using the algorithm

$$DN' = \left(\frac{DN - MIN}{MAX - MIN} \right) 255 \qquad (7.6)$$

where

> DN' = digital number assigned to pixel in output image
> DN = original digital number of pixel in input image
> MIN = minimum value of input image, to be assigned a value of 0 in the output image (60 in our example)
> MAX = maximum value of input image, to be assigned a value of 255 in the output image (158 in our example).

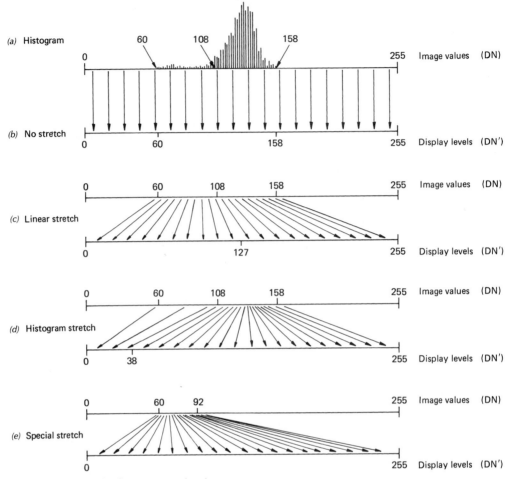

Figure 7.13 Principle of contrast stretch enhancement.

Figure 7.14 illustrates the above algorithm graphically. Note that the values for DN and DN′ must be discrete whole integers. Since the same function is used for all pixels in the image, it is usually calculated for all possible values of DN *before* processing the image. The resulting values of DN′ are then stored in a table (array). To process the image, no additional calculations are necessary. Each pixel's DN is simply used to index a location in the table to find the appropriate DN′ to be displayed in the output image. This process is referred to as a *table lookup* procedure and the list of DN′s associated with each DN is called a *lookup table* (*LUT*). The obvious advantage to the table lookup process is its computational efficiency. All possible values for DN′ are computed only once (for a maximum of 256 times) and the indexing of a location in the table is then all that is required for each pixel in the image.

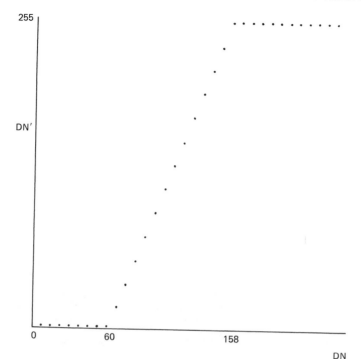

Figure 7.14 Linear stretch algorithm. Each point represents several discrete digital numbers.

One drawback of the linear stretch is that it assigns as many display levels to the rarely occurring image values as it does to the frequently occurring values. For example, as shown in Figure 7.13c, half of the dynamic range of the output device (0 to 127) would be reserved for the small number of pixels having image values in the range 60 to 108. The bulk of the image data (values 109 to 158) are confined to half the output display levels (128 to 255). Although better than the direct display in (b), the linear stretch would still not provide the most expressive display of the data.

To improve on the above situation, a *histogram-equalized stretch* can be applied. In this approach, image values are assigned to the display levels on the basis of their frequency of occurrence. As shown in Figure 7.13d, more display values (and hence more radiometric detail) are assigned to the frequently occurring portion of the histogram. The image value range of 109 to 158 is now stretched over a large portion of the display levels (39 to 255). A smaller portion (0 to 38) is reserved for the infrequently occurring image values of 60 to 108.

For special analyses, specific features may be analyzed in greater radiometric detail by assigning the display range exclusively to a particular range of image values. For example, if water features were represented by a narrow range of values in a scene, characteristics in the water features could be enhanced by stretching this small range to the full display range. As shown in Figure 7.13e,

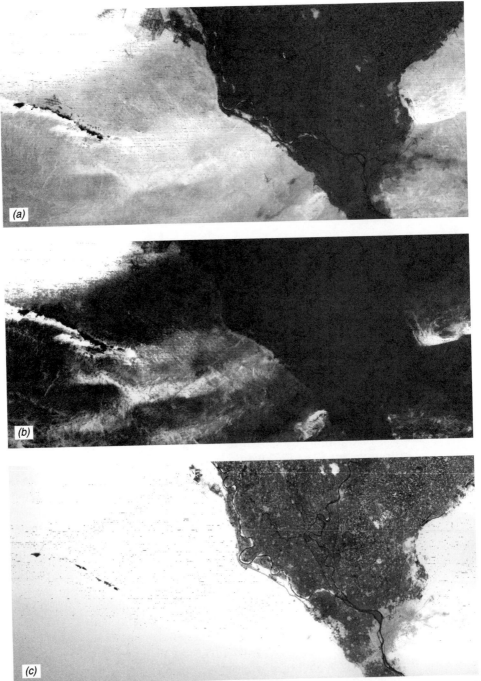

Figure 7.15 Effect of contrast stretching Landsat MSS data acquired over the Nile Delta; (*a*) original image; (*b*) stretch that enhances contrast in bright image areas; (*c*) stretch that enhances contrast in dark image areas. (Courtesy IBM Corp.)

the output range is devoted entirely to the small range of image values between 60 and 92. On the stretched display, minute tonal variations in the water range would be greatly exaggerated. The brighter land features, on the other hand, would be "washed out" by being displayed at a single, bright white level (255).

The visual effect of applying a contrast stretch algorithm is illustrated in Figure 7.15. An original Landsat MSS image covering the Nile Delta in Egypt is shown in (a). The city of Cairo lies close to the apex of the delta near the lower-right edge of the scene. Because of the wide range of image values present in this scene, the original image shows little radiometric detail. That is, features of similar brightness are virtually indistinguishable.

In Figure 7.15b, the brightness range of the desert area has been linearly stretched to fill the dynamic range of the output display. Patterns that were indistinguishable in the low contrast original are now readily apparent in this product. An interpreter wishing to analyze features in the desert region would be able to extract far more information from this display.

Because it reserves all display levels for the bright areas, the desert enhancement shows no radiometric detail in the darker irrigated delta region, which is displayed as black. If an interpreter were interested in analyzing a feature in this area, a different stretch could be applied, resulting in a display as shown in Figure 7.15c. Here, the display levels are devoted solely to the range of values present in the delta region. This rendering of the original image enhances brightness differences in the heavily populated and intensively cultivated delta, at the expense of all information in the bright desert area. Population centers stand out vividly in this display, and brightness differences between crop types are accentuated.

The contrast stretching examples we have illustrated represent only a small subset of the range of possible transformations that can be applied to image data. For example, nonlinear stretches such as sinusoidal transformations can be applied to image data to enhance subtle differences within "homogeneous" features such as forest stands or volcanic flows. Also, we have illustrated only monochromatic stretching procedures. Enhanced color images can be prepared by applying these procedures to separate bands of image data independently and then combining the results into a composite display.

7.5 SPATIAL FEATURE MANIPULATION

Spatial Filtering

In contrast to spectral filters, which serve to block or pass energy over various spectral ranges, spatial filters emphasize or deemphasize image data of various *spatial frequencies*. Spatial frequency refers to the "roughness" of the tonal variations occurring in an image. Image areas of high spatial frequency are tonally "rough." That is, the gray levels in these areas change abruptly over a relatively small number of pixels (e.g., across roads or field borders). "Smooth"

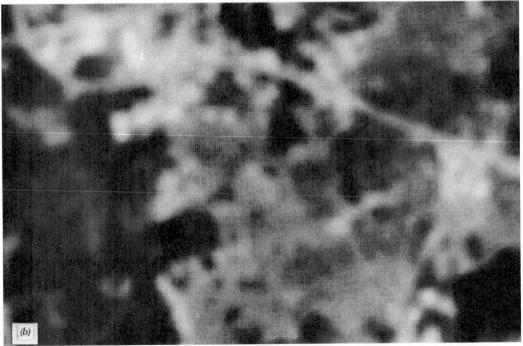

Figure 7.16 Effect of spatial filtering Landsat TM data: (*a*) original image; (*b*) low frequency component image; (*c*) high frequency component image.

518

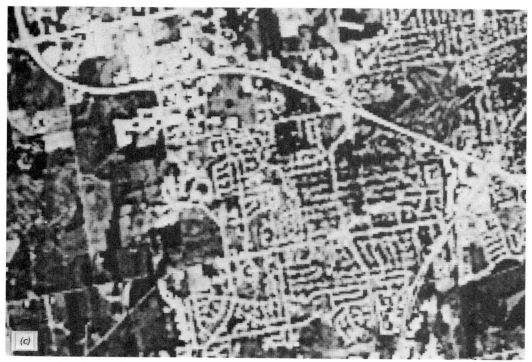

Figure 7.16 *(Continued)*

image areas are those of low spatial frequency, where gray levels vary only gradually over a relatively large number of pixels (e.g., large agricultural fields or water bodies). *Low pass filters* are designed to emphasize low frequency features (large-area changes in brightness) and deemphasize the high frequency components of an image (local detail). *High pass filters* do just the reverse. They emphasize the detailed high frequency components of an image and deemphasize the more general low frequency information.

Spatial filtering is a "local" operation in that pixel values in an original image are modified on the basis of the gray levels of neighboring pixels. For example, a simple low pass filter may be implemented by passing a moving window throughout an original image and creating a second image whose DN at each pixel corresponds to the local average within the moving window at each of its positions in the original image. Assuming a 3×3-pixel window is used, the center pixel's DN in the new (filtered) image would be the average value of the 9 pixels in the original image contained in the window at that point. This process is very similar to that described previously under the topic of noise suppression. (In fact, low pass filters are very useful for reducing random noise.)

A simple high pass filter may be implemented by subtracting a low pass filtered image (pixel by pixel) from the original, unprocessed image. Figure 7.16

illustrates the visual effect of applying this process to an image. The original image is shown in Figure 7.16*a*. Figure 7.16*b* shows the low frequency component image and Figure 7.16*c* illustrates the high frequency component image. Note that the low frequency component image (*b*) reduces deviations from the local average, which smooths or blurs the detail in the original image, reduces the gray-level range, but emphasizes the large-area brightness regimes of the original image. The high frequency component image (*c*) enhances the spatial detail in the image at the expense of the large-area brightness information. Both images have been contrast stretched. (Such stretching is typically required because spatial filtering reduces the gray-level range present in an image.)

Convolution

Spatial filtering is but one special application of the generic image processing operation called *convolution*. Convolving an image involves the following procedures:

1. A moving window is established that contains an array of coefficients or weighting factors. Such arrays are referred to as *operators* or *kernels*, and they are normally an odd number of pixels in size (e.g., 3 × 3, 5 × 5, 7 × 7).

2. The kernel is moved throughout the original image, and the DN at the center of the kernel in a second (convoluted) output image is obtained by multiplying each coefficient in the kernel by the corresponding DN in the original image and adding all the resulting products. This operation is performed for each pixel in the original image.

Figure 7.17 illustrates a 3 × 3-pixel kernel with all of its coefficients equal to 1/9. Convolving an image with this kernel would result in simply averaging the values in the moving window. This is the procedure that was used to prepare the low frequency enhancement shown in Figure 7.16*b*. However, images emphasizing other spatial frequencies may be prepared by simply altering the kernel coefficients used to perform the convolution. Figure 7.18 shows three successively lower frequency enhancements (*b, c,* and *d*) that have been derived from the same original data set (*a*).

The influence convolution may have on an image depends directly upon the size of the kernel used and the values of the coefficients contained within the kernel. The range of kernel sizes and weighting schemes is limitless. For example, by selecting the appropriate coefficients, one can center-weight kernels, make them of uniform weight, or shape them in accordance with a particular statistical model (such as a Gaussian distribution). In short, convolution is a generic image processing operation that has numerous appli-

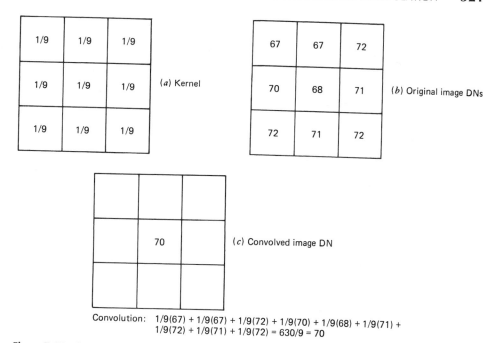

Figure 7.17 Concept of convolution. Shown is a 3 × 3-pixel kernel with all coefficients equal to $\frac{1}{9}$. The central pixel in the convolved image (in this case) contains the average of the DNs within the kernel.

cations in addition to spatial filtering. (Recall the use of "cubic convolution" as a resampling procedure.)

Edge Enhancement

We have seen that high frequency component images emphasize the spatial detail in digital images. That is, these images exaggerate local contrast and are superior to unenhanced original images for portraying linear features or edges in the image data. However, high frequency component images do not preserve the low frequency brightness information contained in original images. Edge-enhanced images attempt to preserve both local contrast and low frequency brightness information. They are produced by "adding back" all or a portion of the gray values in an original image to a high frequency component image of the same scene. Thus, edge enhancement is typically implemented in three steps:

1. A high frequency component image is produced containing the edge information. The kernel size used to produce this image is chosen based on the roughness of the image. "Rough" images suggest small

Figure 7.18 Frequency components of an image resulting from varying the kernel used for convolution: (*a*) original image; (*b–d*) successively lower frequency enhancements.

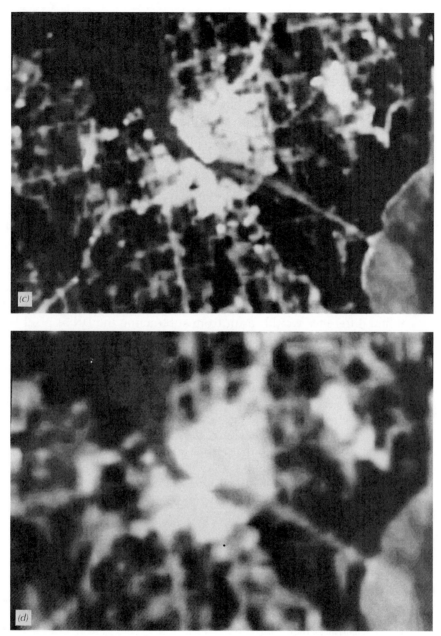

Figure 7.18 *(Continued)*

filter sizes (e.g., 3 × 3 pixels), whereas large sizes (e.g., 9 × 9 pixels) are used with "smooth" images.

2. All or a fraction of the gray level in each pixel of the original scene is added back to the high frequency component image. (The proportion of the original gray levels to be added back may be chosen by the image analyst.)

3. The composite image is contrast stretched. This results in an image containing local contrast enhancement of high frequency features that also preserves the low frequency brightness information contained in the scene.

Directional first differencing is another enhancement technique aimed at emphasizing edges in image data. It is a procedure that systematically compares each pixel in an image to one of its immediately adjacent neighbors and displays the difference in terms of the gray levels of an output image. This process is mathematically akin to determining the first derivative of gray levels with respect to a given direction. The direction used can be horizontal, vertical, or diagonal. In Figure 7.19, a horizontal first difference at pixel A would result from subtracting the DN in pixel H from that in pixel A. A vertical first difference would result from subtracting the DN at pixel V from that in pixel A; a diagonal first difference would result from subtracting the DN at pixel D from that in pixel A.

It should be noted that first differences can be either positive or negative, so a constant such as the display value median (127 for 8-bit data) is normally added to the difference for display purposes. Furthermore, because pixel-to-pixel differences are often very small, the data in the enhanced image often span a very narrow range about the display value median and a contrast stretch must be applied to the output image.

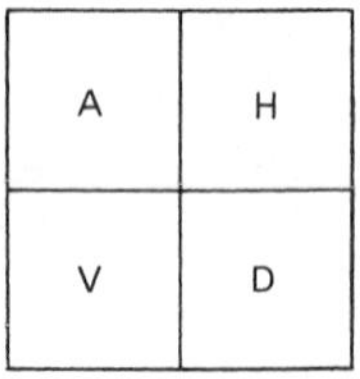

Horizontal first difference = $DN_A - DN_H$

Vertical first difference = $DN_A - DN_V$

Diagonal first difference = $DN_A - DN_D$

Figure 7.19   Primary pixel (A) and reference pixels (H, V, and D) used in horizontal, vertical, and diagonal first differencing, respectively.

First-difference images emphasize those edges normal to the direction of differencing and deemphasize those parallel to the direction of differencing. For example, in a horizontal first-difference image, vertical edges will result in large pixel-to pixel changes in gray level. On the other hand, the vertical first differences for these same edges would be relatively small (perhaps zero). This effect is illustrated in Figure 7.20 where vertical features in the original image (*a*) are emphasized in the horizontal first-difference image (*b*). Horizontal features in the original image are highlighted in the vertical first-difference image (*c*). Features emphasized by the diagonal first difference are shown in (*d*).

Figure 7.21 illustrates yet another form of edge enhancement involving diagonal first differencing. This image was produced by adding the absolute value of the upper-left-to-lower-right diagonal first difference to that of the upper-right-to-lower-left diagonal. This enhancement tends to highlight all edges in the scene.

Fourier Analysis

The spatial feature manipulations we have discussed thus far are implemented in the *spatial domain*—the (*x, y*) coordinate space of images. An alternative coordinate space that can be used for image analysis is the *frequency domain*. In this approach, an image is separated into its various spatial frequency components through application of a mathematical operation known as the *Fourier transform*. A quantitative description of how Fourier transforms are computed is beyond the scope of this discussion. Conceptually, this operation amounts to fitting a continuous function through the discrete DN values if they were plotted along each row and column in an image. The "peaks and valleys" along any given row or column can be described mathematically by a combination of sine and cosine waves with various amplitudes, frequencies, and phases. A Fourier transform results from the calculation of the amplitude and phase for each possible spatial frequency in an image.

After an image is separated into its component spatial frequencies, it is possible to display these values in a two-dimensional scatter plot known as a *Fourier spectrum*. Figure 7.22 illustrates a digital image in (*a*) and its Fourier spectrum in (*b*). The lower frequencies in the scene are plotted at the center of the spectrum and progressively higher frequencies are plotted outward. Features trending horizontally in the original image result in vertical components in the Fourier spectrum; features aligned vertically in the original image result in horizontal components in the Fourier spectrum.

If the Fourier spectrum of an image is known, it is possible to regenerate the original image through the application of an *inverse Fourier transform*. This operation is simply the mathematical reversal of the Fourier transform. Hence, the Fourier spectrum of an image can be used to assist in a number of

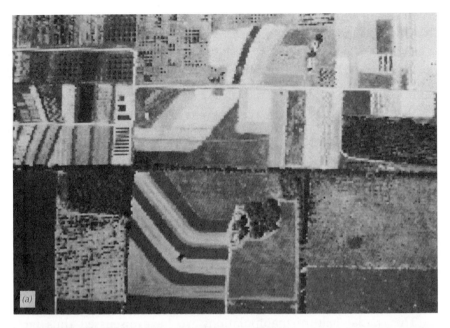

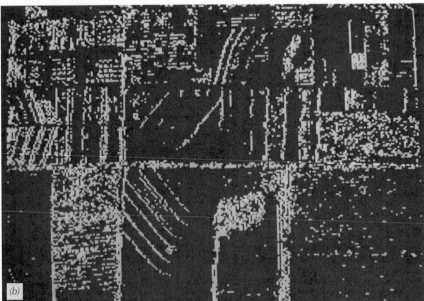

Figure 7.20 Effect of directional first differencing: (*a*) original image; (*b*) horizontal first difference; (*c*) vertical first difference; (*d*) diagonal first difference.

Figure 7.20 *(Continued)*

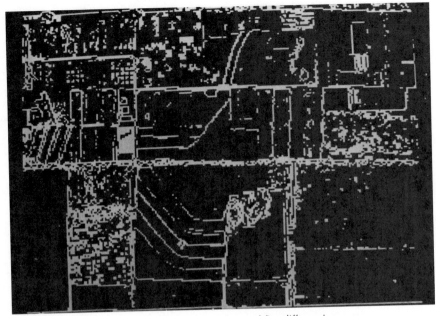

Figure 7.21 Edge enhancement through cross-diagonal first differencing.

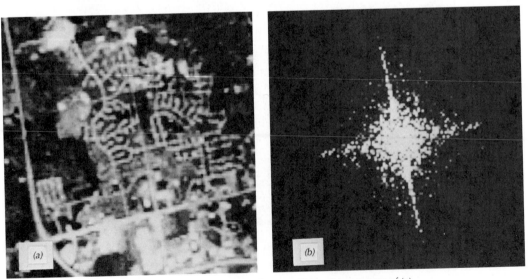

Figure 7.22 Applicaiton of Fourier transform: (*a*) original scene; (*b*) Fourier spectrum of (*a*).

image processing operations. For example, spatial filtering can be accomplished by applying a filter directly on the Fourier spectrum and then performing an inverse transform. This is illustrated in Figure 7.23. In Figure 7.23*a*, a circular high frequency blocking filter has been applied to the Fourier spectrum shown previously in Figure 7.22*b*. Note that this image is a low pass

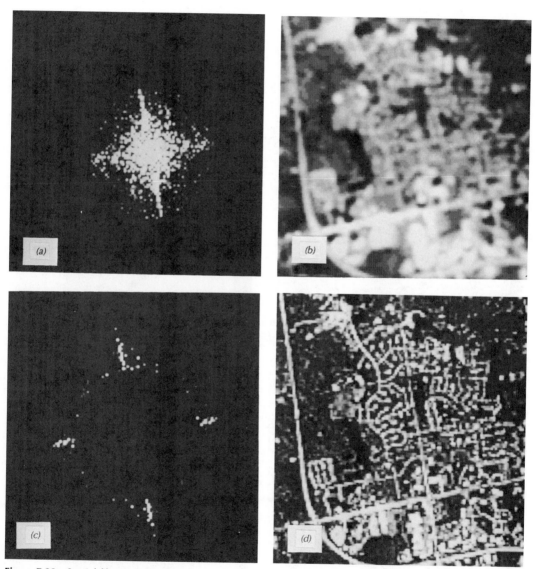

Figure 7.23 Spatial filtering in the frequency domain: (*a*) high frequency blocking filter; (*b*) inverse transform of (*a*); (*c*) low frequency blocking filter; (*d*) inverse transform of (*c*).

filtered version of the original scene. Figures 7.23c and d illustrate the application of a circular low frequency blocking filter (c) to produce a high pass filtered enhancement (d).

Figure 7.24 illustrates another common application of Fourier analysis—the elimination of image noise. Shown in Figure 7.24a is an airborne multispectral scanner image containing substantial noise. The Fourier spectrum

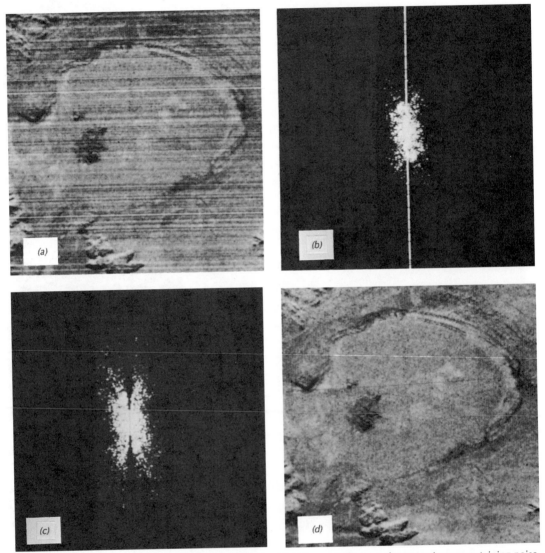

Figure 7.24 Noise elimination in the frequency domain. (a) Airborne multispectral scanner image containing noise. (Courtesy NASA.) (b) Fourier spectrum of (a). (c) Wedge block filter. (d) Inverse transform of (c).

for the image is shown in Figure 7.24*b*. Note that the noise pattern, which occurs in a horizontal direction in the original scene, appears as a band of frequencies trending in the vertical direction in the Fourier spectrum. In Figure 7.24*c* a vertical *wedge block filter* has been applied to the spectrum. This filter passes the lower frequency components of the image but blocks the high frequency components of the original image trending in the horizontal direction. Figure 7.24*d* shows the inverse transform of (*c*). Note how effectively this operation eliminates the noise inherent in the original image.

Fourier analysis is useful in a host of image processing operations in addition to the spatial filtering and image restoration applications we have illustrated in this discussion. However, most image processing is currently implemented in the spatial domain because of the number and complexity of computations required in the frequency domain. (This situation is likely to change with improvements in computer hardware and advances in research on the spatial attributes of digital image data.)

Before leaving the topic of spatial feature manipulation, it should be reemphasized that we have illustrated only a representative subset of the range of possible processing techniques available. Several of the references included at the end of this chapter describe and illustrate numerous other procedures that may be of interest to the reader.

7.6 MULTI-IMAGE MANIPULATION

Spectral Ratioing

Ratio images are enhancements resulting from the division of DN values in one spectral band by the corresponding values in another band. A major advantage of ratio images is that they convey the spectral or color characteristics of image features, regardless of variations in scene illumination conditions. This concept is illustrated in Figure 7.25, which depicts two different land cover types (deciduous and coniferous trees) occurring on both the sunlit and shadowed sides of a ridge line. The DNs observed for each cover type are substantially lower in the shadowed area than in the sunlit area. However, the ratio values for each cover type are nearly identical, irrespective of the illumination condition. Hence, a ratioed image of the scene effectively compensates for the brightness variation caused by the varying topography and emphasizes the color content of the data.

Ratioed images are often useful for discriminating subtle *spectral* variations in a scene that are masked by the *brightness* variations in images from individual spectral bands or in standard color composites. This enhanced discrimination is due to the fact that ratioed images clearly portray the variations in the *slopes* of the spectral reflectance curves between the two bands involved, regardless of the absolute reflectance values observed in the bands.

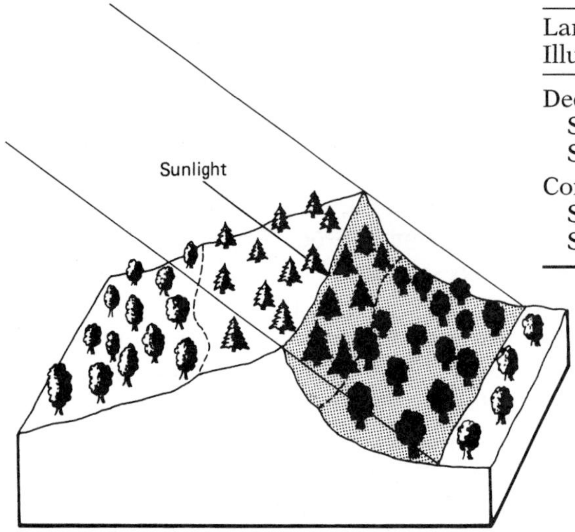

Land Cover/ Illumination	Digital Number		
	Band A	Band B	Ratio (Band A/Band B)
Deciduous			
Sunlit	48	50	0.96
Shadow	18	19	0.95
Coniferous			
Sunlit	31	45	0.69
Shadow	11	16	0.69

Figure 7.25 Reduction of scene illumination effects thorugh spectral ratioing. (Adapted from Sabins, 1997.)

These slopes are typically quite different for various material types in certain bands of sensing. For example, the near-infrared-to-red ratio for healthy vegetation is normally very high. That for stressed vegetation is typically lower (as near-infrared reflectance decreases and the red reflectance increases). Thus a near-infrared-to-red (or red-to-near-infrared) ratioed image might be very useful for differentiating between areas of the stressed and nonstressed vegetation. This type of ratio has also been employed extensively in vegetation indices aimed at quantifying relative vegetation greenness and biomass.

Obviously, the utility of any given spectral ratio depends upon the particular reflectance characteristics of the features involved and the application at hand. The form and number of ratio combinations available to the image analyst also varies depending upon the source of the digital data. The number of possible ratios that can be developed from n bands of data is $n(n - 1)$. Thus, for Landsat MSS data, $4(4 - 1)$, or 12, different ratio combinations are possible (six original and six reciprocal). For the six nonthermal bands of Landsat TM or ETM+ data there are $6(6 - 1)$, or 30, possible combinations.

Figure 7.26 illustrates four representative ratio images generated from TM data. These images depict higher ratio values in brighter tones. Shown in (*a*) is the ratio TM1/TM2. Because these two bands are highly correlated for this scene, the ratio image has low contrast. In (*b*) the ratio TM3/TM4 is depicted so that features such as water and roads, which reflect highly in the red band (TM3) and little in the near-infrared band (TM4), are shown in lighter tones. Features such as vegetation appear in darker tones because of their rel-

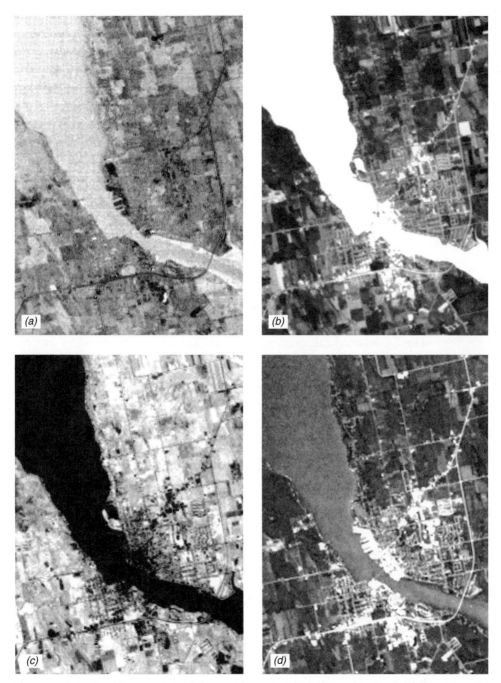

Figure 7.26 Ratioed images derived from midsummer Landsat TM data, near Sturgeon Bay, WI (higher ratio values are displayed in brighter image tones): (*a*) TM1/TM2; (*b*) TM3/TM4; (*c*) TM5/TM2; (*d*) TM3/TM7.

atively low reflectance in the red band (TM3) and high reflectance in the near infrared (TM4). In (c) the ratio TM5/TM2 is shown. Here, vegetation generally appears in light tones because of its relatively high reflectance in the mid-infrared band (TM5) and its comparatively lower reflectance in the green band (TM2). However, note that certain vegetation types do not follow this trend due to their particular reflectance characteristics. They are depicted in very dark tones in this particular ratio image and can therefore be discriminated from the other vegetation types in the scene. Part (d) shows the ratio TM3/TM7. Roads and other cultural features appear in lighter tone in this image due to their relatively high reflectance in the red band (TM3) and low reflectance in the mid-infrared band (TM7). Similarly, differences in water turbidity are readily observable in this ratio image.

Ratio images can also be used to generate false color composites by combining three monochromatic ratio data sets. Such composites have the twofold advantage of combining data from more than two bands and presenting the data in color, which further facilitates the interpretation of subtle spectral reflectance differences. Choosing which ratios to include in a color composite and selecting colors in which to portray them can sometimes be difficult. For example, excluding reciprocals, 20 color combinations are possible when the 6 original ratios of Landsat MSS data are displayed 3 at a time. The 15 original ratios of nonthermal TM data result in 455 different possible combinations.

Various quantitative criteria have been developed to assist in selecting which ratio combinations to include in color composites. The *Optimum Index Factor (OIF)* is one such criterion (Chavez, Berlin, and Sowers, 1982). It ranks all possible three-ratio combinations based on the total variance present in each ratio and the degree of correlation between ratios. That combination containing the most variance and least correlation is assumed to convey the greatest amount of information throughout a scene. A limitation of this procedure is that the best combination for conveying the *overall* information in a scene may not be the best combination for conveying the *specific* information desired by the image analyst. Hence, some trial and error is often necessary in selecting ratio combinations.

Certain caution should be taken when generating and interpreting ratio images. First, it should be noted that such images are "intensity blind." That is, dissimilar materials with different absolute radiances but having similar slopes of their spectral reflectance curves may appear identical. This problem is particularly troublesome when these materials are contiguous and of similar image texture. One way of minimizing this problem is by using a *hybrid color ratio composite*. This product is prepared by displaying two ratio images in two of the primary colors but using the third primary color to display an individual band of data. This restores a portion of the lost absolute radiance information and some of the topographic detail that may be needed to discriminate between certain features. (As we illustrate later, IHS color space transformations can also be used for this purpose.)

Noise removal is an important prelude to the preparation of ratio images since ratioing enhances noise patterns that are uncorrelated in the component images. Furthermore, ratios only compensate for multiplicative illumination effects. That is, division of DNs or radiances for two bands cancels only those factors that are operative equally in the bands and not those that are additive. For example, atmospheric haze is an additive factor that might have to be removed prior to ratioing to yield acceptable results. Alternatively, ratios of between-band differences and/or sums may be used to improve image interpretability in some applications.

The manner in which ratios are computed and displayed will also greatly influence the information content of a ratio image. For example, the ratio between two raw DNs for a pixel will normally be quite different from that between two radiance values computed for the same pixel. The reason for this is that the detector response curves for the two channels will normally have different offsets, which are additive effects on the data. (This situation is akin to the differences one would obtain by ratioing two temperatures using the Fahrenheit scale versus the Celsius scale.) Some trial and error may be necessary before the analyst can determine which form of ratio works best for a particular application.

It should also be noted that ratios can "blow up" mathematically (become equal to infinity) in that divisions by zero are possible. At the same time, ratios less than 1 are common and rounding to integer values will compress much of the ratio data into gray level 0 or 1. Hence, it is important to scale the results of ratio computations somehow and relate them to the display device used. One means of doing this is to employ an algorithm of the form

$$DN' = R \arctan\left(\frac{DN_X}{DN_Y}\right) \tag{7.7}$$

where

$$DN' = \text{digital number in ratio image}$$
$$R = \text{scaling factor to place ratio data in appropriate integer range}$$
$$\arctan(DN_X/DN_Y) = \text{angle (in radians) whose tangent is the ratio of the digital numbers in bands X and Y; if } DN_Y \text{ equals 0, this angle is set to } 90°$$

In the above equation the angle whose tangent is equal to the ratio of the two bands can range from 0° to 90°, or from 0 to approximately 1.571 rad. Therefore, DN' can range from 0 to approximately 1.571R. If an 8-bit display is used, R is typically chosen to be 162.3, and DN' can then range from 0 to 255.

Principal and Canonical Components

Extensive interband correlation is a problem frequently encountered in the analysis of multispectral image data (later illustrated in Figure 7.49). That is, images generated by digital data from various wavelength bands often appear similar and convey essentially the same information. Principal and canonical component transformations are two techniques designed to reduce such redundancy in multispectral data. These transformations may be applied either as an enhancement operation prior to visual interpretation of the data or as a preprocessing procedure prior to automated classification of the data. If employed in the latter context, the transformations generally increase the computational efficiency of the classification process because both principal and canonical component analyses may result in a reduction in the dimensionality of the original data set. Stated differently, the purpose of these procedures is to compress all of the information contained in an original n-band data set into fewer than n "new bands." The new bands are then used in lieu of the original data.

A detailed description of the statistical procedures used to derive principal and canonical component transformations is beyond the scope of this discussion. However, the concepts involved may be expressed graphically by considering a two-band image data set such as that shown in Figure 7.27. In (*a*), a random sample of pixels has been plotted on a scatter diagram according to their gray levels as originally recorded in bands A and B. Superimposed on the band A–band B axis system are two new axes (axes I and II) that are rotated with respect to the original measurement axes and that have their origin at the mean of the data distribution. Axis I defines the direction of the *first*

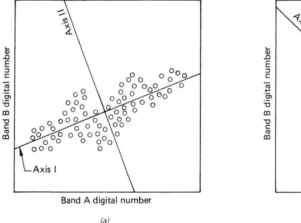

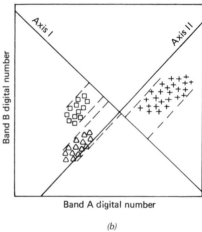

(a)   (b)

Figure 7.27 Rotated coordinate axes used in (*a*) principal component and (*b*) canonical component transformations.

principal component and axis II defines the direction of the *second principal component*. The form of the relationship necessary to transform a data value in the original band A–band B coordinate system into its value in the new axis I–axis II system is

$$DN_I = a_{11}DN_A + a_{12}DN_B \qquad DN_{II} = a_{21}DN_A + a_{22}DN_B \qquad (7.8)$$

where

DN_I, DN_{II} = digital numbers in new (principal component image) coordinate system

DN_A, DN_B = digital numbers in old (original) coordinate system

$a_{11}, a_{12}, a_{21}, a_{22}$ = coefficients for the transformation

In short, the principal component image data values are simply linear combinations of the original data values multiplied by the appropriate transformation coefficients. These coefficients are statistical quantities known as *eigenvectors* or *principal components*. They are derived from the variance/covariance matrix for the original image data.

Hence, a principal component image results from the linear combination of the original data and the eigenvectors on a pixel-by-pixel basis throughout the image. Often, the resulting principal component *image* is loosely referred to as simply a *principal component*. This is theoretically incorrect in that the eigenvalues themselves are the principal components, but we will sometimes not make this distinction elsewhere in this book.

It should be noted in Figure 7.27*a* that the data along the direction of the first principal component (axis I) have a greater variance or dynamic range than the data plotted against either of the original axes (bands A and B). The data along the second principal component direction have far less variance. This is characteristic of all principal component images. In general, the first principal component image (PC1) includes the largest percentage of the total scene variance and succeeding component images (PC2, PC3, . . . , PC*n*) each contain a decreasing percentage of the scene variance. Furthermore, because successive components are chosen to be orthogonal to all previous ones, the data they contain are uncorrelated.

Principal component enhancements are generated by displaying contrast-stretched images of the transformed pixel values. We illustrate the nature of these displays by considering the Landsat MSS images shown in Figure 7.28. This figure depicts the four MSS bands of a scene covering the Sahl al Matran area, Saudi Arabia. Figure 7.29 shows the principal component images for this scene. Some areas of geologic interest labeled in Figure 7.28 are (A) alluvial material in a dry stream valley, (B) flat-lying quaternary and tertiary basalts, and (C) granite and granodiorite intrusion.

Note that in Figure 7.29, PC1 expresses the majority (97.6 percent) of the variance in the original data set. Furthermore, PC1 and PC2 explain virtually all of the variance in the scene (99.4 percent). This compression of

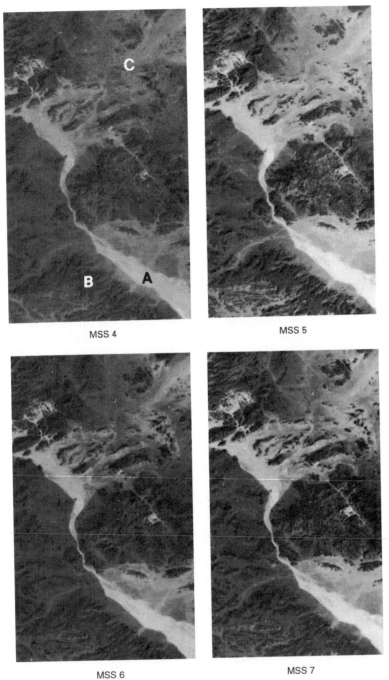

MSS 4

MSS 5

MSS 6

MSS 7

Figure 7.28 Four MSS bands covering the Sahl al Matran area of Saudi Arabia. Note the redundancy of information in these original image displays. (Courtesy NASA.)

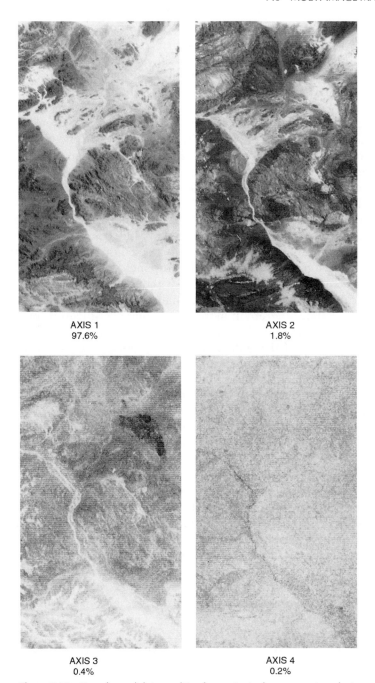

AXIS 1
97.6%

AXIS 2
1.8%

AXIS 3
0.4%

AXIS 4
0.2%

Figure 7.29 Transformed data resulting from principal component analysis of the MSS data shown in Figure 7.28. The percentage of scene variance contained in each axis is indicated. (Courtesy NASA.)

image information in the first two principal component images of Landsat MSS data is typical. Because of this, we refer to the *intrinsic dimensionality* of Landsat MSS data as being effectively 2. Also frequently encountered with Landsat MSS data, the PC4 image for this scene contains virtually no information and tends to depict little more than system noise. However, note that both PC2 and PC3 illustrate certain features that were obscured by the more dominant patterns shown in PC1. For example, a semicircular feature (labeled C in Figure 7.28) is clearly defined in the upper right portion of the PC2 and PC3 images (appearing bright and dark, respectively). This feature was masked by more dominant patterns both in the PC1 image and in all bands of the original data. Also, its tonal reversal in PC2 and PC3 illustrates the lack of correlation between these images.

As in the case of ratio images, principal component images can be analyzed as separate black and white images (as illustrated here), or any three component images may be combined to form a color composite. If used in an image classification process, principal component data are normally treated in the classification algorithm simply as if they were original data. However, the number of components used is normally reduced to the intrinsic dimensionality of the data, thereby making the image classification process much more efficient by reducing the amount of computation required. (For example, Landsat TM or ETM+ data can often be reduced to just three principal component images for classification purposes.)

Principal component enhancement techniques are particularly appropriate where little prior information concerning a scene is available. *Canonical component analysis*, also referred to as multiple discriminant analysis, may be more appropriate when information about particular features of interest is known. Recall that the principal component axes shown in Figure 7.27*a* were located on the basis of a random, undifferentiated sample of image pixel values. In Figure 7.27*b*, the pixel values shown are derived from image areas containing three different analyst-defined feature types (the feature types are represented by the symbols △, □, and +). The canonical component axes in this figure (axes I and II) have been located to maximize the *separability* of these classes while minimizing the variance within each class. For example, the axes have been positioned in this figure such that the three feature types can be discriminated solely on the basis of the first canonical component images (CC1) values located along axis I.

In Figure 7.30, canonical component images are shown for the Landsat MSS scene that was shown in Figure 7.28. Once again, CC1 expresses the highest percentage variation in the data with subsequent component images representing lesser amounts of uncorrelated additional information. These displays, as in cases such as feature C, may enhance subtle features not evident in the original image data. Like principal component data, canonical component data can also be used in image classification. Canonical component images not only improve classification efficiency but also can improve

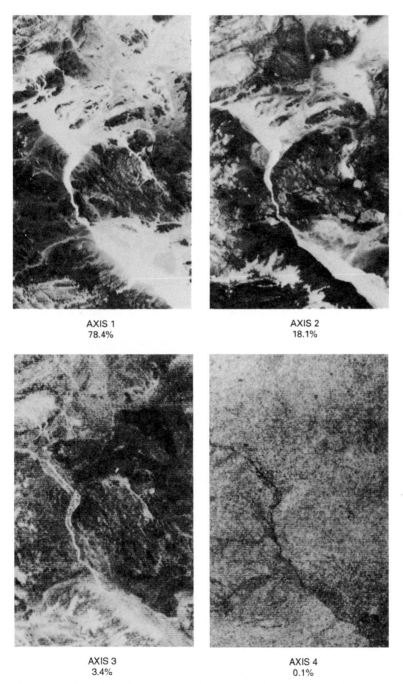

AXIS 1
78.4%

AXIS 2
18.1%

AXIS 3
3.4%

AXIS 4
0.1%

Figure 7.30   Transformed data resulting from canonical component analysis of the MSS data shown in Figure 7.28. The percentage of scene variance contained in each axis is indicated. (Courtesy NASA.)

classification accuracy for the identified features due to the increased spectral separability of classes.

Vegetation Components

Previously (Section 6.16), we introduced the concept of vegetation indices and the use of between-band differences and ratios to produce vegetation index images from AVHRR data. Here we wish to point out that numerous other forms of linear data transformations have been developed for vegetation monitoring, with differing sensors and vegetation conditions dictating different transformations. For example, Kauth and Thomas (1976) derived a linear transformation of the four Landsat MSS bands that established four new axes in the spectral data that can be interpreted as *vegetation components* useful for agricultural crop monitoring. This "tasseled cap" transformation rotates the MSS data such that the majority of information is contained in two components or features that are directly related to physical scene characteristics. Brightness, the first feature, is a weighted sum of all bands and is defined in the direction of the principal variation in soil reflectance. The second feature, greenness, is approximately orthogonal to brightness and is a contrast between the near-infrared and visible bands. Greenness is strongly related to the amount of green vegetation present in the scene. Brightness and greenness together typically express 95 percent or more of the total variability in MSS data and have the characteristic of being readily interpretable features generally applicable from scene to scene (once illumination and atmospheric effects are normalized).

Crist and Cicone (1984) extended the tasseled cap concept to Landsat TM data and found that the six bands of reflected data effectively occupy three dimensions, defining planes of soils, vegetation, and a transition zone between them. The third feature, called wetness, relates to canopy and soil moisture.

Figure 7.31 illustrates the application of the tasseled cap transformation to TM data acquired over north central Nebraska. The northern half of the area is dominated by circular corn fields that have been watered throughout the summer by center-pivot irrigators. The southern half of the area is at the edge of the Nebraska Sand Hills, which are covered by grasslands. Lighter tones in these images correspond to larger values of each component.

Figure 7.32 illustrates another vegetation transformation, namely, the *transformed vegetation index* (*TVI*), applied to the same data set shown in Figure 5.8 and Plate 14. The TVI is computed as

$$TVI = \left[\frac{DN \text{ (near IR)} - DN \text{ (red)}}{DN \text{ (near IR)} + DN \text{ (red)}} + 0.5 \right]^{1/2} \times 100 \qquad (7.9)$$

where

DN (near IR) = digital number in the near-IR band
DN (red) = digital number in the red band

Figure 7.31   Tasseled cap transformation for a late-summer TM image of north central Nebraska. The three components shown here illustrate (a) relative greenness, (b) brightness, and (c) wetness. (Courtesy Institute of Agriculture and Natural Resources, University of Nebraska.)

Using ground reference data, it is frequently possible to "calibrate" TVI values to the green biomass present on a pixel-by-pixel basis. Usually, separate calibration relationships must be established for each cover type present in an image. These relationships may then be used in such applications as "precision crop management" or precision farming to guide the application of

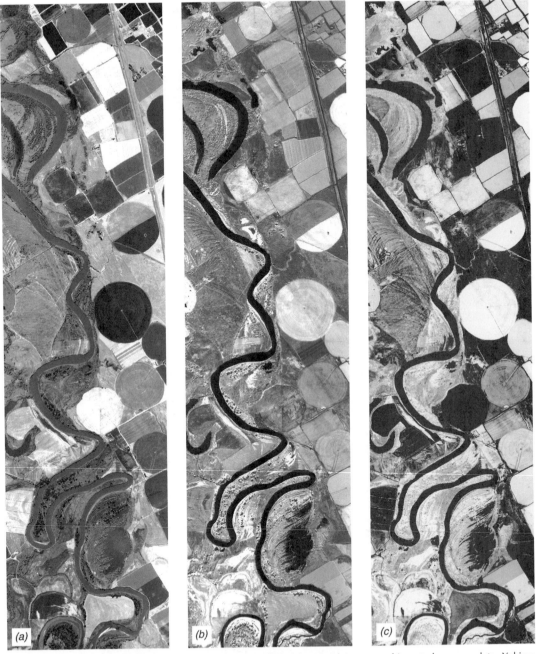

Figure 7.32 Transformed vegetation index (TVI) image derived from mid-August multispectral scanner data, Yakima River valley, WA: (*a*) red band; (*b*) near-IR band; (*c*) TVI image. Same image area as shown in Figure 5.8 and Plate 14. (Courtesy Sensys Technologies, Inc.)

irrigation water, fertilizers, herbicides, and so on (Section 4.6). Similarly, TVI values have been used to aid in making ranch management decision when the TVI data correlate with the estimated level of forage present in pastures contained in an image (Miller et al., 1985).

Indices such as the TVI and NDVI, which are based on the near-infrared and red spectral bands, have been shown to be well correlated not only with crop biomass accumulation, but also with leaf chlorophyll levels, leaf area index values, and the photosynthetically active radiation absorbed by a crop canopy. However, when such biophysical parameters reach moderate to high levels, the *green normalized difference vegetation index* (*GNDVI*) may be a more reliable indicator of crop conditions. The GNDVI is identical in form to the NDVI except that the green band is substituted for the red band.

With the availability of MODIS data on a global basis, several new vegetation indices have been proposed. For example, the *enhanced vegetation index* (*EVI*) has been developed as a modified NDVI with an adjustment factor to minimize soil background influences and a blue band correction of red band data to lessen atmospheric scattering. Additional treatment of this and the scores of other vegetation indices in use is beyond the scope of this discussion. Readers interested in additional information on this subject are encouraged to consult any of the numerous summaries of vegetation indices available in the literature (e.g., Richardson and Everitt, 1992; Running et al., 1994; Lyon et al., 1998; and Jensen, 2000).

Intensity–Hue–Saturation Color Space Transformation

Digital images are typically displayed as additive color composites using the three primary colors: red, green, and blue (RGB). Figure 7.33 illustrates the interrelation among the RGB components of a typical color display device (such as a color monitor). Shown in this figure is the RGB *color cube*, which is defined by the brightness levels of each of the three primary colors. For a display with 8-bit-per-pixel data encoding, the range of possible DNs for each color component is 0 to 255. Hence, there are 256^3 (or 16,777,216) possible combinations of red, green, and blue DNs that can be displayed by such a device. Every pixel in a composited display may be represented by a three-dimensional coordinate position somewhere within the color cube. The line from the origin of the cube to the opposite corner is known as the *gray line* since DNs that lie on this line have equal components of red, green, and blue.

The RGB displays are used extensively in digital processing to display normal color, false color infrared, and arbitrary color composites. For example, a normal color composite may be displayed by assigning TM or

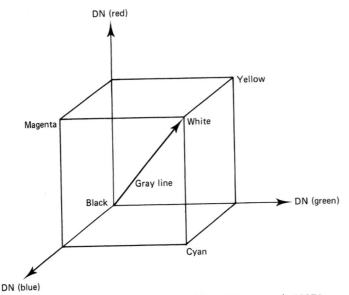

Figure 7.33   The RGB color cube. (Adapted from Schowengerdt, 1997.)

ETM+ bands 1, 2, and 3 to the blue, green, and red components, respectively. A false color infrared composite results when bands 2, 3, and 4 are assigned to these respective components. Arbitrary color composites result when other bands or color assignments are used. Color composites may be contrast stretched on a RGB display by manipulating the contrast in each of the three display channels (using a separate lookup table for each of the three color components).

An alternative to describing colors by their RGB components is the use of the *intensity–hue–saturation* (*IHS*) system. "Intensity" relates to the total brightness of a color. "Hue" refers to the dominant or average wavelength of light contributing to a color. "Saturation" specifies the purity of color relative to gray. For example, pastel colors such as pink have low saturation compared to such high saturation colors as crimson. Transforming RGB components into IHS components *before* processing may provide more control over color enhancements.

Figure 7.34 shows one (of several) means of transforming RGB components into IHS components. This particular approach is called the *hexcone model*, and it involves the *projection* of the RGB color cube onto a plane that is perpendicular to the gray line and tangent to the cube at the corner farthest from the origin. The resulting projection is a hexagon. If the plane of projection is moved from white to black along the gray line, successively smaller color *subcubes* are projected and a series of hexagons of decreasing size re-

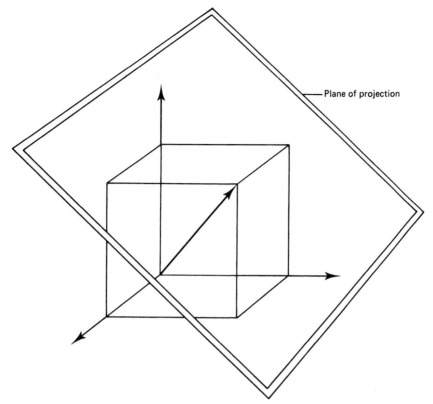

Figure 7.34 Planar projection of the RGB color cube. A series of such projections results when progressively smaller subcubes are considered between white and black.

sult. The hexagon at white is the largest and the hexagon at black degenerates to a point. The series of hexagons developed in this manner define a solid called the *hexcone* (Figure 7.35*a*).

In the hexcone model *intensity* is defined by the distance along the gray line from black to any given hexagonal projection. Hue and saturation are defined at a given intensity, within the appropriate hexagon (Figure 7.35*b*). *Hue* is expressed by the angle around the hexagon, and *saturation* is defined by the distance from the gray point at the center of the hexagon. The farther a point lies away from the gray point, the more saturated the color. (In Figure 7.35*b*, linear distances are used to define hue and saturation, thereby avoiding computations involving trigonometric functions.)

At this point we have established the basis upon which any pixel in the RGB color space can be transformed into its IHS counterpart. Such transformations are often useful as an intermediate step in image enhancement. This

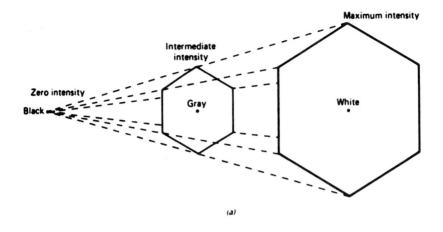

(a)

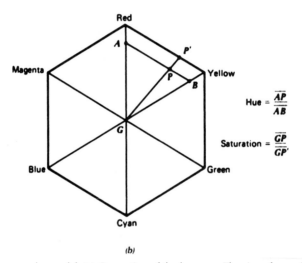

(b)

Figure 7.35 Hexcone color model. (a) Generation of the hexcone. The size of any given hexagon is determined by pixel intensity. (b) Definition of hue and saturation components for a pixel value, P, having a typical, nonzero intensity. (Adapted from Schowengerdt, 1997.)

is illustrated in Figure 7.36. In this figure the original RGB components are shown transformed first into their corresponding IHS components. The IHS components are then manipulated to enhance the desired characteristics of the image. Finally, these modified IHS components are transformed back to the RGB system for final display.

Among the advantages of IHS enhancement operations is the ability to vary each IHS component independently, without affecting the others. For example,

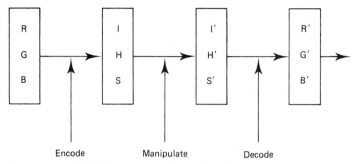

Figure 7.36   IHS/RGB encoding and decoding for interactive image manipulation. (Adapted from Schowengerdt, 1997.)

a contrast stretch can be applied to the intensity component of an image, and the hue and saturation of the pixels in the enhanced image will not be changed (as they typically are in RGB contrast stretches). The IHS approach may also be used to display spatially registered data of varying spatial resolution. For example, high resolution data from one source may be displayed as the intensity component, and low resolution data from another source may be displayed as the hue and saturation components. Such an approach was used to produce Plate 19 (Section 6.15). This plate is a merger of IKONOS 1-m-resolution panchromatic data (used in the intensity component) and 4-m-resolution multispectral data (hue and saturation components). The result is a composite image having the spatial resolution of the 1-m panchromatic data and the color characteristics of the original 4-m multispectral data. Similar procedures are used to merge other sources of same-sensor, multiresolution data sets (e.g., 10-m panchromatic and 20-m multispectral SPOT data, 15-m panchromatic and 30-m ETM+ data). Likewise, IHS techniques are often used to merge data from different sensing systems (e.g., digital orthophotos with satellite data).

One caution to be noted in using IHS transformations to merge multiresolution data is that direct substitution of the panchromatic data for the intensity component may not always produce the best final product in terms of color balance. In such situations, weighted combinations of the panchromatic and multispectral data might be used. The approach used in the production of Plate 19 was to employ a histogram matching operation to match the histogram of the new panchromatic data to that of the intensity component derived from the multispectral RGB data. The modified panchromatic data resulting from this operation were then used in the intensity component of the IHS image prior to transforming the data back to RGB format.

The development and application of various IHS encoding and enhancement schemes are the subject of continuing research. An interesting project in this regard has been the use of IHS transformations to display two bands (bands 1 and 2) of raw AVHRR data in a three-color composite. In such composites, the sum of bands 1 and 2 is used to represent intensity. The band 2/1 ratio is used to define hue, and the difference between bands 1 and 2 is used to define saturation. The resulting image looks very similar to a standard color infrared composite.

Decorrelation Stretching

Decorrelation stretching is a form of multi-image manipulation that is particularly useful when displaying multispectral data that are highly correlated. Data from the NASA Thermal Infrared Multispectral Scanner (TIMS) and other hyperspectral data collected in the same region of the spectrum often fall into this category. Traditional contrast stretching of highly correlated data as R, G, and B displays normally only expands the range of intensities; it does little to expand the range of colors displayed, and the stretched image still contains only pastel hues. For example, no areas in a highly correlated image are likely to have high DNs in the red display channel but low values in the green and blue (which would produce a pure red). Instead, the reddest areas are merely a reddish-gray. To circumvent this problem, decorrelation stretching involves exaggeration of the least correlated information in an image primarily in terms of saturation, with minimal change in image intensity and hue.

As with IHS transformations, decorrelation stretching is applied in a transformed image space, and the results are then transformed back to the RGB system for final display. The major difference in decorrelation stretching is that the transformed image space used is that of the original image's principal components. The successive principal components of the original image are stretched independently along the respective principal component axes (Figure 7.27a). By definition, these axes are statistically independent of one another so the net effect of the stretch is to emphasize the poorly correlated components of the original data. When the stretched data are then transformed back to the RGB system, a display having increased color saturation results. There is usually little difference in the perceived hues and intensities due to enhancement. This makes interpretation of the enhanced image straightforward, with the decorrelated information exaggerated primarily in terms of saturation. Previously pastel hues become much more saturated.

Because decorrelation stretching is based on principal component analysis, it is readily extended to any number of image channels. Recall that the IHS procedure is applied to only three channels at a time.

7.7 IMAGE CLASSIFICATION

The overall objective of image classification procedures is to automatically categorize all pixels in an image into land cover classes or themes. Normally, multispectral data are used to perform the classification and, indeed, the spectral pattern present within the data for each pixel is used as the numerical basis for categorization. That is, different feature types manifest different combinations of DNs based on their inherent spectral reflectance and emittance properties. In this light, a spectral "pattern" is not at all geometric in character. Rather, the term *pattern* refers to the set of radiance measurements obtained in the various wavelength bands for each pixel. *Spectral pattern recognition* refers to the family of classification procedures that utilizes this pixel-by-pixel spectral information as the basis for automated land cover classification.

Spatial pattern recognition involves the categorization of image pixels on the basis of their spatial relationship with pixels surrounding them. Spatial classifiers might consider such aspects as image texture, pixel proximity, feature size, shape, directionality, repetition, and context. These types of classifiers attempt to replicate the kind of spatial synthesis done by the human analyst during the visual interpretation process. Accordingly, they tend to be much more complex and computationally intensive than spectral pattern recognition procedures.

Temporal pattern recognition uses time as an aid in feature identification. In agricultural crop surveys, for example, distinct spectral and spatial changes during a growing season can permit discrimination on multidate imagery that would be impossible given any single date. For example, a field of winter wheat might be indistinguishable from bare soil when freshly seeded in the fall and spectrally similar to an alfalfa field in the spring. An interpretation of imagery from either date alone would be unsuccessful, regardless of the number of spectral bands. If data were analyzed from both dates, however, the winter wheat fields could be readily identified, since no other field cover would be bare in late fall and green in late spring.

As with the image restoration and enhancement techniques we have described, image classifiers may be used in combination in a hybrid mode. Also, there is no single "right" manner in which to approach an image classification problem. The particular approach one might take depends upon the nature of the data being analyzed, the computational resources available, and the intended application of the classified data.

In the remaining discussion we emphasize spectrally oriented classification procedures for land cover mapping. (As stated earlier, this emphasis is based on the relative state of the art of these procedures. They currently form the backbone of most multispectral classification activities.) First, we describe *supervised classification*. In this type of classification the image analyst "supervises" the pixel categorization process by specifying, to the computer algorithm, numerical descriptors of the various land cover types present in a scene. To do this, representative sample sites of known cover type, called *training areas*, are used to compile a numerical "interpretation key" that describes the spectral attributes for each feature type of interest. Each pixel in the data set is then compared numerically to each category in the interpretation key and labeled with the name of the category it "looks most like." As we see in the next section, there are a number of numerical strategies that can be employed to make this comparison between unknown pixels and training set pixels.

Following our discussion of supervised classification we treat the subject of *unsupervised classification*. Like supervised classifiers, the unsupervised procedures are applied in two separate steps. The fundamental difference between these techniques is that supervised classification involves a training step followed by a classification step. In the unsupervised approach the image data are first classified by aggregating them into the natural spectral groupings, or *clusters*, present in the scene. Then the image analyst determines the land cover identity of these spectral groups by comparing the classified image data to ground reference data. Unsupervised procedures are discussed in Section 7.11.

Following our treatment of supervised and unsupervised classification we discuss *hybrid classification* procedures. Such techniques involve aspects of both supervised and unsupervised classification and are aimed at improving the accuracy or efficiency (or both) of the classification process. Hybrid classification is the subject of Section 7.12.

We reiterate that the various classification procedures we discuss in the next several sections are generally applied to multispectral data sets. We defer discussion of hyperspectral image analysis until Section 7.18.

7.8 SUPERVISED CLASSIFICATION

We use a hypothetical example to facilitate our discussion of supervised classification. In this example, let us assume that we are dealing with the analysis of five-channel airborne multispectral scanner data. (The identical procedures would apply to Landsat, SPOT, or virtually any other source of multispectral data.) Figure 7.37 shows the location of a single line of data collected for our hypothetical example over a landscape composed of several cover types. For each of the pixels shown along this line, the multispectral scanner has measured scene radiance in terms of DNs recorded in each of the five spectral bands of sensing: blue, green, red, near infrared, and thermal infrared. Below the scan line, typical DNs measured over six different land cover types are shown. The vertical bars indicate the relative gray values in each spectral band. These five outputs represent a coarse description of the spectral response patterns of the various terrain features along the scan line. If these spectral patterns are sufficiently distinct for each feature type, they may form the basis for image classification.

Figure 7.38 summarizes the three basic steps involved in a typical supervised classification procedure. In the *training stage* (1), the analyst identifies representative training areas and develops a numerical description of the spectral attributes of each land cover type of interest in the scene. Next, in the *classification stage* (2), each pixel in the image data set is categorized into the land cover

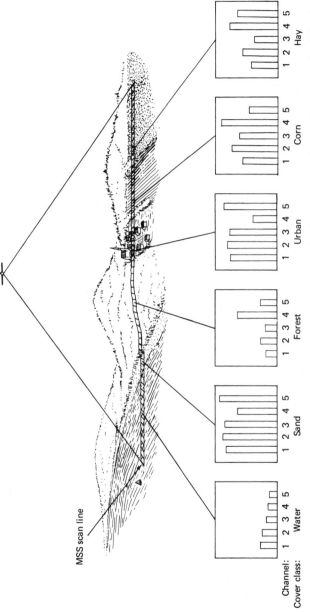

Channel: 1 2 3 4 5   1 2 3 4 5   1 2 3 4 5   1 2 3 4 5   1 2 3 4 5   1 2 3 4 5

Cover class: Water   Sand   Forest   Urban   Corn   Hay

MSS scan line

Figure 7.37 Selected multispectral scanner measurements made along one scan line. Sensor covers the following spectral bands: 1, blue; 2, green; 3, red; 4, near infrared; 5, thermal infrared.

553

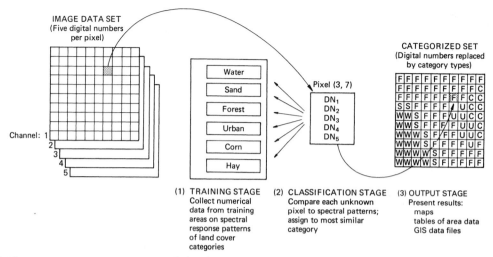

Figure 7.38 Basic steps in supervised classification.

class it most closely resembles. If the pixel is insufficiently similar to any training data set, it is usually labeled "unknown." The category label assigned to each pixel in this process is then recorded in the corresponding cell of an interpreted data set (an "output image"). Thus, the multidimensional image matrix is used to develop a corresponding matrix of interpreted land cover category types. After the entire data set has been categorized, the results are presented in the *output stage* (3). Being digital in character, the results may be used in a number of different ways. Three typical forms of output products are thematic maps, tables of full scene or subscene area statistics for the various land cover classes, and digital data files amenable to inclusion in a GIS. In this latter case, the classification "output" becomes a GIS "input."

We discuss the output stage of image classification in Section 7.14. Our immediate attention is focused on the training and classification stages. We begin with a discussion of the *classification* stage because it is the heart of the supervised classification process—during this stage the spectral patterns in the image data set are evaluated in the computer using predefined decision rules to determine the identity of each pixel. Another reason for treating the classification stage first is because familiarity with this step aids in understanding the requirements that must be met in the training stage.

7.9 THE CLASSIFICATION STAGE

Numerous mathematical approaches to spectral pattern recognition have been developed and extensive discussion of this subject can be found in the various references found at the end of this chapter. Our discussion only "scratches the surface" of how spectral patterns may be classified into categories.

Our presentation of the various classification approaches is illustrated with a two-channel (bands 3 and 4) subset of our hypothetical five-channel multispectral scanner data set. Rarely are just two channels employed in an analysis, yet this limitation simplifies the graphic portrayal of the various techniques. When implemented numerically, these procedures may be applied to any number of channels of data.

Let us assume that we take a sample of pixel observations from our two-channel digital image data set. The two-dimensional digital values, or *measurement vectors*, attributed to each pixel may be expressed graphically by plotting them on a *scatter diagram* (or *scatter plot*), as shown in Figure 7.39. In this diagram, the band 3 DNs have been plotted on the *y* axis and the band 4 DNs on the *x* axis. These two DNs locate each pixel value in the two-dimensional "measurement space" of the graph. Thus, if the band 4 DN for a pixel is 10 and the band 3 DN for the same pixel is 68, the measurement vector for this pixel is represented by a point plotted at coordinate (10, 68) in the measurement space.

Let us also assume that the pixel observations shown in Figure 7.39 are from areas of known cover type (that is, from selected training sites). Each pixel

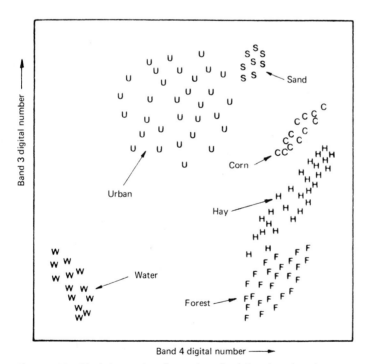

Figure 7.39 Pixel observations from selected training sites plotted on scatter diagram.

value has been plotted on the scatter diagram with a letter indicating the category to which it is known to belong. Note that the pixels within each class do not have a single, repeated spectral value. Rather, they illustrate the natural centralizing tendency—yet variability—of the spectral properties found within each cover class. These "clouds of points" represent multidimensional descriptions of the spectral response patterns of each category of cover type to be interpreted. The following classification strategies use these "training set" descriptions of the category spectral response patterns as interpretation keys by which pixels of unidentified cover type are categorized into their appropriate classes.

Minimum-Distance-to-Means Classifier

Figure 7.40 illustrates one of the simpler classification strategies that may be used. First, the mean, or average, spectral value in each band for each category is determined. These values comprise the *mean vector* for each category. The category means are indicated by +'s in Figure 7.40. By considering the two-channel pixel values as positional coordinates (as they are portrayed in the scatter diagram), a pixel of unknown identity may be classified by computing

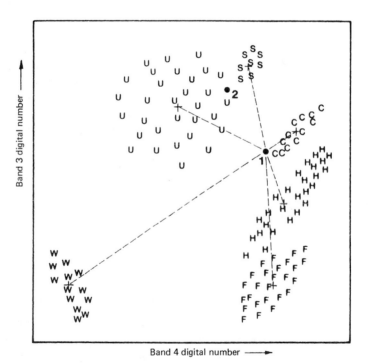

Figure 7.40   Minimum distance to means classification strategy.

the *distance* between the value of the unknown pixel and each of the category means. In Figure 7.40, an unknown pixel value has been plotted at point 1. The distance between this pixel value and each category mean value is illustrated by the dashed lines. After computing the distances, the unknown pixel is assigned to the "closest" class, in this case "corn." If the pixel is farther than an analyst-defined distance from any category mean, it would be classified as "unknown."

The minimum-distance-to-means strategy is mathematically simple and computationally efficient, but it has certain limitations. Most importantly, *it is insensitive to different degrees of variance in the spectral response data.* In Figure 7.40, the pixel value plotted at point 2 would be assigned by the distance-to-means classifier to the "sand" category, in spite of the fact that the greater variability in the "urban" category suggests that "urban" would be a more appropriate class assignment. Because of such problems, this classifier is not widely used in applications where spectral classes are close to one another in the measurement space and have high variance.

Parallelepiped Classifier

We can introduce sensitivity to category variance by considering the *range* of values in each category training set. This range may be defined by the highest and lowest digital number values in each band and appears as a rectangular area in our two-channel scatter diagram, as shown in Figure 7.41. An unknown pixel is classified according to the category range, or *decision region*, in which it lies or as "unknown" if it lies outside all regions. The multidimensional analogs of these rectangular areas are called *parallelepipeds*, and this classification strategy is referred to by that tongue-twisting name. The parallelepiped classifier is also very fast and efficient computationally.

The sensitivity of the parallelepiped classifier to category variance is exemplified by the smaller decision region defined for the highly repeatable "sand" category than for the more variable "urban" class. Because of this, pixel 2 would be appropriately classified as "urban." However, difficulties are encountered when category ranges overlap. Unknown pixel observations that occur in the overlap areas will be classified as "not sure" or be arbitrarily placed in one (or both) of the two overlapping classes. Overlap is caused largely because category distributions exhibiting *correlation* or high *covariance* are poorly described by the rectangular decision regions. Covariance is the tendency of spectral values to vary similarly in two bands, resulting in elongated, slanted clouds of observations on the scatter diagram. In our example, the "corn" and "hay" categories have positive covariance (they slant upward to the right), meaning that high values in band 3 are generally associated with high values in band 4, and low values in band 3 are associated with low values in band 4. The water category in our example exhibits *negative covariance* (its distribution slants down to the right), meaning that high values in band 3 are associated

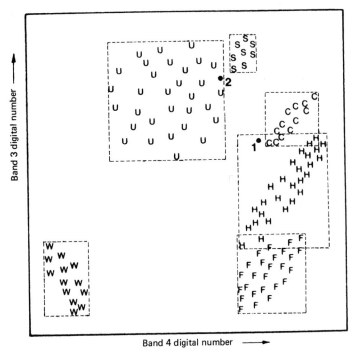

Figure 7.41 Parallelepiped classification strategy.

with low values in band 4. The "urban" class shows a lack of covariance, resulting in a nearly circular distribution on the scatter diagram.

In the presence of covariance, the rectangular decision regions fit the category training data very poorly, resulting in confusion for a parallelepiped classifier. For example, the insensitivity to covariance would cause pixel 1 to be classified as "hay" instead of "corn."

Unfortunately, spectral response patterns are frequently highly correlated, and high covariance is often the rule rather than the exception. The resulting problems can be somewhat alleviated within the parallelepiped classifier by modifying the single rectangles for the various decision regions into a series of rectangles with stepped borders. These borders then describe the boundaries of the elongated distributions more specifically. This approach is illustrated in Figure 7.42.

Gaussian Maximum Likelihood Classifier

The maximum likelihood classifier quantitatively evaluates both the variance and covariance of the category spectral response patterns when classifying an

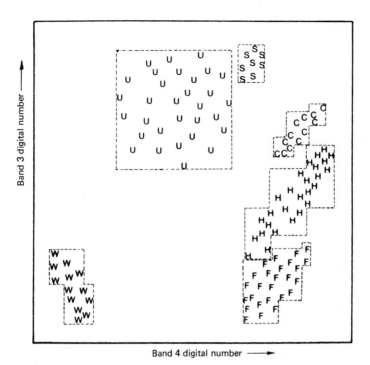

Figure 7.42   Parallelepiped classification strategy employing stepped decision region boundaries.

unknown pixel. To do this, an assumption is made that the distribution of the cloud of points forming the category training data is Gaussian (normally distributed). This *assumption of normality* is generally reasonable for common spectral response distributions. Under this assumption, the distribution of a category response pattern can be completely described by the *mean vector* and the *covariance matrix*. Given these parameters, we may compute the statistical probability of a given pixel value being a member of a particular land cover class. Figure 7.43 shows the probability values plotted in a three-dimensional graph. The vertical axis is associated with the probability of a pixel value being a member of one of the classes. The resulting bell-shaped surfaces are called *probability density functions*, and there is one such function for each spectral category.

The probability density functions are used to classify an unidentified pixel by computing the probability of the pixel value belonging to each category. That is, the computer would calculate the probability of the pixel value occurring in the distribution of class "corn," then the likelihood of its occurring in class "sand," and so on. After evaluating the probability in each category, the pixel would be assigned to the most likely class (highest probability value) or be labeled "unknown" if the probability values are all below a threshold set by the analyst.

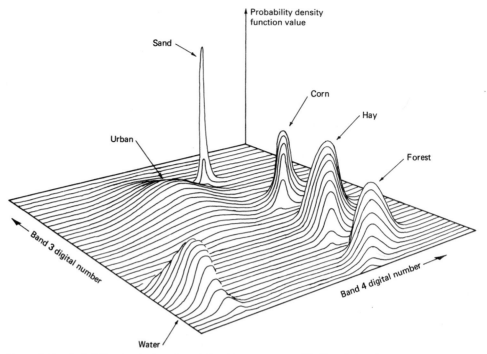

Figure 7.43 Probability density functions defined by a maximum likelihood classifier.

In essence, the maximum likelihood classifier delineates ellipsoidal "equiprobability contours" in the scatter diagram. These decision regions are shown in Figure 7.44. The shape of the equiprobability contours expresses the sensitivity of the likelihood classifier to covariance. For example, because of this sensitivity, it can be seen that pixel 1 would be appropriately assigned to the "corn" category.

An extension of the maximum likelihood approach is the *Bayesian classifier*. This technique applies two weighting factors to the probability estimate. First, the analyst determines the "a priori probability," or the anticipated likelihood of occurrence for each class in the given scene. For example, when classifying a pixel, the probability of the rarely occurring "sand" category might be weighted lightly, and the more likely "urban" class weighted heavily. Second, a weight associated with the "cost" of misclassification is applied to each class. Together, these factors act to minimize the "cost" of misclassifications, resulting in a theoretically optimum classification. In practice, most maximum likelihood classification is performed assuming equal probability of occurrence and cost of misclassification for all classes. If suitable data exist for these factors, the Bayesian implementation of the classifier is preferable.

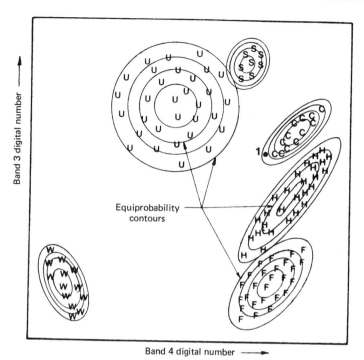

Figure 7.44   Equiprobability contours defined by a maximum likelihood classifier.

The principal drawback of maximum likelihood classification is the large number of computations required to classify each pixel. This is particularly true when either a large number of spectral channels are involved or a large number of spectral classes must be differentiated. In such cases, the maximum likelihood classifier is much slower computationally than the previous techniques.

Several approaches may be taken to increase the efficiency of maximum likelihood classifiers. In the lookup table implementation of such algorithms, the category identity for all possible combinations of digital numbers occurring in an image is determined in advance of actually classifying the image. Hence, the complex statistical computation for each combination is only made once. The categorization of each pixel in the image is then simply a matter of indexing the location of its multichannel gray level in the lookup table.

Another means of optimizing the implementation of the maximum likelihood classifier is to use some method to reduce the dimensionality of the data set used to perform the classification (thereby reducing the complexity of the required computations). As stated earlier, principal or canonical component transformations of the original data may be used for this purpose.

Decision tree, stratified, or *layered* classifiers have also been utilized to simplify classification computations and maintain classification accuracy. These classifiers are applied in a series of steps, with certain classes being separated during each step in the simplest manner possible. For example, water might first be separated from all other classes based on a simple threshold set in a near-infrared band. Certain other classes may require only two or three bands for categorization and a parallelepiped classifier may be adequate. The use of more bands or the maximum likelihood classifier would then only be required for those land cover categories where residual ambiguity exists between overlapping classes in the measurement space.

7.10 THE TRAINING STAGE

Whereas the actual classification of multispectral image data is a highly automated process, assembling the training data needed for classification is anything but automatic. In many ways, the training effort required in supervised classification is both an art and a science. It requires close interaction between the image analyst and the image data. It also requires substantial reference data and a thorough knowledge of the geographic area to which the data apply. Most importantly, the quality of the training process determines the success of the classification stage and, therefore, the value of the information generated from the entire classification effort.

The overall objective of the training process is to assemble a set of statistics that describe the spectral response pattern for each land cover type to be classified in an image. Relative to our earlier graphical example, it is during the training stage that the location, size, shape, and orientation of the "clouds of points" for each land cover class are determined.

To yield acceptable classification results, training data must be both representative and complete. This means that the image analyst must develop training statistics for all *spectral* classes constituting each *information* class to be discriminated by the classifier. For example, in a final classification output, one might wish to delineate an information class called "water." If the image under analysis contains only one water body and if it has uniform spectral response characteristics over its entire area, then only one training area would be needed to represent the water class. If, however, the same water body contained distinct areas of very clear water and very turbid water, a minimum of two spectral classes would be required to adequately train on this feature. If multiple water bodies occurred in the image, training statistics would be required for each of the other spectral classes that might be present in the water-covered areas. Accordingly, the single information class "water" might be represented by four or five spectral classes. In turn, the four or five spectral classes would eventually be used to classify all the water bodies occurring in the image.

By now it should be clear that the training process can become quite involved. For example, an information class such as "agriculture" might contain several crop types and each crop type might be represented by several spectral classes. These spectral classes could stem from different planting dates, soil moisture conditions, crop management practices, seed varieties, topographic settings, atmospheric conditions, or combinations of these factors. *The point that must be emphasized is that all spectral classes constituting each information class must be adequately represented in the training set statistics used to classify an image.* Depending upon the nature of the information classes sought, and the complexity of the geographic area under analysis, it is not uncommon to acquire data from 100 or more training areas to adequately represent the spectral variability in an image.

The location of training areas in an image is normally established by viewing *windows*, or portions of the full scene, in an enlarged format on an interactive color display device. The image analyst normally obtains training sample data by outlining training areas using a reference *cursor*. The cursor may be controlled by any of several means (e.g., a mouse, track ball, or joystick). Figure 7.45 shows the boundaries of several training site polygons that have been delineated in this manner. Note that these polygons have been carefully located to avoid pixels located along the edges between land cover types. The row and column coordinates of the vertices for these polygons are used as the basis for extracting (from the image file) the digital numbers for the pixels located within each training area boundary. These pixel values then form the sample used to develop the statistical description of each training area (mean vector and covariance matrix in the case of the maximum likelihood classifier).

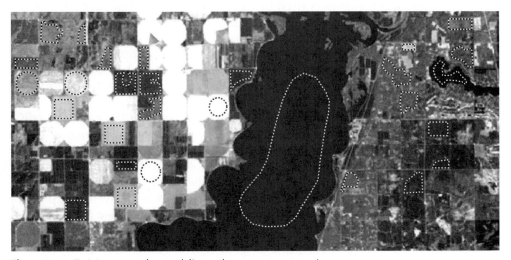

Figure 7.45 Training area polygons delineated on a computer monitor.

An alternative to manually delineating training area polygons is the use of a *seed pixel* approach to training. In this case, the display cursor is placed within a prospective training area and a single "seed" pixel is chosen that is thought to be representative of the surrounding area. Then, according to various statistically based criteria, pixels with similar spectral characteristics that are contiguous to the seed pixel are highlighted on the display and become the training samples for that training area.

Irrespective of how training areas are delineated, when using any statistically based classifier (such as the maximum likelihood method), the theoretical lower limit of the number of pixels that must be contained in a training set is $n + 1$, where n is the number of spectral bands. In our two-band example, *theoretically* only three observations would be required. Obviously, the use of fewer than three observations would make it impossible to appropriately evaluate the variance and covariance of the spectral response values. In practice, a minimum of from $10n$ to $100n$ pixels is used since the estimates of the mean vectors and covariance matrices improve as the number of pixels in the training sets increases. Within reason, the more pixels that can be used in training, the better the statistical representation of each spectral class.

When delineating training set pixels, it is important to analyze several training sites throughout the scene. For example, it would be better to define the training pattern for a given class by analyzing 20 locations containing 40 pixels of a given type than one location containing 800 pixels. Dispersion of the sites throughout the scene increases the chance that the training data will be representative of all the variations in the cover types present in the scene.

The trade-off usually faced in the development of training data sets is that of having sufficient sample size to ensure the accurate determination of the statistical parameters used by the classifier and to represent the total spectral variability in a scene, without going past a point of diminishing returns. In short, one does not want to omit any important spectral classes occurring in a scene, but one also does not want to include redundant spectral classes in the classification process from a computational standpoint. During the process of *training set refinement* the analyst attempts to identify such gaps and redundancies.

As part of the training set refinement process, the overall quality of the data contained in each of the original candidate training areas is assessed and the spectral separability between the data sets is studied. The analyst carefully checks to see if all data sets are essentially normally distributed and spectrally pure. Training areas that inadvertently include more than one spectral class are identified and recompiled. Likewise, extraneous pixels may be deleted from some of the data sets. These might be edge pixels along agricultural field boundaries or within-field pixels containing bare soil rather than the crop trained upon. Training sets that might be merged (or deleted) are identified, and the need to obtain additional training sets for poorly represented spectral classes is addressed.

One or more of the following types of analyses are typically involved in the training set refinement process:

1. **Graphical representation of the spectral response patterns.** The distributions of training area response patterns can be graphically displayed in many formats. Figure 7.46 shows a hypothetical histogram for one of the "hay" category training sites in our five-channel multispectral scanner data set. (A similar display would be available for all training areas.) Histogram output is particularly important when a maximum likelihood classifier is used, since it provides a visual check on the normality of the spectral response distributions. Note in the case of the hay category that the data appear to be normally distributed in all bands except band 2, where the distribution is shown to be bimodal. This indicates that the training site data set chosen by the analyst to represent "hay" is in fact composed of two subclasses with slightly different spectral characteristics. These subclasses may represent two different varieties of hay or different illumination conditions, and so on. In any case, the classification accuracy will generally be improved if each of the subclasses is treated as a separate category.

    Histograms illustrate the distribution of individual categories very well; yet they do not facilitate comparisons between different category types. To evaluate the spectral separation between categories, it is convenient to use some form of *coincident spectral plot*, as shown in Figure 7.47. This plot illustrates, in each spectral band, the mean spectral response of each category (with a letter) and the variance of the distribution (±2 standard deviations shown by black bars). Such plots indicate the overlap between category response patterns. For example, Figure 7.47 indicates that the hay and corn response patterns overlap in all spectral bands. The plot also shows which combination of bands might be best for discrimination because of relative reversals of spectral response (such as bands 3 and 5 for hay/corn separation).

    The fact that the spectral plots for hay and corn overlap in all spectral bands indicates that the categories could not be accurately classified on any *single* multispectral scanner band. However, this does not preclude successful classification when two or more bands are analyzed (such as bands 3 and 4 illustrated in the last section). Because of this, two-dimensional scatter diagrams (as shown in Figures 7.39 to 7.42) provide better representations of the spectral response pattern distributions.

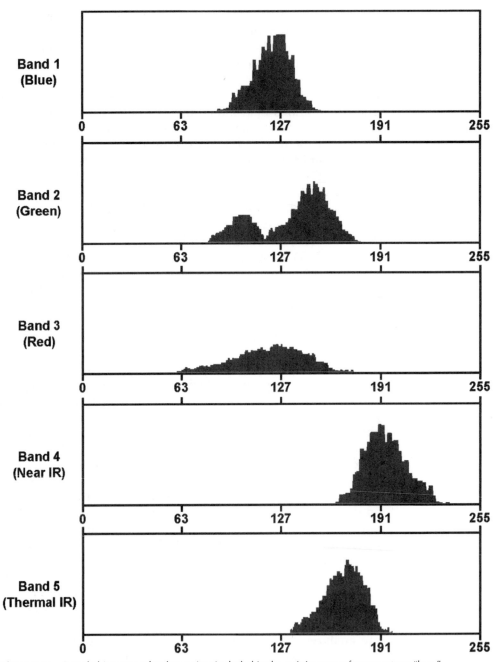

Figure 7.46 Sample histograms for data points included in the training areas for cover type "hay."

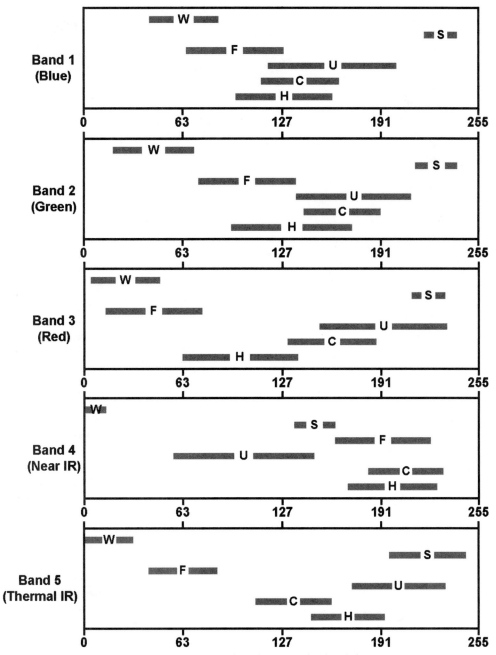

Figure 7.47 Coincident spectral plots for training data obtained in five bands for six cover types.

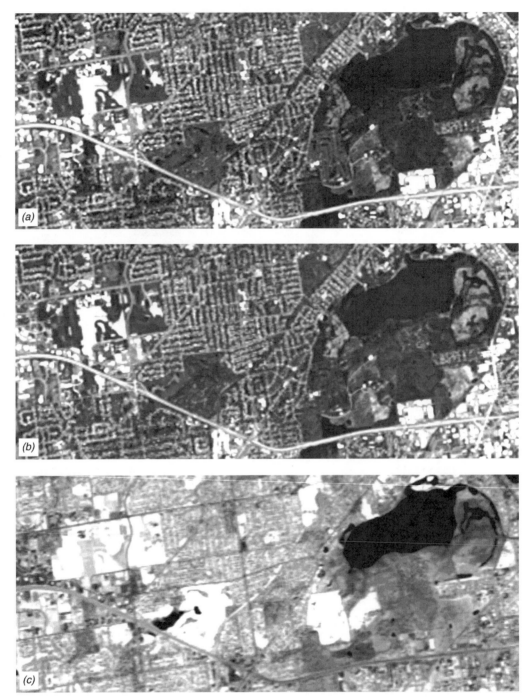

Figure 7.48 SPOT HRV multispectral images of Madison, WI: (*a*) band 1 (green); (*b*) band 2 (red); (*c*) band 3 (near IR).

The utility of scatter diagrams (or scatter plots) is further illustrated in Figures 7.48 to 7.50. Shown in Figure 7.48 are SPOT multispectral HRV images depicting a portion of Madison, Wisconsin. The band 1 (green), band 2 (red), and band 3 (near-IR) images are shown in (a), (b), and (c), respectively. Figure 7.49 shows the histograms for bands 1 and 2 as well as the associated scatter diagram for these two bands. Note that the data in these two bands are highly correlated and a very compact and near-linear "cloud of points" is shown in the scatter diagram.

Figure 7.50 shows the histograms and the scatter diagram for bands 2 and 3. In contrast to Figure 7.49, the scatter diagram in Figure 7.50

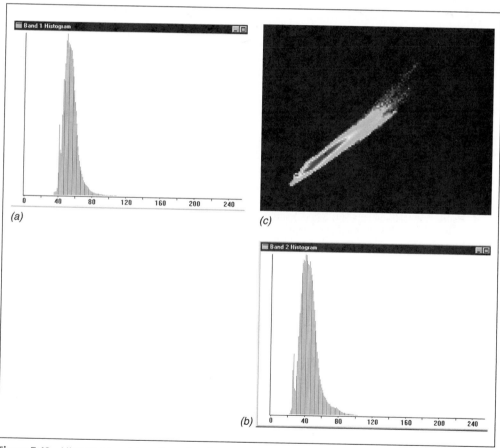

Figure 7.49   Histograms and two-dimensional scatter diagram for the images shown in Figures 7.48a and b: (a) band 1 (green) histogram; (b) band 2 (red) histogram; (c) scatter diagram plotting band 1 (vertical axis) versus band 2 (horizontal axis). Note the high correlation between these two visible bands.

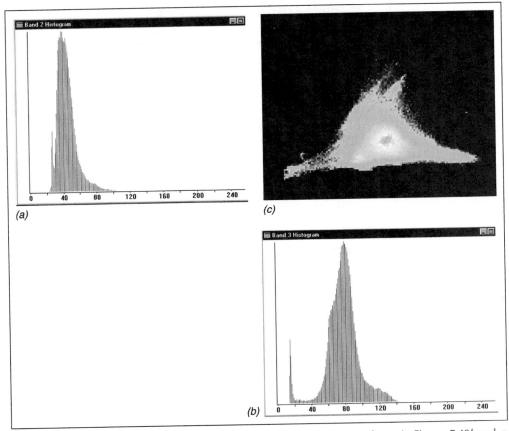

Figure 7.50 Histograms and two-dimensional scatter diagram for the images shown in Figures 7.48*b* and *c*: (*a*) band 2 (red) histogram; (*b*) band 3 (near-IR) histogram; (*c*) scatter diagram plotting band 2 (vertical axis) versus band 3 (horizontal axis). Note the relative lack of correlation between these visible and near-IR bands.

shows that bands 2 and 3 are much less correlated than bands 1 and 2. Whereas various land cover types might overlap one another in bands 1 and 2, they would be much more separable in bands 2 and 3. In fact, these two bands alone may be adequate to perform a generalized land cover classification of this scene.

2. **Quantitative expressions of category separation.** A measure of the statistical separation between category response patterns can be computed for all pairs of classes and can be presented in the form of a matrix. One statistical parameter commonly used for this purpose is *transformed divergence*, a covariance-weighted distance between category means. In general, the larger the transformed divergence, the greater the "statistical distance" between training patterns and the

TABLE 7.1   Portion of a Divergence Matrix Used to Evaluate Pairwise Training Class Spectral Separability

Spectral Class[a]	W1	W2	W3	C1	C2	C3	C4	H1	H2 ...
W1	0								
W2	1185	0							
W3	1410	680	0						
C1	1997	2000	1910	0					
C2	1953	1890	1874	860	0				
C3	1980	1953	1930	1340	1353	0			
C4	1992	1997	2000	1700	1810	1749	0		
H1	2000	1839	1911	1410	1123	860	1712	0	
H2	1995	1967	1935	1563	1602	1197	1621	721	0
⋮	⋮								

[a]W, water; C, corn; H, hay.

higher the probability of correct classification of classes. A portion of a sample matrix of divergence values is shown in Table 7.1. In this example, the maximum possible divergence value is 2000, and values less than 1500 indicate spectrally similar classes. Accordingly, the data in Table 7.1 suggest spectral overlap between several pairs of spectral classes. Note that W1, W2, and W3 are all relatively spectrally similar. However, note that this similarity is all among spectral classes from the same information class ("water"). Furthermore, all the "water" classes appear to be spectrally distinct from the spectral classes of the other information classes. More problematic is a situation typified by the divergence between the H1 and C3 spectral classes (860). Here, a "hay" spectral class severely overlaps a "corn" class.

[Another statistical distance measure of the separability of two spectral classes is the *Jeffries–Matusita (JM) distance*. It is similar to transformed divergence in its interpretation but has a maximum value of 1414.]

3. **Self-classification of training set data.**   Another evaluation of spectral separability is provided by classifying the training set pixels. In such an effort, a preliminary classification of only the training set pixels (rather than the full scene) is made, to determine what percentage of the training pixels are actually classified as expected. These percentages are normally presented in the form of an *error matrix* (to be described in Section 7.16).

It is important to avoid considering an error matrix based on training set values as a measure of *overall* classification accuracy throughout an image. For one reason, certain land cover classes might be inadvertently missed in the training process. Also, the error matrix simply tells us how well the classifier can classify the *training areas* and nothing more. Because the training areas are usually good, homogeneous examples of each cover type, they can be expected to be classified more accurately than less pure examples that may be found elsewhere in the scene. Overall accuracy can be evaluated only by considering *test areas* that are different from and considerably more extensive than the training areas. This evaluation is generally performed after the classification and output stages (as discussed in Section 7.16).

4. **Interactive preliminary classification.** Most modern image processing systems incorporate some provision for interactively displaying how applicable training data are to the full scene to be classified. Often, this involves performing a preliminary classification with a computationally efficient algorithm (e.g., parallelepiped) to provide a visual approximation of the areas that would be classified with the statistics from a given training area. Such areas are typically highlighted in color on the display of the original raw image.

   This is illustrated in Plate 29, which shows a partially completed classification of a subset of the data included in Figure 7.50 (bands 2 and 3). Shown in (*a*) are selected training areas delineated on a color infrared composite of bands 1, 2, and 3 depicted as blue, green, and red, respectively. Part (*b*) shows the histograms and scatter plot for bands 2 and 3. Shown in (*c*) are the parallelepipeds associated with the initial training areas an image analyst has chosen to represent four information classes: water, trees, grass, and impervious surfaces. Part (*d*) shows how the statistics from these initial training areas would classify various portions of the original scene.

5. **Representative subscene classification.** Often, an image analyst will perform a classification of a representative subset of the full scene to eventually be classified. The results of this preliminary classification can then be used interactively on an overlay to the original raw image. Selected classes are then viewed individually or in logical groups to determine how they relate to the original image.

In general, the training set refinement process cannot be rushed with the "maximum efficiency" attitude appropriate in the classification stage. It is normally an iterative procedure in which the analyst revises the statistical descriptions of the category types until they are sufficiently spectrally separable. That is, the original set of "candidate" training area statistics is revised through merger, deletion, and addition to form the "final" set of statistics used in classification.

Training set refinement for the inexperienced data analyst is often a difficult task. Typically, an analyst has little difficulty in developing the statistics for the distinct "nonoverlapping" spectral classes present in a scene. If there are problems, they typically stem from spectral classes on the borders between information classes—"transition" or "overlapping" classes. In such cases, the impact of alternative deletion and pooling of training classes can be tested by trial and error. In this process the sample size, spectral variances, normality, and identity of the training sets should be rechecked. Problem classes that occur only rarely in the image may be eliminated from the training data so that they are not confused with classes that occur extensively. That is, the analyst may accept misclassification of a class that occurs rarely in the scene in order to preserve the classification accuracy of a spectrally similar class that appears over extensive areas. Furthermore, a classification might initially be developed assuming a particular set of detailed information classes will be maintained. After studying the actual classification results, the image analyst might be faced with aggregating certain of the detailed classes into more general ones (for example, "birch" and "aspen" may have to be merged into a "deciduous" class or "corn" and "hay" into "agriculture").

One final note to be made here is that training set refinement is usually the key to improving the accuracy of a classification. However, if certain cover types occurring in an image have inherently similar spectral response patterns, no amount of retraining and refinement will make them spectrally separable! Alternative methods, such as using data resident in a GIS, performing a visual interpretation, or making a field check, must be used to discriminate these cover types. Multitemporal or spatial pattern recognition procedures may also be applicable in such cases. Increasingly, land cover classification involves some merger of remotely sensed data with ancillary information resident in a GIS.

7.11 UNSUPERVISED CLASSIFICATION

As previously discussed, unsupervised classifiers do *not* utilize training data as the basis for classification. Rather, this family of classifiers involves algorithms that examine the unknown pixels in an image and aggregate them into a number of classes based on the natural groupings or clusters present in the image values. The basic premise is that values within a given cover type should be close together in the measurement space, whereas data in different classes should be comparatively well separated.

The classes that result from unsupervised classification are *spectral classes*. Because they are based solely on the natural groupings in the image values, the identity of the spectral classes will not be initially known. The analyst must compare the classified data with some form of reference data (such as larger scale imagery or maps) to determine the identity and informational value of the spectral classes. Thus, in the *supervised* approach we define useful information categories and then examine their spectral separability; in the *unsupervised* approach we determine spectrally separable classes and then define their informational utility.

We illustrate the unsupervised approach by again considering a two-channel data set. Natural spectral groupings in the data can be visually identified by plotting a scatter diagram. For example, in Figure 7.51 we have plotted pixel values acquired over a forested area. Three groupings are apparent in the scatter diagram. After comparing the classified image data with ground reference data, we might find that one cluster corresponds to deciduous trees, one to conifers, and one to stressed trees of both types (indicated by D, C, and S in Figure 7.51). In a supervised approach, we may not have considered training for the "stressed" class. This highlights one of the primary advantages of unsupervised classification: The *classifier* identifies the distinct spectral classes present in the image data. Many of these classes might not be initially apparent to the analyst applying a supervised classifier. Likewise, the spectral classes in a scene may be so numerous that it would be difficult to train on all of them. In the unsupervised approach they are found automatically.

There are numerous *clustering* algorithms that can be used to determine the natural spectral groupings present in a data set. One common form of clustering, called the "K-means" approach, accepts from the analyst the number of clusters to be located in the data. The algorithm then arbitrarily "seeds," or locates, that number of cluster centers in the multidimensional measure-

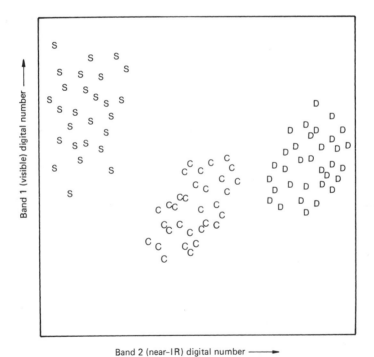

Figure 7.51 Spectral classes in two-channel image data.

ment space. Each pixel in the image is then assigned to the cluster whose arbitrary mean vector is closest. After all pixels have been classified in this manner, revised mean vectors for each of the clusters are computed. The revised means are then used as the basis to reclassify the image data. The procedure continues until there is no significant change in the location of class mean vectors between successive iterations of the algorithm. Once this point is reached, the analyst determines the land cover identity of each spectral class.

A widely used variant on the K-means method for unsupervised clustering is an algorithm called *Iterative Self-Organizing Data Analysis*, or *ISODATA* (Tou and Gonzalez, 1974). This algorithm permits the number of clusters to change from one iteration to the next, by merging, splitting, and deleting clusters. The general process follows that described above for K-means. However, in each iteration, following the assignment of pixels to the clusters, the statistics describing each cluster are evaluated. If the distance between the mean points of two clusters is less than some predefined minimum distance, the two clusters are merged together. On the other hand, if a single cluster has a standard deviation (in any one dimension) that is greater than a predefined maximum value, the cluster is split in two. Clusters with fewer than the specified minimum number of pixels are deleted. Finally, as with K-means, all pixels are then reclassified into the revised set of clusters, and the process repeats, until either there is no significant change in the cluster statistics or some maximum number of iterations is reached.

Another common approach to unsupervised classification is the use of algorithms that incorporate a sensitivity to image "texture" or "roughness" as a basis for establishing cluster centers. Texture is typically defined by the multidimensional variance observed in a moving window passed through the image (e.g., a 3 × 3 window). The analyst sets a variance threshold below which a window is considered "smooth" (homogeneous) and above which it is considered "rough" (heterogeneous). The mean of the first smooth window encountered in the image becomes the first cluster center. The mean of the second smooth window encountered becomes the second cluster center, and so forth. As soon as an analyst-specified maximum number of cluster centers is reached (e.g., 50), the classifier considers the distances between all previously defined cluster centers in the measurement space and merges the two closest clusters, combining their statistics. The classifier continues through the image combining the closest two clusters encountered until the entire image is analyzed. The resulting cluster centers are then analyzed to determine their separability on the basis of an analyst-specified statistical distance. Those clusters separated by less than this distance are combined and their statistics are merged. The final clusters resulting from the analysis are used to classify the image (e.g., with a minimum distance or maximum likelihood classifier).

Data from supervised training areas are sometimes used to augment the results of the above clustering procedure when certain land cover classes are poorly represented in the purely unsupervised analysis. (We discuss other such hybrid approaches in Section 7.12.) Roads and other linear features, for example, may not be represented in the original clustering statistics if these features do not happen to meet the smoothness criteria within the moving window. Likewise, in some unsupervised classifiers the order in which different feature types are encountered can result in poor representation of some classes. For example, the analyst-specified maximum number of classes may be reached in an image long before the moving window passes throughout the scene.

Before ending our discussion of unsupervised classification, we reiterate that the result of such efforts is simply the identification of spectrally distinct classes in image data. The analyst must still use reference data to associate the spectral classes with the cover types of interest. This process, like the training set refinement step in supervised classification, can be quite involved.

Table 7.2 illustrates several possible outcomes of associating spectral classes with information classes for data from a scene covering a forested area. The ideal result would be outcome 1, in which each spectral class is found to be associated uniquely with a feature type of interest to the analyst.

TABLE 7.2 Spectral Classes Resulting from Clustering a Forested Scene

Spectral Class	Identity of Spectral Class	Corresponding Desired Information Category
Possible Outcome 1		
1	Water ⟶	Water
2	Coniferous trees ⟶	Coniferous trees
3	Deciduous trees ⟶	Deciduous trees
4	Brushland ⟶	Brushland
Possible Outcome 2		
1	Turbid water ⟶	Water
2	Clear water ⟶	
3	Sunlit conifers ⟶	Coniferous trees
4	Shaded hillside conifers ⟶	
5	Upland deciduous ⟶	Deciduous trees
6	Lowland deciduous ⟶	
7	Brushland ⟶	Brushland
Possible Outcome 3		
1	Turbid water ⟶	Water
2	Clear water ⟶	
3	Coniferous trees ⟶	Coniferous trees
4	Mixed coniferous/deciduous ⟨	
5	Deciduous trees ⟶	Deciduous trees
6	Deciduous/brushland ⟶	Brushland

This outcome will occur only when the features in the scene have highly distinctive spectral characteristics.

A more likely result is presented in outcome 2. Here, several spectral classes are attributable to each information category desired by the analyst. These "subclasses" may be of little informational utility (sunlit versus shaded conifers) or they may provide useful distinctions (turbid versus clear water and upland versus lowland deciduous). In either case, the spectral classes may be aggregated after classification into the smaller set of categories desired by the analyst.

Outcome 3 represents a more troublesome result in which the analyst finds that several spectral classes relate to more than one information category. For example, spectral class 4 was found to correspond to coniferous trees in some locations and deciduous trees in others. Likewise, class 6 included both deciduous trees and brushland vegetation. This means that these information categories are spectrally similar and cannot be differentiated in the given data set.

As with supervised classification, access to efficient hardware and software is an important factor in determining the *ease* with which an unsupervised classification can be performed. The *quality* of the classification still depends upon the analyst's understanding of the concepts behind the classifiers available and knowledge about the land cover types under analysis.

7.12 HYBRID CLASSIFICATION

Various forms of hybrid classification have been developed to either streamline or improve the accuracy of purely supervised or unsupervised procedures. For example, *unsupervised training areas* might be delineated in an image in order to aid the analyst in identifying the numerous spectral classes that need to be defined in order to adequately represent the land cover information classes to be differentiated in a supervised classification. Unsupervised training areas are image subareas chosen intentionally to be quite different from supervised training areas.

Whereas supervised training areas are located in regions of homogeneous cover type, the unsupervised training areas are chosen to contain numerous cover types at various locations throughout the scene. This ensures that all spectral classes in the scene are represented somewhere in the various subareas. These areas are then clustered independently and the spectral classes from the various areas are analyzed to determine their identity. They are subjected to a pooled statistical analysis to determine their spectral separability and normality. As appropriate, similar clusters representing similar land cover types are combined. Training statistics are developed for the combined classes and used to classify the entire scene (e.g., by a minimum distance or maximum likelihood algorithm).

Hybrid classifiers are particularly valuable in analyses where there is complex variability in the spectral response patterns for individual cover types present. These conditions are quite common in such applications as vegetation mapping. Under these conditions, spectral variability within cover types normally comes about both from variation within cover types per se (species) and from different site conditions (e.g., soils, slope, aspect, crown closure). *Guided clustering* is a hybrid approach that has been shown to be quite effective in such circumstances.

In guided clustering, the analyst delineates numerous "supervised-like" training sets for each cover type to be classified in a scene. Unlike the training sets used in traditional supervised methods, these areas need not be perfectly homogeneous. The data from all the training sites for a given information class are then used in an unsupervised clustering routine to generate several (as many as 20 or more) spectral signatures. These signatures are examined by the analyst; some may be discarded or merged and the remainder are considered to represent spectral subclasses of the desired information class. Signatures are also compared among the different information classes. Once a sufficient number of such spectral subclasses have been acquired for all information classes, a maximum likelihood classification is performed with the full set of refined spectral subclasses. The spectral subclasses are then aggregated back into the original information classes.

Guided clustering may be summarized in the following steps:

1. Delineate training areas for information class X.

2. Cluster all class X training area pixels at one time into spectral subclasses $X_1, \ldots, X_n$ using an automated clustering algorithm.

3. Examine class X signatures and merge or delete signatures as appropriate. A progression of clustering scenarios (e.g., from 3 to 20 cluster classes) should be investigated, with the final number of clusters and merger and deletion decisions based on such factors as (1) display of a given class on the raw image, (2) multidimensional histogram analysis for each cluster, and (3) multivariate distance measures (e.g., transformed divergence or JM distance).

4. Repeat steps 1 to 3 for all additional information classes.

5. Examine all class signatures and merge or delete signatures as appropriate.

6. Perform maximum likelihood classification on the entire image with the full set of spectral subclasses.

7. Aggregate spectral subclasses back to the original information classes.

Bauer et al. (1994) demonstrated the utility of guided clustering in the classification of forest types in northern Minnesota. Based on this success,

Lillesand et al. (1998) made the technique a central part of the Upper Midwest Gap Analysis Program image processing protocol development for classifying land cover throughout the states of Michigan, Minnesota, and Wisconsin. Given the extent and diversity of cover types in this area, the method was found to not only increase classification accuracy relative to either a conventional supervised or unsupervised approach but also increase the efficiency of the entire classification process. Among the advantages of this approach is its ability to help the analyst identify the various spectral subclasses representing an information class "automatically" through clustering. At the same time, the process of labeling the spectral clusters is straightforward because these are developed for one information class at a time. Also, spurious clusters due to such factors as including multiple-cover-type conditions in a single training area can be readily identified (e.g., openings containing understory vegetation in an otherwise closed forest canopy, bare soil in a portion of a crop-covered agricultural field). The method also helps identify situations where mixed pixels might inadvertently be included near the edges of training areas.

7.13 CLASSIFICATION OF MIXED PIXELS

As we have previously discussed (Sections 1.9 and 5.2), mixed pixels result when a sensor's IFOV includes more than one land cover type or feature on the ground. The extent to which mixed pixels are contained in an image is both a function of the spatial resolution of the remote sensing system used to acquire an image and the spatial scale of the surface features in question. For example, if a sensor with a narrow field of view is positioned within a few meters of a healthy soybean crop canopy, the sensor's field of view may be entirely covered by soybean leaves. A lower resolution sensor operating at higher altitude might focus on the same field yet have its field of view occupied by a mixture of soybean leaves, bare soil, and grass. These mixed pixels present a difficult problem for image classification, since their spectral characteristics are not representative of any single land cover type. *Spectral mixture analysis* and *fuzzy classification* are two procedures designed to deal with the classification of mixed pixels. They represent means by which "subpixel classification" is accomplished.

Spectral Mixture Analysis

Spectral mixture analysis involves a range of techniques wherein mixed spectral signatures are compared to a set of "pure" reference spectra (measured in the laboratory, in the field, or from the image itself). The basic assumption is that the spectral variation in an image is caused by mixtures of a limited number of surface materials. The result is an estimate of the approximate proportions of the ground area of each pixel that are occupied by each of the reference classes.

Spectral mixture analysis differs in several ways from other image processing methods for land cover classification. Conceptually, it is a deterministic method rather than a statistical method, since it is based on a physical model of the mixture of discrete spectral response patterns. It provides useful information at the subpixel level, since multiple land cover types can be detected within a single pixel. Many land cover types tend to occur as heterogeneous mixtures even when viewed at very fine spatial scales; thus, this method provides a more realistic representation of the true nature of the surface than would be provided by the assignment of a single dominant class to every pixel.

Many applications of spectral mixture analysis make use of linear mixture models, in which the observed spectral response from an area on the ground is assumed to be a *linear* mixture of the individual spectral signatures of the various land cover types present within the area. These pure reference spectral signatures are referred to as *endmembers*, because they represent the cases where 100 percent of the sensor's field of view is occupied by a single cover type. In this model, the weight for any given endmember signature is the proportion of the area occupied by the class corresponding to the endmember. The input to a linear mixture model consists of a single observed spectral signature for each pixel in an image. The model's output then consists of "abundance" or "fraction" images for each endmember, showing the fraction of each pixel occupied by each endmember.

Linear mixture analysis involves the simultaneous satisfaction of two basic conditions for each pixel in an image. First, the sum of the fractional proportions of all potential endmembers included in a pixel must equal 1. Expressed mathematically,

$$\sum_{i=1}^{N} F_i = F_1 + F_2 + \cdots + F_N = 1 \tag{7.10}$$

where $F_1, F_2, \ldots, F_N$ represent the fraction of each of N possible endmembers contained in a pixel.

The second condition that must be met is that for a given spectral band λ the observed digital number DN_λ for each pixel represents the sum of the DNs that would be obtained from a pixel that is completely covered by a given endmember weighted by the fraction actually occupied by that endmember plus some unknown error. This can be expressed by

$$DN_\lambda = F_1 DN_{\lambda,1} + F_2 DN_{\lambda,2} + \cdots + F_N DN_{\lambda,N} + E_\lambda \tag{7.11}$$

where DN_λ is the composite digital number actually observed in band λ; $F_1, \ldots, F_N$ equal the fractions of the pixel actually occupied by each of the N endmembers; $DN_{\lambda,1}, \ldots, DN_{\lambda,N}$ equal the digital numbers that would be observed if a pixel were completely covered by the corresponding endmember; and E_λ is the error term.

With multispectral data, there would be one version of Eq. 7.11 for each spectral band. So, for B spectral bands, there would be B equations, plus Eq. 7.10. This means that there are $B + 1$ equations available to solve for the various endmember fractions $(F_1, \ldots, F_N)$. If the number of endmember fractions (unknowns) is equal to the number of spectral bands plus 1, the set of equations can be solved simultaneously to produce an exact solution without any error term. If the number of bands $B + 1$ is greater than the number of endmembers N, the magnitude of the error term along with the fractional cover for each endmember can be estimated (using the principles of least squares regression). On the other hand, if the number of endmember classes present in a scene exceeds $B + 1$, the set of equations will not yield a unique solution.

For example, a spectral mixture analysis of a four-band SPOT HRVIR multispectral image could be used to find estimates of the fractional proportions of five different endmember classes (with no estimate of the amount of error), or of four, three, or two endmember classes (in which case an estimate of the error would also be produced). Without additional information, this image alone could not be used in linear spectral mixture analysis to derive fractional cover estimates for more than five endmember classes.

Figure 7.52 shows an example of the output from a linear spectral mixture analysis project in which Landsat TM imagery was used to determine the fractional cover of trees, shrubs, and herbaceous plants in the Steese National Conservation Area of central Alaska. Figure 7.52a shows a single band (TM band 4, near IR), while 7.52b through 7.52d show the resulting output for each of the endmember classes. Note that these output images are scaled such that higher fractional cover values appear brighter while lower fractional cover values appear darker.

One drawback of linear mixture models is that they do not account for certain factors such as multiple reflections, which can result in complex nonlinearities in the spectral mixing process. That is, the observed signal from a pixel may include a mixture of spectral signatures from various endmembers, but it may also include additional radiance reflected multiple times between scene components such as leaves and the soil surface. In this situation, a more sophisticated *nonlinear* spectral mixture model may be required (Borel and Gerstl, 1994). Artificial neural networks (Section 7.17) may be particularly well suited for this task, because they do not require that the input data have a Gaussian distribution and they do not assume that spectra mix linearly (Moody et al., 1996).

Fuzzy Classification

Fuzzy classification attempts to handle the mixed-pixel problem by employing the fuzzy set concept, in which a given entity (a pixel) may have partial

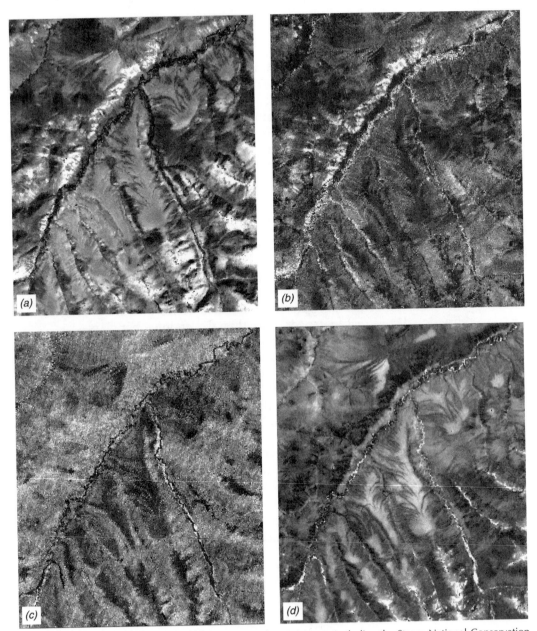

Figure 7.52 Linear spectral mixture analysis of a Landsat TM image including the Steese National Conservation Area of central Alaska: (*a*) band 4 (near IR) of original image; fractional cover images for trees (*b*), shrubs (*c*), and herbaceous plants (*d*). Brighter pixels represent higher fractional cover. (Courtesy Bureau of Land Management–Alaska and Ducks Unlimited, Inc.)

membership in more than one category (Jensen, 1996; Schowengerdt, 1997). One approach to fuzzy classification is *fuzzy clustering*. This procedure is conceptually similar to the K-means unsupervised classification approach described earlier. The difference is that instead of having "hard" boundaries between classes in the spectral measurement space, fuzzy regions are established. So instead of each unknown measurement vector being assigned solely to a single class, irrespective of how close that measurement may be to a partition in the measurement space, *membership grade* values are assigned that describe how close a pixel measurement is to the means of all classes.

Another approach to fuzzy classification is *fuzzy supervised* classification. This approach is similar to application of maximum likelihood classification; the difference being that fuzzy mean vectors and covariance matrices are developed from statistically weighted training data. Instead of delineating training areas that are purely homogeneous, a combination of pure and mixed training sites may be used. Known mixtures of various feature types define the fuzzy training class weights. A classified pixel is then assigned a membership grade with respect to its membership in each information class. For example, a vegetation classification might include a pixel with grades of 0.68 for class "forest," 0.29 for "street," and 0.03 for "grass." (Note that the grades for all potential classes must total 1.)

7.14   THE OUTPUT STAGE

The utility of any image classification is ultimately dependent on the production of output products that effectively convey the interpreted information to its end user. Here the boundaries between remote sensing, computer graphics, digital cartography, and GIS management become blurred. A virtually unlimited selection of output products may be generated. Three general forms that are commonly used include hardcopy graphic products, tables of area statistics, and digital data files.

Graphic Products

Since classified data are in the form of a two-dimensional data array, hardcopy graphic output can be easily computer generated by displaying different colors, tones, or characters for each cell in the array according to its assigned land cover category. A broad range of peripheral equipment can be used for this purpose, including a variety of black and white and color printers, film recorders, and large-format scanners/writers. Printouts can be prepared either in black and white or in color and can vary significantly in color fidelity and geometric precision.

Plate 30 shows a land cover classification of the states of New York and New Jersey that was derived from Landsat TM data. This color output product was prepared on a film recorder as part of a project at the USGS EROS Data Center that will create a generalized and consistent (i.e., "seamless") land cover data layer for the entire conterminous United States. Twenty-eight TM scenes were analyzed to create this classification, and 18 land cover classes are shown at this level of detail.

Tabular Data

Another common form of classification output is a table that lists summary statistics on the areal extent of the cover types present in a scene or in user-defined subscene areas. It is a simple task to derive area statistics from the grid-based interpreted data file. First, the boundary of a region of interest, such as a watershed or a county, is digitized in terms of its image matrix coordinates. Within the boundary, the number of cells in each land cover class is tabulated and multiplied by the ground area covered by a single cell. This process is considerably simpler than manually measuring areas on a map and represents a major advantage of processing land cover data in a digital format.

Digital Information Files

The final general class of output is interpreted data files containing the classification results recorded on some type of computer storage medium. As we illustrate in Section 7.17, the interpreted data in this form may be conveniently input to a GIS for merger with other geographic data files.

7.15   POSTCLASSIFICATION SMOOTHING

Classified data often manifest a salt-and-pepper appearance due to the inherent spectral variability encountered by a classifier when applied on a pixel-by-pixel basis (Figure 7.53a). For example, in an agricultural area, several pixels scattered throughout a corn field may be classified as soybeans, or vice versa. In such situations it is often desirable to "smooth" the classified output to show only the dominant (presumably correct) classification. Initially, one might consider the application of the previously described low pass spatial filters for this purpose. The problem with this approach is that the output from an image classification is an array of pixel locations containing numbers serving the function of *labels*, not *quantities*. That is, a pixel containing land cover 1 may be coded with a 1; a pixel containing land cover 2 may be

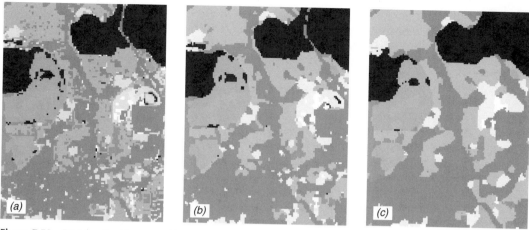

Figure 7.53 Postclassification smoothing: (*a*) original classification; (*b*) smoothed using a 3 × 3 pixel-majority filter; (*c*) smoothed using a 5 × 5-pixel majority filter.

coded with a 2; and so on. A moving low pass filter will not properly smooth such data because, for example, the averaging of class 3 and class 5 to arrive at class 4 makes no sense. In short, postclassification smoothing algorithms must operate on the basis of logical operations, rather than simple arithmetic computations.

One means of classification smoothing involves the application of a *majority filter*. In such operations a moving window is passed through the classified data set and the majority class within the window is determined. If the center pixel in the window is not the majority class, its identity is changed to the majority class. If there is no majority class in the window, the identity of the center pixel is not changed. As the window progresses through the data set, the original class codes are continually used, not the labels as modified from the previous window positions. (Figure 7.53*b* was prepared in this manner, applying a 3 × 3-pixel majority filter to the data shown in Figure7.53*a*. Figure 7.53*c* was prepared by applying a 5 × 5-pixel filter.)

Majority filters can also incorporate some form of class and/or spatial weighting function. Data may also be smoothed more than once. Certain algorithms can preserve the boundaries between land cover regions and also involve a user-specified minimum area of any given land cover type that will be maintained in the smoothed output.

One way of obtaining smoother classifications is to integrate the types of logical operations described above directly into the classification process. This involves the use of spatial pattern recognition techniques that are sensitive to such factors as image texture and pixel context. Compared to purely spectrally based procedures, these types of classifiers have received only limited attention in remote sensing in the past. However, with the continued improvement in the spatial resolution of remote sensing systems and the increasing computational power of image processing systems, such procedures will likely become more common.

7.16 CLASSIFICATION ACCURACY ASSESSMENT

Another area that is continuing to receive increased attention by remote sensing specialists is that of classification accuracy assessment. Historically, the ability to produce digital land cover classifications far exceeded the ability to meaningfully quantify their accuracy. In fact, this problem sometimes precluded the application of automated land cover classification techniques even when their cost compared favorably with more traditional means of data collection. The lesson to be learned here is embodied in the expression "A classification is not complete until its accuracy is assessed."

Congalton and Green (1999) have prepared a thorough overview of the principles and practices currently in use for assessing classification accuracy. Many of the concepts we present here in brief are more fully described in this reference.

Classification Error Matrix

One of the most common means of expressing classification accuracy is the preparation of a classification *error matrix* (sometimes called a *confusion matrix* or a *contingency table*). Error matrices compare, on a category-by-category basis, the relationship between known reference data (ground truth) and the corresponding results of an automated classification. Such matrices are square, with the number of rows and columns equal to the number of categories whose classification accuracy is being assessed.

Table 7.3 is an error matrix that an image analyst has prepared to determine how well a classification has categorized a representative subset of pixels used in the training process of a supervised classification. This matrix stems from classifying the sampled training set pixels and listing the known cover types used for training (columns) versus the pixels actually classified into each land cover category by the classifier (rows).

Several characteristics about classification performance are expressed by an error matrix. For example, one can study the various classification errors of omission (exclusion) and commission (inclusion). Note in Table 7.3 that the training set pixels that are classified into the proper land cover categories are located along the major diagonal of the error matrix (running from upper left to lower right). All nondiagonal elements of the matrix represent errors of omission or commission. Omission errors correspond to nondiagonal column elements (e.g., 16 pixels that should have been classified as "sand" were omit-

TABLE 7.3   Error Matrix Resulting from Classifying Training Set Pixels

	Training Set Data (Known Cover Types)[a]						
	W	S	F	U	C	H	Row Total
Classification data							
W	480	0	5	0	0	0	485
S	0	52	0	20	0	0	72
F	0	0	313	40	0	0	353
U	0	16	0	126	0	0	142
C	0	0	0	38	342	79	459
H	0	0	38	24	60	359	481
Column total	480	68	356	248	402	438	1992

Producer's Accuracy

W = 480/480 = 100%

S = 052/068 = 76%

F = 313/356 = 88%

U = 126/248 = 51%

C = 342/402 = 85%

H = 359/438 = 82%

User's Accuracy

W = 480/485 = 99%

S = 052/072 = 72%

F = 313/353 = 87%

U = 126/142 = 89%

C = 342/459 = 74%

H = 359/481 = 75%

Overall accuracy = (480 + 52 + 313 + 126 + 342 + 359)/1992 = 84%

[a]W, water; S, sand; F, forest; U, urban; C, corn; H, hay.

ted from that category). Commission errors are represented by nondiagonal row elements (e.g., 38 "urban" pixels plus 79 "hay" pixels were improperly included in the "corn" category).

Several other descriptive measures can be obtained from the error matrix. For example, the *overall accuracy* is computed by dividing the total number of correctly classified pixels (i.e., the sum of the elements along the major diagonal) by the total number of reference pixels. Likewise, the accuracies of individual categories can be calculated by dividing the number of correctly classified pixels in each category by either the total number of pixels in the corresponding row or column. What are often termed *producer's accuracies* result from dividing the number of correctly classified pixels in each category (on the major diagonal) by the number of training set pixels used for that category (the column total). This figure indicates how well training set pixels of the given cover type are classified.

User's accuracies are computed by dividing the number of correctly classified pixels in each category by the total number of pixels that were classified in that category (the row total). This figure is a measure of commission error and indicates the probability that a pixel classified into a given category actually represents that category on the ground.

Note that the error matrix in Table 7.3 indicates an overall accuracy of 84 percent. However, producer's accuracies range from just 51 percent ("urban") to 100 percent ("water") and user's accuracies vary from 72 percent ("sand") to 99 percent ("water"). Furthermore, this error matrix is based on training data. *It should be remembered that such procedures only indicate how well the statistics extracted from these areas can be used to categorize the same areas!* If the results are good, it means nothing more than that the training areas are homogeneous, the training classes are spectrally separable, and the classification strategy being employed works well in the training areas. This aids in the training set refinement process, but it indicates little about how the classifier performs elsewhere in a scene. One should expect training area accuracies to be overly optimistic, especially if they are derived from limited data sets. (Nevertheless, training area accuracies are sometimes used in the literature as an indication of overall accuracy. They should not be!)

Sampling Considerations

Test areas are areas of representative, uniform land cover that are different from and considerably more extensive than training areas. They are often located during the training stage of supervised classification by intentionally designating more candidate training areas than are actually needed to develop the classification statistics. A subset of these may then be withheld for the postclassification accuracy assessment. The accuracies obtained in these areas represent at least a first approximation to classification performance throughout the scene. However, being homogeneous, test areas might not provide a valid indication of classification accuracy at the individual pixel level of land cover variability.

One way that would appear to ensure adequate accuracy assessment at the pixel level of specificity would be to compare the land cover classification at every pixel in an image with a reference source. While such "wall-to-wall" comparisons may have value in research situations, assembling reference land cover information for an entire project area is expensive and defeats the whole purpose of performing a remote-sensing-based classification in the first place.

Random sampling of pixels circumvents the above problems, but it is plagued with its own set of limitations. First, collection of reference data for a large sample of randomly distributed points is often very difficult and costly. For example, travel distance and access to random sites might be prohibitive. Second, the validity of random sampling depends on the ability to precisely register the reference data to the image data. This is often difficult to do. One way to overcome this problem is to sample only pixels whose identity is not influenced by potential registration errors (for example, points at least several pixels away from field boundaries).

Another consideration is making certain that the randomly selected test pixels or areas are geographically representative of the data set under analysis. Simple random sampling tends to undersample small but potentially important areas. Stratified random sampling, where each land cover category may be considered a stratum, is frequently used in such cases. Clearly, the sampling approach appropriate for an agricultural inventory would differ from that of a wetlands mapping activity. Each sample design must account for the area being studied and the cover type being classified.

One common means of accomplishing random sampling is to overlay classified output data with a grid. Test cells within the grid are then selected randomly and groups of pixels within the test cells are evaluated. The cover types present are determined through ground verification (or other reference data) and compared to the classification data.

Several papers have been written about the proper sampling scheme to be used for accuracy assessment under various conditions, and opinions vary among researchers. Several suggest the concept of combining both random and systematic sampling. Such a technique may use systematically sampled areas to collect some accuracy assessment data early in a project (perhaps as part of the training area selection process) and random sampling within strata after the classification is complete.

Consideration must also be given to the *sample unit* employed in accuracy assessment. Depending upon the application, the appropriate sample unit might be individual pixels, clusters of pixels, or polygons. Polygon sampling is the most common approach in current use.

Sample size must also weigh heavily in the development and interpretation of classification accuracy figures. Again, several researchers have published recommendations for choosing the appropriate sample size. However, these techniques primarily produce the sample size of test areas or pixels needed to compute the overall accuracy of a classification or of a single category. In general, they are not appropriate for filling in a classification error matrix wherein errors of omission and commission are of interest.

As a broad guideline, it has been suggested that a minimum of 50 samples of each vegetation or land use category be included in the error matrix. Further, "if the area is especially large (i.e., more than a million acres) or the classification has a large number of vegetation or land use categories (i.e., more than 12 categories), the minimum number of samples should be increased to 75 or 100 samples per category" (Congalton and Green, 1999, p. 18). Similarly, the number of samples for each category might be adjusted based on the relative importance of that category for a particular application (i.e., more samples taken in more important categories). Also, sampling might be allocated with respect to the variability within each category (i.e., more samples taken in more variable categories such as wetlands and fewer in less variable categories such as open water).

Evaluating Classification Error Matrices

Once accuracy data are collected (either in the form of pixels, clusters of pixels, or polygons) and summarized in an error matrix, they are normally subject to detailed interpretation and further statistical analysis. For example, a number of features are readily apparent from inspection of the error matrix included in Table 7.4 (resulting from randomly sampled test pixels). First, we can begin to appreciate the need for considering overall, producer's, and user's accuracies simultaneously. In this example, the overall accuracy of the classification is 65 percent. However, if the primary purpose of the classification is to map the locations of the "forest" category, we might note that the producer's accuracy of this class is quite good (84 percent). This would potentially lead one to the conclusion that although the overall accuracy of the classification was poor (65 percent), it is adequate for the purpose of mapping the forest class. The problem with this conclusion is the fact that the user's accuracy for this class is only 60 percent. That is, even though 84 percent of the forested areas have been correctly identified as "forest," only 60 percent of the areas identified as "forest" within the classification are truly of that category. A more careful inspection of

TABLE 7.4   Error Matrix Resulting from Classifying Randomly Sampled Test Pixels

	Reference Data[a]						Row Total
	W	S	F	U	C	H	
Classification data							
W	226	0	0	12	0	1	239
S	0	216	0	92	1	0	309
F	3	0	360	228	3	5	599
U	2	108	2	397	8	4	521
C	1	4	48	132	190	78	453
H	1	0	19	84	36	219	359
Column total	233	328	429	945	238	307	2480

PRODUCER'S ACCURACY

W = 226/233 = 97%

S = 216/328 = 66%

F = 360/429 = 84%

U = 397/945 = 42%

C = 190/238 = 80%

H = 219/307 = 71%

USER'S ACCURACY

W = 226/239 = 94%

S = 216/309 = 70%

F = 360/599 = 60%

U = 397/521 = 76%

C = 190/453 = 42%

H = 219/359 = 61%

Overall accuracy = (226 + 216 + 360 + 397 + 190 + 219)/2480 = 65%

[a]W, water; S, sand; F, forest; U, urban; C, corn; H, hay.

the error matrix shows that there is significant confusion between the "forest" and "urban" classes. Accordingly, although the producer of the classification can reasonably claim that 84 percent of the time an area that was forested was identified as such, a user of this classification would find that only 60 percent of the time will an area visited on the ground that the classification says is "forest" actually be "forest." In fact, the only highly reliable category associated with this classification from both a producer's and a user's perspective is "water."

A further point to be made about interpreting classification accuracies is the fact that even a completely random assignment of pixels to classes will produce percentage correct values in the error matrix. In fact, such a random assignment could result in a surprisingly good apparent classification result. The $\hat{k}$ ("*KHAT*") statistic is a measure of the difference between the actual agreement between reference data and an automated classifier and the chance agreement between the reference data and a random classifier. Conceptually, $\hat{k}$ can be defined as

$$\hat{k} = \frac{\text{observed accuracy} - \text{chance agreement}}{1 - \text{chance agreement}} \tag{7.12}$$

This statistic serves as an indicator of the extent to which the percentage correct values of an error matrix are due to "true" agreement versus "chance" agreement. As true agreement (observed) approaches 1 and chance agreement approaches 0, $\hat{k}$ approaches 1. This is the ideal case. In reality, $\hat{k}$ usually ranges between 0 and 1. For example, a $\hat{k}$ value of 0.67 can be thought of as an indication that an observed classification is 67 percent better than one resulting from chance. A $\hat{k}$ of 0 suggests that a given classification is no better than a random assignment of pixels. In cases where chance agreement is large enough, $\hat{k}$ can take on negative values—an indication of very poor classification performance. (Because the possible range of negative values depends on the specific matrix, the magnitude of negative values should not be interpreted as an indication of relative classification performance).

The KHAT statistic is computed as

$$\hat{k} = \frac{N \sum\limits_{i=1}^{r} x_{ii} - \sum\limits_{i=1}^{r} (x_{i+} \cdot x_{+i})}{N^2 - \sum\limits_{i=1}^{r} (x_{i+} \cdot x_{+i})} \tag{7.13}$$

where

r = number of rows in the error matrix

x_{ii} = number of observations in row i and column i (on the major diagonal)

x_{i+} = total of observations in row i (shown as marginal total to right of the matrix)

x_{+i} = total of observations in column i (shown as marginal total at bottom of the matrix)

N = total number of observations included in matrix

To illustrate the computation of KHAT for the error matrix included in Table 7.4,

$$\sum_{i=1}^{r} x_{ii} = 226 + 216 + 360 + 397 + 190 + 219 = 1608$$

$$\sum_{i=1}^{r} (x_{i+} \cdot x_{+i}) = (239 \cdot 233) + (309 \cdot 328) + (599 \cdot 429)$$

$$+ (521 \cdot 945) + (453 \cdot 238) + (359 \cdot 307) = 1,124,382$$

$$\hat{K} = \frac{2480(1608) - 1,124,382}{(2480)^2 - 1,124,382} = 0.57$$

Note that the KHAT value (0.57) obtained in the above example is somewhat lower than the overall accuracy (0.65) computed earlier. Differences in these two measures are to be expected in that each incorporates different forms of information from the error matrix. The overall accuracy only includes the data along the major diagonal and excludes the errors of omission and commission. On the other hand, KHAT incorporates the nondiagonal elements of the error matrix as a product of the row and column marginal. Accordingly, it is not possible to give definitive advice as to when each measure should be used in any given application. Normally, it is desirable to compute and analyze both of these values.

One of the principal advantages of computing KHAT is the ability to use this value as a basis for determining the statistical significance of any given matrix or the differences among matrices. For example, one might wish to compare the error matrices resulting from different dates of images, classification techniques, or individuals performing the classification. Such tests are based on computing an estimate of the variance of $\hat{k}$ and then using a Z test to determine if an individual matrix is significantly different from a random result and if $\hat{k}$ values from two separate matrices are significantly different from one another. Readers interested in performing such analyses and learning more about accuracy assessment in general are urged to consult the various references on this subject in the Selected Bibliography.

There are three other facets of classification accuracy assessment that we wish to emphasize before leaving the subject. The first relates to the fact that the quality of any accuracy estimate is only as good as the information used to establish the "true" land cover types present in the test sites. To the extent possible, some estimate of the errors present in the reference data should be incorporated into the accuracy assessment process. It is not uncommon to have the accuracy of the reference data influenced by such factors as spatial misregistration, photo interpretation errors, data entry errors, and changes in land cover between the date of the classified image and the date of the reference data. The second point to be made is that the accuracy assessment procedure must be designed to reflect the intended use of the classification. For example, a single pixel misclassified as "wetland" in the midst of a "corn" field might be of little significance in the development of a regional land use plan. However, this same error might be intolerable if the classification forms the basis for land taxation or for enforcement of wetland preservation legislation. Finally, it should be noted that remotely sensed data are normally just a small subset of many possible forms of data resident in a GIS. How errors accumulate through the multiple layers of information in a GIS is the subject of ongoing research.

7.17 DATA MERGING AND GIS INTEGRATION

Many applications of digital image processing are enhanced through the merger of multiple data sets covering the same geographical area. These data sets can be of a virtually unlimited variety of forms. Similarly, the merger of the data may or may not take place in the context of a GIS. For example, one frequently applied form of data merger is the combining of *multiresolution data* acquired by the same sensor. We earlier illustrated (Plate 19) the application of this procedure to IKONOS 1-m panchromatic and 4-m multispectral data.

Also, in Plate 1, we illustrated the merger of automated land cover classification data with soil erodibility and slope information in a GIS environment in order to assist in the process of soil erosion potential mapping. We also illustrated the use of raster remotely sensed imagery as a backdrop for vector overlay data (Figure 1.24). Thus, the reader should already have a basic appreciation for the diversity of forms of data and types of mergers that characterize current spatial analysis procedures. In the remainder of this section, we briefly discuss the following additional data merging operations: multitemporal data merging, change detection procedures, multisensor image merging, merging of image and ancillary data in the image display process, and incorporating GIS data in land cover classification.

We have made the above subdivisions of the topic of data merging and GIS integration purely for convenience in discussing the various procedures involved. As we will see, many of the operations we discuss are extensively used in combination with one another. Similarly, the boundaries between digital image processing and GIS operations have become blurred, and fully integrated spatial analysis systems have become the norm.

Multitemporal Data Merging

Multitemporal data merging can take on many different forms. One such operation is simply combining images of the same area taken on more than one date to create a product useful for visual interpretation. For example, agricultural crop interpretation is often facilitated through merger of images taken early and late in the growing season. In early season images in the upper Midwest, bare soils often appear that later will probably be planted in such crops as corn or soybeans. At the same time, early season images might show perennial alfalfa or winter wheat in an advanced state of maturity. In the late season images, substantial changes in the appearance of the crops present in the scene are typical. Merging various combinations of bands from the two dates to create color composites can aid the interpreter in discriminating the various crop types present.

Plate 31 illustrates two examples of multitemporal NDVI data merging. Shown in (*a*) is the use of this technique to aid in mapping invasive plant species, in this case reed canary grass (*Phalaris arundinacea* L.). This part of the plate consists of a multitemporal color composite of NDVI values derived from Landsat-7 ETM+ images of southern Wisconsin on March 7 (blue), April 24 (green), and October 15 (red). The reed canary grass, which invades native wetland communities, tends to have a relatively high NDVI in the fall (October) compared to the native species and hence appears bright red to pink in the multitemporal composite. It can also be seen that features such as certain agricultural crops also manifest such tones. To eliminate the interpretation of such areas as "false positives" (identifying the areas as reed canary grass when they are not of this cover type) a GIS-derived wetland boundary layer (shown in yellow) has been overlain on the image. In this manner, the image analyst can readily focus solely upon those pink to red areas known to be included in wetlands.

Plate 31*b* represents a multitemporal color composite that depicts the three northernmost lakes shown in 31*a*. In this case, a slightly different set of dates and color assignments have been used to produce the color composite of NDVI values. These include April 24 (blue), October 31 (green), and October 15 (red). It so happened that the timing of the October 31 image corresponded with the occurrence of algal blooms in two of the three lakes shown. These blooms appear as the bright green features within the lakes.

As one would suspect, automated land cover classification is often enhanced through the use of multidate data sets. In fact, in many applications the use of multitemporal data is required to obtain satisfactory cover type discrimination. The extent to which use of multitemporal data improves classification accuracy and/or categorized detail is clearly a function of the particular cover types involved and both the number and timing of the various dates of imagery used.

Various strategies can be employed to combine multitemporal data in automatic land cover classification. One approach is to simply register all spectral bands from all dates of imaging into one master data set for classification. For example, the 6 reflectance (nonthermal) bands of a TM or ETM+ image from one date might be combined with the same 6 bands for an image acquired on another date, resulting in a 12-band data set to be used in the classification. Alternatively, principal components analysis can be used to reduce the dimensionality of the combined data set prior to classification. For example, the first three principal components for each image could be computed separately and then merged to create a final 6-band data set for classification. The 6-band image can be stored, manipulated, and classified with much greater efficiency than the original 12-band image.

Another means of dealing with multitemporal data for crop classification is the *multitemporal profile* approach. In this approach, classification is based on physical modeling of the time behavior of each crop's spectral response pattern. It has been found that the time behavior of the greenness of annual crops is sigmoidal (Figure 7.54), whereas the greenness of the soils (G_0) in a given region is nearly constant. Thus, the greenness at any time t can be modeled in terms of the peak greenness G_m, the time of peak greenness t_p, and the width σ of the profile between its two inflection points. (The inflection points, t_1 and t_2, are related to the rates of change in greenness early in the growing season and at the onset of senescence.) The features G_m, t_p, and σ account for more than 95 percent of the information in the original data and can therefore be used for classification instead of the original spectral response patterns. These three features are important because they not only reduce the

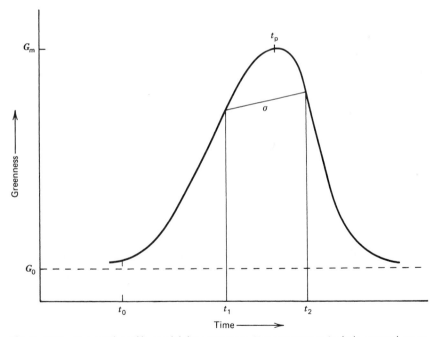

Figure 7.54 Temporal profile model for greenness. Key parameters include spectral emergence date (t_0), time (t_p) of peak greenness (G_m), and width of the profile (σ). (Adapted from Bauer, 1985, after Badhwar, 1985.)

dimensionality of the original data but also provide variables directly relatable to agrophysical parameters.

Change Detection Procedures

Change detection involves the use of multitemporal data sets to discriminate areas of land cover change between dates of imaging. The types of changes that might be of interest can range from short term phenomena such as snow cover or floodwater to long term phenomena such as urban fringe development or desertification. Ideally, change detection procedures should involve data acquired by the same (or similar) sensor and be recorded using the same spatial resolution, viewing geometry, spectral bands, radiometric resolution, and time of day. Often *anniversary dates* are used to minimize sun angle and seasonal differences. Accurate spatial registration of the various dates of imagery is also a requirement for effective change detection. Registration to within $\frac{1}{4}$ to $\frac{1}{2}$ pixel is generally required. Clearly, when misregistration is greater than one pixel, numerous errors will result when comparing the images.

The reliability of the change detection process may also be strongly influenced by various environmental factors that might change between image dates. In addition to atmospheric effects, such factors as lake level, tidal stage, wind, or soil moisture condition might also be important. Even with the use of anniversary dates of imagery, such influences as different planting dates and season-to-season changes in plant phenology must be considered.

One way of discriminating changes between two dates of imaging is to employ *postclassification comparison*. In this approach, two dates of imagery are independently classified and registered. Then an algorithm can be employed to determine those pixels with a change in classification between dates. In addition, statistics (and change maps) can be compiled to express the specific nature of the changes between the dates of imagery. Obviously, the accuracy of such procedures depends upon the accuracy of each of the independent classifications used in the analysis. The errors present in each of the initial classifications are compounded in the change detection process.

Another approach to change detection using spectral pattern recognition is simply the *classification of multitemporal data sets*. In this alternative, a single classification is performed on a combined data set for the two dates of interest. Supervised or unsupervised classification is used to categorize the land cover classes in the combined image. The success of such efforts depends upon the extent to which "change classes" are significantly different spectrally from the "nonchange" classes. Also, the dimensionality and complexity of the classification can be quite great, and if all bands from each date are used, there may be substantial redundancy in their information content.

Principal components analysis is sometimes used to analyze multidate image composites for change detection purposes. In this approach, two (or more) images are registered to form a new multiband image containing various bands from each date. Several of the uncorrelated principal components computed from the combined data set can often be related to areas of change. One disadvantage to this process is that it is often difficult to interpret and identify the specific nature of the changes involved.

Plate 32 illustrates the application of multidate principal components analysis to the process of assessing tornado damage from "before" and "after" images of the tornado's path of destruction. The "before" image shown in (a), is a Landsat-7 ETM+ composite of bands, 1, 2, and 5 shown as blue, green, and red, respectively. The "after" image, shown in (b), was acquired 32 days later than the image shown in (a), on the day immediately following the tornado. While the damage path from the tornado is fairly discernable in (b), it is most distinct in the principal component image shown in (c). This image depicts the second principal component image computed from the 6-band composite formed by registering bands, 1, 2, and 5 from both the "before" and "after" images.

Temporal image differencing is yet another common approach to change detection. In the image differencing procedure, DNs from one date are simply subtracted from those of the other. The difference in areas of no change will be very small (approaching zero), and areas of change will manifest larger negative or positive values. If 8-bit images are used, the possible range of values for the difference image is -255 to $+255$, so normally a constant (e.g., 255) is added to each difference image value for display purposes.

Temporal image ratioing involves computing the ratio of the data from two dates of imaging. Ratios for areas of no change tend toward 1 and areas of change will have higher or lower ratio values. Again, the ratioed data are normally scaled for display purposes (Section 7.6). One of the advantages to the ratioing technique is that it tends to normalize the data for changes in such extraneous factors as sun angle and shadows.

Whether image differencing or ratioing is employed, the analyst must find a meaningful "change–no change threshold" within the data. This can be done by compiling a histogram for the differenced or ratioed image data and noting that the change areas will reside within the tails of the distribution. A variance from the mean can then be chosen and tested empirically to determine if it represents a reasonable threshold. The threshold can also be varied interactively in most image analysis systems so the analyst can obtain immediate visual feedback on the suitability of a given threshold.

In lieu of using raw DNs to prepare temporal difference or ratio images, it is often desirable to correct for illumination and atmospheric effects and to transform the image data into physically meaningful quantities such as radiances or reflectances (Section 7.2). Also, the images may be prepared using spatial filtering or transformations such as principal components or vegetation components. Likewise, linear regression procedures may be used to compare the two dates of imagery. In this approach a linear regression model is applied to predict data values for date 2 based on those of date 1. Again, the analyst must set a threshold for detecting meaningful change in land cover between the dates of imaging.

Change vector analysis is a change detection procedure that is a conceptual extension of image differencing. Figure 7.55 illustrates the basis for this approach in two dimensions. Two spectral variables (e.g., data from two bands, two vegetation components) are plotted at dates 1 and 2 for a given pixel. The vector connecting these two data sets describes both the magnitude and direction of spectral change between dates. A threshold on the magnitude can be established as the basis for determining areas of change, and the direction of the spectral change vector often relates to the type of change. For example, Figure 7.55*b* illustrates the differing directions of the spectral change vector for vegetated areas that have been recently cleared versus those that have experienced regrowth between images.

One of the more efficient approaches to delineating change in multidate imagery is use of a *change-versus-no-change binary mask to guide multidate classification*. This method begins with a traditional classification of one image as a reference (time 1). Then, one of the spectral bands from this date is registered to the same band in a second date (time 2). This two-band data set is then analyzed using one of the earlier described algebraic operations (e.g., image differencing or ratioing). A threshold is then set to separate areas that have changed between dates from those that have not. This forms the basis for creating a binary mask of change-versus-no-change areas. This mask is then applied to the multiband image acquired at time 2 and only the areas of change are then classified for time 2. A traditional postclassifica-

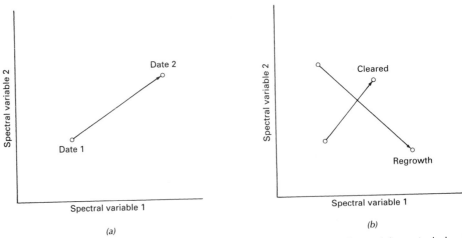

Figure 7.55 Spectral change vector analysis: (*a*) spectral change vector observed for a single land cover type; (*b*) length and direction of spectral change vectors for hypothetical "cleared" and "regrowth" areas.

tion comparison is then performed in the areas known to have changed between dates.

Research continues on the development of software tools to assist in multidate change detection. Representative of such tools is a technique proposed by Walkey (1997) that consists of a set of interactive, iterative steps to aid an analyst in delineating true land cover changes from incidental scene-to-scene changes. The technique is based on the simple notion that a two-dimensional scatter plot of a band at time 1 versus the same band at time 2 would result in an elongated ellipse oriented at 45° in the spectral measurement space (Figure 7.56b). The ellipse, rather than a straight line (Figure 7.56a), results from the natural variability of the landscape even in the absence of land cover change. Also, extraneous effects such as changing atmospheric conditions or sensor response drift can cause the "stable spectral space ellipse" to depart from an exact 45° orientation (Figures 7.56c and d).

Figure 7.56e illustrates a situation where certain features in the two-date data set have experienced a land cover change. The small ellipse above the stable spectral space ellipse represents those pixels that have gotten brighter between the two dates, and the ellipse below the stable spectral space ellipse corresponds to those areas that have gotten darker between dates. By displaying such a scatter plot, it is possible to interactively describe new *delta transformation* axes with a transformed band 1 aligned in the direction of the stable spectral space ellipse and a transformed band 2 located at 90° from band 1 (Figure 7.56f). The transformed band 2 axis then defines a change axis that the analyst can use to set a pair of change-versus-no-change thresholds—one in the positive (brighter) direction and one in the negative (darker) direction. Hence, the analyst can then display separately images of the positive-change component, the negative-change component, and the stable component.

Delta transformations can be established for each spectral band available, such that the two change images plus the stable component of each band pair can be generated. The options for display of the images are numerous. For example, the various change and stable components can be displayed individually in black and white or in the blue, green, and red planes of a color monitor in various combinations.

It should also be emphasized that the scatter plots shown in Figure 7.56 have been greatly simplified in order to describe the delta transformation process. The scatter plots for actual image data can be more complex and the analyst must be very careful in delineating the direction of "stable spectral space." Again, this can be done interactively and iteratively until an acceptable change-versus-no-change axis is determined.

Figure 7.57 illustrates the application of the delta transformation process to images acquired over a portion of the Nicolet National Forest in northeastern Wisconsin. Shown in (a) is a Landsat TM band 5 (mid-IR) image acquired in 1984, and its counterpart acquired in 1993 is shown in (b). Note that between dates several areas of the forest have become brighter due to

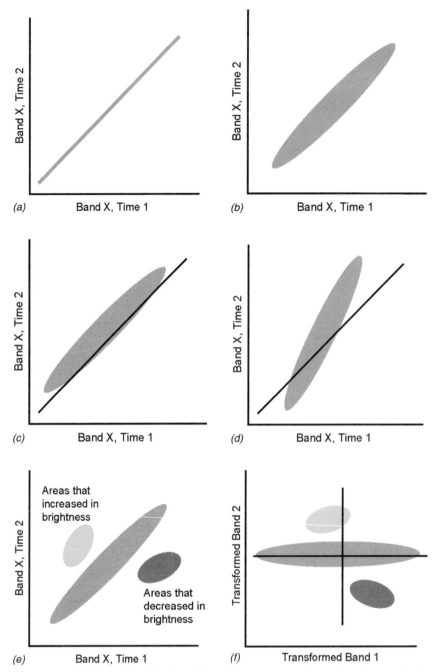

Figure 7.56 Conceptual basis for the delta transformation: (*a*) plot of a band vs. itself with no spectral change whatsoever; (*b*) stable spectral space ellipse showing natural variability in the landscape between dates; (*c*) effect of uniform atmospheric haze differences between dates; (*d*) effect of sensor drift between dates; (*e*) pixels that appear brighter (above) or darker (below) than the stable spectral space ellipse due to land cover change; (*f*) delta transformation with transformed band 2 defining the image change axis. (After Walkey, 1997.)

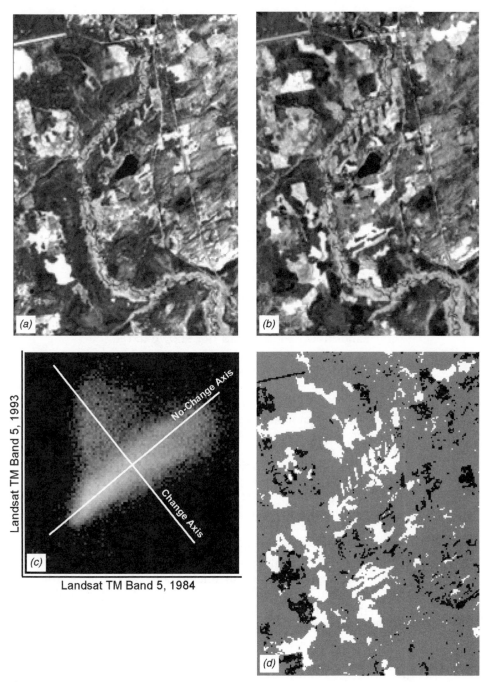

Figure 7.57 Application of delta transformation techniques to Landsat TM band 5 (mid-IR) images of the Nicolet National Forest: (*a*) 1984 image; (*b*) 1993 image; (*c*) two-date scatter plot, including "no-change" and "change" axes; (*d*) areas of forest cutting between image dates and forest regeneration between image dates shown as light and dark areas, respectively.

forest cutting activity between the dates, and several areas appear darker due to forest regeneration over the same time period.

Figure 7.57c depicts the two-date scatter plot for the images shown in (a) and (b). The direction of the stable spectral space and the land cover change axis are also shown. Figure 7.57d illustrates those areas that have changed between dates due to forest cutting and regeneration as lighter and darker areas, respectively.

Multisensor Image Merging

Plate 33 illustrates yet another form of data merging, namely, the combination of image data from more than one type of sensor. This image depicts an agricultural area in southcentral Wisconsin. The following sources of image data comprise this IHS-enhanced composite image:

1. SPOT HRV band 2 (red), shown in blue.
2. SPOT HRV band 3 (near IR), shown in green.
3. Landsat TM band 5 (mid IR), shown in red.
4. Digital orthophotograph, used in the intensity component of the IHS transformation.
5. GIS overlay of the boundaries of selected farm fields.

This image demonstrates how multisensor image merging often results in a composite image product that offers greater interpretability than an image from any one sensor alone. For example, Plate 33 affords the spatial information content of the digital orthophoto data (2 m) and the red and near-infrared spectral data from the SPOT HRV, as well as the mid-infrared information content of the Landsat TM (band 5). The composite image is aimed at exploiting the advantages of each of the data sources used in the merger process.

In addition to the merger of digital photographic and multispectral scanner data, multisensor image merging has been extensively used to combine multispectral scanner and radar image data. Such combinations take advantage of the spectral resolution of the multispectral scanner data in the optical wavelengths and the radiometric and "sidelighting" characteristics of the radar data.

Merging of Image Data with Ancillary Data

Probably one of the most important forms of data merger employed in digital image processing is the registration of image data with "nonimage," or ancillary, data sets. This latter type of data set can vary, ranging from soil type to elevation data to assessed property valuation. The only requirement is that the ancillary data be amenable to accurate geocoding so that they can be registered with the image data to a common geographic base. Usually, although not a necessity, the merger is made in a GIS environment.

Digital elevation models (DEMs) have been combined with image data for a number of different purposes. Figure 7.58 illustrates the merger of DEM and image data to produce *synthetic stereoscopic images*. Shown in this figure is a synthetic stereopair generated by introducing simulated parallax into a Landsat MSS image. Whereas standard Landsat images exhibit only a fixed, weak stereoscopic effect in the relatively small areas of overlap between orbit passes, the synthetic image can be viewed in stereo over its entirety and with

Figure 7.58 Synthetic stereopair generated from a single Landsat MSS image and a digital elevation model, Black Canyon of the Gunnison, CO. Maximum canyon depth is 850 m. Scale 1 : 400,000. (Courtesy USGS.)

an analyst-specified degree of vertical exaggeration. These images are produced in a manner similar to the process used to produce stereomates for orthophotographs (Section 3.9). That is, the elevation at each pixel position is used to offset the pixel according to its relative elevation. When this distorted image is viewed stereoscopically with the original scene, a three-dimensional effect is perceived. Such images are particularly valuable in applications where landform analysis is central to the interpretation process. The technique is also useful for restoring the topographic information lost in the preparation of spectral ratio images.

Another common use of DEM data is in the production of perspective-view images, such as Figure 7.59, a merger of Landsat TM and DEM data, which shows Mount Fuji, the highest peak in Japan.

Merging topographic information and image data is often useful in image classification. For example, topographic information is often important in forest-type mapping in mountainous regions. In such situations, species that have very similar spectral characteristics might occupy quite different elevation ranges, slopes, or aspects. Thus, the topographic information might serve as another "channel" of data in the classification directly or as a postclassification basis upon which to discriminate between only the spectrally similar classes in an image. In either case, the key to improving the classification is being able to define and model the various associations between the cover types present in a scene and their habitats.

Incorporating GIS Data in Automated Land Cover Classification

Obviously, topographic information is not the only type of ancillary data that might be resident in a GIS and useful as an aid in image classification. For example, data as varied as soil types, census statistics, ownership boundaries, and zoning districts have been used extensively in the classification process. The basic premise of any such operation is that the accuracy and/or the categorical detail of a classification based on image *and* ancillary data will be an improvement over a classification based on either data source alone.

Ancillary data are often used to perform *geographic stratification* of an image prior to classification. As with the use of topographic data, the aim of this process is to subdivide an image into a series of relatively homogeneous geographic areas (strata) that are then classified separately. The basis of stratification need not be a single variable (e.g., upland versus wetland, urban versus rural) but can also be such factors as landscape units or ecoregions that combine several interrelated variables (e.g., local climate, soil type, vegetation, landform).

There are an unlimited number of data sources and ways of combining them in the classification process. Similarly, the ancillary data can be used either prior to, during, or after the image classification process (or even some

Figure 7.59 Perspective-view image of Mount Fuji, Japan, produced by combining Landsat TM data (bands 2, 3, and 4, shown in black and white) and digital elevation model (DEM) data. (Copyright © RESTEC/NIHON University, 1991.) (Figure 8.45 shows a radar image of this area as well.)

combination of these choices might be employed in a given application). The particular sources of data used and how and when they are employed are normally determined through the formulation of *multisource image classification decision rules* developed by the image analyst. These rules are most often formulated on a case-by-case basis through careful consideration of the form, quality, and logical interrelationship among the data sources available. For example, the "roads" in a land cover classification might be extracted from current digital line graph (DLG) data rather than from an image source. Similarly, a particular spectral class might be labeled "alfalfa" or "grass" in different locations of a classification depending upon whether it occurs within an area zoned as agricultural or residential.

Space limits us from providing numerous examples of how user-defined classification decision rules are designed and implemented. In lieu of such a discussion, Plate 34 is presented to illustrate the general principles involved. Shown in Plate 34 is a "composite" (lower right) land cover classification performed in the vicinity of Fox Lake, Wisconsin, which is located in the east central portion of the state. The composite classification was produced from five separate data sources:

1. A supervised classification of the scene using a TM image acquired in early May (upper left).

2. A supervised classification of the scene using a TM image acquired in late June (upper right).

3. A supervised classification of both dates combined using a principal components analysis (not shown).

4. A wetlands GIS layer prepared by the Wisconsin Department of Natural Resources (DNR) (lower left).

5. A DNR-supplied road DLG (lower left).

Used alone, none of the above data sources provided the classification accuracy or detail needed for the purpose of monitoring the wildlife habitat characteristics of the area depicted in the plate. However, when all of the data were integrated in a GIS, the data analyst was able to develop a series of postclassification decision rules utilizing the various data sources in combination. Simply put, these decision rules were based on the premise that certain cover types were better classified in one classification than the others. In such cases, the optimal classification for that category was used for assigning that cover type to the composite classification. For example, the water class was classified with nearly 100 percent accuracy in the May classification. Therefore, all pixels having a classification of water in the May scene were assigned to that category in the composite classification.

Other categories in the above example were assigned to the composite classification using other decision rules. For example, early attempts to automatically discriminate roads in any of the TM classifications suggested that this class would be very poorly represented on the basis of the satellite data. Accordingly, the road class was dropped from the training process and the roads were simply included as a DLG overlay to the composite classification.

The wetland GIS layer was used in yet another way, namely to aid in the discrimination between deciduous upland and deciduous wetland vegetation. None of TM classifications could adequately distinguish between these two classes. Accordingly, any pixel categorized as deciduous in the May TM data was assigned either to the upland or wetland class in the composite classification based on whether that pixel was outside or within a wetland according to the wetland GIS layer. A similar procedure was used to discriminate between the grazed upland and grazed wetland classes in the principal components classification.

Several of the land cover categories in this example were discriminated using rules that involved comparison among the various classifications for a given pixel in each of the three preliminary classifications. For example, hay was classified well in the May scene and the principal components image but with some errors of commission into the grazed upland, cool season grass, and old field categories. Accordingly, a pixel was classified as hay if it was hay in the May or principal components classification but at the same time was not classified as grazed upland or cool season grass in the principal components classification or as old field in the June classification. Similarly, pixels were assigned to the oats class in the composite classification if they were originally classified as oats, corn, peas, or beans in the May scene but oats in the June scene.

Table 7.5 lists a representative sample of the various decision rules used in the composite classification depicted in Plate 34. They are presented to illustrate the basic manner in which GIS data and spatial analysis techniques can be combined with digital image processing to improve the accuracy and categorical detail of land cover classifications. The integration of remote sensing, GIS, and "expert system" techniques for such purposes is an active area of current research. Indeed, these combined technologies are resulting in the development of increasingly "intelligent" information systems.

Another area of current research is the use of *artificial neural networks* in image classification. Such systems are "self-training" in that they adaptively construct linkages between a given pattern of input data and particular outputs. Neural networks can be used to perform traditional image classification tasks (Foody et al., 1995) and are also increasingly used for more complex operations such as spectral mixture analysis (Moody et al., 1996). For image classification, neural networks do not require that the training class data have a Gaussian statistical distribution, a requirement that is held by maximum likelihood algorithms. This allows neural networks to be used with a much wider range of types of input data than could be used in a traditional maximum likelihood classification process. In addition, once they have been fully trained, neural networks can perform image classification relatively rapidly,

TABLE 7.5 Basis for Sample Decision Rules Used for Composite Classification Shown in Plate 34

Sample Class in Composite Classification	Classification			GIS Data	
	May	June	PC	Wetlands	Roads
Water	Yes				
Roads					Yes
Deciduous upland	Yes			Outside	
Deciduous wetland	Yes			Inside	
Grazed upland			Yes	Outside	
Grazed wetland			Yes	Inside	
Hay	Yes		Yes		
		Old field—no	Grazed upland—no Cool season grass—no		
Oats	Oats—yes Corn—yes Peas—yes Beans—yes	Oats—yes			
Peas	Oats—yes Corn—yes Peas—yes Beans—yes		Peas—yes		
Beans	Oats—yes Corn—yes Peas—yes Beans—yes	Beans—yes			
Reed canary grass			Yes		
Warm season grass	Yes				
Cool season grass			Yes		

although the training process itself can be quite time consuming. In the following discussion we will focus on back-propagation neural networks, the type most widely used in remote sensing applications, although other types of neural networks have been described.

A neural network consists of a set of three or more layers, each made up of multiple nodes. These nodes are somewhat analogous to the neurons in a biological neural network and thus are sometimes referred to as neurons. The network's layers include an input layer, an output layer, and one or more hidden layers. The nodes in the input layer represent variables used as input to the neural network. Typically, these might include spectral bands from a remotely sensed image, textural features or other intermediate products derived from such images, or ancillary data describing the region to be analyzed. The nodes in the output layer represent the range of possible output categories to be produced by the network. If the network is being used for image classification, there will be one output node for each class in the classification system.

Between the input and output layers are one or more hidden layers. These consist of multiple nodes, each linked to many nodes in the preceding layer and to many nodes in the following layer. These linkages between nodes are represented by weights, which guide the flow of information through the network. The number of hidden layers used in a neural network is arbitrary. Generally speaking, an increase in the number of hidden layers permits the network to be used for more complex problems but reduces the network's ability to generalize and increases the time required for training.

Figure 7.60 shows an example of a neural network that is used to classify land cover based on a combination of spectral, textural, and topographic information. There are seven nodes in the input layer, as follows: nodes 1 to 4 correspond to the four spectral bands of a multispectral scanner image, node 5 corresponds to a textural feature that is calculated from a radar image, and nodes 6 and 7 correspond to terrain slope and aspect, calculated from a digital elevation model. After the input layer, there are two hidden layers, each with nine nodes. Finally, the output layer consists of six nodes, each corresponding to a land cover class (water, sand, forest, urban, corn, and hay). When given any combination of input data, the network will produce the output class that is most likely to result from that set of inputs, based on the network's analysis of previously supplied training data.

If multiple sources of input data are used, the range of values in each data set may differ. In the example shown in Figure 7.60, the four multispectral bands might have digital number values ranging from 0 to 255, while the terrain slope might be expressed as a percentage and the terrain aspect might be in degrees. Before these various data sets can be provided to the neural network, they each must be rescaled to fit within the same range. Generally, all input data are scaled to fit within the range from 0 to 1.

Applying a neural network to image classification makes use of an iterative training procedure in which the network is provided with matching sets of input and output data. Each set of input data represents an example of a pattern to be learned, and each corresponding set of output data represents the desired output that should be produced in response to the input. During the training process, the network autonomously modifies the weights on the linkages between each pair of nodes in such a way as to reduce the discrepancy between the desired output and the actual output.

It should be noted that a back-propagation neural network is not guaranteed to find the ideal solution to any particular problem. During the training process the network may develop in such a way that it becomes caught in a "local minimum" in the output error field, rather than reaching the absolute minimum error. Alternatively, the network may begin to oscillate between two slightly different states, each of which results in approximately equal error (Paola and Schowengerdt, 1995). A variety of strategies have been proposed to

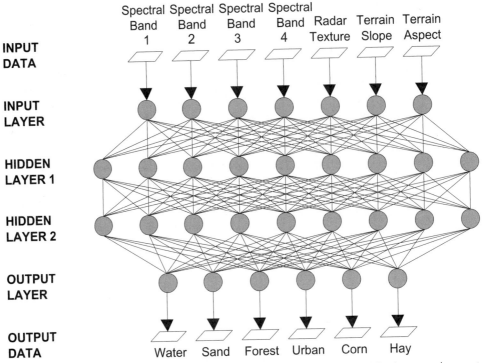

Figure 7.60 Example of an artificial neural network with one input layer, two hidden layers, and one output layer.

help push neural networks out of these pitfalls and enable them to continue development toward the absolute minimum error. Again, the use of artificial neural networks in image classification is the subject of continuing research.

7.18 HYPERSPECTRAL IMAGE ANALYSIS

The hyperspectral sensors discussed in Sections 5.14 and 6.15 differ from other optical sensors in that they typically produce contiguous, high resolution radiance spectra rather than discrete measurements of average radiance over isolated, wide spectral bands. As a result, these sensors can potentially provide vast amounts of information about the physical and chemical composition of the surface under observation as well as insight into the characteristics of the atmosphere between the sensor and the surface. While most multispectral sensors merely *discriminate* among various earth surface features, hyperspectral sensors afford the opportunity to *identify* and *determine many characteristics about* such features. However, these sensors have their disadvantages as well, including an increase in the volume of data to be processed, a tendency toward relatively poor signal-to-noise ratios, and an increased susceptibility to the effects of unwanted atmospheric interference if such effects are not corrected for. As a result, the image processing techniques that are used to analyze hyperspectral imagery differ somewhat from those discussed previously for multispectral sensors. In general, hyperspectral image analysis requires more attention to issues of atmospheric correction and relies more heavily on physical and biophysical models rather than on purely statistical techniques such as maximum likelihood classification.

Atmospheric Correction of Hyperspectral Images

Atmospheric constituents such as gases and aerosols have two types of effects on the radiance observed by a hyperspectral sensor. The atmosphere absorbs (or attenuates) light at particular wavelengths, thus decreasing the radiance that can be measured. At the same time, the atmosphere also scatters light into the sensor's field of view, thus adding an extraneous source of radiance that is unrelated to the properties of the surface. The magnitude of absorption and scattering will vary from place to place (even within a single image) and time to time depending on the concentrations and particle sizes of the various atmospheric constituents. The end result is that the "raw" radiance values observed by a hyperspectral sensor cannot be directly compared to either laboratory spectra or remotely sensed hyperspectral imagery acquired at other times or places. Before such comparisons can be performed, an atmospheric correction process must be used to compensate for the transient effects of atmospheric absorption and scattering (Green, 1990).

One significant advantage of hyperspectral sensors is that the contiguous, high resolution spectra that they produce contain a substantial amount of information about atmospheric characteristics at the time of image acquisition. In some cases, atmospheric models can be used with the image data themselves to compute quantities such as the total atmospheric column water vapor content (Gao and Goetz, 1990) and other atmospheric correction parameters. Alternatively, ground measurements of atmospheric transmittance or optical depth, obtained by instruments such as sunphotometers, may also be incorporated into the atmospheric correction models. Such methods are beyond the scope of this discussion.

Hyperspectral Image Analysis Techniques

Many of the techniques used for the analysis of remotely sensed hyperspectral imagery are derived from the field of spectroscopy, wherein the molecular composition of a particular material is related to the distinctive patterns in which the material absorbs and reflects light at individual wavelengths. Once a hyperspectral image has been corrected for the effects of atmospheric absorption and scattering, the reflectance "signature" of each pixel can be compared to previously acquired spectra for known material types. Many "libraries" of spectral reference data have been collected in the laboratory and in the field, representing primarily minerals, soils, and vegetation types. (See Section 5.14 and Appendix B.)

Numerous approaches have been taken to compare remotely sensed hyperspectral image data to known reference spectra. At the simplest level, individual wavelength-specific absorption features can be identified in the image and compared to similar features in the reference spectra. This can be done by selecting one spectral band that occurs in the "trough," or low point, of an absorption feature and two "bracket" bands located on either side of the absorption feature. The default spectral reflectance that would be observed if the absorption feature were not present is calculated by interpolating between the two bracket bands, and the value measured in the trough band is then subtracted from this default value (or divided by this default value) to obtain an estimate of the strength of the absorption that is occurring. This process can be applied to both the image spectra and the reference spectra in the same fashion. Drawbacks of this method include its sensitivity to noise in the image data and its inability to deal with multiple adjacent or overlapping absorption features (Kruse, Kieren-Young, and Boardman, 1990).

With the increase in computational power that has occurred in recent years, the above method has evolved to the direct comparison of entire spectral signatures, rather than individual absorption features within a signature. Spectrum ratioing consists of dividing every reflectance value in the reference spectrum by the corresponding value in the image spectrum. If the average deviation from 1.0 across all wavelengths falls within some small level of tolerance, the image spectrum for that pixel is considered to match the reference spectrum. It should be noted that this method will tend to fail if the average brightness of the image spectrum is higher or lower than the average brightness of the reference spectrum. This would occur, for example, on slopes that are facing away from the sun and are topographically shaded; in this case, at all wavelengths the observed image spectrum would have lower values than would the corresponding reference spectrum.

Another technique for comparing image and reference spectra is referred to as *spectral angle mapping (SAM)* (Kruse et al., 1993). This technique is based on the idea that an observed reflectance spectrum can be considered as a vector in a multidimensional space, where the number of dimensions equals the number of spectral bands. If the overall illumination increases or decreases (perhaps due to the presence of a mix of sunlight and shadows), the length of this vector will increase or decrease, but its angular orientation will remain constant. This is shown in Figure 7.61*a*, where a two-band "spectrum" for a particular material will lie somewhere along a line passing through the origin of a two-dimensional space. Under low illumination conditions, the

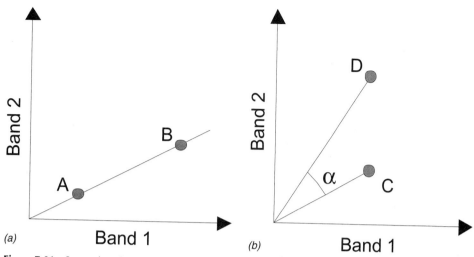

Figure 7.61 Spectral angle mapping concept. (*a*) For a given feature type, the vector corresponding to its spectrum will lie along a line passing through the origin, with the magnitude of the vector being smaller (A) or larger (B) under lower or higher illumination, respectively. (*b*) When comparing the vector for an unknown feature type (C) to a known material with laboratory-measured spectral vector (D), the two features match if the angle α is smaller than a specified tolerance value. (After Kruse et al., 1993.)

length of the vector will be short, and the point will be located closer to the origin of the multidimensional space (point *A* in Figure 7.61*a*). If the illumination is increased, the length of the vector will increase, and the point will move farther from the origin (point *B*).

To compare two spectra, such as an image pixel spectrum and a library reference spectrum, the multidimensional vectors are defined for each spectrum and the angle between the two vectors is calculated. If this angle is smaller than a given tolerance level, the spectra are considered to match, even if one spectrum is much brighter than the other (farther from the origin) overall. For example, Figure 7.61*b* shows vectors for a pair of two-band spectra. If angle α is acceptably small, then the spectrum corresponding to vector *C* (derived from an unknown pixel in the image) will be considered to match the spectrum of vector *D* (derived from a laboratory measurement).

It should be noted that the examples shown in Figure 7.61*a* and *b* are for a hypothetical sensor with only 2 spectral bands. Needless to say, for a real sensor with 200 or more spectral bands, the dimensionality of the space in which the vectors are located becomes too large for human visualization.

These and other techniques for comparing image and reference spectra can be further developed by the incorporation of prior knowledge about the likely characteristics of materials on the ground. This leads to the area of expert systems, whereby the existing knowledge of expert scientists in a particular field, such as geology, is incorporated as a set of constraints for the results of the spectral analysis process. The development of these systems requires an extensive effort on the part of both the "expert" and the programmer. Despite this difficulty, such techniques have been used successfully, particularly in mineralogical applications (Kruse, Lefkoff, and Dietz, 1993).

As discussed earlier in Section 7.13, spectral mixture analysis provides a powerful, physically based method for estimating the proportions of multiple surface components within a single pixel. As might be expected, this analytical technique is particularly well suited to hyperspectral imagery. Since the maximum number of endmembers that can be included in a spectral mixture analysis is directly proportional to the number of spectral bands +1, the vastly increased dimensionality of a hyperspectral sensor effectively removes the sensor-related limit on the number of endmembers available. The fact that the number of channels far exceeds the likely number of endmembers for most applications readily permits the exclusion from the analysis of any bands with low signal-to-noise ratios or with significant atmospheric absorption effects. The endmember spectra may be determined by examining image pixels for known areas on the landscape or by adopting reference spectra from laboratory or field measurements. The end result of the process is an estimate of the proportion of each pixel in the image that is represented by each endmember. Spectral mixture analysis is fast becoming one of the most widely used methods for extracting biophysical information from remotely sensed hyperspectral images.

Derivative analysis, another method derived from the field of spectroscopy, has also been adopted for use with hyperspectral imagery. This approach makes use of the fact that the derivative of a function tends to emphasize changes irrespective of the mean level. The advantage of spectra derivatives is their ability to locate and characterize subtle spectral details. A common disadvantage of this method is its extreme sensitivity to noise. For this reason, the derivative computation is typically coupled with spectral smoothing, the most common method being the optimized Savitsky–Golay procedure (Savitsky and Golay, 1964). When the smoothing and the derivative are well matched to a spectral feature, the method is very effective in isolating and characterizing spectral details independent of the magnitude of the signal (Tsai and Philpot, 1999). Hence, radiance data that are uncalibrated for atmospheric effects may be analyzed in this manner.

It is also possible to apply traditional image classification methods such as maximum likelihood classification to hyperspectral data. However, there are significant disadvantages to this approach. First, the increased number of spectral bands results in a vast increase in the computational load for the statistical classification algorithms. In addition, maximum likelihood classification requires that every training set must include at least one more pixel than there are bands in the sensor, and more often it is desirable to have the number of pixels per training set be between 10 times and 100 times as large as the number of sensor bands, in order to reliably derive class-specific covariance matrices. Obviously, this means that the shift from a 7-band Landsat ETM+ image to a 224-band AVIRIS hyperspectral image will result in a major increase in the number of pixels required for each training set. Finally, the use of a statistical image analysis technique, as opposed to the physically based methods that have been discussed previously, represents a potential loss of the valuable information that is provided in the continuous reflectance spectrum about the physical and chemical characteristics of the features under analysis.

Various efforts are underway to develop software toolkits to cover the full sequence of hyperspectral image analysis activities. One example is the "Tetracorder" algorithm (formerly Tricorder), developed at the USGS Spectroscopy Laboratory in Denver, Colorado. Using this system, an image can be converted to apparent surface reflectance, then subjected to multiple analyses by comparison to reference spectra. A typical sequence of activities that might be performed would be as follows (after Clark and Swayze, 1995):

Step 1. For a particular pixel or region in an image, identify the spectral features that are present, such as chlorophyll absorption and clay mineral absorption features.

Step 2. Using an unmixing algorithm, determine the proportion of the area covered by plants.

Step 3. Using the plant spectrum, determine the water content of the leaves, remove this component from the plant spectrum, and calculate the lignin–nitrogen ratio.

Step 4. Derive a soil spectrum by removing the plant spectrum from the original spectrum and search for various minerals present.

For other applications, such as water quality monitoring, the steps shown above would be replaced with similar analyses that are relevant to the application of interest. It should be emphasized that the successful use of software systems such as Tetracorder require the user to be familiar with the principles of spectroscopy as well as with the potential material types and environmental conditions that are likely to be found in the area under investigation.

7.19 BIOPHYSICAL MODELING

Digital remote sensing data have been used extensively in the realm of quantitative biophysical modeling. The intent of such operations is to relate quantitatively the data recorded by a remote sensing system to biophysical features and phenomena measured on the earth's surface. For example, remotely sensed data might be used in applications as varied as crop yield estimation, defoliation measurement, biomass prediction, water depth determination, and pollution concentration estimation.

Three basic approaches can be employed to relate digital remote sensing data to biophysical variables. In *physical modeling*, the data analyst attempts to account mathematically for all known parameters affecting the radiometric characteristics of the remote sensing data (e.g., earth–sun distance, solar elevation, atmospheric effects, sensor gain and offset, viewing geometry). Alternatively, *empirical modeling* can be employed. In this approach, the quantitative relationship between the remote sensing data and ground-based data is calibrated by interrelating known points of coincident observation of the two (e.g., field-based measurement of forest defoliation conditions precisely at the time a satellite image is acquired). Statistical regression procedures are often used in this process. The third approach to biophysical modeling is simply to employ some *combination of physical and empirical techniques* (e.g., conversion of DNs to absolute radiance values prior to relating these values to a ground-based measurement).

Plate 35 illustrates the application of biophysical modeling in the context of using Landsat-7 ETM+ data to estimate lake water clarity. In this application, water clarity is measured in the field in terms of *Secchi disk* readings. A Secchi disk is simply a white (or black and white) circular (20.3-cm diameter) metal disk that is attached to a rope on which depth markings are delineated. The disk is lowered from a boat until it just disappears from view, and this depth is noted using the marks on the rope. The disk is then lowered slightly more, and then it is raised until it just reappears, and a second depth reading is taken. The two depth readings are then averaged to yield a final transparency measurement. This very quick and easy measurement is often well correlated with various other water quality measurements (e.g., concentration of chlorophyll *a* or suspended materials in the water).

In Wisconsin, some 600 volunteers distributed across the state regularly take Secchi disk readings coincident with the overpass of Landsat (and other) satellites. These field measurements are then interrelated (via multiple regression) with the corresponding digital data collected by the satellite system. (This relationship then is used to predict the transparency of lakes appearing in the satellite image that have not been sampled on the ground.) With over 15,000 lakes in Wisconsin, this approach permits substantial, cost-effective extension of the geographic reach of purely surface-based monitoring efforts.

Plate 35 illustrates a small sample of how the above statewide effort is applied locally. Shown in (*a*) is a normal color composite (band 1, 2, and 3 shown as blue, green, and red, respectively) of a small portion of a Landsat-7 ETM+ image acquired over an area in southeast Wisconsin. A band 4 image (near IR) was used as the basis for creating a binary mask to separate water from land. (A vector GIS hydrography layer could also have been used for this purpose.) In (*b*), such a mask has been applied to portray the normal color composite data for the water only. Note the relatively large variation in tone and brightness in this image. It has been shown that in the case of most Wisconsin lakes, the ratio of the radiance measured in band 1 (blue) to that measured in band 3 (red) is an excellent predictor of Secchi disk depth. This is shown in (c), which is a plot of the natural logarithm of numerous Secchi disk readings made in a full Landsat-7 image versus the band 1/band 3 ratio measured over the corresponding area in each field-sampled lake. Usually, the field measurements of Secchi disk depth are made in the deepest, darkest central portion of each lake. A combination of unsupervised classification and a histogram analysis procedure is used to isolate lake pixels corresponding to these deep, dark areas. This procedure eliminates pixels affected by the presence of such features as small islands, aquatic vegetation, and bottom reflectance in shallow areas. Multiple regression is used to relate the resulting image data to the field observations of all sampled lakes. This regression model is then applied to all lakes in the scene. This is shown in Figure 7.62d where pixel-by-pixel colors correspond to the various Secchi disk readings predicted by the model shown in 7.62c. At the time of this writing (2002), the above process had been successfully employed for statewide surveys of lake transparency both in Wisconsin and Minnesota (Chipman et al., 2004), and its application in the state of Michigan had just begun.

Increasingly, remotely sensed data are being used in concert with GIS techniques in the context of *environmental modeling*. The objective of environmental modeling is to simulate the processes functioning within environmental systems and to predict and understand their behavior under altered conditions. Over the past several decades, improved understanding of environmental systems, in conjunction with increased remote sensing and GIS capabilities, has permitted highly accurate and detailed spatial representations of the behavior of environmental systems (e.g., hydrologic systems, terrestrial ecosystems). Remotely sensed data normally form only one source of input to environmental models. GIS procedures permit the bringing together of models of numerous different processes, often using data from different sources, with different structures, formats, and levels of accuracy.

Environmental modeling using remote sensing and GIS is an active area of continuing research and development. Progress to date suggests that these combined technologies will play an increasingly important role in effective resource management, environmental risk assessment, and predicting and analyzing the impacts of global environmental change. This combination of technologies is at the heart of the EOS/ESE program (Section 6.18).

7.20 SCALE EFFECTS

One of the most important considerations in the use of remotely sensed data in environmental modeling is that of spatial scale. The definition of the term *scale* can be problematic when specialists from varying disciplines are involved in modeling activities. First, there is a distinction to be made between temporal scale and spatial scale. The former refers to the frequency with

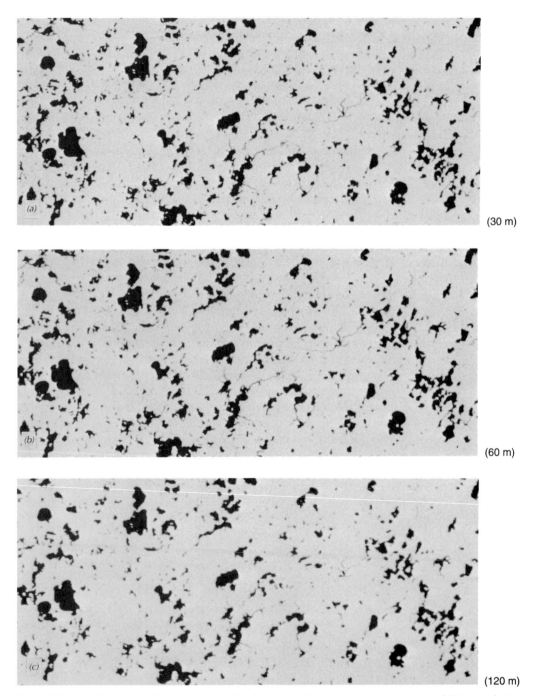

(30 m)

(60 m)

(120 m)

Figure 7.62 Land vs. water binary masks resulting from analyzing successive aggregations of 30-m-resolution Landsat TM (band 4) data. (Adapted from Benson and Mackenzie, 1995.)

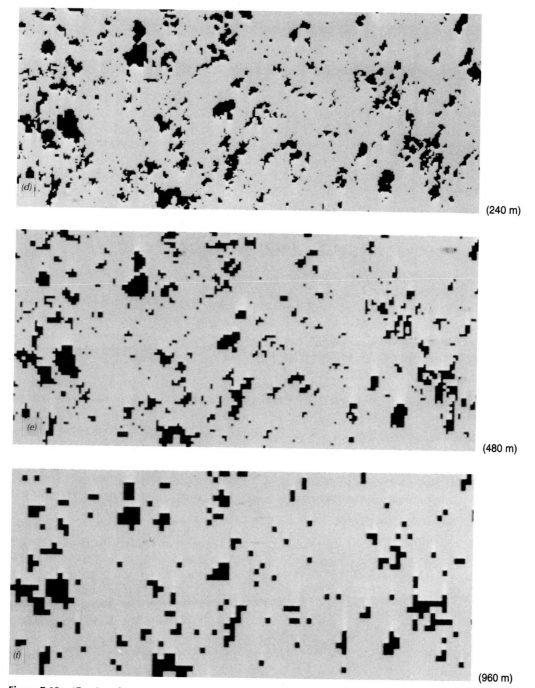

(240 m)

(480 m)

(960 m)

Figure 7.62 (Continued)

which a measurement or observation is made. The latter has various, often confusing, meanings to individuals from different backgrounds. For example, a remote sensing specialist uses the term scale to refer to the relationship between the size of a feature on a map or image to the corresponding dimensions on the ground. In contrast, an ecologist typically uses the term *spatial scale* to infer two characteristics of data collection: *grain*, the finest spatial resolution within which data are collected, and *extent*, the size of the study area. In essence, these terms are synonymous with what we have defined in this book as spatial resolution and area of coverage, respectively.

Another potential source of confusion in terminology associated with spatial scale is the varying use of the adjectives "small" and "large." Again, to the remote sensing specialist, small and large scale refer to the relative relationship between map and image dimensions and their counterparts on the earth's surface. Under this convention, small scale infers relatively coarser spatial portrayal of features than does the term large scale. The ecologist (or other scientist) will usually use these terms with the reverse meaning. That is, a small scale study in this context would relate to an analysis performed in relatively great spatial detail (usually over a small area). Similarly, a large scale analysis would encompass a large area (usually with less spatial detail).

Again, in this book *grain* and *spatial resolution* are used synonymously. What is important to recognize is that the appropriate resolution to be used in any particular biophysical modeling effort is both a function of the spatial structure of the environment under investigation and the kind of information desired. For example, the spatial resolution required in the study of an urban area is usually much different than that needed to study an agricultural area or the open ocean.

Figure 7.62 illustrates the effects of changing the spatial resolution (grain) with which an area in northeastern Wisconsin is recorded in the near infrared. This area contains numerous lakes of various sizes, and Figure 7.62 shows a series of binary masks produced to portray the lakes (dark areas) versus the land (light areas) in the scene. Part (*a*) was produced using data from the Landsat TM (band 4) acquired at a resolution of 30 m. The remaining parts of this figure were generated by simulating successively coarser grain sizes through the application of a pixel aggregation algorithm. This "majority rules" algorithm involved clumping progressively larger numbers of the 30-m pixels shown in (*a*) into progressively larger synthesized ground resolution cells. The grain sizes so developed were 60, 120, 240, 480, and 960 m, shown in (*b*) to (*f*), respectively.

After the above masks were produced, three basic landscape parameters were extracted from each: (1) the percentage of the scene covered by water, (2) the number of lakes included, and (3) the mean surface area of the lakes. Figure 7.63 includes graphs of the results of this data extraction process. Note how sensitive all of the landscape parameters are to changes in grain. As the ground resolution cell size increases from 30 to 960 m, the percentage of the

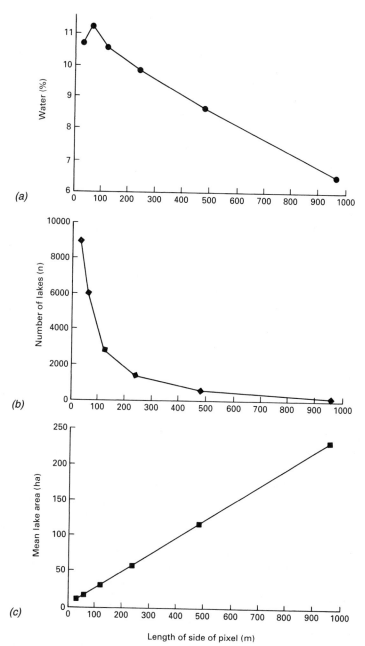

Figure 7.63 Influence of grain (spatial resolution) on measurement of selected landscape parameters for the scene included in Figure 7.62. Aggregation of 30-m-resolution Landsat TM pixels was used to simulate progressively larger ground resolution cell sizes. (*a*) Percentage of water. (*b*) Number of lakes. (*c*) Mean lake area. (Adapted from Benson and Mackenzie, 1995.)

scene masked into the water class at first increases slightly and then continually decreases. At the same time, the number of lakes decreases in a nonlinear fashion and the mean lake surface area increases nearly linearly.

The above example, involving the quantification of only three very basic landscape characteristics, illustrates the interrelationships among the spatial resolution of the sensor, the spatial structure of the environment under investigation, and the nature of the information sought in any given image processing operation. These three factors, as well as the specific techniques used to extract information from digital imagery, constantly interact. This must always be kept in mind as one selects the appropriate spatial resolution and analysis techniques for any given setting and application.

7.21 IMAGE TRANSMISSION AND COMPRESSION

The volume of digital data collected by currently operational earth resources satellites and airborne remote sensing systems is already extremely large. In the near future, with the launch of additional new satellites in the EOS series and other public-sector and commercial satellite programs, the volume of data collected will be immense. Storing, transmitting, and distributing these data will continue to represent formidable tasks even with continual increases in available computational power. For this reason, the subject of image compression is attracting a great deal of attention within the computer science and image processing communities. In particular, the goal of distributing useful digital image data over the Internet has sparked much research into methods for decomposing imagery into multiple components, compressing those components individually for transmission, and re-creating an approximate or exact version of the original image upon reception.

Traditional image transmission methods generally transfer the image data on a line-by-line basis, starting at the top of the image and working toward the bottom. If the image is large and the medium over which the image is being transmitted is slow (such as a modem-limited network connection), the user will have to wait a considerable time to get any information about the appearance of the majority of the image, although the top of the image will be provided in complete detail early in the transmission process. This is not necessarily the most useful arrangement. In many cases, it may be preferable to transmit a lower resolution version of the entire image first before proceeding to transmit the full-resolution data for any part of the image. *Progressive image transmission* refers to this general process of transmitting an image as a series of approximations, each increasingly close to the original image (Sayood, 1996).

For example, consider a case where a user is interested in examining a series of remotely sensed images that are stored in a remote archive, with the purpose of identifying one or more images that are largely cloud free over a particular study site. Using traditional image transmission techniques, the user will have no information about whether the bottom half of each image is cloud free until after the top half of the image has been received in full detail. Using a progressive image transmission technique, however, a lower resolution approximation of the entire image would be received quite quickly, with increasing levels of detail following. This may allow the user to readily establish that a particular image is largely covered by clouds and to move on to examining the next image rather than waiting for the full level of detail to be transmitted.

It should be noted that any method for improving image transmission efficiency is dependent upon both the sender and the receiver employing the same techniques for encoding and decoding the image data. This becomes particularly important when advanced strategies are used for improving transmission efficiency. One such advanced compression strategy makes use of *wavelet transforms* (Mallat, 1989), whereby the raw imagery is decomposed into a series of wavelets that are then independently compressed and transmitted. The receiver then recomposes the original image, starting with the coarsest levels of detail and progressing to finer levels of detail. If desired, the original image can be re-created exactly, but often a certain amount of loss of fine detail is tolerated for the sake of a dramatic decrease in the volume of data that must be transferred.

The wavelet transform is related to the Fourier transform (discussed in Section 7.5). In the case of the Fourier transform, the image was represented by a set of spatial frequencies, consisting of sine or cosine functions of varying wavelengths. In the case of wavelets, a "mother wavelet" is selected, and a series of wavelet functions are then derived from it by "dilation and translation," so as to fit the spatial patterns in each row of the image. The process is then repeated for each column of the image. In this context, "dilation" refers to the stretching or compression of the wavelet, so as to represent higher or lower spatial frequencies, while translation refers to the repositioning of the leading edge of the wavelet. Since each wavelet function is based on the same mother wavelet, the only pieces of information that need to be stored are the dilation and translation coefficients for each wavelet function. For the very high frequency components of the image, the wavelet coefficients may be discarded, resulting in a relatively small loss in detail in exchange for a potentially substantial decrease in the volume of data to be transmitted.

Numerous wavelet functions have been described, of which those by Daubechies (1992) are among the most widely used for image compression. In general, these wavelet transforms will result in image qualities that are comparable to that which can be obtained by other compression techniques that require two to four times as much computational effort. It should also be noted that there may be useful applications of wavelet transforms in areas other than image compression and transmission. For example, several studies have investigated the use of wavelet transforms for reducing the high frequency "speckle" that is present in radar images (Dong et al., 1998; Fukuda and Hirosawa, 1998).

7.22 CONCLUSION

In concluding this discussion, we must reiterate that this chapter has only been a general introduction to digital image processing. This subject is extremely broad, and the procedures we have discussed are only a representative sample of the types of image processing operations useful in remote sensing. The scope of digital image processing and its application in spatial analysis in general are virtually unlimited.

SELECTED BIBLIOGRAPHY

Abeyta, A.M., and J. Franklin, "The Accuracy of Vegetation Stand Boundaries Derived from Image Segmentation in a Desert Environment," *Photogrammetric Engineering and Remote Sensing*, vol. 64, no. 1, 1998, pp. 59–66.

Afek, Y., and A. Brand, "Mosaicking of Orthorectified Aerial Images," *Photogrammetric Engineering and Remote Sensing*, vol. 64, no. 2, 1998, pp. 115–125.

American Society of Photogrammetry (ASP), *Manual of Remote Sensing*, 2nd ed., ASP, Falls Church, VA, 1983.

American Society for Photogrammetry and Remote Sensing (ASPRS), *GAP Analysis Program Proceedings*, ASPRS, Bethesda, MD, 1996.

American Society for Photogrammetry and Remote Sensing (ASPRS), *Earth Observing Platforms and Sensors*, Manual of Remote Sensing, 3rd ed. (A Series), CD-ROM, Version 1.1, ASPRS, Bethesda, MD, 1997.

American Society for Photogrammetry and Remote Sensing (ASPRS), *Accuracy Assessment of Remote Sensing-Derived Change Detection*, ASPRS, Bethesda, MD, 1999.

American Society for Photogrammetry and Remote Sensing (ASPRS), *Digital Elevation Model Technologies and Applications: The DEM Users Manual*, ASPRS, Bethesda, MD, 2001.

Ammenberg, P., et al., "Bio-Optical Modelling Combined with Remote Sensing to Assess Water Quality," *International Journal of Remote Sensing*, vol. 23, no. 8, 2002, pp. 1621–1638.

Arlinghaus, S.L., and D.A. Griffith, *Practical Handbook of Spatial Statistics*, Lewis Publishers, Boca Raton, FL, 1996.

Ashton, E.A., and A. Schaum, "Algorithms for the Detection of Sub-Pixel Targets in Multispectral Imagery," *Photogrammetric Engineering and Remote Sensing*, vol. 64, no. 7, 1998, pp. 723–731.

Atkinson, P.M., and N.J. Tate, eds., *Advances in Remote Sensing and GIS Analysis*, Wiley, Hoboken, N.J., 2000.

Atkinson, P.M., and A.R.L. Tatnall, "Neural Networks in Remote Sensing," *International Journal of Remote Sensing*, vol. 18, no. 4, 1997, pp. 699–709.

Avard, M.M., F.R. Schiebe, and J.H. Everitt, "Quantification of Chlorophyll in Reservoirs of the Little Washita River Watershed Using Airborne Video," *Photogrammetric Engineering and Remote Sensing*, vol. 66, no. 2, 2000, pp. 213–218.

Avery, T.E., and G.L. Berlin, *Fundamentals of Remote Sensing and Airphoto Interpretation*, 6th ed., Prentice Hall, Upper Saddle River, NJ, 2004.

Baber, M.L., D. Wood, and R.A. McBride, "Use of Digital Image Analysis and GIS to Assess Regional Soil Compaction Risk," *Photogrammetric Engineering and Remote Sensing*, vol. 62, no. 12, 1996, pp. 1397–1404.

Badhwar, G.D., "Classification of Corn and Soybeans Using Multitemporal Thematic Mapper Data," *Remote Sensing of Environment*, vol. 16, 1985, pp. 175–181.

Barandela, R., and M. Juarez, "Supervised Classification of Remotely Sensed Data with Ongoing Learning

Capability," *International Journal of Remote Sensing*, vol. 23, no. 22, 2002, pp. 4965–4970.

Bardossy, A., and L. Samaniego, "Fuzzy Rule-Based Classification of Remotely Sensed Imagery," *IEEE Transactions on Geoscience and Remote Sensing*, vol. 40, no. 2, 2002, pp. 362–374.

Barnsley, J.J., and S.L. Barr, "Inferring Urban Land Use from Satellite Sensor Images Using Kernel-Based Spatial Reclassification," *Photogrammetric Engineering and Remote Sensing*, vol. 62, no. 8, 1996, pp. 949–958.

Batson, R.M., K. Edwards, and E.M. Eliason, "Synthetic Stereo and Landsat Pictures," *Photogrammetric Engineering and Remote Sensing*, vol. 44, no. 4, 1978, pp. 503–505.

Bauer, M.E., "Spectral Inputs to Crop Identification and Condition Assessment," *Proceedings of the IEEE*, vol. 73, no. 6, 1985, pp. 1071–1085.

Bauer, M.E., et al., "Satellite Inventory of Minnesota Forest Resources," *Photogrammetric Engineering and Remote Sensing*, vol. 60, no. 3, 1994, pp. 287–298.

Baxes, G.A., *Digital Image Processing*, Prentice-Hall, Englewood Cliffs, NJ, 1984.

Beaudemin, M., and K.B. Fung, "On Statistical Band Selection for Image Visualization," *Photogrammetric Engineering and Remote Sensing*, vol. 67, no. 5, 2001, pp. 571–574.

Benson, B.J., and M.D. MacKenzie, "Effects of Sensor Spatial Resolution on Landscape Structure Parameters," *Landscape Ecology*, vol. 10, no. 2, 1995, pp. 113–120.

Bischof, H., and A. Leonardis, "Finding Optimal Neural Networks for Land Use Classification," *IEEE Transactions on Geoscience and Remote Sensing*, vol. 36, no. 1, 1998, pp. 337–341.

Bolstad, P. V., and T. M. Lillesand, "Rapid Maximum Likelihood Classification," *Photogrammetric Engineering and Remote Sensing*, vol. 57, no. 1, 1991, pp. 67–74.

Borel, C.C., and S.A.W. Gerstl, "Nonlinear Spectral Mixing Models for Vegetative and Soil Surfaces," *Remote Sensing of Environment*, vol. 47, 1994, pp. 403–416.

Bouman, B.A.M., "Accuracy of Estimating the Leaf Area Index from Vegetation Indices Derived from Crop Reflectance Characteristics, a Simulation Study," *International Journal of Remote Sensing*, vol. 13, no. 16, 1992, pp. 3069–3084.

Bronge, L.B., and B. Näsland-Landenmark, "Wetland Classification for Swedish CORINE Land Cover Adopting a Semi-Automatic Interactive Approach," *Canadian Journal of Remote Sensing*, vol. 28, no. 2, 2002, pp. 139–155.

Brown, M., H.G. Lewis, and S.R. Gunn, "Linear Spectral Mixture Models and Support Vector Machines for Remote Sensing," *IEEE Transactions on Geoscience and Remote Sensing*, vol. 38, no. 5, 2000, pp. 2346–2360.

Bruzzone, L., et al., "Multisource Classification of Complex Rural Areas by Statistical and Neural-Network Approaches," *Photogrammetric Engineering and Remote Sensing*, vol. 63, no. 5, 1997, pp. 523–533.

Bruzzone, L., and D.F. Prieto, "Automatic Analysis of the Difference Image for Unsupervised Change Detection," *IEEE Transactions on Geoscience and Remote Sensing*, vol. 38, no. 3, 2000, pp. 1171–1182.

Cairns, S.H., K.L. Dickson, and S.F. Atkinson, "An Examination of Measuring Selected Water Quality Trophic Indicators with SPOT Satellite HRV Data," *Photogrammetric Engineering and Remote Sensing*, vol. 63, no. 3, 1997, pp. 263–265.

Campbell, J.B., *Introduction to Remote Sensing*, 3rd ed., Guilford, New York, 2002.

Canadian Journal of Remote Sensing, Special Issue on Remote Sensing in BOREAS, vol. 23, no. 2, 1997.

Canters, F., "Evaluating the Uncertainty of Area Estimates Derived from Fuzzy Land-Cover Classification," *Photogrammetric Engineering and Remote Sensing*, vol. 63, no. 4, 1997, pp. 403–414.

Carbone, G.J., S. Narumalani, and M. King, "Application of Remote Sensing and GIS Technologies with Physiological Crop Models," *Photogrammetric Engineering and Remote Sensing*, vol. 62, no. 2, 1996, pp. 171–179.

Carlotto, M.J., "Spectral Shape Classification of Landsat Thematic Mapper Imagery," *Photogrammetric Engineering and Remote Sensing*, vol. 64, no. 9, 1998, pp. 905–913.

Carlson, T.N., and G.A. Sanchez-Azofeifa, "Satellite Remote Sensing of Land Use Changes In and Around San Jose, Costa Rica," *Remote Sensing of Environment*, vol. 70, no. 3, 1999, pp. 247–256.

Carmel, Y., et al., "Combining Location and Classification Error Sources for Estimating Multi-Temporal Database Accuracy," *Photogrammetric Engineering and Remote Sensing*, vol. 67, no. 7, 2001, pp. 865–872.

Carpenter, G.A., et al., "A Neural Network Method for Efficient Vegetation Mapping," *Remote Sensing of Environment*, vol. 70, no. 3, 1999, pp. 326–338.

Chalifoux, S., F. Cavayas, and J.T. Gray, "Map-Guided Approach for the Automatic Detection on Landsat TM Images of Forest Stands Damaged by the Spruce Budworm," *Photogrammetric Engineering and Remote Sensing*, vol. 64, no. 6, 1998, pp. 629–635.

Chavez, P.S., Jr., "Image-Based Atmospheric Corrections Revisited and Improved," *Photogrammetric Engineering and Remote Sensing*, vol. 62, no. 9, 1996, pp. 1025–1036.

Chavez, P.S., Jr., G.L. Berlin, and L.B. Sowers, "Statistical Method for Selecting Landsat MSS Ratios," *Journal of Applied Photographic Engineering*, no. 8, 1982, pp. 23–30.

Chavez, P.S., Jr., S.C. Sides, and J.A. Anderson, "Comparison of Three Different Methods to Merge Multiresolution and Multispectral Data: Landsat TM and SPOT Panchromatic," *Photogrammetric Engineering and Remote Sensing*, vol. 57, no. 3, 1991, pp. 295–303.

Chen, D.M., and D. Stow, "The Effect of Training Strategies on Supervised Classification at Different Spatial Resolutions," *Photogrammetric Engineering and Remote Sensing*, vol. 68, no. 11, 2002, pp. 1155–1162.

Cheng, T.D., *Canopy Biomass Measurement of Individual Agricultural Fields with Landsat Imagery*, Ph.D. Thesis, Department of Forest Science, Texas A&M University, College Station, TX, 1985.

Cheng, T., "Fuzzy Objects: Their Changes and Uncertainties," *Photogrammetric Engineering and Remote Sensing*, vol. 68, no. 1, 2002, pp. 41–49.

Chibani, Y., and A. Houacine, "The Joint Use of HIS Transform and Redundant Wavelet Decomposition for Fusing Multispectral and Panchromatic Images," *International Journal of Remote Sensing*, vol. 23, no. 18, 2002, pp. 3821–3833.

Chipman, J.W., et al., "Mapping Lake Water Clarity with Landsat Images in Wisconsin, USA," *Canadian Journal of Remote Sensing*, vol. 30, no. 1, 2004, in press.

Clark, R.N., and G.A. Swayze, "Mapping Minerals, Amorphous Materials, Environmental Materials, Vegetation, Water, Ice, and Snow, and Other Materials: The USGS Tricorder Algorithm," *Summaries of the Fifth Annual JPL Airborne Earth Science Workshop*, JPL Publication 95-1, Jet Propulsion Laboratory, Pasadena, CA, 1995, pp. 39–40.

Collins, J.B., and C.E. Woodcock, "An Assessment of Several Linear Change Detection Techniques for Mapping Forest Mortality Using Multitemporal Landsat TM Data," *Remote Sensing of Environment*, vol. 56, no. 1, 1996, pp. 66–77.

Congalton, R.G., and K. Green, *Assessing the Accuracy of Remotely Sensed Data: Principles and Practices*, Lewis Publishers, Boca Raton, FL, 1999.

Conrac Division, *Raster Graphics Handbook*, Conrac Corp., Covina, CA, 1980.

Cook, A.E., and J.E. Pinder, "Relative Accuracy of Rectifications Using Coordinates Determined from Maps and the Global Positioning System," *Photogrammetric Engineering and Remote Sensing*, vol. 62, no. 1, 1996, pp. 73–77.

Coppin, P., et al., "Operational Monitoring of Green Biomass Change for Forest Management," *Photogrammetric Engineering and Remote Sensing*, vol. 67, no. 5, 2001, pp. 603–611.

Cortijo, F.J., and N.P. De La Blanca, "Improving Classical Contextual Classifications," *International Journal of Remote Sensing*, vol. 19, no. 8, 1998, pp. 1591–1613.

Cosentino, B.L., *Satellite Remote Sensing Techniques in Support of Natural Resource Monitoring: A View Towards Statewide Land Cover Mapping*, M.S. Thesis, University of Wisconsin–Madison, 1992.

Couteron, P., "Quantifying Change in Patterned Semi-Arid Vegetation by Fourier Analysis of Digitized Aerial Photographs," *International Journal of Remote Sensing*, vol. 23, no. 17, 2002, pp. 3407–3425.

Crist, E.P., and R.C. Cicone, "Application of the Tasseled Cap Concept to Simulated Thematic Mapper Data," *Photogrammetric Engineering and Remote Sensing*, vol. 50, no. 3, 1984, pp. 343–352.

Dai, X., "The Effects of Image Misregistration on the Accuracy of Remotely Sensed Change Detection," *IEEE Transactions on Geoscience and Remote Sensing*, vol. 36, no. 5, 1998, pp. 1566–1577.

Dai, X.L., and S. Khorram, "Remotely Sensed Change Detection Based on Artificial Neural Network," *Photogrammetric Engineering and Remote Sensing*, vol. 65, no. 10, 1999, pp. 1187–1194.

Daubechies, I., *Ten Lectures on Wavelets*, Society for Industrial and Applied Mathematics, Philadelphia, 1992.

Debeir, O., et al., "Textural and Contextual Land-Cover Classification Using Single and Multiple Classifier Systems," *Photogrammetric Engineering and Remote Sensing*, vol. 68, no. 6, 2002, pp. 597–605.

Dechka, J.A., et al., "Classification of Wetland Habitat and Vegetation Communities Using Multi-temporal

Ikonos Imagery in Southern Saskatchewan," *Canadian Journal of Remote Sensing*, vol. 28, no. 5, 2002, pp. 679–685.

Dekker, A.G., R.J. Vos, and S.W.M. Peters, "Analytical Algorithms for Lake Water TSM Estimation for Retrospective Analyses of TM and SPOT Sensor Data," *International Journal of Remote Sensing*, vol. 23, no. 1, 2002, pp. 15–35.

Demetriades-Shah, T.H., M.D. Steven, and J.A. Clark, "High Resolution Derivatives Spectra in Remote Sensing," *Remote Sensing of Environment*, vol. 33, 1990, pp. 55–64.

Deppe, F., "Forest Area Estimation Using Sample Surveys and Landsat MSS and TM Data," *Photogrammetric Engineering and Remote Sensing*, vol. 64, no. 4, 1998, pp. 285–292.

de Vasconcelos, M.J.P., et al., "Spatial Prediction of Fire Ignition Probabilities: Comparing Logistic Regression and Neural Networks," *Photogrammetric Engineering and Remote Sensing*, vol. 67, no. 1, 2001, pp. 73–81.

Dhakal, A.S., et al., "Detection of Areas Associated with Flood and Erosion Caused by a Heavy Rainfall Using Multitemporal Landsat TM Data," *Photogrammetric Engineering and Remote Sensing*, vol. 68, no. 3, 2002, pp. 233–239.

Dikshit, O., and D.P. Roy, "Empirical Investigation of Image Resampling Effects upon the Spectral and Textural Supervised Classification of a High Spatial Resolution Multispectral Image," *Photogrammetric Engineering and Remote Sensing*, vol. 62, no. 9, 1996, pp. 1085–1092.

Dong, Y., et al., "Speckle Suppression Using Recursive Wavelet Transforms," *International Journal of Remote Sensing*, vol. 19, no. 2, 1998, pp. 317–330.

Drake, N.A., S. Mackin, and J.J. Settle, "Mapping Vegetation, Soils, and Geology in Semiarid Shrublands Using Spectral Matching and Mixture Modeling of SWIR AIRIS Imagery," *Remote Sensing of Environment*, vol. 68, no. 1, 1999, pp. 12–25.

Duda, T., and M. Canty, "Unsupervised Classification of Satellite Imagery: Choosing a Good Algorithm," *International Journal of Remote Sensing*, vol. 23, no. 11, 2002, pp. 2193–2212.

Duhaime, R.J., P.V. August, and W.R. Wright, "Automated Vegetation Mapping Using Digital Orthophotography," *Photogrammetric Engineering and Remote Sensing*, vol. 63, no. 11, 1997, pp. 1295–1302.

Dymond, C.C., D.J. Mladenoff, and V. Radeloff, "Phenological Differences in Tasseled Cap Indices Improve Deciduous Forest Classification," *Remote Sensing of Environment*, vol. 80, no. 3, 2002, pp. 460–472.

Ediriwickrema, J., and S. Khorram, "Hierarchical Maximum-Likelihood Classification for Improved Accuracies," *IEEE Transactions on Geoscience and Remote Sensing*, vol. 35, no. 4, 1997, pp. 810–816.

Edwards, T.C., G.G. Moisen, and D.R. Cutler, "Assessing Map Accuracy in a Remotely Sensed, Ecoregion-Scale Cover Map," *Remote Sensing of Environment*, vol. 63, no. 1, 1998, pp. 73–83.

Ekstrand, S., "Landsat TM-Based Forest Damage Assessment: Correction for Topographic Effects," *Photogrammetric Engineering and Remote Sensing*, vol. 62, no. 2, 1996, pp. 151–161.

El-Manadili, Y., and K. Novak, "Precision Rectification of SPOT Imagery Using the Direct Linear Transformation Model," *Photogrammetric Engineering and Remote Sensing*, vol. 62, no. 1, 1996, pp. 67–72.

Elmore, A.J., et al., "Quantifying Vegetation Change in Semiarid Environments: Precision and Accuracy of Spectral Mixture Analysis and the Normalized Difference Vegetation Index," *Remote Sensing of Environment*, vol. 73, no. 1, 2000, pp. 87–102.

Evans, C., et al., "Segmenting Multispectral Landsat TM Images into Field Units," *IEEE Transactions on Geoscience and Remote Sensing*, vol. 40, no. 5, 2002, pp. 1054–1064.

Fardanesh, M.T., and O.K. Ersoy, "Classification Accuracy Improvement of Neural Network Classifiers by Using Unlabeled Data," *IEEE Transactions on Geoscience and Remote Sensing*, vol. 36, no. 3, 1998, pp. 1020–1025.

Fassnacht, K.S., et al., "Estimating the Leaf Area Index of North Central Wisconsin Forests Using the Landsat Thematic Mapper," *Remote Sensing of Environment*, vol. 61, no. 2, 1997, pp. 229–245.

Ferro, C.J.S., and T.A. Warner, "Scale and Texture in Digital Image Classification," *Photogrammetric Engineering and Remote Sensing*, vol. 68, no. 1, 2002, pp. 51–63.

Fonseca, L.M.G., and B.S. Manjunath, "Registration Techniques for Multisensor Remotely Sensed Imagery," *Photogrammetric Engineering and Remote Sensing*, vol. 62, no. 9, 1996, pp. 1049–1056.

Foody, G.M., "Relating the Land-Cover Classification of Mixed Pixels to Artificial Neural Network Classification Output," *Photogrammetric Engineering and Remote Sensing*, vol. 62, no. 5, 1996a, pp. 491–499.

Foody, G.M., "Approaches for the Production and Evaluation of Fuzzy Land Cover Classifications from Remotely-Sensed Data," *International Journal of Remote Sensing*, vol. 17, no. 7, 1996b, pp. 1317–1340.

Foody, G.M., "The Continuum of Classification Fuzziness in Thematic Mapping," *Photogrammetric Engineering and Remote Sensing*, vol. 65, no. 4, 1999, pp. 443–451.

Foody, G.M., "Monitoring the Magnitude of Land-Cover Change Around the Southern Limits of the Sahara," *Photogrammetric Engineering and Remote Sensing*, vol. 67, no. 7, 2001, pp. 841–847.

Foody, G.M., M.B. McCulloch, and W.B. Yates, "The Effect of Training Set Size and Composition on Neural Network Classification," *International Journal of Remote Sensing*, vol. 16, no. 9, 1995, pp. 1707–1723.

Foschi, P.G., and D.K. Smith, "Detecting Subpixel Woody Vegetation in Digital Imagery Using Two Artificial Intelligence Approaches," *Photogrammetric Engineering and Remote Sensing*, vol. 63, no. 5, 1997, pp. 493–500.

Franklin, S.E., et al., "An Integrated Decision Tree Approach (IDTA) to Mapping Landcover Using Satellite Remote Sensing in Support of Grizzly Bear Habitat Analysis in the Alberta Yellowhead Ecosystem," *Canadian Journal of Remote Sensing*, vol. 27, no. 6, 2001a, pp. 579–592.

Franklin, S.E., A.J. Maudie, and M.B. Lavigne, "Using Spatial Co-Occurrence Texture to Increase Forest Structure and Species Composition Classification Accuracy," *Photogrammetric Engineering and Remote Sensing*, vol. 67, no. 7, 2001b, pp. 849–855.

Franklin, S.E., et al., "Large-Area Forest Structure Change Detection: An Example," *Canadian Journal of Remote Sensing*, vol. 28, no. 4, 2002a, pp. 588–592.

Franklin, S.E., et al., "Evidential Reasoning with Landsat TM, DEM, and GIS Data for Landcover Classification in Support of Grizzly Bear Habitat Mapping," *International Journal of Remote Sensing*, vol. 23, no. 21, 2002b, pp. 4633–4652.

Friedl, M.A., and C.E. Brodley, "Decision-Tree Classification of Land Cover from Remotely Sensed Data," *Remote Sensing of Environment*, vol. 61, no. 3, 1997, pp. 399–409.

Frizzelle, B.G., and A. Moody, "Mapping Continuous Distributions of Land Cover: A Comparison of Maximum-Likelihood Estimation and Artificial Neural Networks," *Photogrammetric Engineering and Remote Sensing*, vol. 67, no. 6, 2001, pp. 693–705.

Frohn, R.C., *Remote Sensing for Landscape Ecology*, Lewis Publishers, Boca Raton, FL, 1998.

Fukuda, S., and H. Hirosawa, "Suppression of Speckle in Synthetic Aperture Radar Images Using Wavelets," *International Journal of Remote Sensing*, vol. 19, no. 2, 1998, pp. 507–519.

Furby, S.L., and N.A. Campbell, "Calibrating Images from Different Dates to 'Like-Value' Digital Counts," *Remote Sensing of Environment*, vol. 77, no. 2, 2001, pp. 186–196.

Gahegan, M., and J. Flack, "A Model to Support the Integration of Image Understanding Techniques within a GIS," *Photogrammetric Engineering and Remote Sensing*, vol. 62, no. 5, 1996, pp. 483–490.

Gao, B., and Goetz, A.F.H., "Column Atmospheric Water Vapor and Vegetation Liquid Water Retrievals from Airborne Imaging Spectrometer Data," *Journal of Geophysical Research*, vol. 95, no. D4, 1990, pp. 3549–3564.

Gerylo, G.R., et al., "Empirical Relations Between Landsat TM Spectral Response and Forest Stands Near Fort Simpson, Northwest Territories, Canada," *Canadian Journal of Remote Sensing*, vol. 28, no. 1, 2002, pp. 67–79.

Gibson, P., and C. Power, *Introductory Remote Sensing: Digital Image Processing and Applications*, Taylor & Francis, New York, 2000.

Gilabert, M.A., et al., "A Generalized Soil-Adjusted Vegetation Index," *Remote Sensing of Environment*, vol. 82, nos. 2–3, 2002, pp. 303–310.

Gitelson, A.A., et al., "Novel Algorithms for Remote Estimation of Vegetation Fraction," *Remote Sensing of Environment*, vol. 80, no. 1, 2002a, pp. 76–87.

Gitelson, A.A., et al., "Vegetation and Soil Lines in Visible Spectral Space: A Concept and Technique for Remote Estimation of Vegetation Fraction," *International Journal of Remote Sensing*, vol. 23, no. 13, 2002b, pp. 2537–2562.

Gobron, N., et al., "Advanced Vegetation Indices Optimized for Up-Coming Sensors: Design, Performance, and Applications," *IEEE Transactions on Geoscience and Remote Sensing*, vol. 38, no. 6, 2000, pp. 2489–2505.

Gomez, O., and E. Medina, "The Vegetation Distribution Patterns from a Landsat-4 TM Image Along the San Juan River (Monagas State), Venezuela," *International Journal of Remote Sensing*, vol. 19, no. 6, 1998, pp. 1017–1020.

Gong, P., "Integrated Analysis of Spatial Data from Multiple Sources: Using Evidential Reasoning and Artificial Neural Network Techniques for Geological Mapping," *Photogrammetric Engineering and Remote Sensing*, vol. 62, no. 5, 1996, pp. 513–523.

Gopal, S., C.E. Woodcock, and A.H. Strahler, "Fuzzy Neural Network Classification of Global Land Cover from a 1° AVHRR Data Set," *Remote Sensing of Environment*, vol. 67, no. 2, 1999, pp. 230–243.

Goward, S.N., et al., "Normalized Difference Vegetation Index Measurements from the Advanced Very High Resolution Radiometer," *Remote Sensing of Environment*, vol. 35, 1991, pp. 257–277.

Green, R., "Retrieval of Reflectance from Calibrated Radiance Imagery Measured by the Airborne Visible/Infrared Imaging Spectrometer (AVIRIS) for Lithological Mapping of the Clark Mountains, California," *Proceedings of the Second Airborne Visible/Infrared Imaging Spectrometer (AVIRIS) Workshop*, Jet Propulsion Laboratory, Pasadena, CA, 1990, pp. 167–175.

Gross, H.N., and J.R. Schott, "Application of Spectral Mixture Analysis and Image Fusion Techniques for Image Sharpening," *Remote Sensing of Environment*, vol. 63, no. 2, 1998, pp. 85–94.

Guch, R., "Urban Growth Detection Using Texture Analysis on Merged Landsat TM and SPOT-P Data," *Photogrammetric Engineering and Remote Sensing*, vol. 68, no. 12, 2002, pp. 1283–1288.

Guindon, B., "Computer-Based Aerial Image Understanding: A Review and Assessment of Its Application to Planimetric Information Extraction from Very High Resolution Satellite Images," *Canadian Journal of Remote Sensing*, vol. 23, no. 1, 1997, pp. 38–47.

Guindon, B., "A Framework for the Development and Assessment of Object Recognition Modules from High-Resolution Satellite Images," *Canadian Journal of Remote Sensing*, vol. 26, no. 4, 2000, pp. 334–348.

Guindon, B., "Application of Perceptual Grouping Concepts to the Recognition of Residential Buildings in High Resolution Satellite Images," *Canadian Journal of Remote Sensing*, vol. 27, no. 3, 2001, pp. 264–275.

Guindon, B., and C.M. Edmonds, "Large-Area Land-Cover Mapping Through Scene-Based Classification Compositing," *Photogrammetric Engineering and Remote Sensing*, vol. 68, no. 6, 2002, pp. 589–596.

Hall, F.G., et al., "Radiometric Rectification: Toward a Common Radiometric Response Among Multidate, Multisensor Images," *Remote Sensing of Environment*, vol. 35, no. 1, 1991, pp. 11–27.

Hardin, P.J., and J.M. Shumway, "Statistical Significance and Normalized Confusion Matrices," *Photogrammetric Engineering and Remote Sensing*, vol. 63, no. 6, 1997, pp. 735–740.

Harris, J.R., et al., "Mapping Altered Rocks Using Landsat TM and Lithogeochemical Data: Sulphurets-Brucejack Lake District, British Columbia, Canada," *Photogrammetric Engineering and Remote Sensing*, vol. 64, no. 4, 1998, pp. 309–322.

Hayes, D.J., and S.A. Sader, "Comparison of Change-Detection Techniques for Monitoring Tropical Forest Clearing and Vegetation Regrowth in a Time Series," *Photogrammetric Engineering and Remote Sensing*, vol. 67, no. 9, 2001, pp. 1067–1075.

Hazel, G.G., "Object-Level Change Detection in Spectral Imagery," *IEEE Transactions on Geoscience and Remote Sensing*, vol. 39, no. 3, 2001, pp. 553–561.

Hepinstall, J.A., and S.A. Sader, "Using Bayesian Statistics, Thematic Mapper Satellite Imagery, and Breeding Bird Survey Data to Model Bird Species Probability of Occurrence in Maine," *Photogrammetric Engineering and Remote Sensing*, vol. 63, no. 10, 1997, pp. 1231–1237.

Heuvelink, G., *Error Propagation in Environmental Modelling with GIS*, Taylor & Francis, New York, 1998.

Hill, J., and D. Peter, eds., *The Use of Remote Sensing for Land Degradation and Desertification Monitoring in the Mediterranean Basin*, European Commission, Luxembourg, 1996.

Hobbs, R.J., and H.A. Mooney eds., *Remote Sensing of Biosphere Functioning*, Springer, New York, 1990.

Hsieh, P.F., L.C. Lee, and N. Chen, "Effect of Spatial Resolution on Classification Errors of Pure and Mixed Pixels in Remote Sensing," *IEEE Transactions on Geoscience and Remote Sensing*, vol. 39, no. 12, 2001, pp. 2657–2663.

Hu, B.X., et al., "Surface Albedos and Angle-Corrected NDVI from AVHRR Observations of South America," *Remote Sensing of Environment*, vol. 71, no. 2, 2000, pp. 119–132.

Huang, C., et al., "Derivation of a Tasselled Cap Transformation Based on Landsat 7 At-Satellite Reflectance," *International Journal of Remote Sensing*, vol. 23, no. 8, 2002, pp. 1741–1748.

Huang, K.Y., "A Synergistic Automatic Clustering Technique (SYNERACT) for Multispectral Image Analysis," *Photogrammetric Engineering and Remote Sensing*, vol. 68, no. 1, 2002, pp. 33–40.

Huang, X., and J.R. Jensen, "A Machine-Learning Approach to Automated Knowledge-Based Building for Remote Sensing Image Analysis with GIS Data," *Photogrammetric Engineering and Remote Sensing*, vol. 63, no. 10, 1997, pp. 1185–1194.

Hudak, A.T., and C.A. Wessman, "Textural Analysis of Historical Aerial Photography to Characterize Woody Plant Encroachment in South African Savanna," *Remote Sensing of Environment*, vol. 66, no. 3, 1998, pp. 317–330.

Huguenin, R.L., et al., "Subpixel Classification of Bald Cypress and Tupelo Gum Trees in Thematic Mapper Imagery," *Photogrammetric Engineering and Remote Sensing*, vol. 63, no. 6, 1997, pp. 717–725.

Hung, M.C., and M.K. Ridd, "A Subpixel Classifier for Urban Land-Cover Mapping Based on a Maximum-Likelihood Approach and Expert System Rules," *Photogrammetric Engineering and Remote Sensing*, vol. 68, no. 11, 2002, pp. 1173–1180.

International Journal of Remote Sensing, Special Issue on Neural Networks in Remote Sensing, vol. 18, no. 4, 1997.

Jacobsen, A., et al., "Spectral Identification of Plant Communities for Mapping of Semi-Natural Grasslands," *Canadian Journal of Remote Sensing*, vol. 26, no. 5, 2000, pp. 370–383.

Jahne, B., *Digital Image Processing: Concepts, Algorithms and Scientific Applications*, Springer, New York, 1991.

Jain, A.K., *Fundamentals of Digital Image Processing*, Prentice Hall, Upper Saddle River, NJ, 1989.

Jensen, J.R., *Introductory Digital Image Processing: A Remote Sensing Perspective*, 2nd ed., Prentice Hall, Upper Saddle River, NJ, 1996.

Jensen, J.R., *Remote Sensing of the Environment: An Earth Resource Perspective*, Prentice Hall, Upper Saddle River, NJ, 2000.

Jia, X.P., and J.A. Richards, "Progressive Two-Class Decision Classifier for Optimization of Class Discriminations," *Remote Sensing of Environment*, vol. 63, no. 3, 1998, pp. 289–297.

Jia, X., and J.A. Richards, "Segmented Principal Components Transformation for Efficient Hyperspectral Remote-Sensing Image Display and Classification," *IEEE Transactions on Geoscience and Remote Sensing*, vol. 37, no. 1, 1999, pp. 538–542.

Johnson, R.D., and E.S., Kasischke, "Change Vector Analysis: A Technique for the Multispectral Monitoring of Land Cover and Condition," *International Journal of Remote Sensing*, vol. 19, no. 3, 1998, pp. 411–426.

Joria, P.E., and J.C. Jorgenson, "Comparison of Three Methods for Mapping Tundra with Landsat Digital Data," *Photogrammetric Engineering and Remote Sensing*, vol. 62, no. 2, 1996, pp. 163–169.

Kalkhan, M.A., R.M. Reich, and R.L. Czaplewski, "Variance Estimates and Confidence Intervals for the Kappa Measure of Classification Accuracy," *Canadian Journal of Remote Sensing*, vol. 23, no. 3, 1997, pp. 210–216.

Kandun, P., et al., "Influence of Flood Conditions and Vegetation Status on the Radar Backscatter of Wetland Ecosystems," *Canadian Journal of Remote Sensing*, vol. 27, no. 6, 2002, pp. 651–662.

Kardoulas, N.G., A.C. Bird, and A.L. Lawan, "Geometric Correction of SPOT and Landsat Imagery: A Comparison of Map- and GPS-Derived Control Points," *Photogrammetric Engineering and Remote Sensing*, vol. 62, no. 10, 1996, pp. 1173–1177.

Kauth, R.J., and G.S. Thomas, "The Tasselled Cap—A Graphic Description of Spectral-Temporal Development of Agricultural Crops as Seen by Landsat," *Proceedings: 2nd International Symposium on Machine Processing of Remotely Sensed Data*, Purdue University, West Lafayette, IN, 1976.

Kavzoglu, T., and P.M. Mather, "The Role of Feature Selection in Artificial Neural Network Applications," *International Journal of Remote Sensing*, vol. 23, no. 15, 2002, pp. 2919–2937.

Kerekes, J.P., and J.E. Baum, "Spectral Imaging System Analytical Model for Subpixel Object Detection," *IEEE Transactions on Geoscience and Remote Sensing*, vol. 40, no. 5, 2002, pp. 1088–1101.

Kiema, J.B.K., "Texture Analysis and Data Fusion in the Extraction of Topographic Objects from Satellite Imagery," *International Journal of Remote Sensing*, vol. 23, no. 4, 2002, pp. 767–776.

Kimes, D.S., et al., "Attributes of Neural Networks for Extracting Continuous Vegetation Variables from Optical and Radar Measurements," *International Journal of Remote Sensing*, vol. 19, no. 14, 1998, pp. 2639–2663.

Kloiber, S.M., et al., "A Procedure for Regional Lake Water Clarity Assessment Using Landsat Multispectral Data," *Remote Sensing of Environment*, vol. 82, no. 1, 2002, pp. 38–47.

Kloditz, C., et al., "Estimating the Accuracy of Coarse Scale Classification Using High Scale Information,"

Photogrammetric Engineering and Remote Sensing, vol. 64, no. 2, 1998, pp. 127–133.

Knick, S.T., J.R. Rotenberry, and T.J. Zarriello, "Supervised Classification of Landsat Thematic Mapper Imagery in a Semi-Arid Rangeland by Nonparametric Discriminant Analysis," *Photogrammetric Engineering and Remote Sensing,* vol. 63, no. 1, 1997, pp. 79–86.

Koch, M., and F. El-Baz, "Identifying the Effects of the Gulf War on the Geomorphic Features of Kuwait by Remote Sensing and GIS," *Photogrammetric Engineering and Remote Sensing,* vol. 64, no. 7, 1998, pp. 739–747.

Koponen, S., et al., "Lake Water Quality Classification with Airborne Hyperspectral Spectrometer and Simulated MERIS Data," *Remote Sensing of Environment,* vol. 79, no. 1, 2002, pp. 51–59.

Koukoulas, S., and G.A. Blackburn, "Introducing New Indices for Accuracy Evaluation of Classified Images Representing Semi-Natural Woodland Environments," *Photogrammetric Engineering and Remote Sensing,* vol. 67, no. 4, 2001, pp. 499–510.

Koutsias, N., M. Karteris, and E. Chuvico, "The Use of Intensity-Hue-Saturation Transformation of Landsat-5 Thematic Mapper Data for Burned Land Mapping," *Photogrammetric Engineering and Remote Sensing,* vol. 66, no. 7, 2000, pp. 829–839.

Kruse, F., K. Kieren-Young, and J. Boardman, "Mineral Mapping at Cuprite, Nevada with a 63-Channel Imaging Spectrometer," *Photogrammetric Engineering and Remote Sensing,* vol. 56, no. 1, 1990, pp. 83–92.

Kruse, F.A., A.B. Lefkoff, and J.B. Dietz, "Expert System-Based Mineral Mapping in Northern Death Valley, California/Nevada Using the Airborne Visible/Infrared Imaging Spectrometer (AVIRIS)," *Remote Sensing of Environment,* vol. 44, 1993, pp. 309–336.

Kruse, F., et al., "The Spectral Image Processing System (SIPS)—Interactive Visualization and Analysis of Imaging Spectrometer Data," *Remote Sensing of Environment,* vol. 44, 1993, pp. 145–163.

Lambin, E.F., "Change Detection at Multiple Temporal Scales: Seasonal and Annual Variations in Landscape Variables," *Photogrammetric Engineering and Remote Sensing,* vol. 62, no. 8, 1996, pp. 931–938.

Lathrop, R.G., Jr., and T.M. Lillesand, "Use of Thematic Mapper Data to Assess Water Quality in Green Bay and Central Lake Michigan," *Photogrammetric Engineering and Remote Sensing,* vol. 52, no. 5, 1986, pp. 671–680.

Lathrop, R.G., Jr., and T.M. Lillesand, "Calibration of Thematic Mapper Thermal Data for Water Surface Temperature Mapping: Case Study on the Great Lakes," *Remote Sensing of Environment,* vol. 22, 1987, pp. 297–307.

Lawrence, R.L., and W.J. Ripple, "Calculating Change Curves for Multitemporal Satellite Imagery: Mount St. Helens 1980–1995," *Remote Sensing of Environment,* vol. 67, no. 3, 1999, pp. 309–319.

Lawrence, R.L., and A. Wright, "Rule-Based Classification Systems Using Classification and Regression Tree (CART) Analysis," *Photogrammetric Engineering and Remote Sensing,* vol. 67, no. 10, 2001, pp. 1137–1142.

Levesque, J., K. Staenz, and T. Szeredi, "The Impact of Spectral Band Characteristics on Unmixing of Hyperspectral Data for Monitoring Mine Tailings Site Rehabilitation," *Canadian Journal of Remote Sensing,* vol. 26, no. 3, 2000, pp. 231–240.

Lewis, M., "Spectral Characterization of Australian Arid Zone Plants," *Canadian Journal of Remote Sensing,* vol. 28, no. 2, 2002, pp. 219–230.

Li, X., and A.G.O. Yeh, "Urban Simulation Using Principal Components Analysis and Cellular Automata for Land-Use Planning," *Photogrammetric Engineering and Remote Sensing,* vol. 68, no. 4, 2002, pp. 341–351.

Liang, S.L., et al., "Atmospheric Correction of Landsat ETM+ Land Surface Imagery—Part I: Methods," *IEEE Transactions on Geoscience and Remote Sensing,* vol. 39, no. 11, 2001, pp. 2490–2498.

Lillesand, T.M., et al., *Upper Midwest Gap Analysis Program Image Processing Protocol,* USGS Environmental Management Technical Center, Onalaska, WI, 1998.

Liu, X., and R.G. Lathrop, "Urban Change Detection Based on an Artificial Neural Network," *International Journal of Remote Sensing,* vol. 23, no. 12, 2002, pp. 2513–2518.

Liu, X., A.K. Skidmore, and H. Van Oosten, "Integration of Classification Methods for Improvement of Land-Cover Map Accuracy," *ISPRS Journal of Photogrammetry and Remote Sensing,* vol. 56, no. 4, 2002, pp. 257–268.

Lo, C.P., and L.J. Watson, "Influence of Geographic Sampling Methods on Vegetation Map Accuracy Evaluation in a Swampy Environment," *Photogrammetric Engineering and Remote Sensing,* vol. 64, no. 12, 1998, pp. 1189–1200.

Loughlin, W.P., "Principal Component Analysis for Alteration Mapping," *Photogrammetric Engineering and Remote Sensing*, vol. 57, no. 9, 1991, pp. 1163–1169.

Loyarte, M.M.G., "Detecting Spatial and Temporal Patterns in NDVI Time Series Using Histograms," *Canadian Journal of Remote Sensing*, vol. 28, no. 2, 2002, pp. 275–290.

Lunetta, R.S., and M.E. Balogh, "Application of Multi-Temporal Landsat 5 TM Imagery for Wetland Identification," *Photogrammetric Engineering and Remote Sensing*, vol. 65, no. 11, 1999, pp. 1303–1310.

Lunetta, R.S., and C.D. Elvidge, *Remote Sensing Change Detection: Environmental Monitoring Methods and Applications*, Ann Arbor Press, Ann Arbor, MI, 1999.

Lunetta, R.S., et al., "North American Landscape Characterization Dataset Development and Data Fusion Issues," *Photogrammetric Engineering and Remote Sensing*, vol. 64, no. 8, 1998, pp. 821–829.

Lymburner, L., P.J. Beggs, and C.R. Jacobson, "Estimation of Canopy-Average Surface-Specific Leaf Area Using Landsat TM Data," *Photogrammetric Engineering and Remote Sensing*, vol. 66, no. 2, 2000, pp. 183–191.

Lyon, J.G., et al., "A Change Detection Experiment Using Vegetation Indices," *Photogrammetric Engineering and Remote Sensing*, vol. 64, no. 2, 1998, pp. 143–150.

Macleod, R.D., and R.G. Congalton, "Quantitative Comparison of Change-Detection Algorithms for Monitoring Eelgrass from Remotely Sensed Data," *Photogrammetric Engineering and Remote Sensing*, vol. 64, no. 3, 1998, pp. 207–216.

Mallat, S.G., "A Theory for Multiresolution Signal Decomposition: The Wavelet Representation," *IEEE Transactions on Pattern Analysis and Machine Intelligence*, vol. 11, 1989, pp. 674–693.

Masek, J.G., F.E. Lindsay, and S.N. Goward, "Dynamics of Urban Growth in the Washington DC Metropolitan Area, 1973–1996, from Landsat Observations," *International Journal of Remote Sensing*, vol. 21, no. 18, 2000, pp. 3473–3486.

Mather, P.M., *Computer Processing of Remotely-Sensed Images: An Introduction*, 2nd ed., Wiley, New York, 1999.

Matsushita, B., and M. Tamura, "Integrating Remotely Sensed Data with an Ecosystem Model to Estimate Net Primary Productivity in East Asia," *Remote Sensing of Environment*, vol. 81, no. 1, 2002, pp. 58–66.

McDonald, A.J., F.M. Gemmell, and P.E. Lewis, "Investigation of the Utility of Spectral Vegetation Indices for Determining Information on Coniferous Forests," *Remote Sensing of Environment*, vol. 66, no. 3, 1998, pp. 250–272.

McGwire, K.C., "Cross-Validated Assessment of Geometric Accuracy," *Photogrammetric Engineering and Remote Sensing*, vol. 62, no. 10, 1996, pp. 1179–1187.

Mesev, V., "The Use of Census Data in Urban Image Classification," *Photogrammetric Engineering and Remote Sensing*, vol. 64, no. 5, 1998, pp. 431–438.

Mickelson, J.G., D.L. Civco, and J.A. Silander, "Delineating Forest Canopy Species in the Northeastern United States Using Multi-Temporal TM Imagery," *Photogrammetric Engineering and Remote Sensing*, vol. 64, no. 9, 1998, pp. 891–904.

Miller, L.D., et al., "Assessing Forage Conditions in Individual Ranch Pastures Using Thematic Mapper Imagery and an IBM Personal Computer," *Proceedings: 10th William T. Pecora Memorial Symposium*, American Society for Photogrammetry and Remote Sensing, Fort Collins, CO, 1985.

Moody, A., S. Gopal, and A.H. Strahler, "Artificial Neural Network Response to Mixed Pixels in Coarse-Resolution Satellite Data." *Remote Sensing of Environment*, vol. 58, 1996, pp. 329–343.

Morisette, J.T., and S. Khorram, "Exact Binomial Confidence Interval for Proportions," *Photogrammetric Engineering and Remote Sensing*, vol. 64, no. 4, 1998, pp. 281–283.

Morisette, J.T., S. Khorram, and T. Mace, "Land-Cover Change Detection Enhanced with Generalized Linear Models," *International Journal of Remote Sensing*, vol. 20, no. 14, 1999, pp. 2703–2721.

Mowrer, H.T., and R.G. Congalton, *Quantifying Spatial Uncertainty in Natural Resources*, Ann Arbor Press, Chelsea, MI, 2000.

Muchoney, D.M., and A.H. Strahler, "Pixel- and Site-based Calibration and Validation Methods for Evaluating Supervised Classification of Remotely Sensed Data," *Remote Sensing of Environment*, vol. 81, nos. 2–3, 2002, pp. 290–299.

Muller, S.V., et al., "Accuracy Assessment of a Land-Cover Map of the Kupaaruk River Basin, Alaska: Considerations for Remote Regions," *Photogrammetric Engineering and Remote Sensing*, vol. 64, no. 6, 1998, pp. 619–628.

Myint, S.W., and N.D. Walker, "Quantification of Surface Suspended Sediments along a River Dominated Coast with NOAA AVHRR and SeaWiFS Measurements: Louisiana, USA," *International Journal of Remote Sensing*, vol. 23, no. 16, 2002, pp. 3229–3249.

Nag, P., et al., *Digital Remote Sensing*, Concept Publishing, New Delhi, 1998.

National Center for Geographic Information and Analysis, *Environmental Modeling with GIS*, Oxford University Press, Oxford, England, 1993.

Nielsen, A.A., K. Conradsen, and J.J. Simpson, "Multivariate Alternation Detection (MAD) and MAF Postprocessing in Multispectral, Bitemporal Image Data: New Approaches to Change Detection Studies," *Remote Sensing of Environment*, vol. 64, no. 1, 1998, pp. 1–19.

Oetter, D.R., et al., "Land Cover Mapping in an Agricultural Setting Using Multiseasonal Thematic Mapper Data," *Remote Sensing of Environment*, vol. 76, no. 2, 2001, pp. 139–155.

Oindo, B.O., "Predicting Mammal Species Richness and Abundance Using Multi-Temporal NDVI," *Photogrammetric Engineering and Remote Sensing*, vol. 68, no. 6, 2002, pp. 623–629.

Oki, K., et al., "Subpixel Classification of Alder Trees Using Multitemporal Landsat Thematic Mapper Imagery," *Photogrammetric Engineering and Remote Sensing*, vol. 68, no. 1, 2002, pp. 77–82.

Paola, J.D., and R.A. Schowengerdt, "A Review and Analysis of Backpropagation Neural Networks for Classification of Remotely-Sensed Multi-Spectral Imagery," *International Journal of Remote Sensing*, vol. 16, no. 16, 1995, pp. 3033–3058.

Paola, J.D., and R.A. Schowengerdt, "The Effect of Neural-Network Structure on a Multispectral Land-Use/Land-Cover Classification," *Photogrammetric Engineering and Remote Sensing*, vol. 63, no. 5, 1997, pp. 535–544.

Peddle, D.R., F.G. Hall, and E.F. LeDrew, "Spectral Mixture Analysis and Geometric-Optical Reflectance Modeling of Boreal Forest Biophysical Structure," *Remote Sensing of Environment*, vol. 67, no. 3, 1999, pp. 288–297.

Pekkarinen, A., "A Method for the Segmentation of Very High Spatial Resolution Images of Forested Landscapes," *International Journal of Remote Sensing*, vol. 23, no. 14, 2002, pp. 2817–2836.

Pellikka, P., "Illumination Compensation for Aerial Video Images to Increase Land Cover Classification Accuracy in Mountains," *Canadian Journal of Remote Sensing*, vol. 22, no. 4, 1996, pp. 368–381.

Pereira, J.M.C., "A Comparative Evaluation of NOAA/AVHRR Vegetation Indexes for Burned Surface Detection and Mapping," *IEEE Transactions on Geoscience and Remote Sensing*, vol. 37, no. 1, 1999, pp. 217–226.

Peters, A.J., et al., "Drought Monitoring with NDVI-Based Standardized Vegetation Index," *Photogrammetric Engineering and Remote Sensing*, vol. 68, no. 1, 2002, pp. 71–75.

Philpot, W.D., "The Derivative Ratio Algorithm: Avoiding Atmospheric Effects in Remote Sensing," *IEEE Transactions on Geoscience and Remote Sensing*, vol. 29, no. 2, 1991, pp. 350–357.

Photogrammetric Engineering and Remote Sensing, Special Issue on Image Processing, vol. 56, no. 1, 1990a.

Photogrammetric Engineering and Remote Sensing, Special Issue on Knowledge-Based Expert Systems, vol. 56, no. 6, 1990b.

Photogrammetric Engineering and Remote Sensing, Special Issue on Integration of Remote Sensing and GIS, vol. 57, no. 6, 1991.

Photogrammetric Engineering and Remote Sensing, Special Issue on Geostatistics and Scaling of Remote Sensing and Spatial Data, vol. 65, no. 1, 1999a.

Photogrammetric Engineering and Remote Sensing, Special Issue on the South Florida Vegetation Mapping Project, vol. 65, no. 2, 1999b.

Photogrammetric Engineering and Remote Sensing, Special Issue on Decision Support Systems, vol. 66, no. 10, 2000.

Pohl, C., and J.L. van Genderen, "Multisensor Image Fusion in Remote Sensing: Concepts, Methods, and Applications," *International Journal of Remote Sensing*, vol. 19, no. 5, 1998, pp. 823–854.

Polzer, P., *Assessment of Classification Accuracy Improvement Using Multitemporal Satellite Data: Case Study in the Glacial Habitat Restoration Area in East Central Wisconsin*, M.S. Thesis, University of Wisconsin–Madison, 1992.

Prakash, A., et al., "Data Fusion for Investigating Land Subsidence and Coal Fire Hazards in a Coal Mining Area," *International Journal of Remote Sensing*, vol. 22, no. 6, 2001, pp. 921–932.

Price, J., "Examples of High Resolution Visible to Near-Infrared Reflectance Spectra and a Standardized Collection for Remote Sensing Studies," *International Journal of Remote Sensing*, vol. 16, no. 6, 1995, pp. 993–1000.

Pugh, S.A., and R.G. Congalton, "Applying Spatial Autocorrelation Analysis to Evaluate Error in New England Forest-Cover-Type Maps Derived from Landsat Thematic Mapper Data," *Photogrammetric Engineering and Remote Sensing*, vol. 67, no. 5, 2001, pp. 613–620.

Purevdorj, T., et al., "Relationships between Percent Vegetation Cover and Vegetation Indices," *International Journal of Remote Sensing*, vol. 19, no. 18, 1998, pp. 3519–3536.

Qari, M.Y.H.T., "Application of Landsat TM Data to Geological Studies, Al-Khabt Area, Southern Arabian Shield," *Photogrammetric Engineering and Remote Sensing*, vol. 57, no. 4, 1991, pp. 421–429.

Quattrochi, D.A., and M.F. Goodchild, *Scale in Remote Sensing and GIS*, Lewis Publishers, Boca Raton, FL, 1997.

Radeloff, V., J. Hill, and W. Mehl, *Forest Mapping from Space: Enhanced Satellite Data Processing by Spectral Mixture Analysis and Topographic Corrections*, Joint Research Centre, European Commission, Brussels, Belgium, 1997.

Radeloff, V.C., D.J. Mladenoff, and M.S. Boyce, "Detecting Jack Pine Budworm Defoliation Using Spectral Mixture Analysis: Separating Effects from Determinants," *Remote Sensing of Environment*, vol. 69, no. 2, 1999, pp. 156–169.

Read, J.M., and N.S.N. Lam, "Spatial Methods for Characterizing Land Cover and Detecting Land-Cover Changes for the Tropics," *International Journal of Remote Sensing*, vol. 23, no. 12, 2002, pp. 2457–2474.

Reese, H.M., et al., "Statewide Land Cover Derived from Multiseasonal Landsat TM Data: A Retrospective of the WISCLAND Project," *Remote Ssensing of Environment*, vol. 82, nos. 2–3, 2002, pp. 224–237.

Remote Sensing of Environment, Special Issue on Physical Measurements and Signatures in Remote Sensing, vol. 41, nos. 2/3, 1992.

Remote Sensing of Environment, Special Issue on Biophysical Remote Sensing, vol. 79, nos. 2/3, 2002.

Richardson, J.A., and X. Jia, *Remote Sensing Digital Image Analysis: An Introduction*, Springer, New York, 1999.

Richardson, A.J., and J.H. Everitt, "Using Spectral Vegetation Indices to Estimate Rangeland Productivity," *GeoCarto International*, vol. 1, 1992, pp. 63–77.

Ridd, M.K., and J.J. Liu, "A Comparison of Four Algorithms for Change Detection in an Urban Environ-ment," *Remote Sensing of Environment*, vol. 63, no. 2, 1998, pp. 95–100.

Riera, J.L., et al., "Analysis of Large-Scale Spatial Heterogeneity in Vegetation Indices Among North American Landscapes," *Ecosystems*, vol. 1, 1998, pp. 268–282.

Robinson, G.D., H.N. Gross, and J.R. Schott, "Evaluation of Two Applications of Spectral Mixing Models to Image Fusion," *Remote Sensing of Environment*, vol. 71, no. 3, 2000, pp. 272–281.

Roderick, M., R. Smith, and S. Cridland, "The Precision of the NDVI Derived from AVHRR Observations," *Remote Sensing of Environment*, vol. 56, no. 1, 1996, pp. 57–65.

Roderick, M., R. Smith, and G. Lodwick, "Calibrating Long-Term AVHRR-Derived NDVI Imagery," *Remote Sensing of Environment*, vol. 58, no. 1, 1996, pp. 1–12.

Rosin, P.L., "Robust Pixel Unmixing," *IEEE Transactions on Geoscience and Remote Sensing*, vol. 39, no. 9, 2001, pp. 1978–1983.

Rowe, N.C., and L.L. Grewe, "Change Detection for Linear Features in Aerial Photographs Using Edge-Finding," *IEEE Transactions on Geoscience and Remote Sensing*, vol. 39, no. 7, 2001, pp. 1608–1612.

Rumelhart, D.E., G.E. Hinton, and R.J. Williams, "Learning Internal Representations by Error Propagation," in Rumelhart, D.E. and J.L. McClelland (eds.), *Parallel Distributed Processing: Explorations in the Microstructure of Cognition*, vol. 1, MIT Press, Cambridge, MA, 1986, pp. 318–362.

Running, S.W., et al., "Terrestrial Remote Sensing Science and Algorithms Planned for EOS/MODIS," *International Journal of Remote Sensing*, vol. 15, no. 17, 1994, pp. 3587–3620.

Russ, J. C., *The Image Processing Handbook*, 2nd ed., CRC Press, Boca Raton, FL, 1995.

Russ, J. C., *The Image Processing Handbook*, 4th ed., CRC Press, Boca Raton, FL, 2002.

Ryherd, S., and C. Woodcock, "Combining Spectral and Texture Data in the Segmentation of Remotely Sensed Images," *Photogrammetric Engineering and Remote Sensing*, vol. 62, no. 2, 1996, pp. 181–194.

Sabins, F.F., Jr., *Remote Sensing—Principles and Interpretation*, 3rd ed., W. H. Freeman, New York, 1997.

Sanchez, J., and M.P. Canton, *Space Image Processing*, CRC Press, Boca Raton, FL, 1998.

Sanchez, J., and M.P. Canton, *Space Image Processing*, CRC Press, Boca Raton, FL, 1999.

Sanden, E.M., C.M. Britton, and J.H. Everitt, "Total Ground-Cover Estimates from Corrected Scene Brightness Measurements," *Photogrammetric Engineering and Remote Sensing*, vol. 62, no. 2, 1996, pp. 147–150.

Savitsky, A., and M.J.E. Golay, "Smoothing and Differentiation of Data by Simiplified Least Squares Procedures," *Analytical Chemistry*, vol. 36, no. 8, 1964, pp. 1627–1639.

Sayood, K., *Introduction to Data Compression*, Morgan Kaufman, San Francisco, 1996.

Schetselaar, E.M., "On Preserving Spectral Balance in Image Fusion and Its Advantages for Geological Image Interpretation," *Photogrammetric Engineering and Remote Sensing*, vol. 67, no. 8, 2001, pp. 925–934.

Schott, J.R., *Remote Sensing: The Image Chain Approach*, Oxford University Press, New York, 1997.

Schowengerdt, R.A., *Remote Sensing Models and Methods for Image Processing*, 2nd ed., Academic, New York, 1997.

Seong, J.C., and E.L. Usery, "Fuzzy Image Classification for Continental-Scale Multitemporal NDVI Series Images Using Invariant Pixels and an Image Stratification Method," *Photogrammetric Engineering and Remote Sensing*, vol. 67, no. 3, 2001, pp. 287–294

Shao, G., D. Liu, and G. Zhao, "Relationships of Image Classification Accuracy and Variation of Landscape Statistics," *Canadian Journal of Remote Sensing*, vol. 27, no. 1, 2001, pp. 33–43.

Sharma, G., *Digital Color Imaging Handbook*, CRC Press, Boca Raton, FL, 2002.

Sharma, K.M.S., and A. Sarkar, "Modified Contextual Classification Technique for Remote Sensing Data," *Photogrammetric Engineering and Remote Sensing*, vol. 64, no. 4, 1998, pp. 273–280.

Skidmore, A.K., *Environmental Modelling with GIS and Remote Sensing*, Taylor & Francis, New York, 2002.

Skidmore, A.K., et al., "An Operational GIS Expert System for Mapping Forest Soils," *Photogrammetric Engineering and Remote Sensing*, vol. 62, no. 5, 1996, pp. 501–511.

Skidmore, A.K., et al., "Performance of a Neural Network: Mapping Forests Using GIS and Remotely Sensed Data," *Photogrammetric Engineering and Remote Sensing*, vol. 63, no. 5, 1997, pp. 501–514.

Smith, J.H., et al., "Impacts of Patch Size and Land-Cover Heterogeneity on Thematic Image Classification Accuracy," *Photogrammetric Engineering and Remote Sensing*, vol. 68, no. 1, 2002, pp. 65–70.

Sohl, T.L., "Change Analysis in the United Arab Emirates: An Investigation of Techniques," *Photogrammetric Engineering and Remote Sensing*, vol. 65, no. 4, 1999, pp. 475–484.

Sohn, Y., and R.M. McCoy, "Mapping Desert Shrub Rangeland Using Spectral Unmixing and Modeling Spectral Mixtures with TM Data," *Photogrammetric Engineering and Remote Sensing*, vol. 63, no. 6, 1997, pp. 707–716.

Sohn, Y.S., E. Moran, and F. Gurri, "Deforestation in North-Central Yucatan (1985–1995): Mapping Secondary Succession of Forest and Agricultural Land Use in Sotuta Using the Cosine of the Angle Concept." *Photogrammetric Engineering and Remote Sensing*, vol. 65, no. 8, 1999, pp. 947–958.

Song, C., et al., "Classification and Change Detection Using Landsat TM Data: When and How to Correct Atmospheric Effects," *Remote Sensing of Environment*, vol. 75, no. 2, 2001, pp. 230–244.

Steele, M., J.C. Winne, and R.L. Redmond, "Estimation and Mapping of Misclassification Probabilities for Thematic Land Cover Maps," *Remote Sensing of Environment*, vol. 66, 1998, pp. 192–202.

Stehman, S., "Estimating the Kappa Coefficient and Its Variance under Stratified Random Sampling," *Photogrammetric Engineering and Remote Sensing*, vol. 62, no. 4, 1996, pp. 401–407.

Stehman, S.V., "Selecting and Interpreting Measures of Thematic Classification Accuracy," *Remote Sensing of Environment*, vol. 62, no. 1, 1997, pp. 77–89.

Stehman, S.V., "Practical Implications of Design-Based Sampling Inference for Thematic Map Accuracy Assessment," *Remote Sensing of Environment*, vol. 72, no. 1, 2000, pp. 35–45.

Stehman, S.V., "Statistical Rigor and Practical Utility in Thematic Map Accuracy Assessment," *Photogrammetric Engineering and Remote Sensing*, vol. 67, no. 6, 2001, pp. 727–734.

Stehman, S.V., and R.L. Czaplewski, "Design and Analysis for Thematic Map Accuracy Assessment: Fundamental Principles," *Remote Sensing of Environment*, vol. 64, no. 3, 1998, pp. 331–344.

Stein, A., et al., "Integrating Spatial Statistics and Remote Sensing," *International Journal of Remote Sensing*, vol. 19, no. 9, 1998, pp. 1793–1814.

Stein, A., et al., *Spatial Statistics for Remote Sensing*, Kluwer Academic, Boston, 1999.

Stern, A.J., P.C. Doraiswamy, and P.W. Cook, "Spring Wheat Classification in an AVHRR Image by Signature Extension from a Landsat TM Classified Image," *Photogrammetric Engineering and Remote Sensing*, vol. 67, no. 2, 2001, pp. 207–211.

Stow, D.A., D. Collins, and D. McKinsey, "Land Use Change Detection Based on Multi-Date Imagery from Different Satellite Sensor Systems," *Geocarto International*, vol. 5, no. 3, 1990, pp. 3–12.

Stuckens, J., P.R. Coppin, and M.E. Bauer, "Integrating Contextual Information with Per-Pixel Classification for Improved Land Cover Classification," *Remote Sensing of Environment*, vol. 71, no. 3, 2000, pp. 282–296.

Teillet, P.M., K. Staenz, and D.J. Williams, "Effects of Spectral, Spatial, and Radiometric Characteristics on Remote Sensing Vegetation Indices of Forested Regions," *Remote Sensing of Environment*, vol. 61, no. 1, 1997, pp. 139–149.

Thenkabail, P.S., R.B. Smith, and E. De Pauw, "Hyperspectral Vegetation Indices and Their Relationships with Agricultural Crop Characteristics," *Remote Sensing of Environment*, vol. 71, no. 2, 2000, pp. 158–182.

Theseira, M.A., G. Thomas, and C.A.D. Sannier, "An Evaluation of Spectral Mixture Modelling Applied to a Semi-Arid Environment," *International Journal of Remote Sensing*, vol. 23, no. 4, 2002, pp. 687–700.

Todd, S.W., and R.M. Hoffer, "Responses of Spectral Indices to Variations in Vegetation Cover and Soil Background," *Photogrammetric Engineering and Remote Sensing*, vol. 64, no. 9, 1998, pp. 915–921.

Tokola, T., S. Löfman, and A. Erkkilä, "Relative Calibration of Multitemporal Landsat Data for Forest Cover Change Detection," *Remote Sensing of Environment*, vol. 68, no. 1, 1999, pp. 1–11.

Tompkins, S., et al., "Optimization of Endmembers for Spectral Mixture Analysis," *Remote Sensing of Environment*, vol. 59, 1997, pp. 472–489.

Tou, J.T., and R.C. Gonzalez, *Pattern Recognition Principlels*, Addison-Wesley, London, 1974.

Toutin, T., "SPOT and Landsat Stereo Fusion for Data Extraction over Mountainous Areas," *Photogrammetric Engineering and Remote Sensing*, vol. 64, no. 2, 1998, pp. 109–113.

Treitz, P., and P. Howarth, "High Spatial Resolution Remote Sensing Data for Forest Ecosystem Classification: An Examination of Spatial Scale," *Remote Sensing of Environment*, vol. 72, no. 3, 2000, pp. 268–289.

Tsai, F., and W.D. Philpot, "Derivative Analysis of Hyperspectral Data," *Remote Sensing of Environment*, vol. 66, 1999, pp. 41–51.

Tsai, F., and W.D. Philpot, "A Derivative-Aided Hyperspectral Image Analysis System for Land-Cover Classification," *IEEE Transactions on Geoscience and Remote Sensing*, vol. 40, no. 2, 2002, pp. 416–425.

Tso, B., and P.M. Mather, *Classification Methods for Remotely Sensed Data*, Taylor & Francis, New York, 2001.

Turner, M.G., et al., "Effects of Changing Spatial Scale on the Analysis of Landscape Pattern," *Landscape Ecology*, vol. 3, 1989, pp. 153–162.

Urban, D.L., R.V. O'Neill, and H.H. Shugart, Jr., "Landscape Ecology," *Bioscience*, vol. 37, 1987, pp. 119–127.

VanDeusen, P.C., "Unbiased Estimates of Class Proportions from Thematic Maps," *Photogrammetric Engineering and Remote Sensing*, vol. 62, no. 4, 1996, pp. 409–412.

VanDeventer, A.P., et al., "Using Thematic Mapper Data to Identify Contrasting Soil Plains and Tillage Practices," *Photogrammetric Engineering and Remote Sensing*, vol. 63, no. 1, 1997, pp. 87–93.

Verbyla, D.L., *Satellite Remote Sensing of Natural Resources*, CRC Press, Boca Raton, FL, 1995.

Verhoeye, J., and R. De Wulf, "Land Cover Mapping at Sub-Pixel Scales Using Linear Optimization Techniques," *Remote Sensing of Environment*, vol. 79, no. 1, 2002, pp. 96–104.

Vogelmann, J.E., T. Sohl, and S.M. Howard, "Regional Characterization of Land Cover Using Multiple Sources of Data," *Photogrammetric Engineering and Remote Sensing*, vol. 64, no. 1, 1998, pp. 45–57.

Vrabel, J., "Multispectral Imagery Band Sharpening Study," *Photogrammetric Engineering and Remote Sensing*, vol. 62, no. 9, 1996, pp. 1075–1083.

Walkey, J.A., *Development of a Change Detection Tool for Image Analysts*, M.S. Thesis, University of Wisconsin-Madison, 1997.

Wang, J., K.P. Price, and P.M. Rich, "Spatial Patterns of NDVI in Response to Precipitation and Temperature in the Central Great Plains," *International Journal of Remote Sensing*, vol. 22, no. 18, 2001, pp. 3827–3844.

Wang, J., and Q. Zhang, "Applicability of a Gradient Profile Algorithm for Road Network Extraction— Sensor, Resolution and Background Considerations," *Canadian Journal of Remote Sensing*, vol. 26, no. 5, 2000, pp. 428–439.

Warner, T.A., and M. Shank, "An Evaluation of the Potential for Fuzzy Classification of Multispectral Data Using Artificial Neural Networks," *Photogrammetric Engineering and Remote Sensing*, vol. 63, no. 11, 1997, pp. 1285–1294.

Wayman, J.P., et al., "Landsat TM-Based Forest Area Estimation Using Iterative Guided Spectral Class Rejection," *Photogrammetric Engineering and Remote Sensing*, vol. 67, no. 10, 2001, pp. 1155–1166.

Wickham, J.D., et al., "Sensitivity of Selected Landscape Pattern Metrics to Land-Cover Misclassification and Differences in Land-Cover Composition," *Photogrammetric Engineering and Remote Sensing*, vol. 63, no. 4, 1997, pp. 397–402.

Williams, E.A., and D.E. Jelinski, "On Using the NOAA AVHRR Experimental Calibrated Biweekly Global Vegetation Index," *Photogrammetric Engineering and Remote Sensing*, vol. 62, no. 8, 1996, pp. 959–960.

Wilson, P.A., "Rule-Based Classification of Water in Landsat MSS Images Using the Variance Filter," *Photogrammetric Engineering and Remote Sensing*, vol. 63, no. 5, 1997, pp. 485–491.

Woodcock, C.E., S. Gopal, and W. Albert, "Evaluation of the Potential for Providing Secondary Labels in Vegetation Maps," *Photogrammetric Engineering and Remote Sensing*, vol. 62, no. 4, 1996, pp. 393–399.

Woodruff, D.L., et al., "Remote Estimation of Water Clarity in Optically Complex Estuarine Waters," *Remote Sensing of Environment*, vol. 68, no. 1, 1999, pp. 41–52.

Yamamoto, T., H. Hanaizumi, and S. Chino, "A Change Detection Method for Remotely Sensed Multispectral and Multitemporal Images Using 3-D Segmentation," *IEEE Transactions on Geoscience and Remote Sensing*, vol. 39, no. 5, 2001, pp. 976–985.

Yin, Z.S., and T.H.L. Williams, "Obtaining Spatial and Temporal Vegetation Data from Landsat MSS and AVHRR/NOAA Satellite Images for a Hydrologic Model," *Photogrammetric Engineering and Remote Sensing*, vol. 63, no. 1, 1997, pp. 69–77.

Yocky, D.A., "Multiresolution Wavelet Decomposition Image Merger of Landsat Thematic Mapper and SPOT Panchromatic Data," *Photogrammetric Engineering and Remote Sensing*, vol. 62, no. 9, 1996, pp. 1067–1074.

Young, S.S., and C.Y. Wang, "Land-Cover Change Analysis of China Using Global-Scale Pathfinder AVHRR Landcover (PAL) Data, 1982–92," *International Journal of Remote Sensing*, vol. 22, no. 8, 2001, pp. 1457–1477.

Zhang, J., and G.M. Foody, "Fully-Fuzzy Supervised Classification of Sub-Urban Land Cover from Remotely Sensed Imagery: Statistical and Artificial Neural Network Approaches," *International Journal of Remote Sensing*, vol. 22, no. 4, 2001, pp. 615–628.

Zhao, G., and A.L. Maclean, "A Comparison of Canonical Discriminant Analysis and Principal Component Analysis for Spectral Transformation," *Photogrammetric Engineering and Remote Sensing*, vol. 66, no. 7, 2000, pp. 841–847.

Zhou, J., and D. L. Civco, "Using Genetic Learning Neural Networks for Spatial Decision Making in GIS," *Photogrammetric Engineering and Remote Sensing*, vol. 62, no. 11, 1996, pp. 1287–1295.

Zhu, Z.L., et al., "Accuracy Assessment for the US Geological Survey Regional Land-Cover Mapping Program: New York and New Jersey Region," *Photogrammetric Engineering and Remote Sensing*, vol. 66, no. 12, 2000, pp. 1425–1435.

8 MICROWAVE AND LIDAR SENSING

8.1 INTRODUCTION

An increasing amount of valuable environmental and resource information is being acquired by sensors that operate in the *microwave* portion of the electromagnetic spectrum. In the context of the sensors we have discussed thus far, microwaves are not "micro" at all. That is, the microwave portion of the spectrum includes wavelengths within the approximate range of 1 mm to 1 m. Thus, the longest microwaves are about 2,500,000 times longer than the shortest light waves!

There are two distinctive features that characterize microwave energy from a remote sensing standpoint:

1. Microwaves are capable of penetrating the atmosphere under virtually all conditions. Depending on the wavelengths involved, microwave energy can "see through" haze, light rain and snow, clouds, and smoke.

2. Microwave reflections or emissions from earth materials bear no direct relationship to their counterparts in the visible or thermal portions of the spectrum. For example, surfaces that appear "rough" in the visible portion of the spectrum may be "smooth" as seen by mi-

crowaves. In general, microwave responses afford us a markedly different "view" of the environment—one far removed from the views experienced by sensing light or heat.

In this chapter we discuss both airborne and spaceborne, as well as *active* and *passive*, microwave sensing systems. Recall that the term "active" refers to a sensor that supplies its own source of energy or illumination. *Radar* is an active microwave sensor, and it is the major focus of attention in this chapter. To a lesser extent, we also treat the passive counterpart to radar, the *microwave radiometer*. This device responds to the extremely low levels of microwave energy that are naturally emitted and/or reflected from ambient sources (such as the sun) by terrain features.

It should be recognized that practical resource management experience with radar and passive microwave systems is relatively limited compared to photographic or scanning systems. However, with the increasing availability of spaceborne radar data, the applications of microwave sensing have increased substantially.

We conclude this chapter with a brief introduction to *lidar* remote sensing. Like radar, lidar sensors are active remote sensing systems. However, they use pulses of laser light, rather than microwave energy, to illuminate the terrain. As with radar, the outlook for applications of such systems is an extremely promising one.

8.2 RADAR DEVELOPMENT

The word *radar* is an acronym for *radio detection and ranging*. As its name implies, radar was developed as a means of using radio waves to detect the presence of objects and to determine their distance and sometimes their angular position. The process entails transmitting short bursts, or pulses, of microwave energy in the direction of interest and recording the strength and origin of "echoes" or "reflections" received from objects within the system's field of view.

Radar systems may or may not produce images, and they may be ground based or mounted in aircraft or spacecraft. A common form of nonimaging radar is the type used to measure vehicle speeds. These systems are termed *Doppler radar* systems because they utilize Doppler frequency shifts in the transmitted and returned signals to determine an object's velocity. Doppler frequency shifts are a function of the relative velocities of a sensing system and a reflector. For example, we perceive Doppler shifts in sound waves as a change in pitch, as in the case of a passing car horn or train whistle. The Doppler shift principle is often used in analyzing the data generated from imaging radar systems.

Another common form of radar is the *plan position indicator (PPI)* system. These systems have a circular display screen on which a radial sweep indicates the position of radar "echoes." Essentially PPI radar images a continuously updated plan-view map of objects surrounding its rotating antenna. These systems are common in weather forecasting, air traffic control, and navigation applications. However, PPI systems are not appropriate to most remote sensing applications because they have rather poor spatial resolution.

Airborne and spaceborne radar remote sensing is done with systems that use an antenna fixed below the aircraft (or spacecraft) and pointed to the side. Such systems are termed *side-looking radar (SLR)*, or *side-looking airborne radar (SLAR)* in the case of airborne systems. Side-looking radar systems produce continuous strips of imagery depicting extensive ground areas that parallel the aircraft flight line.

Side-looking airborne radar was first developed for military reconnaissance purposes in the late 1940s. It became an ideal military reconnaissance system, not only because it affords nearly an all-weather operating capability, but also because it is an active, day-or-night imaging system. The military genesis of SLAR has had two general impacts on its subsequent application to civilian remote sensing uses. First, there was a time lag between military development, declassification, and civilian application. Less obvious, but nonetheless important, is the fact that military SLAR systems were developed to look at military targets. Terrain features that "cluttered" SLAR imagery and masked objects of military importance were naturally not of interest in original system designs. However, with military declassification and improvement in nonmilitary capabilities, SLR has evolved into a powerful tool for acquiring natural resource data.

Although SLR acquisition and analysis techniques have been developed to a high degree of sophistication, it should be pointed out that the application of radar technology to earth resource sensing is still in an active state of advancement. What determines the overall "radar reflectivity" of various earth resources features under various conditions is still not always known precisely. Even though much is yet to be learned about how radar signals interact with the natural environment, productive applications of existing radar technology have been many and varied. While the use of radar is continuing to increase in general, it is particularly extensive in regions where persistent cloud cover limits the acquisition of imagery in the optical portion of the spectrum.

The first large-scale project for mapping terrain with side-looking airborne radar was a complete survey of the Darien province of Panama. This survey was undertaken in 1967 and resulted in images used to produce a mosaic of a 20,000-km^2 ground area. Prior to that time, this region had never been photographed or mapped in its entirety because of persistent (nearly perpetual) cloud cover. The success of the Panama radar mapping project led to the application of radar remote sensing throughout the world. Since the early 1970s, extensive radar mapping programs have been conducted by several governments as well as by mining and petroleum companies.

In 1971, a radar survey was begun in Venezuela that resulted in the mapping of nearly 500,000 km² of land. This project resulted in improvements in the accuracy of the location of the boundaries of Venezuela with its neighboring countries. It also permitted a systematic inventory and mapping of the country's water resources, including the discovery of the previously unknown source of several major rivers. Likewise, improved geologic maps of the country were produced.

Also beginning in 1971, was Project Radam (standing for *Radar of the Amazon*), a reconnaissance survey of the Amazon and the adjacent Brazilian northeast. At that time, this was the largest radar mapping project ever undertaken. By the end of 1976, more than 160 radar mosaic sheets covering an area in excess of 8,500,000 km² had been completed. Scientists used these radar mosaics as base maps in a host of studies, including geologic analysis, timber inventory, transportation route location, and mineral exploration. Large deposits of important minerals were discovered after intensive analysis was made of newly discovered features shown by radar. Mapping of previously uncharted volcanic cones, and even large rivers, resulted from this project. In such remote and cloud-covered areas of the world, radar imagery is a prime source of inventory information about potential mineral resources, forestry and range resources, water supplies, transportation routes, and sites suitable for agriculture. Such information is essential to planning sustainable development in such ecologically sensitive areas.

Radar imagery has also been used extensively to monitor the surface of the oceans to determine wind, wave, and ice conditions. Radar data have also been used to study (indirectly) ocean bottom contours under some conditions. Numerous other applications of radar have been demonstrated in the areas of geologic mapping, mineral exploration, flood inundation mapping, and small-scale thematic mapping.

Radar remote sensing from space began with the launch of *Seasat* in 1978 and continued with the Shuttle Imaging Radar (SIR) and Soviet Cosmos experiments in the 1980s. Beginning in 1991, three radar satellites were launched within an 11-month period. These were the *Almaz-1*, *ERS-1*, and *JERS-1* systems launched by the former Soviet Union, European Space Agency, and Japan, respectively. These systems were followed by Canada's first *Radarsat* satellite in 1995. Hence, the 1990s initiated a new era in terms of the broad-scale availability of radar data collected from space. This trend continued into the new century with the globe-spanning Shuttle Radar Topography Mission and ESA's *Envisat*, the most advanced satellite radar system yet launched. Several sophisticated new radar satellite systems are scheduled for deployment in the next 5 years, beginning with those developed by Canada (*Radarsat-2*) and Japan (*ALOS*). The field of spaceborne radar remote sensing thus continues to be characterized by rapid technological advances, an expanding range of sources of data, and a high level of international participation.

8.3 SIDE-LOOKING RADAR SYSTEM OPERATION

The basic operating principle of a SLR system is shown in Figure 8.1. Microwave energy is transmitted from an antenna in very short bursts or pulses. These high energy pulses are emitted over a time period on the order of microseconds (10^{-6} sec). In Figure 8.1, the propagation of one pulse is shown by indicating the wavefront locations at successive increments of time. Beginning with the solid lines (labeled 1 through 10), the transmitted pulse moves radially outward from the aircraft in a constrained (or narrow) beam. Shortly after time 6, the pulse reaches the house, and a reflected wave (dashed line) is shown beginning at time 7. At time 12, this return signal reaches the antenna and is registered at that time on the antenna response graph (Figure 8.1*b*). At time 9, the transmitted wavefront is reflected off the tree, and this "echo" reaches the antenna at time 17. Because the tree is less reflective of radar waves than the house, a weaker response is recorded in Figure 8.1*b*.

By electronically measuring the return time of signal echoes, the range, or distance, between the transmitter and reflecting objects may be determined.

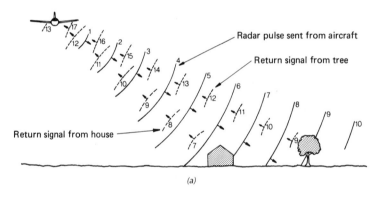

(a)

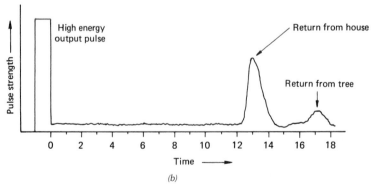

(b)

Figure 8.1 Operating principle of SLR: (*a*) propagation of one radar pulse (indicating the wavefront location at time intervals 1–17); (*b*) resulting antenna return.

Since the energy propagates in air at approximately the velocity of light c, the slant range, $\overline{SR}$, to any given object is given by

$$\overline{SR} = \frac{ct}{2} \qquad (8.1)$$

where
$\overline{SR}$ = slant range (direct distance between transmitter and object)
c = speed of light (3×10^8 m/sec)
t = time between pulse transmission and echo reception

(Note that the factor 2 enters into the equation because the time is measured for the pulse to travel the distance both to and from the target, or twice the range.) This principle of determining distance by electronically measuring the transmission–echo time is central to imaging radar systems.

One manner in which SLR images are created is illustrated in Figure 8.2. As the aircraft advances, the antenna (1) is continuously repositioned along the flight direction at the aircraft velocity V_a. The antenna is switched from a transmitter to a receiver mode by a synchronizer switch (2). A portion of each transmitted pulse (3) is returned (as an echo) from terrain features occurring along a single antenna beamwidth. Shown in (4) is the return signal from one line of data. Return signals (echoes) are received by the airborne antenna, processed, and then recorded (5). Spaceborne systems operate on the same general principle.

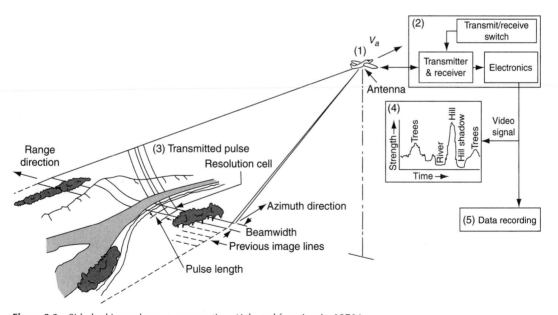

Figure 8.2   Side-looking radar system operation. (Adapted from Lewis, 1976.)

Figure 8.3 Seasat SLR image of Appalachian Mountains of Pennsylvania, L band, midsummer. Scale 1 : 575,000. (Courtesy NASA/JPL/Caltech.)

Figure 8.3, a satellite radar image of an area of folded sedimentary rocks in the Appalachian Mountains of Pennsylvania, illustrates the appearance of radar images. Here, the "sidelighting" nature of SLR images obtained as illustrated in Figure 8.2 is apparent. In Figure 8.3, the signals from the radar system in the spacecraft were transmitted toward the bottom of the page, and the signals received by the radar system are those reflected back toward the top of the page. Note that the topographic slopes of the linear hills and valleys associated with the folded sedimentary rocks that face the spacecraft return strong signals, whereas the flatter areas and slopes facing away from the spacecraft return weaker signals. Note also that the river seen at upper left is dark toned because of specular reflection away from the sensor. (Section 8.8 describes earth surface feature characteristics influencing radar returns in some detail.)

Figure 8.4 illustrates the nomenclature typically used to describe the geometry of radar data collection. A radar system's *look angle* is the angle from

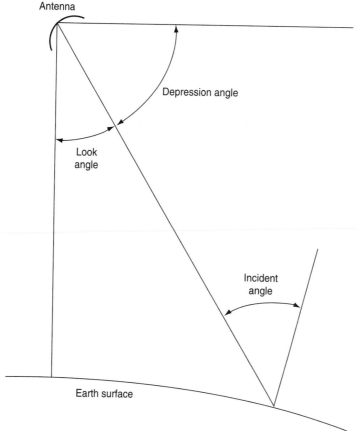

Figure 8.4   Nomenclature for the geometry of radar data collection.

nadir to a point of interest on the ground. The complement of the look angle is called the *depression angle*. The *incident angle* is the angle between the incident radar beam at the ground and the normal to the earth's surface at the point of incidence. (The incident angle is often referred to using the grammatically incorrect term "incidence angle" in the literature. The two terms are synonymous.) In the case of airborne imaging over flat terrain, the incident angle is approximately equal to the look angle. In radar imaging from space, the incident angle is slightly greater than the look angle due to earth curvature. The *local incident angle* is the angle between the incident radar beam at the ground and the normal to the ground surface at the point of incidence. The incident angle and the local incident angle are the same only in the case of level terrain.

The ground resolution cell size of an SLR system is controlled by two independent sensing system parameters: *pulse length* and *antenna beamwidth*. The pulse length of the radar signal is determined by the length of time that the antenna emits its burst of energy. As can be seen in Figure 8.5, the signal pulse length dictates the spatial resolution in the direction of energy propagation. This direction is referred to as the *range* direction. The width of the antenna beam determines the resolution cell size in the flight, or *azimuth*, direction. We consider each of these elements controlling radar spatial resolution separately.

Range Resolution

For an SLR system to image separately two ground features that are close to each other in the range direction, it is necessary for the reflected signals

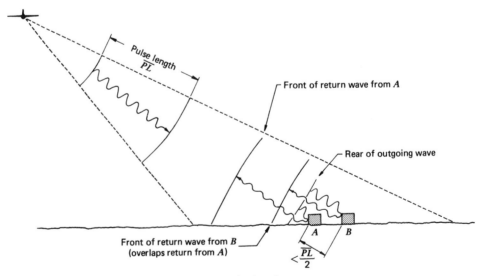

Figure 8.5 Dependence of range resolution on pulse length.

from all parts of the two objects to be received separately by the antenna. Any time overlap between the signals from two objects will cause their images to be blurred together. This concept is illustrated in Figure 8.5. Here a pulse of length $\overline{PL}$ (determined by the duration of the pulse transmission) has been transmitted toward buildings A and B. Note that the slant-range distance (the direct sensor-to-target distance) between the buildings is less than $\overline{PL}/2$. Because of this, the pulse has had time to travel to B and have its echo return to A while the end of the pulse at A continues to be reflected. Consequently, the two signals are overlapped and will be imaged as one large object extending from building A to building B. If the slant-range distance between A and B were anything greater than $\overline{PL}/2$, the two signals would be received separately, resulting in two separate image responses. Thus, the slant-range resolution of an SLR system is independent of the distance from the aircraft and is equal to half the transmitted pulse length.

Although the *slant-range* resolution of an SLR system does not change with distance from the aircraft, the corresponding *ground*-range resolution does. As shown in Figure 8.6, the ground resolution in the range direction varies inversely with the cosine of the depression angle. This means that the ground-range resolution becomes smaller with increases in the slant-range distance.

Accounting for the depression angle effect, the ground resolution in the range direction R_r is found from

$$R_r = \frac{c\tau}{2\cos\theta_d} \tag{8.2}$$

where τ is the pulse duration.

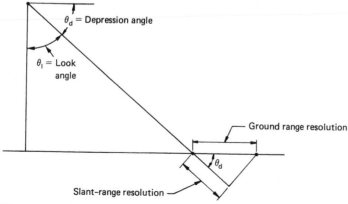

Figure 8.6   Relationship between slant-range resolution and ground-range resolution.

EXAMPLE 8.1

A given SLAR system transmits pulses over a duration of 0.1 μsec. Find the range resolution of the system at a depression angle of 45°.

Solution
From Eq. 8.2

$$R_r = \frac{(3 \times 10^8 \text{ m/sec})(0.1 \times 10^{-6} \text{ sec})}{2 \times 0.707} = 21 \text{ m}$$

Azimuth Resolution

As shown in Figure 8.7, the resolution of an SLR system in the azimuth direction, R_a, is determined by the angular *beamwidth* β of the antenna and the ground range $\overline{GR}$. As the antenna beam "fans out" with increasing distance from the aircraft, the azimuth resolution deteriorates. Objects at points A and B in Figure 8.6 would be resolved (imaged separately) at $\overline{GR}_1$

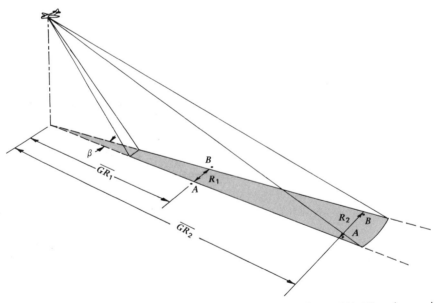

Figure 8.7 Dependence of azimuth resolution (R_a) on antenna beamwidth (β) and ground range ($\overline{GR}$).

but not at $\overline{GR}_2$. That is, at distance $\overline{GR}_1$, A and B result in separate return signals. At $\overline{GR}_2$, distance, A and B would be in the beam simultaneously and would not be resolved.

Azimuth resolution R_a is given by

$$R_a = \overline{GR} \cdot \beta \tag{8.3}$$

EXAMPLE 8.2

A given SLAR system has a 1.8-mrad antenna beamwidth. Determine the azimuth resolution of the system at ranges of 6 and 12 km.

Solution
From Eq. 8.3

$$R_{a_{6\ km}} = (6 \times 10^3 \text{ m})(1.8 \times 10^{-3}) = 10.8 \text{ m}$$

and

$$R_{a_{12\ km}} = (12 \times 10^3 \text{ m})(1.8 \times 10^{-3}) = 21.6 \text{ m}$$

The beamwidth of the antenna of an SLR system is directly proportional to the wavelength of the transmitted pulses, λ, and inversely proportional to the length of the antenna, $\overline{AL}$. That is,

$$\beta = \frac{\lambda}{\overline{AL}} \tag{8.4}$$

For any given wavelength, the effective antenna beamwidth can be controlled by one of two different means: (1) by controlling the *physical* length of the antenna or (2) by synthesizing a virtual antenna length. Those systems wherein beamwidth is controlled by the physical antenna length are called *brute force, real aperture,* or *noncoherent* radars. As expressed by Eq. 8.4, the antenna in a brute force system must be many wavelengths long for the antenna beamwidth to be narrow. For example, to achieve even a 10-mrad beamwidth with a 5-cm-wavelength radar, a 5-m antenna is required $[(5 \times 10^{-2} \text{ m})/(10 \times 10^{-3}) = 5 \text{ m}]$. To obtain a resolution of 2 mrad, we would need an antenna 25 m long! Obviously, antenna length requirements of brute force systems present considerable logistical problems if fine resolution at long range is the objective.

Brute force systems enjoy relative simplicity of design and data processing. Because of resolution problems, however, their operation is often restricted to relatively short range, low altitude operation, and the use of relatively short wavelengths. These restrictions are unfortunate because short-range, low altitude operation limits the area of coverage obtained by the system and short wavelengths experience more atmospheric attenuation and dispersion.

8.4 SYNTHETIC APERTURE RADAR

The deficiencies of brute force operation are overcome in *synthetic aperture radar (SAR)* systems. These systems employ a short physical antenna, but through modified data recording and processing techniques, they synthesize the effect of a very long antenna. The result of this mode of operation is a very narrow effective antenna beamwidth, even at far ranges, without requiring a physically long antenna or a short operating wavelength.

At the detailed level, the operation of SAR systems is quite complex. Suffice it to say here that these systems operate on the principle of using the sensor motion along track to transform a single physically short antenna into an array of such antennas that can be linked together mathematically as part of the data recording and processing procedures (Elachi, 1987b). This concept is shown in Figure 8.8. The "real" antenna is shown in several successive positions along the flight line. These successive positions are treated mathemati-

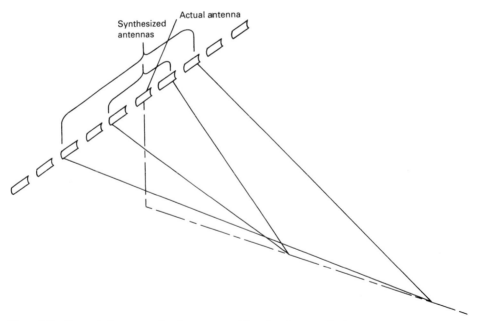

Figure 8.8   Concept of an array of real antenna positions forming a synthetic aperture.

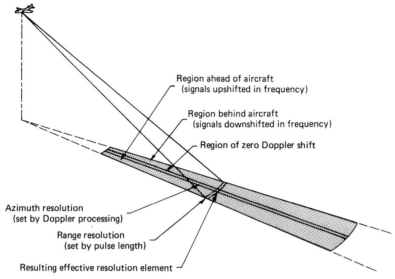

Figure 8.9 Determinants of resolution in synthetic aperture SLR.

cally (or electronically) as if they were simply successive elements of a single, long synthetic antenna. Points on the ground at near range are viewed by proportionately fewer antenna elements than those at far range, meaning effective antenna length increases with range. This results in essentially constant azimuth resolution irrespective of range. Through this process, antennas as long as several kilometers can be synthesized with spaceborne synthetic aperture systems.

Figure 8.9 illustrates yet another approach to explaining how synthetic aperture systems operate, namely discriminating only the near-center return signals from the real antenna beamwidth by detecting Doppler frequency shifts. Recall that a Doppler shift is a change in wave frequency as a function of the relative velocities of a transmitter and a reflector. Within the wide antenna beam, returns from features in the area ahead of the aircraft will have upshifted (higher) frequencies resulting from the Doppler effect. Conversely, returns from the area behind the aircraft will have downshifted (lower) frequencies. Returns from features near the centerline of the beamwidth will experience little or no frequency shift. By processing the return signals according to their Doppler shifts, a very small effective beamwidth can be generated.

Figure 8.10 illustrates the variation with distance of the ground resolution cell size of real aperture systems (*a*) versus synthetic aperture systems (*b*). Note that as the distance from the aircraft increases, the azimuth resolution size increases with real aperture systems and remains constant with synthetic

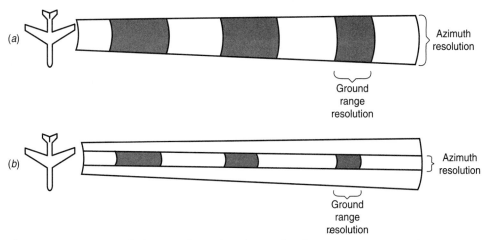

Figure 8.10 Variation with distance of spatial resolution of real aperture (*a*) versus synthetic aperture (*b*) SLR systems.

aperture systems and that the ground range resolution size decreases with both systems.

A final point about SAR systems is that both *unfocused* and *focused* systems exist. Again the details of these systems are beyond our immediate concern. The interesting point about these systems is that the theoretical resolution of *unfocused* systems is a function of wavelength and range, not antenna length. The theoretical resolution of a *focused* system is a function of antenna length, regardless of range or wavelength. More particularly, the resolution of a focused synthetic aperture system is approximately one-half the actual antenna length. That is, the *shorter* the antenna, the finer the resolution. In theory, the resolution for a 1-m antenna would be about 0.5 m, whether the system is operated from an aircraft or a spacecraft! Radar system design is replete with trade-offs among operating range, resolution, wavelength, antenna size, and overall system complexity. (For additional technical information on SAR systems, see Raney, 1998.)

8.5 GEOMETRIC CHARACTERISTICS OF SIDE-LOOKING RADAR IMAGERY

The geometry of SLR imagery is fundamentally different from that of both photography and scanner imagery. This difference basically results because SLR is a *distance* rather than an *angle* measuring system. The influences this has on image geometry are many and varied. Here we limit our discussion to treatment of the following geometric elements of SLR image acquisition and interpretation: *scale distortion, relief displacement,* and *parallax.*

Slant-Range Scale Distortion

Side-looking radar systems use one of two types of image recording systems. A *slant-range* image recording system involves a constant sweep rate across each line. Consequently, the spacing between return signals on slant-range imagery is directly proportional to the time interval between echoes from adjacent terrain features. This interval is directly proportional to the *slant,* rather than *horizontal,* distance between the sensor and any given object. In *ground-range* image recording systems, the sweep often incorporates a hyperbolic timing correction in which the spacing between image points is approximately proportional to the horizontal ground distance between terrain features.

Figure 8.11 illustrates the characteristics of slant-range and ground-range image recording. The symbols A, B, and C represent objects of equal size that are equally separated in the near, middle, and far range. The respective ground ranges to the points are $\overline{GR}_A$, $\overline{GR}_B$, and $\overline{GR}_C$. Based directly on the signal return time, the slant-range image shows unequal distances between the features as well as unequal widths for the features. The result is a varying image scale that is at a minimum in the near range and progresses hyperbolically to a maximum at the far range. Therefore, on a slant-range presentation, object width $A_1 < B_1 < C_1$ and distance $\overline{AB} < \overline{BC}$. Applying a hyperbolic correction, a ground-range image of essentially constant scale can be formed with width $A = B = C$ and distance $\overline{AB} = \overline{BC}$. For a given swath width, the change in scale across an image decreases with increasing flying height. Thus, satellite systems have less scale change across an image than do airborne systems.

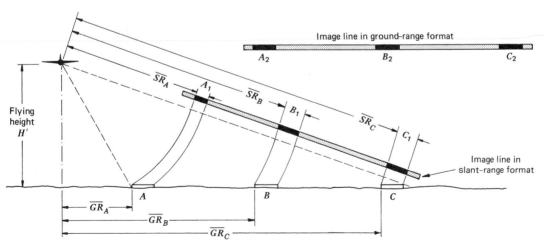

Figure 8.11   Slant-range versus ground-range image format. (Adapted from Lewis, 1976.)

Obviously, the scale distortions inherent in slant-range imagery preclude its direct use for accurate planimetric mapping. However, approximate ground range, $\overline{GR}$, can be derived from slant range, $\overline{SR}$, and flying height, H', under the assumption of flat terrain. From Figure 8.11 it can be seen that

$$\overline{SR}^2 = H'^2 + \overline{GR}^2$$

so

$$\overline{GR} = (\overline{SR}^2 - H'^2)^{1/2} \tag{8.5}$$

Therefore, a ground-range distance can be calculated from an image slant-range distance if the flying height is known. The assumption of flat terrain should be noted, however, and it should be pointed out that flight parameters also affect both range and azimuth scales. The range scale will vary with changes in aircraft altitude, and the azimuth scale will be dependent on precise synchronization between the aircraft ground speed and the data recording system.

Maintaining consistent scale in the collection and recording of SLR data is a complex task. Whereas scale in the range (or across-track) direction is determined by the speed of light, scale in the azimuth (or along-track) direction is determined by the speed of the aircraft or spacecraft. To reconcile and equalize these independent scales, strict control of data collection parameters is needed. In most airborne systems, this is provided by an *inertial navigation and control system*. This device guides the aircraft at the appropriate flying height along the proper course. Angular sensors measure aircraft roll, crab, and pitch and maintain a constant angle of the antenna beam with respect to the line of flight. Inertial systems also provide the output necessary to synchronize the data recording with the aircraft ground speed. Spaceborne systems provide a more stable flight platform.

Relief Displacement

As in line scanner imagery, relief displacement in SLR images is one dimensional and perpendicular to the flight line. However, in contrast to scanner imagery and photography, the *direction* of relief displacement is reversed. This is because radar images display ranges, or distances, from terrain features to the antenna. When a vertical feature is encountered by a radar pulse, the top of the feature is often reached before the base. Accordingly, return signals from the top of a vertical feature will often reach the antenna before returns from the base of the feature. This will cause a vertical feature to "lay over" the closer features, making it appear to lean toward the nadir. This radar *layover effect*, most severe at near range (steeper incident angles), is compared to photographic relief displacement in Figure 8.12.

Terrain slopes facing the antenna at near range are often displayed with a dramatic layover effect. This occurs whenever the terrain slope is steeper

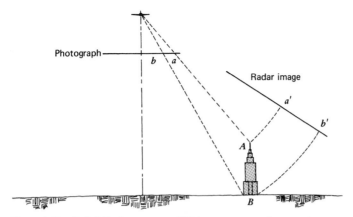

Figure 8.12 Relief displacement on SLR images versus photographs.

than a line perpendicular to the direction of the radar pulse, expressed by its look angle. This condition is met by the left sides of features A and B in Figure 8.13. As such, the tops of these slopes will be imaged before their bases, causing layover. It can be seen in the image representations that the amount of layover displacement is greatest at short range, where the look angle is smaller.

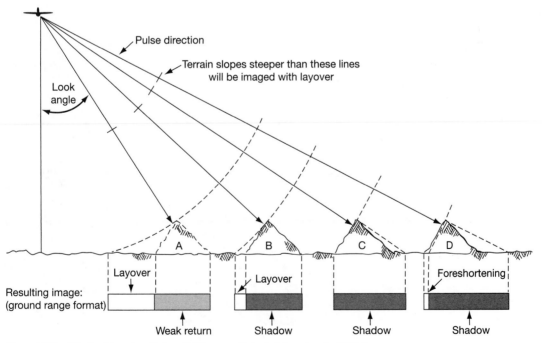

Figure 8.13 Effects of terrain relief on SLR images. (Adapted from Lewis, 1976.)

When the slope facing the antenna is less steep than the line perpendicular to the look direction, as in feature D in Figure 8.13, no layover occurs. That is, the radar pulse reaches the base of the feature before the top. The slopes of the surfaces will not be presented in true size, however. As shown in feature D, the size of the sloped surface is compressed on the image. This *foreshortening effect* becomes more severe as the slope's steepness approaches perpendicularity to the look direction. In feature C, the front slope is precisely perpendicular to the look direction, and it can be seen that the image of the front slope has been foreshortened to zero length.

Figure 8.14, a satellite radar image of the west coast of Vancouver Island, British Columbia, illustrates layover very well. In this image, the effect of layover is very prominent because of the relatively steep look direction of the radar system and the mountainous terrain contained in this image (the satellite track was to the left of the image).

The look angle and terrain slope also affect the phenomenon of *radar shadow*. Slopes facing away from the radar antenna will return weak signals

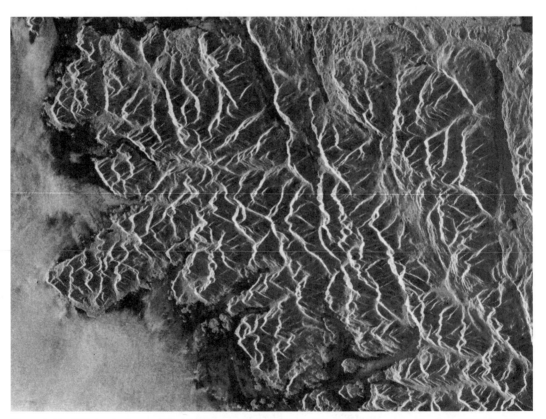

Figure 8.14   ERS-1 radar image, C band, Vancouver Island, British Columbia, midsummer. Scale 1 : 625,000. (Copyright © ESA. Courtesy Canada Centre for Remote Sensing.)

or no signal at all. In Figure 8.13, the right side of feature *A* faces away from the aircraft, but it is less steep than the look direction and will therefore be illuminated by the radar pulse. This illumination, however, will be very slight and the resulting return signals will be weak, causing a fairly dark image area. In feature *B*, its right side is parallel to the look direction and will therefore not be illuminated. As a result, the antenna will receive no return echoes from that side, and the corresponding image area will be totally black. When a slope faces away from the aircraft and is steeper than the look direction, as in features *C* and *D*, the area of nonillumination will extend beyond the sloped area, masking down-range features in a radar shadow. As shown in Figure 8.13, the shadow length increases with range because of the increase in look angle. Thus, a feature that casts an extensive shadow at far range *(D)* could be completely illuminated at close range *(A)*. Note in Figure 8.14 that radar shadows increase toward the right side of the image (farther distance from the satellite).

Parallax

When an object is imaged from two different flight lines, differential relief displacements cause image parallax on SLR images. This allows images to be viewed stereoscopically. Stereo radar images can be obtained by acquiring data from flight lines that view the terrain feature from opposite sides (Figure 8.15*a*). However, because the radar sidelighting will be reversed on

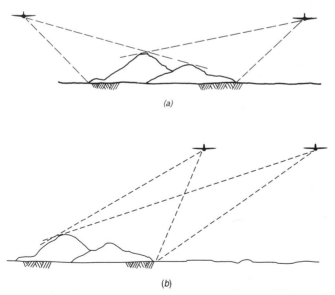

Figure 8.15 Flight orientations to produce parallax on SLR images: (*a*) opposite-side configuration; (*b*) same-side configuration.

the two images in the stereopair, stereoscopic viewing is somewhat difficult using this technique. Accordingly, stereo radar imagery is often acquired from two flight lines at the same altitude on the same side of the terrain feature. In this case, the direction of illumination and the sidelighting effects will be similar on both images (Figure 8.15*b*). It is also possible to acquire stereo radar imagery in the same-side configuration by using different flying heights on the same flight line and, therefore, varying the antenna look angle.

Figure 8.16 shows a stereopair of space radar images of volcanic terrain in Chile that were acquired from two laterally offset flight lines at the same altitude on the same side of the volcano. This resulted in the two different look angles (45° and 54°) that were used for data collection. Although the stereo convergence angle is relatively small (9°), the stereo perception of the imagery is excellent because of the ruggedness of the terrain. The

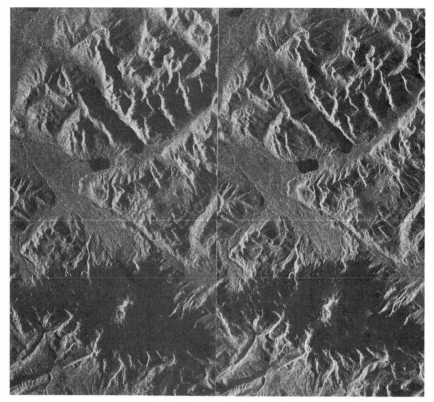

Figure 8.16 Shuttle imaging radar stereopair (SIR-B), Michinmahuida Volcano, Chiloe Province, Chile. Scale 1 : 350,000. The data for this stereopair were collected at two incident angles from the same altitude, with same-side illumination. (Courtesy NASA/JPL/ Caltech.)

volcano near the bottom of the figure is Michinmahuida volcano; it rises 2400 m above the surrounding terrain. The snow-covered slopes of this volcano appear dark toned, because of absorption of the radar signal by the snow.

In addition to providing a stereoscopic view, image parallax may be measured and used to compute approximate feature heights. As with aerial photography, parallax is determined by measuring mutual image displacements on the two images forming a stereomodel. Such measurements are part of the science of *radargrammetry*, a field beyond the scope of our interest in this text. (For further information on radargrammetry, see Leberl, 1990, 1998.)

8.6   TRANSMISSION CHARACTERISTICS OF RADAR SIGNALS

The two primary factors influencing the transmission characteristics of the signals from any given radar system are the wavelength and the polarization of the energy pulse used. Table 8.1 lists the common wavelength bands used in pulse transmission. The letter codes for the various bands (e.g., **K, X, L**) were originally selected arbitrarily to ensure military security during the early stages of radar development. They have continued in use as a matter of convenience, and various authorities designate the various bands in slightly different wavelength ranges.

Naturally, the wavelength of a radar signal determines the extent to which it is attenuated and/or dispersed by the atmosphere. Serious atmospheric effects on radar signals are confined to the shorter operating wavelengths (less than about 4 cm). Even at these wavelengths, under most operating conditions the atmosphere only slightly attenuates the signal. As one would anticipate, attenuation generally increases as operating wavelength decreases, and

TABLE 8.1   Radar Band Designations

Band Designation	Wavelength λ (cm)	Frequency $\nu = c\lambda^{-1}$ [MHz (10^6 cycles sec^{-1})]
K_a	0.75–1.1	40,000–26,500
K	1.1–1.67	26,500–18,000
K_u	1.67–2.4	18,000–12,500
X	2.4–3.75	12,500–8,000
C	3.75–7.5	8000–4000
S	7.5–15	4000–2000
L	15–30	2000–1000
P	30–100	1000–300

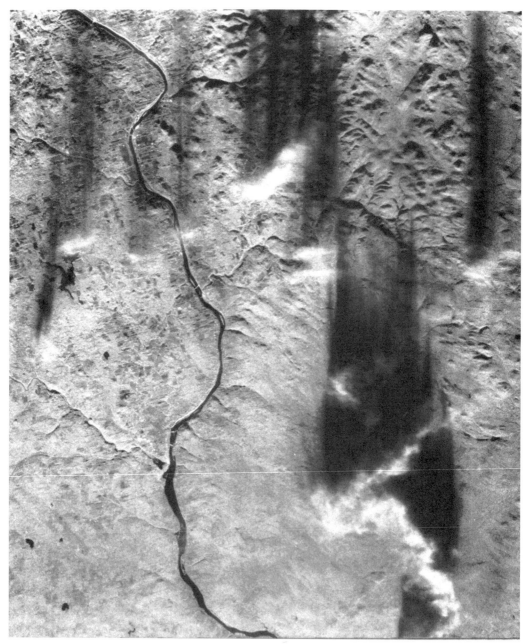

Figure 8.17 X-band airborne side-looking radar data acquired near Woodstock, New Brunswick, Canada, illustrating an unusual shadow effect created by severe rainfall activity and radar signal attenuation. (Courtesy Agriculture and Agri-Food Canada, Fredericton.)

the influence of clouds and rain is variable. Whereas radar signals are relatively unaffected by clouds, echoes from heavy precipitation can be considerable. Precipitation echoes are proportional, for a single drop, to the quantity D^6/λ^4, where D is the drop diameter. With the use of short wavelengths, radar reflection from water droplets is substantial enough to be used in PPI systems to distinguish regions of precipitation. For example, rain and clouds can affect radar signal returns when the radar wavelength is 2 cm or less. At the same time, the effect of rain is minimal with wavelengths of operation greater than 4 cm. With K- and X-band radar, rain may attenuate or scatter radar signals significantly.

Figure 8.17 illustrates an unusual shadow effect created by severe rainfall activity and demonstrates the wavelength dependence of radar systems. In this X-band airborne radar image, the bright "cloudlike" features are due to backscatter from rainfall. The dark regions "behind" these features (especially well seen at lower right) can be created by one of two mechanisms. One explanation is that they are "hidden" because the rain completely attenuates the incident energy, preventing illumination of the ground surface. Alternatively, a portion of the energy penetrates the rain during radar signal transmission but is completely attenuated after backscatter. This is termed *two-way attenuation*. In either case, much less energy is returned to the receiving antenna. When the signal is not completely attenuated, some backscatter is received (upper left).

Another effect of rainfall on radar images is that rain occurring at the time of data acquisition can change the physical and dielectric properties of the surface soil and vegetation, thus affecting backscatter. Figure 8.18 shows an area in Alberta, Canada, with dense forest and clearcut areas. The forested areas present a rougher surface to the radar waves and appear lighter toned than the smoother clearcut areas. Figure 8.18*a* was imaged under dry conditions, and the tonal differences between the forested and clearcut areas are very clear. Figure 8.18*b* was imaged during a period of heavy rain, and the tonal differences between the forested and clearcut areas are subdued.

Irrespective of wavelength, radar signals can be transmitted and/or received in different modes of *polarization*. That is, with polarimetric radar systems, the signal can be filtered in such a way that its electrical wave vibrations are restricted to a single plane perpendicular to the direction of wave propagation. (Unpolarized energy vibrates in all directions perpendicular to that of propagation.) A radar signal can be transmitted in either a horizontal (H) or a vertical (V) plane. Likewise, it can be received in either a horizontal or a vertical plane. Thus, we have the possibility of dealing with four different combinations of signal transmission and reception: H send, H receive (HH); H send, V receive (HV); V send, H receive (VH); and V send, V receive (VV). *Like-polarized* imagery results from the HH or

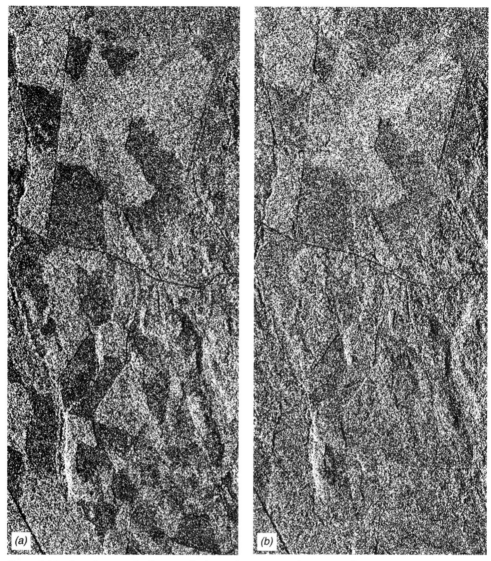

Figure 8.18 Spaceborne radar images of a forested area near Whitecourt, Alberta, Canada, illustrating the effects of heavy rain on radar returns. C band, HH polarization. Figure 8.18*a* was imaged in October under dry conditions, and the distinction between the light-toned forested areas and the darker toned clearcut areas is very clear. Figure 8.18*b* was imaged in August during a period of heavy rain, and the tonal differences between the forested and clearcut areas are much less. (Radarsat images copyright © Canadian Space Agency 1996. Courtesy Canada Centre for Remote Sensing.)

VV combinations. *Cross-polarized* imagery is obtained from HV or VH combinations. Systems that simultaneously collect data in HH, HV, VH, and VV combinations are said to have *quadrature polarization*. *Circular polarization*, where the plane of wave vibrations rotates as the waves propagate, can also be utilized. Since various objects modify the polarization of the energy they reflect to varying degrees, the mode of signal polarization influences how the objects look on the resulting imagery. We illustrate this in Section 8.8. (For further information on radar polarization, see Boerner, 1998.)

For long wavelengths (P band) at high altitude (greater than 500 km), the ionosphere can have significant effects on the transmission of radar signals. These effects occur in at least two ways. First, passage through the ionosphere can result in a propagation delay, which leads to errors in the measurement of the slant range. Second, there is a phenomenon known as *Faraday rotation*, whereby the plane of polarization is rotated somewhat, in direct proportion to the amount of ionospheric activity from the planet's magnetic field. These factors can cause significant problems for polarimetric, long wavelength, orbital SAR systems. (For further information on these effects, see Curlander and McDonough, 1991.)

8.7   OTHER RADAR IMAGE CHARACTERISTICS

Other characteristics that affect the appearance of radar images are *radar image speckle* and *radar image range brightness variation*. These factors are described below.

Radar Image Speckle

Radar images, such as those illustrated here, contain some degree of *speckle*. Microwave signals returning from a given location on the earth's surface can be in phase or out of phase by varying degrees when received by the sensor. This produces a seemingly random pattern of brighter and darker pixels in radar images, giving them a distinctly grainy appearance (or speckle).

Speckle can be reduced through the application of image processing techniques, such as averaging neighboring pixel values, or by special filtering and averaging techniques but cannot be completely eliminated. One technique useful for reducing speckle is *multiple-look processing*. In this procedure, several independent images of the same area, produced by using different portions of the synthetic aperture, are averaged together to

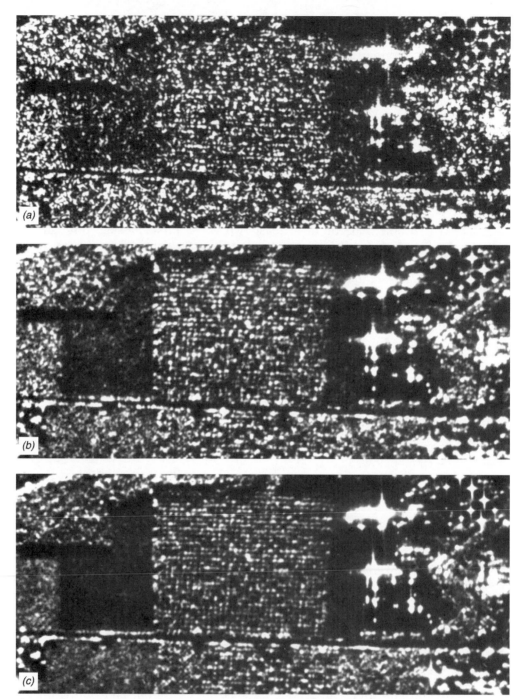

Figure 8.19 An example of multilook processing and its effect on image speckle: (*a*) 1 look; (*b*) 4 looks; (*c*) 16 looks. X-band airborne SAR radar image. Note that speckle decreases as the number of looks increases. These images were specially processed such that the image resolution is the same for all three parts of the figure; 16 times as much data were required to produce the image in (*c*) as the image in (*a*). (From American Society for Photogrammetry and Remote Sensing, 1998. Images copyright © John Wiley & Sons.)

produce a smoother image. The number of statistically independent images being averaged is called the *number of looks,* and the amount of speckle is inversely proportional to the square root of this value. Given that the input data characteristics are held constant, the size of the resolution cell of the output image is directly proportional to the number of looks. For example, a four-look image would have a resolution cell four times larger than a one-look image and a speckle standard deviation one-half that of a one-look image. Hence, both the number of looks and the resolution of a system contribute to the overall quality of a radar image. For further information on number of looks versus other image characteristics, such as speckle and resolution, see Raney (1998) and Lewis, Henderson, and Holcomb (1998).

Figure 8.19 illustrates the appearance of radar images of the same scene with different numbers of looks. The amount of speckle is much less with 4 looks (*b*) than with 1 look (*a*), and even less with 16 looks (*c*). These images were specially processed such that the image resolution is the same for all three parts of the figure; 16 times as much data were required to produce the image in (*c*) as the image in (*a*).

Radar Image Range Brightness Variation

Synthetic aperture radar images often contain a systematic gradient in image brightness across the image in the range direction. This is principally caused by two geometric factors. First, the size of the ground resolution cell decreases from near range to far range, reducing the strength of the return signal. Second, and more significantly, backscatter is inversely related to the local incident angle (i.e., as the local incident angle increases, backscatter decreases), which is in turn related to the distance in the range direction. As a result, radar images will tend to become darker with increasing range. This effect is typically more severe for airborne radar systems than for spaceborne systems because the range of look angles is larger for the airborne systems with a lower flying height (for the same swath width). To some degree, mathematical models can be used to compensate for this effect, resulting in images without visible range brightness illumination effects. Some SAR systems (e.g., SIR-C) correct for the first of these geometric factors (decreasing ground resolution cell size), but not the second (increasing local incident angle), the effects of which are more complex.

Figure 8.20 is an airborne SAR image with a difference in look angle from near to far range of about 14°. Figure 8.20*a* has no compensation for range-related brightness variation, while a simple empirical model has been used to compensate for this effect in Figure 8.20*b*.

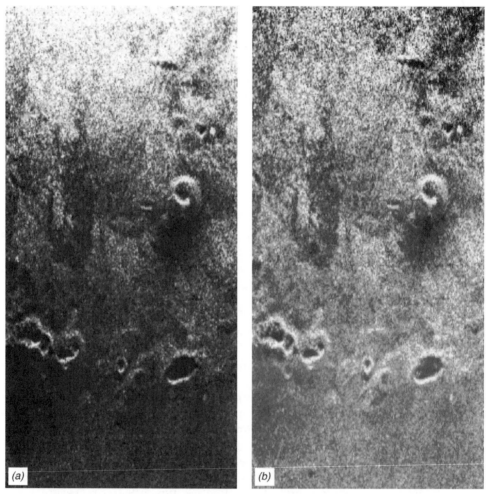

Figure 8.20 Airborne SAR radar images, Hualalai volcano, Hawaii: (*a*) without compensation for range brightness falloff; (*b*) with compensation for range brightness falloff. The difference in look angle from near range (top of the image) to far range (bottom of the image) is about 14°. (Courtesy NASA/JPL/Caltech.)

8.8 RADAR IMAGE INTERPRETATION

Side-looking radar image interpretation has been successful in many fields of application. These include, for example, mapping major rock units and surficial materials, mapping geologic structure (folds, faults, and joints), mapping vegetation types (natural vegetation and crops), determining sea ice types, and mapping surface drainage features (streams and lakes).

Because of its sidelighted character, SLR imagery superficially resembles aerial photography taken under low sun angle conditions. However, a host of earth surface feature characteristics work together with the wavelength, incident angle, and polarization of radar signals to determine the intensity of radar returns from various objects. These factors are many, varied, and complex. Although several theoretical models have been developed to describe how various objects reflect radar energy, most practical knowledge on the subject has been derived from empirical observation. It has been found that the primary factors influencing objects' return signal intensity are their geometric and electrical characteristics; these are described below. The effects of radar signal polarization are illustrated, and radar wave interactions with soil, vegetation, water and ice, and urban areas are also described.

Geometric Characteristics

Again, of the most readily apparent features of radar imagery is its "side-lighted" character. This arises through variations in the relative sensor/terrain geometry for differing terrain orientations, as was illustrated in Figure 8.13. Variations in local incident angle result in relatively high returns from slopes facing the sensor and relatively low returns, or no returns, from slopes facing away from the sensor.

In Figure 8.21, the return-strength-versus-time graph has been positioned over the terrain such that the signals can be correlated with the feature that produced them. Above the graph is the corresponding image line, in which the signal strength has been converted schematically to brightness values. The response from this radar pulse initially shows a high return from the slope facing the sensor. This is followed by a duration of no return signal from areas blocked from illumination by the radar wave. This radar shadow

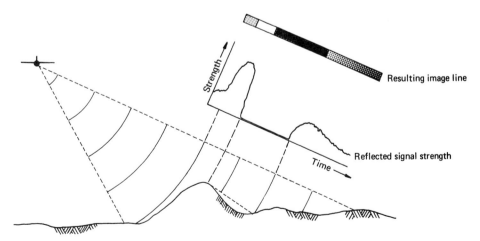

Figure 8.21 Effect of sensor/terrain geometry on SLR imagery. (Adapted from Lewis, 1976.)

is completely black and sharply defined, unlike shadows in photography that are weakly illuminated by energy scattered by the atmosphere. Note that radar shadows can be seen in several radar images in this chapter. Following the shadow, a relatively weak response is recorded from the terrain that is not oriented toward the sensor.

Radar backscatter and shadow areas are affected by different surface properties over a range of local incident angles. As a generalization, for local incident angles of 0° to 30°, radar backscatter is dominated by topographic slope. For angles of 30° to 70°, surface roughness dominates. For angles greater than 70°, radar shadows dominate the image.

Figure 8.22 illustrates radar reflection from surfaces of varying roughness and geometry. The *Rayleigh criterion* states that surfaces can be considered "rough," and act as diffuse reflectors (Figure 8.22a), if the root-mean-square (rms) height of the surface variations exceeds one-eighth of the wavelength of sensing ($\lambda/8$) divided by the cosine of the local incident angle (Sabins, 1997). Such surfaces scatter incident energy in all directions and return a significant portion of the incident energy to the radar antenna. Surfaces are considered "smooth" by the Rayleigh criterion, and act as specular reflectors (Figure 8.22b), when their rms height variation is less than approximately $\lambda/8$ divided by the cosine of the local incident angle. Such surfaces reflect most of the energy away from the sensor, resulting in a very low return signal.

The Rayleigh criterion does not consider that there can be a category of surface relief that is intermediate between definitely rough and definitely smooth surfaces. A *modified Rayleigh criterion* is used to typify such situations. This criterion considers rough surfaces to be those where the rms

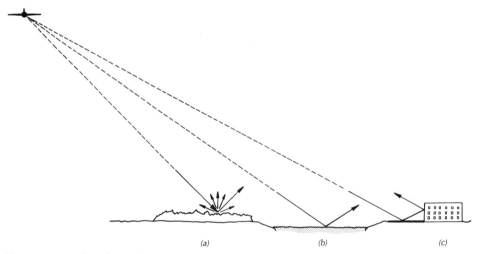

(a)   *(b)*   *(c)*

Figure 8.22 Radar reflection from various surfaces: (*a*) diffuse reflector; (*b*) specular reflector; (*c*) corner reflector.

TABLE 8.2   Definition of Synthetic Aperture Radar Roughness Categories for Three Local Incident Angles[a]

Roughness Category	Root-Mean-Square Surface Height Variation (cm)		
	K_a Band (λ = 0.86 cm)	X Band (λ = 3.2 cm)	L Band (λ = 23.5 cm)
(a) Local incidence angle of 20°			
Smooth	<0.04	<0.14	<1.00
Intermediate	0.04–0.21	0.14–0.77	1.00–5.68
Rough	>0.21	>0.77	>5.68
(b) Local incidence angle of 45°			
Smooth	<0.05	<0.18	<1.33
Intermediate	0.05–0.28	0.18–1.03	1.33–7.55
Rough	>0.28	>1.03	>7.55
(c) Local incidence angle of 70°			
Smooth	<0.10	<0.37	<2.75
Intermediate	0.10–0.57	0.37–2.13	2.75–15.6
Rough	>0.57	>2.13	>15.6

Source: Adapted from Sabins, 1997.

[a]The table is based on a modified Rayleigh criterion.

height is greater than $\lambda/4.4$ divided by the cosine of the local incident angle and smooth when the rms height variation is less than $\lambda/25$ divided by the cosine of the local incident angle (Sabins, 1997). Intermediate values are considered to have intermediate roughnesses. Table 8.2 lists the surface height variations that can be considered smooth, intermediate, and rough for several radar bands for local incident angles of 20°, 45°, and 70°. (Values for other wavelength bands and/or incident angles can be calculated from the information given above.)

Figure 8.23 graphically illustrates how the amount of diffuse versus specular reflection for a given surface roughness varies with wavelength, and Table 8.3 describes how rough various surfaces appear to radar pulses of various wavelengths using the modified Rayleigh criterion described above. It should be noted that some features, such as corn fields, might appear rough when seen in both the visible and the microwave portion of the spectrum. Other surfaces, such as roadways, may be diffuse reflectors in the visible region but specular reflectors of microwave energy. In general, radar images manifest many more specular surfaces than do photographs.

The shape and orientation of objects must be considered as well as their surface roughness when evaluating radar returns. A particularly bright response results from a *corner reflector*, as illustrated in Figure 8.22c. In this case, adjacent smooth surfaces cause a double reflection that yields a very

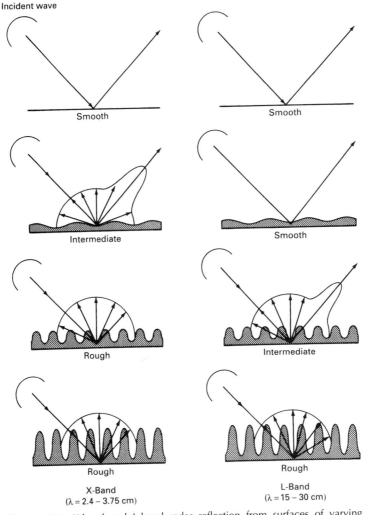

Figure 8.23 X-band and L-band radar reflection from surfaces of varying roughness. (Modified from diagram by Environmental Research Institute of Michigan.)

high return. Because corner reflectors generally cover only small areas of the scene, they typically appear as bright spots on the image.

Electrical Characteristics

The electrical characteristics of terrain features work closely with their geometric characteristics to determine the intensity of radar returns. One

TABLE 8.3   Synthetic Aperture Radar Roughness at a Local Incident Angle of 45°

Root-Mean-Square Surface Height Variation (cm)	K_a Band (λ = 0.86 cm)	X Band (λ = 3.2 cm)	L Band (λ = 23.5 cm)
0.05	Smooth	Smooth	Smooth
0.10	Intermediate	Smooth	Smooth
0.5	Rough	Intermediate	Smooth
1.5	Rough	Rough	Intermediate
10.0	Rough	Rough	Rough

Source: Adapted from Sabins, 1997.

measure of an object's electrical character is the *complex dielectric constant*. This parameter is an indication of the reflectivity and conductivity of various materials.

In the microwave region of the spectrum, most natural materials have a dielectric constant in the range 3 to 8 when dry. On the other hand, water has a dielectric constant of approximately 80. Thus, the presence of moisture in either soil or vegetation can significantly increase radar reflectivity. In fact, changes in radar signal strength from one material to another are often linked to changes in moisture content much more closely than they are to changes in the materials themselves. Because plants have large surface areas and often have a high moisture content, they are particularly good reflectors of radar energy. Plant canopies with their varying complex dielectric constants and their microrelief often dominate the texture of radar image tones.

It should be noted that the dielectric constant of vegetation changes with atmospheric conditions. Clouds limit incident radiation on the earth's surface, changing the water content of the surface vegetation. In particular, clouds decrease or stop vegetation transpiration, which in turn changes the water potential, dielectric constant, and radar backscatter of vegetation.

Metal objects also give high returns, and metal bridges, silos, railroad tracks, and poles usually appear as bright features on radar images. Figure 8.24*a* is a side-looking airborne radar image showing a small urban area (population 6000) adjacent to the Mississippi River. In the urban area, note the high return from the larger buildings of the central business district, which act as corner reflectors, as well as from the metallic bridges crossing the river. The river water acts as a specular reflector and has a very dark image tone. The urban area is located on a river terrace with flat topography. Note the rectangular field patterns resulting from image tone differences caused by differing amounts of diffuse reflection from different crops on this terrace. Figure 8.24*b* shows agricultural land in an area of thick loess (wind blown silt) with moderate relief. Strip farming (to control soil erosion) can be

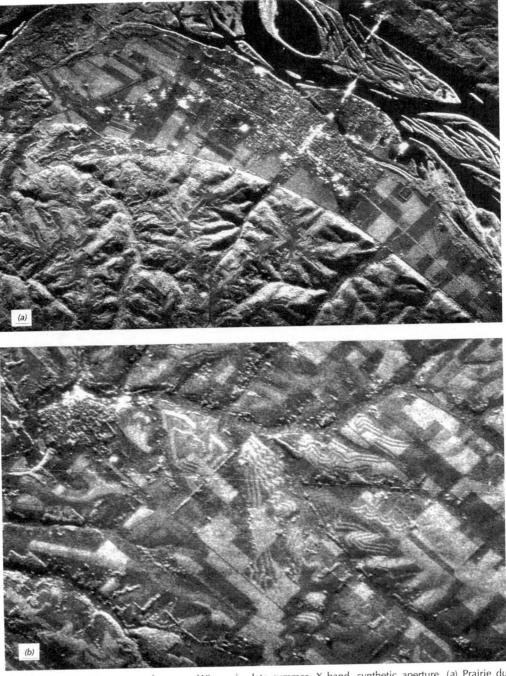

Figure 8.24 SLAR images, southwestern Wisconsin, late summer, X band, synthetic aperture. (a) Prairie du Chien and vicinity. 1 : 80,000. (b) Rural area near Prairie du Chien. 1 : 50,000. (Courtesy Strategic Air Command and Goodyear Aerospace.)

seen on this image as alternations of light-toned and dark-toned crops resembling contour lines.

Effect of Polarization

Figures 8.25 and 8.26 illustrate the effect of radar signal polarization on the resulting images. Figure 8.25 is a side-looking airborne radar image of an area of folded sedimentary rocks obtained using both HH and HV polarizations. A large synclinal mountain is seen in the upper left and center portions of the scene. Note that the slopes facing the top of the page have a lighter tone than those facing the bottom. This is because the flight line was to the top of this image and slopes facing that direction produce a greater return signal. Despite the fact that the topography of this bedrock hill is so strikingly exhibited on this radar image, the radar signal is principally returned from vegetation surfaces. The banding that can be seen around the synclinal mountain is due to an alternation of bedrock types, principally sandstone and shale. Some of the banding results from differences in shading because of topography, and some results from differences in vegetation type and vigor over the different rock formations. Note also the dark tone of the lake at right center and the various rivers in this scene caused by specular reflection of the radar signal. The cross-polarized signal (HV), in this case, results in an image with less image contrast, showing fewer distinctions among vegetation types, than the like-polarized (HH) image. Because the complex manner in which radar signals interact with and return from features is dependent on slope orientation, surface roughness, vegetation cover, and soil and vegetation water content, it is not always possible to predict whether HH or HV images will have a higher information content for a particular application. As shown in the next illustration, there can be conditions where the cross-polarized image has a greater information content than the like-polarized image.

Figure 8.26 is a side-looking airborne radar image showing an area with a variety of soil and rock conditions obtained using both HH and HV polarizations. At the top of the image are dissected bedrock hills with sufficient relief to exhibit considerable shadowing. Below the bedrock hills is a light-toned area of alluvial materials washed down from the bedrock hills. Some braiding of the distributary stream channels can be seen, especially on the HV image. Basaltic lava flows can be seen at C and D. The "Sunshine Basalt" flow (C) issued from Sunshine Crater (20 mm above and to the right of the letter C on this image). The younger "Pisgah Basalt" flow (D) issued from Pisgah Crater, located just outside the lower right-hand corner of this image. The Sunshine Basalt flow has a much darker tone on the HV image than the Pisgah Basalt flow, whereas both have nearly the same tone on the HH image. The small alluvial fan at A has a much lighter tone than the adjacent Sunshine Basalt flow on the HV image but nearly the same tone on the HH image. This lighter tone

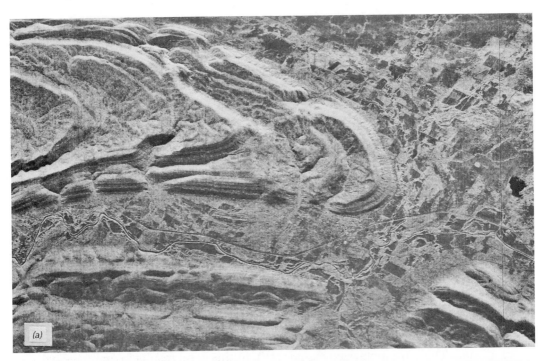

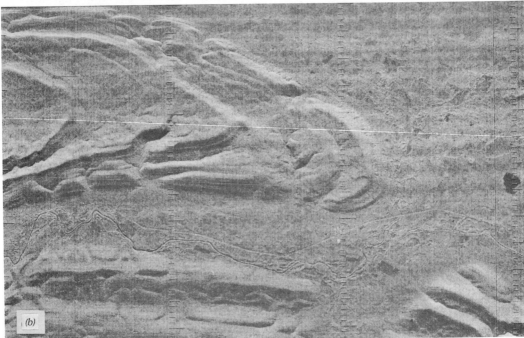

Figure 8.25 SLAR images, Ouachita Mountains, OK, K band, real aperture (scale 1:160,000): (a) HH polarization; (b) HV polarization. (Courtesy Westinghouse Electric Corp.)

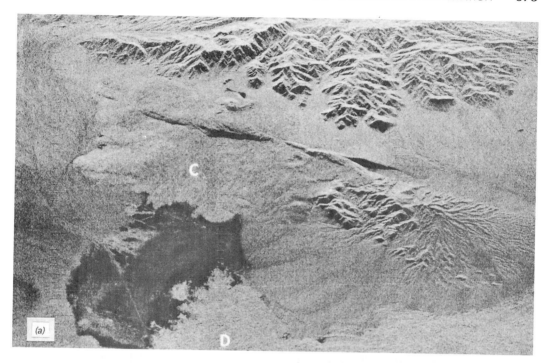

Figure 8.26 SLAR images, Sunshine Crater Area, CA, K band, real aperture (scale 1 : 75,000): (*a*) HH polarization; (*b*) HV polarization. (Courtesy Westinghouse Electric Corp.)

is principally due to the greater density of vegetation on the alluvial fan than on the adjacent Sunshine Basalt flow. The contrast in tones at *B* on the HV image represents the boundary between the lighter toned alluvial materials and the darker toned Sunshine Basalt flows. Note that this boundary is not visible on the HH image. The large dark-toned area at lower left is a playa (dry lakebed) with a mud-cracked clay surface that acts as a specular reflector. The light-toned line running across the playa is a gravel road. Note that the playa is dark toned on both images and the gravel road is light toned.

Soil Response

Because the dielectric constant for water is at least 10 times that for dry soil, the presence of water in the top few centimeters of bare (unvegetated) soil can be detected in radar imagery. Soil moisture and surface wetness conditions become particularly apparent at longer wavelengths. Soil moisture normally limits the penetration of radar waves to depths of a few centimeters. However, signal penetrations of several meters have been observed under extremely dry soil conditions with L-band radar.

Figure 8.27 shows a comparison of Landsat TM (*a*) and spaceborne synthetic aperture radar (*b*) imagery of the Sahara Desert near Safsaf Oasis in southern Egypt. This area, located approximately 190 km west of the lakes shown in Figures 4.26 and 4.27, is largely covered by a thin layer of windblown sand, which obscures many of the underlying bedrock and drainage features. Field studies in the area have shown that L-band (23 cm) radar signals can penetrate up to 2 m through this sand, providing imagery of subsurface geologic features. The dark, braided patterns in (*b*) represent former drainage channels from an ancient river valley, now filled with sand more than 2 m deep. While some of these channels are believed to date back tens of millions of years, other most likely formed during intervals within the past half-million years when the region experienced a wetter climate. Archaeologists working in this area have found stone tools used by early humans more than 100,000 years ago. Other features visible in the radar imagery primarily relate to bedrock structures, which include sedimentary rocks, gneisses, and other rock types. Very few of these features are visible in the Landsat imagery, due to the obscuring sand cover.

Vegetation Response

Radar waves interact with a vegetation canopy as a group of volume scatterers composed of a large number of discrete plant components (leaves, stems, stalks, limbs, etc.). In turn, the vegetation canopy is underlain by soil that may cause surface scattering of the energy that penetrates the vegetation canopy. When the radar wavelengths approximate the mean size of plant components, volume scattering is strong, and if the plant canopy is dense,

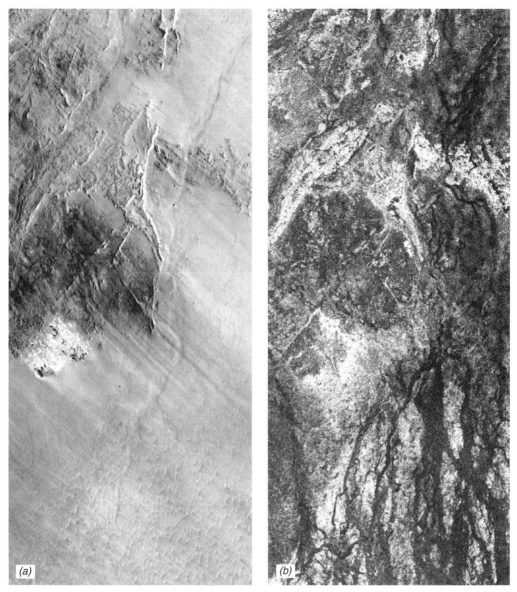

Figure 8.27 Sahara Desert near Safsaf Oasis, souithern Egypt: (*a*) Landsat TM image; (*b*) SIR-C image, L band, HH polarization, 45° incident angle. North is to the upper left. Scale 1 : 170,000. (Courtesy NASA/JPL/Caltech.)

there will be strong backscatter from the vegetation. In general, shorter wavelengths (2 to 6 cm) are best for sensing crop canopies (corn, soybeans, wheat, etc.) and tree leaves. At these wavelengths, volume scattering predominates and surface scattering from the underlying soil is minimal. Longer wavelengths (10 to 30 cm) are best for sensing tree trunks and limbs.

In addition to plant size and radar wavelength, many other factors affect radar backscatter from vegetation. Recall that vegetation with a high moisture content returns more energy than dry vegetation. Also, more energy is returned from crops having their rows aligned in the azimuth direction than from those aligned in the range direction of radar sensing.

Figure 8.28 shows a spaceborne SLR image of a relatively flat area that includes a portion of the Canada/United States border, which appears as an inclined straight line near the bottom of the image. Different agricultural practices emphasize this line, with rangeland predominating in Alberta and wheat fields (at various stages of growth) in Montana. Also clearly evident is the Milk River and its tributaries flowing southeast out of Canada and into the United States.

Figure 8.29 shows a spaceborne SLR image of an agricultural area located in southwest Manitoba. The light-toned circular features are center-pivot irrigation areas with crops that have a much higher moisture content than the nonirrigated crops in the scene. The greater moisture content of the irrigated crops increases the dielectric constant, which in turn increases the reflectivity of the crop surface. That is, leaves with a high moisture content reflect radar waves more strongly than dry leaves of the same plant species. The light-toned

Figure 8.28 ERS-1 radar image, C band, Alberta (Canada)/Montana (United States) border, midsummer. Scale 1 : 700,000. (Copyright © ESA. Courtesy Canada Centre for Remote Sensing.)

Figure 8.29 ERS-1 radar image, C band, southwest Manitoba, midsummer. Scale 1:180,000. (Copyright © ESA. Courtesy Canada Centre for Remote Sensing.)

area at upper right of Figure 8.29 is a marsh (wetland with grasses, sedges, cattails, and rushes). The brightness of this feature is due both to the increased roughness of the vegetation present and to its increased moisture content. Often, vegetated areas that are flooded can cause a corner reflector effect. Each stalk of vegetation forms a right angle with the calm water. Combined, these can produce a bright radar return, which is a useful indicator of water standing beneath a vegetation canopy (see also Figure 8.32).

Figure 8.30 illustrates the effect of wavelength on the appearance of airborne SAR images. Here, the scene is imaged with three different wave-

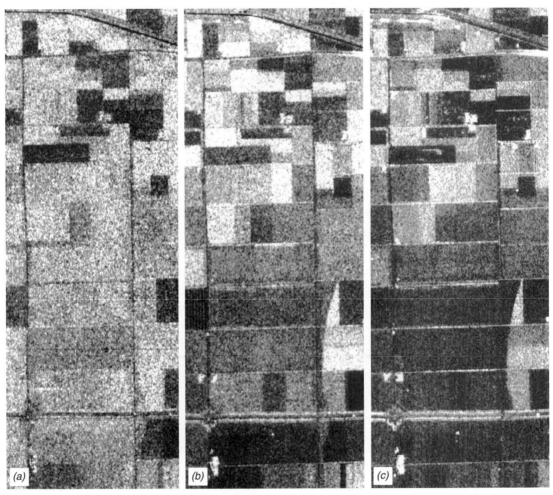

Figure 8.30 Airborne SAR images of an agricultural area in the Netherlands: (a) C band (3.75–7.5 cm); (b) L band (15–30 cm); (c) P band (30–100 cm). HH polarization. (From American Society for Photogrammetry and Remote Sensing, 1998. Images copyright © John Wiley & Sons.)

lengths. Most crop types reflect differently in all three wavelength bands, with generally lighter tones in the C band and darker tones in the P band. Many crop types in this image could be identified by comparing the relative amounts of backscatter in the three different bands.

Figure 8.31 shows a C-band (*a*) and an L-band (*b*) image of an area in northern Wisconsin that is mostly forested, containing many lakes. The look direction is toward the bottom of the images, and the effect of reduced backscatter with increasing range (as described in Section 8.7) can be seen (the near-range ground areas are generally lighter toned than the far-range ground areas). Because of specular reflection from their smooth surfaces, the lakes appear dark throughout both images. A tornado scar can be seen as a dark-toned linear feature running through the center of Figure 8.31*b*, from upper left to lower right. The tornado occurred 10 years before the date of this image. It destroyed many buildings and felled most of the trees in its

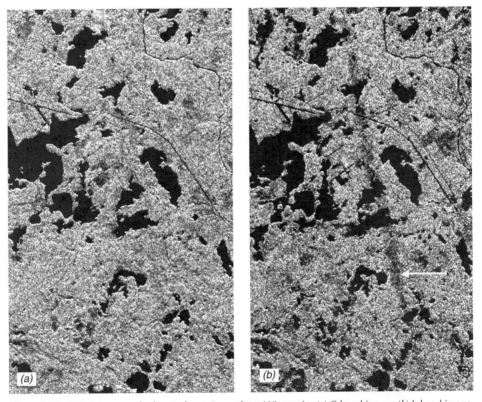

Figure 8.31 SIR-C images of a forested area in northern Wisconsin: (*a*) C-band image; (*b*) L-band image. Scale 1 : 150,000. Note the dark-toned lakes throughout the image and the tornado scar that is visible only in the L-band image. (Courtesy NASA/JPL/Caltech and UW-Madison Environmental Remote Sensing Center.)

path. After the tornado damage, timber salvage operations removed most of the fallen trees, and young trees were established. At the time of acquisition of these spaceborne radar images, the young growing trees in the area of the tornado scar appear rough enough in the C-band (6-cm) image that they blend in with the larger trees in the surrounding forested area. In the L-band (24-cm) image, they appear smoother than the surrounding forested area and the tornado scar can be seen as a dark-toned linear feature.

Incident angle also has a significant effect on radar backscatter from vegetation. Figure 8.32 shows spaceborne SAR images of a forested area in northern Florida and further illustrates the effect of radar imaging at multiple incident angles on the interpretability of radar images. The terrain is flat, with a mean elevation of 45 m. Sandy soils overlay weathering limestone; lakes are sinkhole lakes. Various land cover types can be identified in Figure 8.32*b* by their tone, texture, and shape. Water bodies (W) have a dark tone and smooth texture. Clear-cut areas (C) have a dark tone with a faint mottled texture and rectangular to angular shapes. The powerline right-of-way (P) and roads (R) appear as dark-toned, narrow, linear swaths. Pine forest (F), which covers the majority of this image, has a medium tone with a mottled texture. Cypress-tupelo swamps (S), which consist mainly of deciduous species, have a light tone and a mottled texture. However, the relative tones of the forested areas vary considerably with incident angle. For example, the cypress-tupelo swamp areas are dark toned at an incident angle of 58° and cannot be visually distinguished from the pine forest. These same swamps are somewhat lighter toned than the pine forest at an incident angle of 45° and much lighter toned than the pine forest at an incident angle of 28°. The very high radar return from these swamps on the 28° image is believed to be caused by specular reflection from the standing water in these areas acting in combination with reflection from the tree trunks, resulting in a complex corner reflector effect (Hoffer, Mueller, and Lozano-Garcia, 1985). This effect is more pronounced at an incident angle of 28° than at larger incident angles because the penetration of radar waves through the forest canopy is greater at the smaller angle.

Water and Ice Response

Smooth water surfaces act as specular reflectors of radar waves and yield no returns to the antenna, but rough water surfaces return radar signals of varying strengths. Experiments conducted with the Seasat radar system (L-band system with look angles of 20° to 26°, as described later in Section 8.11) showed that waves with a wavelength greater than 100 m could be detected when wave heights were greater than about 1 m and surface wind speeds exceeded about 2 m/sec (Fu and Holt, 1982). It was also found that waves moving in the range direction (moving toward or away from the radar system) could be detected more readily than waves moving in the azimuth direction.

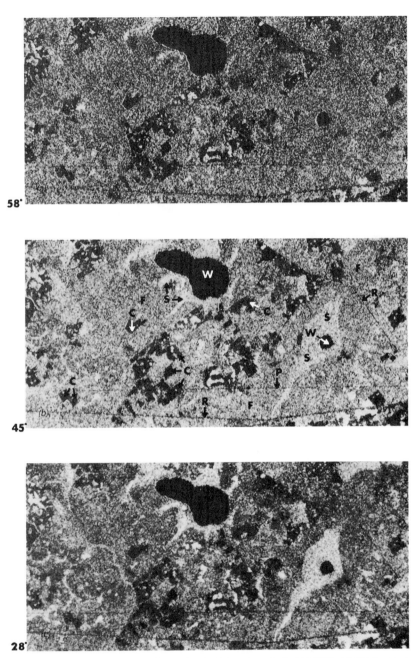

Figure 8.32 SIR-B images, northern Florida, L band (scale 1:190,000): (*a*) 58° incident angle, October 9; (*b*) 45° incident angle, October 10; (*c*) 28° incident angle, October 11. C = clear-cut area; F = pine forest; P = powerline right-of-way; R = road; S = cypress-tupelo swamp; W = open water. (Courtesy Department of Forestry and Natural Resources, Purdue University, and NASA/JPL/Caltech.)

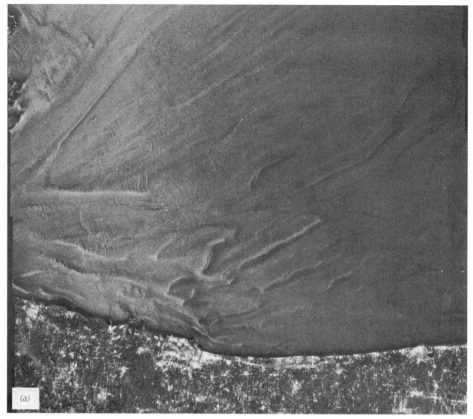

Figure 8.33 English Channel near the Strait of Dover: (a) Seasat SAR image, L band, midsummer; (b) map showing ocean bottom contours in meters. (Courtesy NASA/JPL/Caltech.)

Radar images from space have revealed interesting patterns that have been shown to correlate with ocean bottom configurations. Figure 8.33a is a spaceborne SAR image of the English Channel near the Strait of Dover. Here, the channel is characterized by tidal variations of up to 7 m and reversing tidal currents with velocities at times over 1.5 m/sec. Also, there are extensive sand bars on both sides of the strait and along the coasts of France and England. The sand bars in the channel are long, narrow ridges from 10 to 30 m in depth, with some shallower than 5 m. Together with the high volume of ship traffic, these sand bars make navigation in the channel hazardous. By comparing this image with Figure 8.33b, it can be seen that the surface patterns on the radar image follow closely the sand bar patterns present in the area. Tidal currents at the time of image acquisition were 0.5 to 1.0 m/sec, generally in a northeast-to-southwest direction. The more prominent patterns are visible over bars 20 m or less in depth.

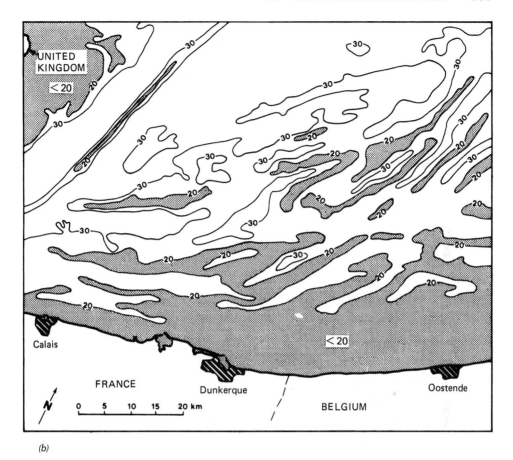

(b)

Figure 8.33   *(Continued)*

Radar backscatter from ice is dependent on the dielectric properties and spatial distribution of the ice. In addition, such factors as ice age, surface roughness, internal geometry, temperature, and snow cover also affect radar backscatter. X- and C-band radar systems have proven useful in determining ice types and, by inference, ice thickness. L-band radar is useful for showing the total extent of ice, but it is often not capable of discriminating ice type and thickness. (See Figure 8.49 for radar images of several forms of ice.)

Urban Area Response

As illustrated in Figure 8.34, urban areas typically appear light toned in SLR images because of their many corner reflectors.

Figure 8.34 Airborne SAR image, Las Vegas, NV, X band, HH polarization. North is to the top, the look direction is from the right. Scale 1 : 250,000. (From American Society for Photogrammetry and Remote Sensing, 1998. Images copyright © John Wiley & Sons.)

Figure 8.35, an airborne SAR image of Sun City, Arizona, illustrates the effect of urban building orientation on radar reflection. The "corner reflection" from buildings located on the part of the circular street system where the wide faces of the houses (front and rear) face the direction from which the radar waves have originated provides the strongest radar returns. At right angles to this direction, there is again a relatively strong radar return where the sides of the houses face the direction from which the radar waves have originated. This effect is sometimes called the *cardinal effect*, a term that has survived from the early days of radar remote sensing. At that time, it was noted that reflections from urban areas, often laid out according to the cardinal directions of a compass, caused significantly larger returns when the linear features were illuminated at an angle orthogonal to their orientation,

Figure 8.35   Airborne SAR image, Sun City, AZ, X band. The look direction is from the top of the image. Scale 1 : 28,000. (From American Society for Photogrammetry and Remote Sensing, 1998. Images copyright © John Wiley & Sons.)

hence the name cardinal effect (Raney, 1998). Other earth surface features also respond with a similar effect. For example, the orientation of row crops affects their response, as described in Section 8.8, and the orientation of ocean waves also strongly affects their response.

Summary

In summary, as a generalization, larger SLR return signals are received from slopes facing the aircraft, rough objects, objects with a high moisture content, metal objects, and urban and other built-up areas (resulting from corner reflections). Surfaces acting as diffuse reflectors return a weak to moderate signal and may often have considerable image texture. Low returns are received from surfaces acting as specular reflectors, such as smooth water, pavements, and playas (dry lakebeds). No return is received from radar "shadow" areas.

8.9 INTERFEROMETRIC RADAR

As discussed in Section 8.5, the presence of differential relief displacement in overlapping radar images acquired from different flight lines produces image parallax. This is analogous to the parallax present in aerial photographs or electro-optical scanner data. Just as photogrammetry can be used to measure surface topography and feature heights in optical images, radargrammetry can be used to make similar measurements in radar images.

In recent years, much attention has been paid to an alternative method for topographic mapping with radar. *Imaging radar interferometry* is based on analysis of the phase of the radar signals as received by two antennas located at different positions in space. As shown in Figure 8.36, the radar signals returning from a single point P on the earth's surface will travel slant-range distances r_1 and r_2 to antennas A_1 and A_2, respectively. The difference between

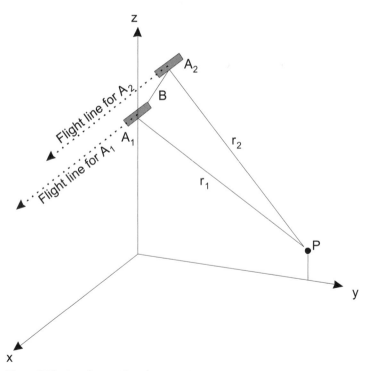

Figure 8.36 Interferometric radar geometry.

lengths r_1 and r_2 will result in the signals being out of phase by some phase difference ϕ, ranging from 0 to 2π radians. If the geometry of the *interferometric baseline* (*B*) is known with a high degree of accuracy, this phase difference can be used to compute the elevation of point *P*.

Figure 8.37 illustrates the interferometric determination of earth's surface elevations. Figure 8.37*a* is a SAR image of a large volcano. Figure 8.37*b* shows an *interferogram*, which displays the phase difference values for each pixel of an interferometric radar data set. The resulting interference pattern consists of a series of stripes, or *fringes*, that represent differences in surface height and sensor position. When the effect of sensor position is removed, a *flattened interferogram* is produced in which each fringe corresponds to a particular elevation range (Figure 8.37*c*).

There are several different approaches to collecting interferometric radar data. In the simplest case, referred to as *single-pass interferometry*, two antennas are located on a single aircraft or satellite platform. One antenna acts as both a transmitter and receiver, while the second antenna acts only as a

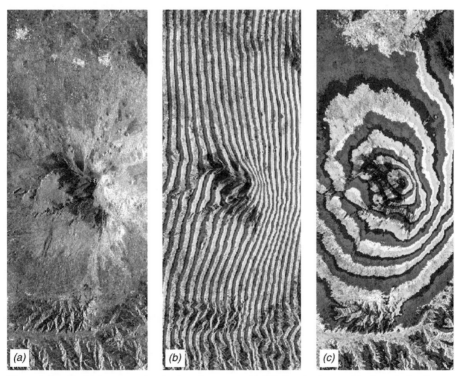

Figure 8.37 Radar image and radar interferograms: (*a*) SIR-C radar image, Mt. Etna, Italy; (*b*) raw interferogram; (*c*) flattened interferogram showing elevation ranges. Parts (*b*) and (*c*) are black and white reproductions of color interferograms. (Courtesy NASA/JPL/Caltech and USGS EROS Data Center.)

receiver. In this case, as shown in Figure 8.36, the interferometric baseline is the physical distance between the two antennas. Alternatively, in *repeat-pass interferometry*, an aircraft or satellite with only a single radar antenna makes two or more passes over the area of interest, with the antenna acting as both a transmitter and receiver on each pass. The interferometric baseline is then the distance between the two flight lines or orbital tracks. It is generally desirable to have the sensor pass as close as possible to its initial position, to keep this baseline small. For airborne repeat-pass interferometry, the flight lines should generally be separated by no more than tens of meters, while for spaceborne systems this distance can be as much as hundreds or thousands of meters.

In repeat-pass interferometry, the position and orientation of objects on the surface may change substantially between passes, particularly if the passes are separated by an interval of days or weeks. This results in a situation known as *temporal decorrelation* in which precise phase matching between the two signals is degraded. For example, in a forested area the individual leaf elements at the top of the canopy may change position due to wind action over the course of a single day. For a short wavelength system such as an X-band SAR, which is highly sensitive to individual leaves and other small features, this decorrelation may limit the use of repeat-pass interferometry substantially. In more arid landscapes, where vegetation is sparse, temporal decorrelation will be less of a problem. Likewise, longer wavelength interferometric radar systems will tend to be less affected by temporal decorrelation than will systems using shorter wavelengths. The use of single-pass interferometry avoids the problem of surface change between passes, so that decorrelation is not a problem.

In some cases, repeat-pass interferometry can actually be used to study surface changes that have occurred between the two passes. In addition to the "before" and "after" images, this approach—known as *differential interferometry*—also requires prior knowledge about the underlying topography. This can be in the form of an existing digital elevation model (DEM), but errors in the DEM will lead to incorrect estimates of surface change. A better approach is to acquire one interferometric image pair from the period before the surface change occurs, for use in combination with a third image acquired after the change. In either case, the phase difference between the before and after images can be corrected to account for topography, with the residual phase differences then representing changes in the position of features on the surface. If the interferometric correlation between the two images is high, these changes can be accurately measured to within a small fraction of the radar system's wavelength—often to less than 1 cm. With a single pair of images, surface changes are measured only as line-of-sight displacements, meaning that only the degree to which a point moved toward or away from the radar look direction can be measured. If two sets of interferometric image pairs are available from different look directions, such as from the ascending and descending segments of a satellite's orbit, the two-dimensional movement of the surface can be derived.

This approach works best for changes that affect large areas in a spatially correlated manner, such as the entire surface of a glacier moving downhill, as opposed to changes that occur in a spatially disjointed manner, such as the growth of trees in a forest. Plate 36 illustrates the use of interferometric radar to monitor continuing ground uplift caused by magma accumulation at depth. Scientists from the USGS Cascades Volcano Observatory, in connection with other agencies, have confirmed the slow uplift of a broad area centered about 5 km west of South Sister volcano in the central Oregon Cascade Range. The radar interferogram shown in Plate 36 was produced using radar data from the European Space Agency's ERS satellites. In this interferogram, each full color band from blue to red represents about 2.8 cm of ground movement in the direction of the radar satellite. (No information is available for the uncolored areas, where forest vegetation, or other factors, hinders the acquisition of useful radar data.) The four concentric bands show that the ground surface moved toward the satellite by as much as 10 cm between August 1996 and October 2000. Surface uplift caused by magma accumulation at depth can be a precursor to volcanic activity at the earth's surface.

Another example of differential interferometry is illustrated in Plate 37. This differential interferogram, derived from ERS-1 and ERS-2 images, shows surface displacement between April 1992 and December 1997 in the area around Las Vegas, Nevada. For most of the past century, pumping of groundwater from an underground aquifer for domestic and commercial consumption has caused land subsidence at a rate of several centimeters per year in Las Vegas, with significant damage to the city's infrastructure. In recent years, artificial recharge of groundwater has been used in an attempt to reduce subsidence. Analysis of the interferometric radar imagery in combination with geologic maps shows that the spatial extent of subsidence is controlled by geologic structures (faults, indicated by white lines in Plate 37) and sediment composition (clay thickness). The maximum detected subsidence during the 1992 to 1997 period was 19 cm. Other potential applications of differential radar interferometry include monitoring the movement of glaciers and ice sheets, measuring displacement across faults after earthquakes, and detecting land subsidence due to oil extraction, mining, and other activities.

8.10   RADAR REMOTE SENSING FROM SPACE

Several satellites have provided operational radar remote sensing from space, including the Almaz-1 system of the former Soviet Union, the ERS-1, ERS-2, and Envisat systems of the European Space Agency, Japan's JERS-1 system, and Canada's Radarsat system.

Among the precursors to these systems were the experimental spaceborne systems Seasat-1 and three Shuttle Imaging Radar systems (SIR-A, SIR-B, and SIR-C). More recently, the Shuttle Radar Topography Mission (SRTM) employed the SIR-C antenna for a brief but highly productive operational program to map global topography using radar interferometry.

In general, images acquired at small incident angles (less than 30°) emphasize variations in surface slope, although geometric distortions due to layover and foreshortening in mountainous regions can be severe. Images with large incident angles have reduced geometric distortion and emphasize variations in surface roughness, although radar shadows increase.

A limitation in the use of airborne radar imagery is the large change in incident angle across the image swath. In these circumstances, it is often difficult to distinguish differences in backscatter caused by variations in incident angle from those actually related to the structure and composition of the surface materials present in an image. Spaceborne radar images overcome this problem because they have only small changes in incident angle. This makes their interpretation less difficult.

Because radar is an active sensor that can gather data both day and night, radar images may be acquired during both a south-to-north (ascending) orbital direction and a north-to-south (descending) orbital direction. A "right-looking" sensor will face east during an ascending orbit and west during a descending orbit. (The opposite is true for a left-looking sensor.)

Some spaceborne systems (e.g., Radarsat and Envisat) use a *ScanSAR* imaging mode in which the radar beam is electronically steered back and forth in the range direction to cover a wider area. In effect, the beam illuminates two or more separate swaths in alternation, with the far-range side of the first swath being contiguous with the near-range side of the second swath, and so on. The multiple swaths are then processed to form a single, wide image. The disadvantage of ScanSAR mode is that the spatial resolution is reduced.

An additional imaging mode, referred to as *Spotlight*, involves steering the radar beam in azimuth rather than in range, in order to dwell on a given site for a longer period of time. As the satellite approaches the target area, the beam is directed slightly forward of the angle at which it is normally transmitted; then, while the satellite moves past, the beam swings back to continue covering the target area. Through an extension of the synthetic aperture principle, this Spotlight mode allows a finer resolution to be achieved by acquiring more "looks" over the target area from a longer segment of the orbit path. This increase in resolution comes at the expense of continuous coverage because while the antenna is focusing on the target area it is missing the opportunity to image other portions of the ground swath. The SIR-C radar mission provided the first tests of both ScanSAR and Spotlight modes from space.

8.11   SEASAT-1

Seasat-1 was the first of a proposed series of satellites oriented toward oceanographic research. The Seasat-1 satellite was launched on June 27, 1978, into an 800-km near-polar orbit. The satellite was designed to provide alternating day and night coverage each 36 hr. Approximately 95 percent of the earth's oceans were to be covered by the system. Unfortunately, prime power system failure 99 days after launch limited the image data produced by the satellite.

An important "first" realized with Seasat-1 was a spaceborne L-band (23.5-cm) SLR system with HH polarization. It was designed to generate imagery across a 100-km swath with a look angle of 20° to 26° and four-look 25 m resolution in both range and azimuth. Table 8.4 summarizes these characteristics (as well as those of the SIR systems).

Although the primary rationale for placing the imaging radar system on board Seasat was its potential for monitoring the global surface wave field and polar sea ice conditions, the resultant images of the oceans revealed a much wider spectrum of oceanic and atmospheric phenomena, including internal waves, current boundaries, eddies, fronts, bathymetric features, storms, rainfalls, and windrows. Seasat was also operated over the world's land areas, and many excellent images illustrating applications to geology, water resources, land cover mapping, agricultural assessment, and other land-related uses were obtained.

Figures 8.3 and 8.33 (described previously) are examples of Seasat-1 imagery.

TABLE 8.4   Characteristics of Major Experimental Synthetic Aperture Radar Systems

Characteristic	Seasat-1	SIR-A	SIR-B	SIR-C
Launch date	June 1978	November 1981	October 1984	April 1994 October 1994
Length of mission	99 days	3 days	8 days	10 days
Nominal altitude, km	800	260	225	225
Wavelength band	L band	L band	L band	X band (X-SAR) C and L bands (SIR-C)
Polarization	HH	HH	HH	HH, HV, VV, VH (X band HH only)
Look angle	20–26° (fixed)	47–53° (fixed)	15–60° (variable)	15–60° (variable)
Swath width, km	100	40	10–60	15–90
Azimuth resolution, m	25	40	25	25
Range resolution, m	25	40	15–45	15–45

8.12 SHUTTLE IMAGING RADAR

A number of early spaceborne radar experiments were conducted using SIR systems—SIR-A in 1981 and SIR-B in 1984. Two SIR-C missions were conducted in 1994. Table 8.4 summarizes the characteristics of these three systems.

SIR-A

The SIR-A experiments were conducted from the Space Shuttle during November 1981. This was the second flight of the Space Shuttle, and the first scientific payload ever flown. The SIR-A system possessed many of the same characteristics as the radar system onboard Seasat. The principal difference between these two was that the SIR-A antenna (9.4 m long) illuminated the earth's surface at a larger look angle (47° to 53°) than Seasat. As with Seasat, an L-band (23.5-cm) system with HH polarization was used. The swath width imaged was 40 km and resolution was 40 m in both range and azimuth. SIR-A obtained imagery over 10 million km^2 of the earth's surface and acquired radar images of many tropical, arid, and mountainous regions for the first time.

Figure 8.38 shows a Landsat MSS band 7 (near-IR) image (*a*) and a SIR-A image (*b*) of an area in the Widyan region of Saudi Arabia and Iraq. This is an area of rugged terrain composed of extensively dissected carbonate rocks (several types of limestone and dolomite). Numerous dry river channels in the area form the regional drainage network. The dry river beds are covered with smooth, dry layers of wind-deposited silt and sand that produce very low radar returns. The outcropping carbonate rocks adjacent to the river beds have rough angular surfaces that produce strong radar returns. The contrast between the light-toned carbonate rocks and the dark-toned dry river beds provides ready discrimination of the dendritic drainage pattern on the radar image. The Landsat image of this area shows little contrast between the river beds and the adjacent carbonate bedrock.

Figure 8.39 is a SIR-A image showing villages, roads, and cultivated fields in eastern China. Each of the hundreds of white spots on this image is a village. The agricultural crops common in this area are winter wheat, kaoliang, corn, and millet. The dark linear and winding features with white lines on each side are rivers and drainageways located between levees.

SIR-B

The SIR-B experiments were conducted from the Space Shuttle during October 1984. Again, an L-band system with HH polarization was used. The prin-

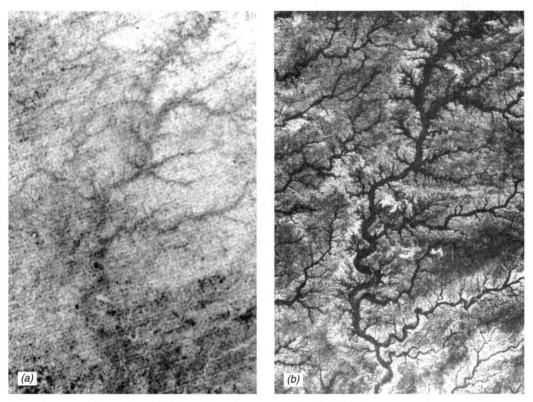

Figure 8.38 Widyan region of Saudi Arabia and Iraq: (*a*) Landsat MSS band 7 image (0.8–1.1 μm), July 1973; (*b*) SIR-A image, L band, November 1981 (scale 1 : 485,000). (Courtesy NASA/JPL/Caltech.)

cipal difference between the SIR-A and SIR-B radar systems is that SIR-B was equipped with an antenna that could be tilted mechanically to beam its radar signals toward the earth at varying look angles (ranging from 15° to 60°). This provided the opportunity for scientific studies aimed at assessing the effect of various incident angles on radar returns. In addition, it provided the opportunity for the acquisition of stereo radar images. The azimuth resolution of SIR-B was 25 m. The range resolution varied from 15 m at a look angle of 60° to 45 m at a look angle of 15°.

Figure 8.40 shows SIR-B images of Mt. Shasta, a 4300-m-high volcano in northern California. These images illustrate the effect of incident angle on elevation displacement. In Figure 8.40*a*, having an incident angle of 60°, the peak of the volcano is imaged near its center. In Figure 8.40*b*, having an incident angle of 30°, the peak is imaged near the top of the figure (the look direction was from top to bottom in this figure). Several light-toned tongues of lava can be seen on the flanks of this strato volcano. The surface of the young lava flow seen at upper left in this radar image consists of unvegetated angular

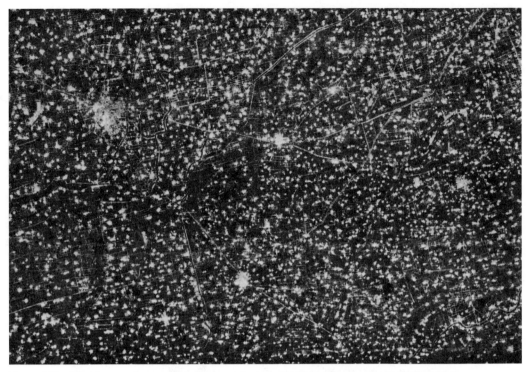

Figure 8.39 SIR-A image of eastern China, L band. Scale 1 : 530,000. (Courtesy NASA/JPL/Caltech.)

chunks of basalt $\frac{1}{3}$ to 1 m in size that present a very rough surface to the L-band radar waves. The other, somewhat darker toned lava flows on the flanks of Mt. Shasta are older, have weathered more, and are more vegetated.

The image parallax associated with data from two different incident angles can be used not only to generate stereopairs of radar data but also to prepare topographic maps and digital elevation models of the imaged terrain. For example, Figure 8.41 shows 12 perspective views of Mt. Shasta that were generated from a digital elevation model extracted from the radar data shown in Figures 8.40a and b. Image parallax was achieved by same-side illumination using the same altitude but different look angles (Section 8.5). Through computer modeling, perspective views from various vantage points around the mountain were calculated for this display.

SIR-C

SIR-C missions were conducted in April and October 1994. SIR-C was designed to explore multiple-wavelength radar sensing from space. Onboard

SIR-C were L-band (23.5-cm) and C-band (5.8-cm) systems from NASA and an X-band (3.1-cm) system (known as X-SAR) from a consortium of the German Space Agency (DARA) and the Italian Space Agency (ASI). The multiple wavelength bands available allowed scientists to view the earth in up to three wavelength bands, either individually or in combination. SIR-C look angles were variable in one-degree increments from 15° to 60°, and four polarizations were available (HH, HV, VV, and VH).

The scientific emphasis in selecting sites for SIR-C image acquisition was on studying five basic themes—oceans, ecosystems, hydrology, geology, and rain and clouds. Ocean characteristics studied included large surface and internal waves, wind motion at the ocean surface, and ocean current motion, as well as sea ice characteristics and distribution. Ecosystem characteristics studied included land use, vegetation type and extent, and the effects of fires, flooding, and clear cutting. Hydrologic studies focused on water and wetland conditions, soil moisture patterns, and snow and glacier cover. Geologic applications included mapping geologic structures (including those buried under dry sand), studying soil erosion, transportation and deposition, and monitoring active volcanoes. Also under study was the attenuation of radar signals at X-band and C-band wavelengths by rain and clouds.

Figure 8.42 is an example of SIR-C imagery. In this black-and-white reproduction of a three-band, multipolarization image, the sinuous dark area crossing the image from lower right to upper left is the Missouri River. The narrow light-toned area on both sides of the river channel is its natural levee. The darker toned band several times wider than the river is its low-lying floodplain. The heavily dissected uplands on both sides of the floodplain can be seen at lower left and upper right, covering more than one-half of the image.

NASA's Jet Propulsion Laboratory (JPL) has released many color-composite SIR-C images from around the world (see Appendix B for website address). Three different wavelength–polarization combinations are used to produce these images, with one combination displayed as blue, one as green, and one as red. If the different wavelength–polarization images show reflection from different features with different intensities, then these features are displayed with different colors. This can be seen in Plate 38, which shows a volcano-dominated landscape in central Africa; parts of Rwanda, Uganda, and the Democratic Republic of the Congo (formerly Zaire) are each present in this image. In this image, C-band data with HH polarization are displayed as blue, C-band data with HV polarization are displayed as green, and L-band data with HV polarization are displayed as red. The volcano at top center is Karisimba, 4500 m high. The green band on the lower slopes of Karisimba volcano, to the right of its peak, is an area of bamboo forest, one of the world's few remaining natural habitats for mountain gorillas. Just right of the center of the image is Nyiragongo volcano, an active volcano 3465 m high. The lower portion of the image is dominated by Nyamuragira volcano,

Figure 8.40   SIR-B images, Mt. Shasta, CA, L band, mid-fall (scale 1 : 240,000): (*a*) 60° incident angle; (*b*) 30° incident angle. Note the severe layover of the mountain top in (*b*). (Courtesy NASA/JPL/Caltech.)

Figure 8.41   Perspective views of Mt. Shasta, CA, generated from the SIR-B data shown in Figure 8.40*a* and *b*. (Courtesy NASA/JPL/Caltech.)

Figure 8.42 SIR-C image, Missouri River near Glasgow, MO. Black and white reproduction of a three-band color composite image (L band HH, L band HV, and L band HH + HV). North is to the bottom of the image. Scale 1 : 220,000. (Courtesy NASA/JPL/Caltech.)

3053 m high, and the many lava flows (purple in this image) that issue from its flanks.

Plate 39 shows SIR-C imagery of a portion of Yellowstone National Park in Wyoming. Yellowstone was the world's first national park and is known for its geological features, including geysers and hot springs. The park and the surrounding region also provide habitat for populations of grizzly bears, elk, and bison. In 1988, massive forest fires burned across some 3200 km² within the park. The intensity of the burn varied widely, leaving a complex mosaic of heavily burned, lightly burned, and unburned areas that will continue to dominate the park's landscape for decades. The effects of these fires can be clearly seen in the L_{VH}-band SIR-C image (*a*), acquired on October 2, 1994.

Unburned lodgepole pine forest returned a strong response in the L_{VH}-band and thus appear relatively bright in this image. Areas of increasing burn intensity have proportionately less above-ground forest biomass present; these areas produced less L_{VH} backscatter and appear darker. Yellowstone Lake, near the bottom of the image, appears black due to specular reflection and negligible L_{VH} backscatter.

Plate 39*b* shows a map of above-ground forest biomass, derived from the SIR-C data shown in (*a*) and from field measurements by the Yellowstone National Biological Survey. Colors in the map indicate the amount of biomass, ranging from brown (less than 4 tons per hectare) to dark green (nonburned forest with a biomass of greater than 35 tons per hectare). Rivers and lakes are shown in blue. The ability of long-wavelength and cross-polarized radar systems to estimate forest biomass may provide a valuable tool for natural resource managers and scientists, whether in the wake of natural disasters such as fires and windstorms, or in the course of routine forest inventory operations.

Due to the experimental nature of the SIR-C system, not all of the imagery from the two missions was processed. The raw data are archived at the USGS EROS Data Center and are currently being processed on demand. Unlike the subsequent Shuttle Radar Topography Mission (Section 8.17), which collected data over the majority of the earth's land surface, SIR-C data were only collected over isolated areas of research interest. Low resolution browse images for all SIR-C data acquisitions can be viewed on the EROS Data Center website. Precision processing of the full resolution data can be ordered at low cost.

8.13 ALMAZ-1

The Soviet Union (just prior to its dissolution) became the first country to operate an earth-orbiting radar system on a commercial basis with the launch of *Almaz-1* on March 31, 1991. Previously, the Soviet Union had operated *Cosmos-1870*, an experimental spaceborne radar system for a period of 2 years, and had collected a large volume of radar images over both land areas and the world's oceans. Although these images were not made generally available during the period of the system's operation, Cosmos-1870 served as a prototype to validate the design and operational capabilities of the Almaz-1 satellite.

Almaz-1 returned to earth on October 17, 1992, after operating for about 18 months. Other Almaz missions were planned prior to the dissolution of the Soviet Union.

Almaz-1 was launched with a nominal altitude of 300 km and an orbit that ranged from approximately 73° N latitude to 73° S latitude. About halfway through its lifetime, the orbital altitude of Almaz-1 was changed from 300 to 360 km in an attempt to prolong its lifetime in orbit. Depending on the region of interest, Almaz-1 provided repeated image coverage at intervals of 1 to 3 days.

The primary sensor on board Almaz-1 was a SAR system operating in the S-band spectral region (10 cm wavelength) with HH polarization. The look angle for the system could be varied by rolling the satellite. The look angle range of 20° to 70° was divided into a standard range of 32° to 50° and two experimental ranges of 20° to 32° and 50° to 70°. The effective spatial resolution varied from 10 to 30 m, depending on the range and azimuth of the area imaged. The data swaths were approximately 350 km wide. Onboard tape recorders were used to record all data until they were transmitted in digital form to a ground receiving station.

8.14 ERS-1, ERS-2, AND ENVISAT

The European Space Agency (ESA) launched its first remote sensing satellite, *ERS-1*, on July 17, 1991, and its successor *ERS-2* on April 21, 1995. Both had a projected life span of at least 3 years; ERS-1 was retired from service on March 10, 2000, and ERS-2 is still in operation. The characteristics of both systems were essentially the same. They were positioned in sun-synchronous orbits at an inclination of 98.5° and a nominal altitude of 785 km. During the 1995 to 2000 period, a particular focus of the two satellites was tandem operation for repeat-pass radar interferometry.

On March 1, 2002, ESA launched *Envisat*. This large satellite platform carries a number of instruments, including the ocean monitoring system MERIS (Section 6.17), and an advanced imaging radar system. Following the launch, Envisat was maneuvered into an orbit matching that of ERS-2, just 30 min ahead of ERS-2 and covering the same ground track to within 1 km. North American data from ERS-1, ERS-2, and Envisat have been provided by receiving stations located in Prince Albert, Saskatchewan, Gatineau, Quebec, and Fairbanks, Alaska.

Sensors Onboard ERS-1 and ERS-2

ERS-1 and ERS-2 carry three principal sensors: (1) a C-band active microwave instrumentation (AMI) module, (2) a Ku-band radar altimeter (a nadir looking instrument for altitude, wind speed, and significant wave height measurements), and (3) an along-track scanning radiometer (a passive instrument consisting of an infrared radiometer and a microwave sounder). We limit this discussion to the AMI, which has a SAR system that operates in either an Image Mode or a Wave Mode and a microwave scatterometer that is used in a Wind Mode. In the Image Mode, the AMI produces SAR data over a 100-km, right-looking swath with a four-look resolution of approximately 30 m, VV polarization, and a 23° look angle (beam center). The high power and data rates for the system preclude onboard storage so the system is operated only while in line-of-sight communication with an ERS ground station.

The AMI Wave Mode is used to measure the radar reflectivity of the ocean surface as it is influenced by waves, yielding information on wavelength and direction of ocean wave systems. In the Wind Mode (a nonimaging mode) sea surface wind speeds and directions are measured.

Examples of ERS Images

The ERS radar systems operate in a shorter wavelength (C) band than many of the operational spaceborne radar systems and also employ a VV polarization, chosen to enhance the reflectivity of the oceanic returns. They also employ a relatively steep incident angle (23°). Thus, ERS images can be expected to look somewhat different than images of the same area obtained with most other operational spaceborne radar systems.

Figure 8.43 is an ERS-1 radar image of the Strait of Gibraltar showing internal waves (wavelength around 2 km) moving from the Atlantic Ocean to the Mediterranean Sea. Internal waves are usually created by the presence of two different layers of water combined with a current effect. In the case of the Strait of Gibraltar, the two layers are caused by different salinities, whereas the current is caused by the tide.

Figure 8.44 is an ERS-1 radar image showing an oil slick (lower right) in the Mediterranean Sea off the coast of France. Oil films have a dampening effect on waves, and the oil-coated smoother water has greater specular reflection than the surrounding water, thus appearing darker on this radar image. Figures 8.14, 8.28, and 8.29 (described previously) are also examples of ERS images.

Sensors Onboard Envisat

As mentioned previously, Envisat carries several sensors, including MERIS (Section 6.17) and the *Advanced Synthetic Aperture Radar (ASAR)* system. The ASAR system operates in the C band, with wavelengths similar to the SAR system onboard ERS-1 and ERS-2. ASAR can be programmed to function in a variety of modes, two of which (Image Mode and Wave Mode) were also implemented on ERS-1 and -2. In Image Mode, ASAR generates four-look high resolution (30-m) images and operates in one of seven predetermined swath configurations, with swath widths ranging from 58 to 109 km and look angles ranging from 14° to 45°. Either HH or VV polarization can be used in Image Mode. Wave Mode is designed to measure radar backscatter from ocean wave action intermittently over large areas. It uses the same swaths and polarizations as Image Mode but only collects data over small samples of the ocean at

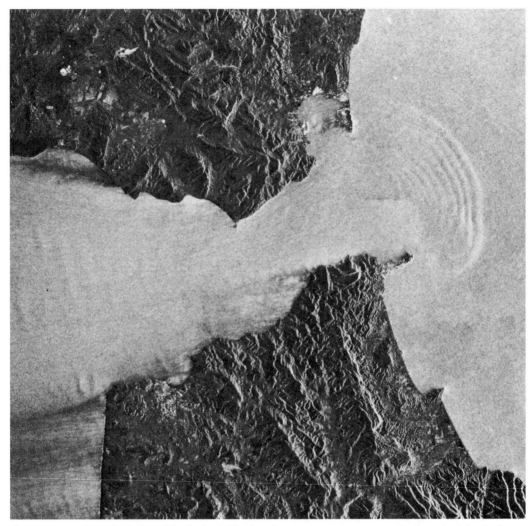

Figure 8.43 ERS-1 radar image of the Strait of Gibraltar showing internal waves moving from the Atlantic Ocean (at left) to the Mediterranean Sea (at right). Scale 1 : 650,000. (Courtesy ESA/ERS-1.)

regular intervals along the swath, rather than collecting a continuous strip of data. This intermittent operation provides a low data rate, such that the data can be stored on board the satellite, rather than being downlinked immediately to the ground station.

Other ASAR modes are based on the ScanSAR technique discussed in Section 8.10. Wide Swath Mode provides images at 150 m resolution, while Global Monitoring Mode uses a coarser resolution of 1 km. Both of these modes cover a 405-km swath with either HH or VV polarization. The final ASAR mode, Al-

Figure 8.44   ERS-1 radar image showing an oil slick (darker area at lower right) in the Mediterranean Sea off the coast of France. Scale 1 : 475,000. (Courtesy ESA/ERS-1.)

ternating Polarization Mode, represents a modified ScanSAR technique that alternates between polarizations over a single swath, rather than alternating between near- and far-range swaths. This mode provides 30-m-resolution dual-polarization imagery, with one of three polarization combinations (HH and VV, VV and VH, or HH and HV).

8.15   JERS-1 AND ALOS

Developed by the National Space Development Agency of Japan, the *JERS-1* satellite was launched on February 11, 1992, and operated until October 12, 1998. It included both a four-band optical sensor (OPS; described earlier in Section 6.15) and an L-band (23-cm) SAR operating with HH polarization. The radar system had a three-look ground resolution of 18 m and covered a swath width of 75 km at a look angle of 35°. The JERS-1 was in a sun-synchronous orbit at an inclination of 98° and an altitude of 568 km. The repeat period for the orbit was 44 days.

Figure 8.45 is a JERS-1 radar image of Mt. Fuji, Japan. The snow-covered peak of Mt. Fuji is dark toned, as was the snow-covered peak of Michinmahuida Volcano that was shown in Figure 8.16, a SIR-B stereopair. Several lakes appear black, and recently clearcut areas appear dark toned, as contrasted with the lighter toned forested areas. Developed areas appear light toned.

The National Space Development Agency of Japan plans to launch an *Advanced Land Observing Satellite (ALOS)* in the summer of 2004. Among the systems onboard this satellite is a *Phased Array L-band Synthetic Aperture Radar (PALSAR)* system. Operating at the L-band, PALSAR will have a cross-

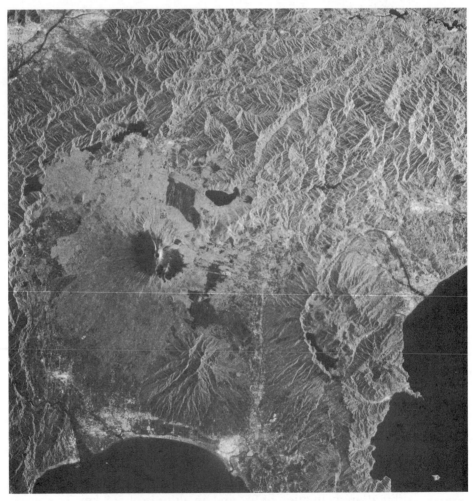

Figure 8.45 JERS-1 radar image showing Mt. Fuji and vicinity, Japan. Scale 1 : 450,000. (NASDA, 1992. Courtesy Remote Sensing Technology Center of Japan.)

track pointing capability over a range of look angles from 18° to 51°. In its fine resolution mode, PALSAR will use an HH or VV single polarization in its standard configuration, with a two-look spatial resolution of 10 m. However, a dual polarization mode will also be available (either HH and HV, or VV and VH) at a spatial resolution of 20 m. In its ScanSAR mode, PALSAR will have a swath width of up to 350 km, with a resolution of about 100 m in both azimuth and range directions, and a polarization of either HH or VV.

8.16   RADARSAT

Radarsat-1, launched on November 28, 1995, is the first Canadian remote sensing satellite. It was developed by the Canadian Space Agency in cooperation with the United States, provincial governments, and the private sector. Canada is responsible for the design, control, and operations of the overall system, while NASA provided the launch services. Radarsat-1 had an initial expected lifetime of 5 years, and as of the time of this writing (2002), it is still fully operational. It will be followed by *Radarsat-2*, scheduled for launch in 2004. Radarsat data for North America are received at the Prince Albert, Saskatchewan, Gatineau, Quebec, and Fairbanks, Alaska, ground stations. The data are distributed commercially worldwide by *Radarsat International* (*RSI*), an arrangement that will continue for Radarsat-2.

The orbit for Radarsat is sun synchronous and at an altitude of 798 km and inclination of 98.6°. The orbit period is 100.7 min and the repeat cycle is 24 days. Radarsat is a right-looking sensor, facing east during the ascending orbit and west during the descending orbit. Because the antenna can be operated at various look angles and swath widths, the system provides 1-day repeat coverage over the high Arctic and approximately 3-day repeat coverage at midlatitudes.

The Radarsat SAR is a C-band (5.6-cm) system with HH polarization. The system can be operated in a variety of beam selection modes providing various swath widths, resolutions, and look angles. In the ScanSAR mode, either two, three, or four single beams are used during data collection. Table 8.5 and Figure 8.46 summarize the modes in which the system operates.

The primary applications for which Radarsat has been designed include ice reconnaissance, coastal surveillance, land cover mapping, and agricultural and forestry monitoring. The near-real-time monitoring of sea ice is important for reducing the navigational risks of Arctic ships. Other uses include disaster monitoring (e.g., oil spill detection, landslide identification, flood monitoring), snow distribution mapping, wave forecasting, ship surveillance in offshore economic zones, and measurement of soil moisture. The different operating modes of the system allow both broad monitoring programs to be conducted as well as more detailed investigations using the fine resolution mode.

TABLE 8.5 Radarsat-1 Beam Selection Modes

Beam Mode	Number of Beam Positions	Swath Width (km)	Look Angle (deg)	Resolution[a] (m)	Number of Looks
Standard	7	100	20–49	25	4
Wide	3	150–165	20–39	30	4
Fine	5+	45	37–48	8	1
Extended high	6	75	50–60	25	4
Extended low	1	170	10–23	35	4
ScanSAR narrow	2	305	20–46	50	2–4
ScanSAR wide	1	510	20–49	100	4–8

[a]Resolution values are approximate. Azimuth and range resolution values differ, and range resolution varies with range (which in turn varies with look angle).

Figure 8.47 shows an area of forest clearcuts in Alberta, Canada, imaged by Radarsat-1 at two different incident angles. Clearcut areas in northern Canada generally have a rough surface (with varying topography, standing tree trunks, new trees, and slash piles), and it has been found that they are difficult to distinguish from forested areas when incident angles are low. The clearcut areas are clearly visible (as dark-toned patches) in (*b*), where the incident angle varies from 45° to 49°, and much less visible in (*a*), where the incident angle varies from 20° to 27°.

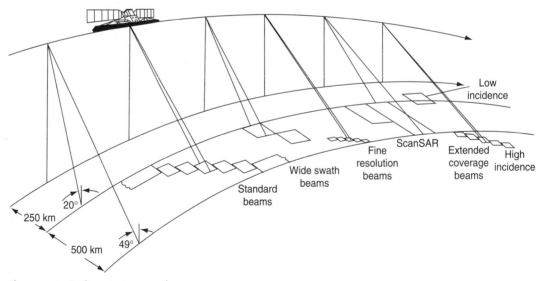

Figure 8.46 Radarsat imaging modes.

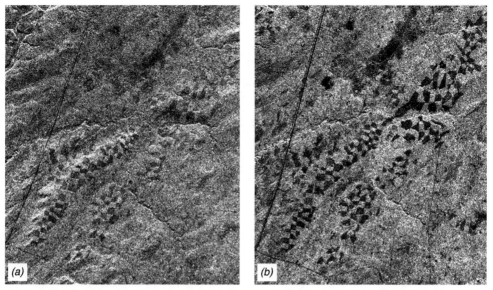

Figure 8.47 Radarsat-1 images illustrating the effect of incident angle on the appearance of clearcuts in a forested area near Whitecourt, Alberta, Canada: (*a*) Incident angle 20 to 27°; (*b*) Incident angle 45 to 49°. Winter image, cold, dry snow ground cover. (Radarsat data copyright © Canadian Space Agency 1996. Received by the Canada Centre for Remote Sensing. Processed and distributed by RADARSAT International. Courtesy Canada Centre for Remote Sensing.)

Figure 8.48 shows flooding of the Red River, Manitoba, Canada, in May 1996. The broad, dark area from lower right to the top of the image is smooth, standing water. The lighter toned areas to the left and upper right are higher, nonflooded, ground. Where standing water is present under trees or bushes, corner reflection takes place and the area appears very light toned. This is especially evident near the Red River. (See also Figure 8.32 for an example of this effect.) The town of Morris can be identified as a light toned rectangle in the flooded area. The town is protected by a levee and, as a result, has not flooded. Other, smaller areas that have not flooded (but are surrounded by water) can also be seen in this image.

Figure 8.49 illustrates the use of Radarsat data for ice type identification in the Gulf of St. Lawrence, Canada. Because radar returns vary with the salinity, surface roughness, and surface wetness of ice, image interpretations can be made to assess ice conditions. Also, images obtained over a period of days and weeks can be used to monitor ice movement. Thin first-year ice floes (A) are distinguishable by their shape and tone relative to the surrounding features. Lower ice salinity and a relatively smooth surface result in low to moderate radar returns (relatively dark-toned areas on the image). "Brash ice" (B) found between first-year ice floes consists of floating ice made up of ice fragments from other

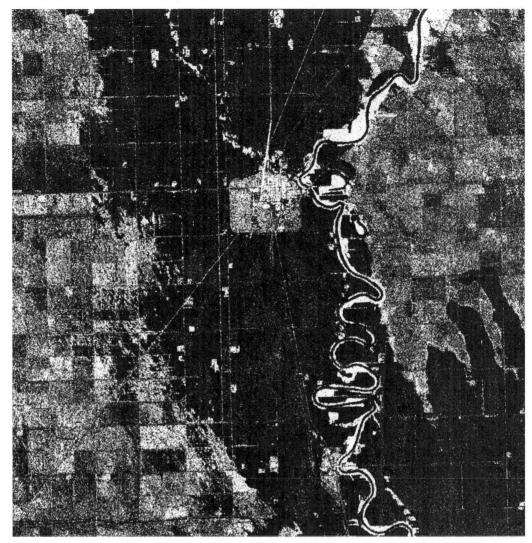

Figure 8.48 Radarsat-1 image showing flooding of the Red River, Manitoba, Canada, May 1996. Standard beam mode, incident angle 30 to 37°. Scale 1:135,000. (Radarsat data copyright © Canadian Space Agency 1996. Received by the Canada Centre for Remote Sensing. Processed and distributed by RADARSAT International. Enhanced and interpreted by the Canada Centre for Remote Sensing.)

forms. This ice type returns moderate to strong amounts of energy and appears light toned. Shown at (C) is a pressure ridge between thin first-year ice floes. These ridges act as corner reflectors and appear light toned, if large enough. A crack, or "lead," within a first-year ice floe is shown at (D). This feature exhibits strong radar returns because of the presence of brash ice within the crack.

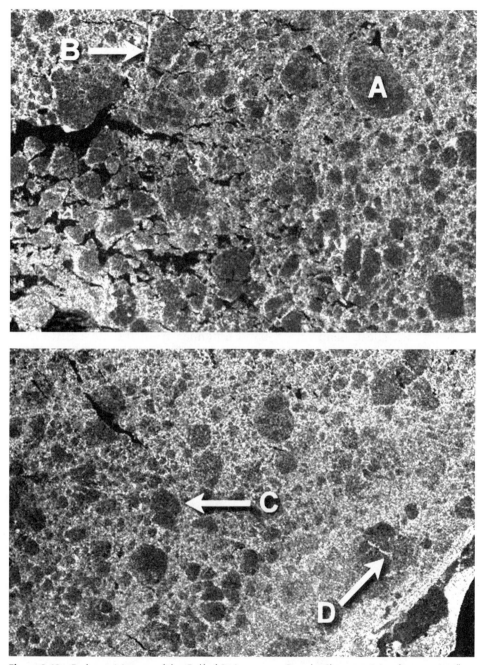

Figure 8.49 Radarsat-1 images of the Gulf of St. Lawrence, Canada. Shown at A is a first-year ice floe; at B is "brash ice"; at C is a pressure ridge; and at D is a crack or "lead" within first-year ice. (Radarsat data copyright © Canadian Space Agency 1996. Courtesy Canada Centre for Remote Sensing and Canadian Ice Service.)

As mentioned earlier, Radarsat-2 is scheduled for launch in 2004. This satellite will provide data continuity to Radarsat-1 users and data for new commercial applications tailored to market needs. Its SAR is a C-band system with HH, VV, HV, and VH polarization options. It has beam selection modes similar to Radarsat-1 but with an increased number of modes available. The various modes support swath widths from 10 to 500 km, look angles from 10° to 60°, resolutions varying from 3 to 100 m, and number of looks varying from 1 to 10.

8.17   SHUTTLE RADAR TOPOGRAPHY MISSION

The *Shuttle Radar Topography Mission (SRTM)* was a joint project of the National Imagery and Mapping Agency (NIMA) and NASA to map the world in three dimensions. During a single Space Shuttle mission on February 11 to 22, 2000, SRTM collected single-pass radar interferometry data covering 119.51 million km^2 of the earth's surface, including over 99.9 percent of the land area between 60° N and 56° S latitude. This represents approximately 80 percent of the total land surface worldwide and is home to nearly 95 percent of the world's population.

The C-band and X-band antennas from the 1994 SIR-C/X-SAR shuttle missions (Section 8.12) were used for data collection. To provide an interferometric baseline suitable for data acquisition from space, a 60-m-long rigid mast was extended when the shuttle was in orbit, with a second pair of C-band and X-band antennas located at the end of the mast. The primary antennas in the shuttle's payload bay were used to send and receive data, while the outboard antennas on the mast operated only in receiving mode. The extendible mast can be seen during testing prior to launch in Figure 8.50. In

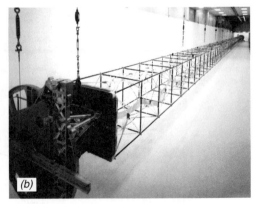

Figure 8.50 The SRTM extendable mast during prelaunch testing: (*a*) view of the mast emerging from the canister in which it is stowed during launch and landing; (*b*) the mast fully extended. (Courtesy NASA/JPL/NIMA.)

Figure 8.50*a* most of the mast is stowed within the canister visible in the background. In Figure 8.50*b* the mast is extended to its full length. The two outboard antennas were not installed on the mast at the time of these tests; during the mission they were mounted on the triangular plate visible at the end of the mast in Figure 8.50. An artist's rendition of the shuttle in orbit during SRTM is provided in Figure 8.51. This illustration shows the position and orientation of the various SRTM components, including the main antenna inside the shuttle's payload bay; the canister for storage of the mast, the mast itself, and the outboard antennas at the end of the mast.

The system collected 12 terabytes of raw data during the 11-day mission, a volume of data that would fill over 15,000 CD-ROMs. Processing this volume of data has taken several years and is not yet complete as of this writing (2002). Plans call for the elevation data to be released as each continent is completed; data for the continental United States were released in 2002 and are being distributed by the U.S. Geological Survey. The SRTM processor produces digital elevation models with a pixel spacing of 1 arcsecond of latitude and longitude (about 30 m). The data for areas outside the United States will be aggregated prior to public distribution and will have a pixel spacing of 3 arcseconds (about 90 m). The absolute horizontal and vertical accuracy of the data are better than 20 and 16 m, respectively. In addition to the elevation data, the SRTM processor produces orthorectified radar image products and maps showing the expected level of error in the elevation model.

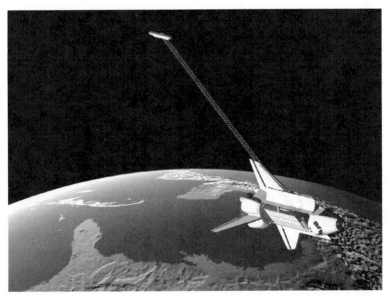

Figure 8.51 Artist's rendition of the shuttle in orbit during SRTM, showing the positions of the main antenna inside the payload bay, the canister, the mast, and the outboard antennas. (Courtesy NASA/JPL/NIMA.)

Figure 8.52 Perspective view of a DEM of the Los Angeles area, derived from SRTM interferometric radar data. A Landsat-7 ETM+ image has been draped over the DEM to show land cover patterns. (Courtesy NASA.)

Figure 8.52 shows a perspective view of a DEM for the Los Angeles metropolitan area. The DEM was derived from interferometric analysis of SRTM imagery, and a Landsat-7 ETM+ image was draped over the DEM. This view is dominated by the San Gabriel Mountains, with Santa Monica and the Pacific Ocean in the lower right and the San Fernando Valley to the left.

Technical problems during the shuttle mission caused 50,000 km^2 of the targeted land area to be omitted by SRTM. These omitted areas represent less than 0.01 percent of the land area intended for coverage. All the omitted areas were located within the United States, where topographic data are already available from other sources. As the first large-scale effort to utilize single-pass radar interferometry for topographic mapping from space, the project has proved to be highly successful. The resulting topographic data and radar imagery represent a unique and highly valuable resource for future geospatial applications.

8.18 SPACEBORNE RADAR SYSTEM SUMMARY

Given the number and diversity of recent past, currently operational, and planned future radar satellites, it is difficult to readily compare the operating

TABLE 8.6   Characteristics of Major Past Operational Spaceborne Synthetic Aperture Radar Systems

Characteristic	Almaz-1 (Soviet Union)	ERS-1 (ESA)	JERS-1 (Japan)
Launch date	March 31, 1991	July 17, 1991	February 11, 1992
End of operations	October 17, 1992	March 10, 2000	October 12, 1998
Altitude, km	300 or 360	785	568
Wavelength band	S band	C band	L band
Polarization modes	Single	Single	Single
Polarization	HH	VV	HH
Look angle	20–70°	23°	35°
Swath width, km	350	100	75
Resolution, m	10–30	30	18

characteristics of the various systems. Tables 8.6, 8.7, and 8.8 are presented to assist in this process. Note that there has been a general trend toward increasing design sophistication, including multiple polarizations, multiple look angles, and multiple combinations of resolution and swath width. However, none of these systems operates at more than one wavelength band—a fact that testifies to the technical challenges involved in designing spaceborne multiwavelength radar systems. In fact, all currently available spaceborne radar systems only operate at the C band. This places significant limits on the range of potential applications of spaceborne radar systems, a circumstance that will not be changed until the scheduled launch of the L-band ALOS PALSAR system in 2004.

TABLE 8.7   Characteristics of Major Current Operational Spaceborne Synthetic Aperture Radar Systems

Characteristic	ERS-2 (ESA)	Radarsat-1 (Canada)	Envisat (ESA)
Launch date	April 21, 1995	November 28, 1995	March 1, 2002
Altitude, km	785	798	785
Wavelength band	C band	C band	C band
Polarization modes	Single	Single	Single, dual
Polarization(s)	VV	HH	HH, VV, HV, VH
Look angle	23°	10–60°	14–45°
Swath width, km	100	45–500	58–405
Resolution, m	30	8–100	30–1000

TABLE 8.8   Characteristics of Planned Future Spaceborne Synthetic Aperture Radar Systems

Characteristic	Radarsat-2 (Canada)	ALOS PALSAR (Japan)
Scheduled launch	2004	2004
Altitude, km	798	692
Wavelength band	C band	L band
Polarization modes	Single, dual, quad	Single, dual
Polarization(s)	HH, VV, HV, VH	HH, VV, HV, VH
Look angle	10–60°	10–51°
Swath width, km	10–500	Up to 350
Resolution, m	3–100	10–100

8.19   PLANETARY EXPLORATION

Synthetic aperture radar is especially well suited for obtaining images of planetary bodies that may be too cloud covered for sensing by optical means. Below we describe missions to Venus and to Saturn's moon Titan.

Radar Remote Sensing of Venus

Launched from the space shuttle *Atlantis* on May 4, 1989, the *Magellan* spacecraft has provided remarkable radar images of the planet Venus. A typical elliptical polar orbit of Magellan around the planet took approximately 3 hr and 15 min and dipped from an altitude of 2100 km above the poles to an altitude of 175 km above the equator. From this orbit, the S-band SAR aboard the spacecraft collected successive image swaths 16,000 km long and 25 km wide, with a multilook spatial resolution of 75 m. The radar was able to penetrate Venus' inhospitable atmosphere, which includes thick clouds of sulfuric acid. The data from the mission are not only advancing the scientific understanding of Venus but are also helping scientists understand more clearly the geologic forces that shaped Earth. Often referred to as Earth's sister planet, Venus is nearly identical to Earth in size and density. Magellan unveiled a new view of Venus' striking volcanism, tectonics, and meteoroid impact processes.

Figure 8.53 is a Magellan image mosaic showing the largest (280-km-diameter) impact crater known to exist on Venus. Named Mead, after American anthropologist Margaret Mead, this multiring crater has a light-toned inner floor that is interpreted as resulting from considerable infilling of the original crater cavity by impact melt and/or by volcanic lavas.

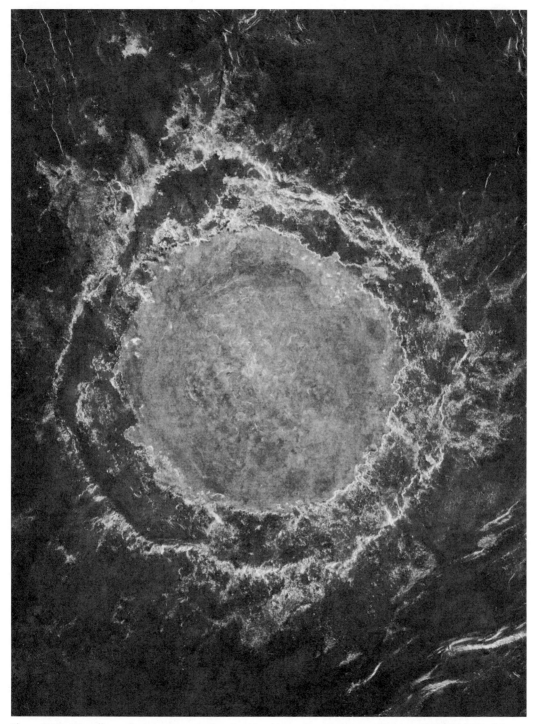

Figure 8.53 Magellan radar image mosaic showing Mead impact crater, Venus. (Courtesy NASA/JPL/Caltech.)

Figure 8.54 is a stereopair of radar images of the Geopert-Meyer crater on Venus, named for the twentieth-century Polish physicist and Nobel laureate. This crater was imaged from the same side at look angles of 28° and 15°. This crater is 35 km in diameter and can be seen near the center of the stereopair. The crater lies above an escarpment at the edge of a ridge belt. Analysis of stereopairs allows planetary scientists to resolve details of topographic relationships on Venusian craters, volcanoes, mountain belts, and fault zones.

Figure 8.55 is a perspective view image of Venus' surface created by merging Magellan radar data with radar altimetry data collected during the mission. Shown is Sapas Mons, from a viewpoint located 243 km to the south. Sapas Mons appears at the center of the image, with two peaks that ascend to

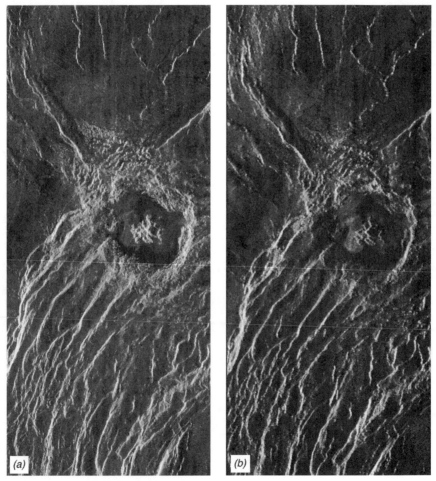

Figure 8.54 Magellan same-side radar stereopair of the crater Geopert-Meyer, Venus: (*a*) 28° incident angle; (*b*) 15° incident angle. (Courtesy NASA/JPL/Caltech.)

Figure 8.55 Magellan perspective view radar image of Sapas Mons, Venus. The vertical scale in this perspective has been exaggerated 10 times. (Courtesy NASA/JPL/Caltech.)

a height of 3 km above the mean surface. Lava flows extend for hundreds of kilometers from the large volcano onto the smooth plains.

Radar Remote Sensing of Saturn's Moon Titan

Launched from Kennedy Space Center on October 15, 1997, NASA's *Cassini* spacecraft is expected to reach Saturn in June 2004. One of several remote sensing systems that will be used to explore Saturn's moon, Titan, is a synthetic aperture radar system that operates in the Ku band (2.2 cm) and has a resolution of 0.35 to 1.7 km. This system has the scientific objectives of studying the geologic features and topography of the solid surface of Titan and to acquire data on non-Titan objects (e.g., Saturn's rings) as conditions permit.

8.20   PASSIVE MICROWAVE SENSING

Operating in the same spectral domain as radar, passive microwave systems yield yet another "look" at the environment—one quite different from that of radar. Being passive, these systems do not supply their own illumination but rather sense the naturally available microwave energy within their field of view. They operate in much the same manner as thermal radiometers and scanners. In fact, passive microwave sensing principles and sensing instrumentation parallel those of thermal sensing in many respects. As with thermal sensing, blackbody radiation theory is central to the conceptual understanding of passive microwave sensing. Again as in thermal sensing, passive microwave sensors exist in the form of both radiometers and scanners. However, passive microwave sensors incorporate antennas rather than photon detection elements.

Most passive microwave systems operate in the same spectral region as the shorter wavelength radar (out to 30 cm). As shown in Figure 8.56, passive microwave sensors operate in the low energy tail of the 300 K blackbody radiation curve typifying terrestrial features. In this spectral region, all objects in

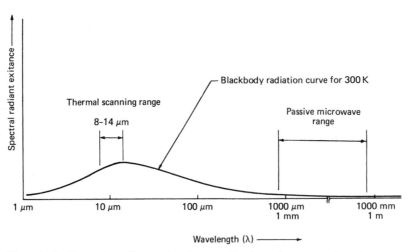

Figure 8.56   Comparison of spectral regions used for thermal versus passive microwave sensing.

the natural environment emit microwave radiation, albeit faintly. This includes terrain elements and the atmosphere. In fact, passive microwave signals are generally composed of a number of source components—some emitted, some reflected, and some transmitted. This is illustrated in Figure 8.57. Over any given object, a passive microwave signal might include (1) an emitted component related to the surface temperature and material attributes of the object, (2) an emitted component coming from the atmosphere, (3) a surface-reflected component from sunlight and skylight, and (4) a transmitted component having a subsurface origin. In short, the intensity of remotely sensed passive microwave radiation over any given object is dependent not only on the object's temperature and the incident radiation but also on the emittance, reflectance, and transmittance properties of the object. These properties in turn are influenced by the object's surface electrical, chemical, and textural characteristics, its bulk configuration and shape, and the angle from which it is viewed.

Because of the variety of its possible sources and its extremely weak magnitude, the signal obtained from various ground areas is "noisy" compared to that which cameras, scanners, or radars provide. The interpretation of this signal is thus much more complex than that of the other sensors discussed. In spite of the difficulties, the utility of passive microwave systems ranges from measuring atmospheric temperature profiles on the one hand to analyzing subsurface variations in soil, water, and mineral content on the other.

The technology of passive microwave sensors has largely been adapted from concepts used in the field of radio astronomy. Major problem areas in the design of such systems include system sensitivity, absolute accuracy, spectral selectivity, and response directionality. Each application involves a particular set

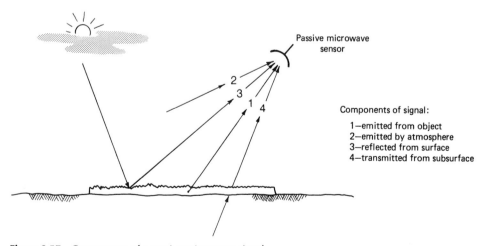

Figure 8.57 Components of a passive microwave signal.

of constraints of system cost, size, weight, power, reliability, operational simplicity, and signal interpretability. Both airborne and satellite systems exist. Here, we consider the basic configuration of airborne radiometers and scanners.

Microwave Radiometers

The basic configuration of a typical microwave radiometer system is shown in Figure 8.58. Scene energy is collected at the antenna. A microwave switch permits rapid, alternate sampling between the antenna signal and a calibration temperature reference signal. The low strength antenna signal is amplified and compared with that of the internal reference signal. The difference between the antenna signal and the reference signal is electronically detected and input to some mode of readout and recording. (It should be noted that we have greatly simplified the operation of a microwave radiometer and that many variations of the design illustrated here exist.)

Common to all radiometer designs is the trade-off between antenna beamwidth and system sensitivity. Because of the very low levels of radiation available to be passively sensed in the microwave region, a comparatively

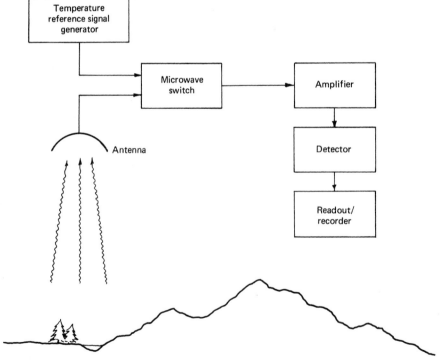

Figure 8.58 Block diagram of a passive microwave radiometer.

large antenna beamwidth is required to collect enough energy to yield a detectable signal. Consequently, passive microwave radiometers are characterized by low spatial resolution.

Microwave radiometers, like thermal radiometers, are nonimaging, profiling devices. Their output is normally digitally recorded on a magnetic medium. During daylight operation, photography can be concurrently acquired to provide a visual frame of reference for the profile data. Normally, the radiometer output is expressed in terms of *apparent antenna temperature*. That is, the system is calibrated in terms of the temperature that a blackbody located at the antenna must reach to radiate the same energy as was actually collected from the ground scene.

Passive Microwave Scanners

To afford the advantages of image output, a number of scanning microwave radiometer systems have been developed. Conceptually, a scanning radiometer operates on the same principle as a profiling system except that its antenna field of view is scanned transverse to the direction of flight. This may be performed mechanically, electronically, or by using a multiple antenna array.

Figure 8.59 shows three segments of a strip of imagery acquired with a scanning passive microwave radiometer, or scanner. The image covers a transect running from Coalinga, California, visible at the western (left) end, to Tulare Lake (dry, now in agriculture) at the eastern (right) end in California's San Joaquin Valley. (Note that the image has the tonal and geometric appearance of thermal scanner imagery. However, in this image bright areas are radiometrically "cold" and dark areas are "warm.") Agricultural fields are visible along the length of the transect. The striping in several of the fields is due to irrigation. The darker fields are natural vegetation or dry bare soil. Radiance measurements made using this type of data have been found to relate quite systematically to the moisture content of the top 50 mm of the soil (Estes, Mel, and Hooper, 1977).

Applications of Passive Microwave Sensing

Like radar imaging, the field of passive microwave sensing has been developed relatively recently, and the interpretation of passive microwave data is still not fully understood. However, there are certain very positive characteristics inherent in this form of remote sensing. As with radar, passive microwave systems can be operated day or night, under virtually all weather conditions. By the appropriate selection of operating wavelength, systems can either look *through* or look *at* the atmosphere. That is, a number of atmospheric "windows" and "walls" exist in microwave regions, principally due to

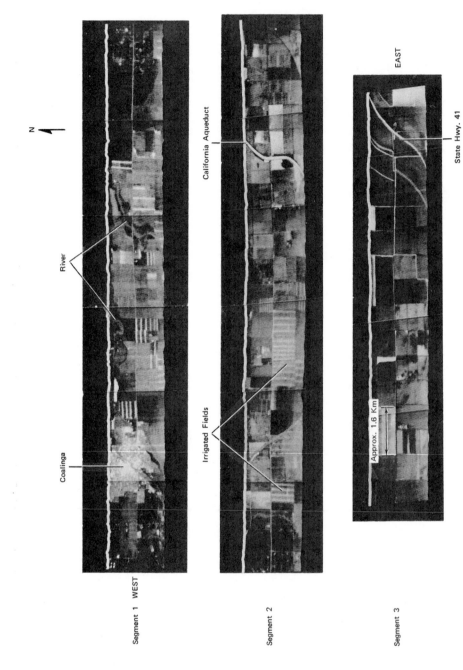

N

Segment 1 WEST

Coalinga

River

Segment 2

Irrigated Fields

California Aqueduct

Segment 3

Approx. 1.6 Km

State Hwy. 41

EAST

Figure 8.59   Passive microwave image transect, Coalinga to Tulare Lake, CA, midsummer, 760 m flying height. (Courtesy Geography Remote Sensing Unit, University of California—Santa Barbara, and Naval Weapons Center, China Lake, CA.)

selective absorption by water vapor and oxygen. Meteorologists are now using selected wavelength microwave sensing to measure atmospheric temperature profiles and to determine the atmospheric distribution of water and ozone.

Passive microwave sensing has strong utility in the field of oceanography. These applications range from monitoring sea ice, currents, and winds to detecting oil pollution in even trace amounts. Though currently sparse in number, investigations related to the utility of passive microwave sensing in hydrology have shown potential for providing information on snow melt conditions, soil temperature, and soil moisture over large areas.

The coarse resolution of passive microwave systems does not preclude their value for synoptic surveillance of many earth resource features that occur over large areas. In fact, from satellite altitudes, the concept of gross scale worldwide resource monitoring with such systems is a present reality. Other useful applications involve the study of features that lie beneath a soil overburden. *Multispectral* microwave radiometry appears to be a means by which we can peer through this overburden. This has obvious potential implications in the field of geology in terms of delineating geologic structure, material changes, subsurface voids, and so on.

8.21 LIDAR

Lidar (which stands for *light detection and ranging*), like radar, is an active remote sensing technique. This technology involves the use of pulses of laser light directed toward the ground and measuring the time of pulse return. The return time for each pulse back to the sensor is processed to calculate the variable distances between the sensor and the various surfaces present on (or above) the ground.

The use of lidar for accurate determination of terrain elevations began in the late 1970s. Initial systems were profiling devices that obtained elevation data only directly under the path of an aircraft. These initial laser terrain systems were complex and not necessarily suited for cost-effective terrain data acquisition over large areas, so their utilization was limited. (Among the primary limitations was the fact that neither airborne GPS nor Inertial Measurement Units were yet available for accurate georeferencing of the raw laser data.) One of the more successful early applications of lidar was the determination of accurate water depths. In this situation the first reflected return records the water surface, closely followed by a weaker return from the bottom of the water body. The depth of the water can then be calculated from the differential travel time of the pulse returns (Figure 8.60).

The advantages of using lidar to supplement or replace traditional photogrammetric methodologies for terrain and surface feature mapping stimulated the development of high-performance scanning systems. Among their

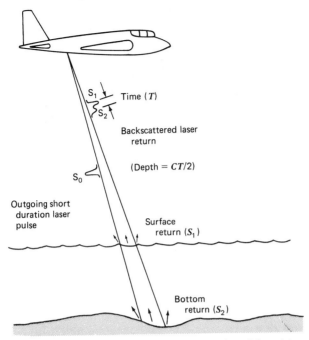

Figure 8.60 Principle of lidar bathymetry. (Adapted from Measures, 1984.)

advantages, these systems afford the opportunity to collect terrain data about steep slopes and shadowed areas (such as the Grand Canyon) and inaccessible areas (such as large mud flats and ocean jetties).

Modern lidar acquisition begins with a "photogrammetric aircraft" equipped with Airborne GPS (for x, y, z sensor location), an Inertial Measuring Unit (for measuring the angular orientation of the sensor with respect to the ground), a rapidly pulsing (20,000 to 50,000 pulses/sec) laser, a highly accurate clock, substantial onboard computer support, reliable electronics, and robust data storage. Flight planning for lidar acquisition requires special considerations. Flights are conducted using a digital flight plan without ground visibility, often at night. Individual flight lines are planned with sufficient overlap (30 to 50 percent) to assure that data gaps do not occur in steep terrain. Areas with dense vegetation cover usually require a narrow field of view, so that most of the lidar pulses are pointing nearly straight down. The distance between sampling points or *post spacing* is derived from the altitude and speed of the aircraft, the scanning angle, and the scan rate. Figure 8.61 portrays the operation of an airborne lidar scanning system.

As with any airborne GPS activity, the lidar system requires a surveyed ground base location to be established in the project area as well as differen-

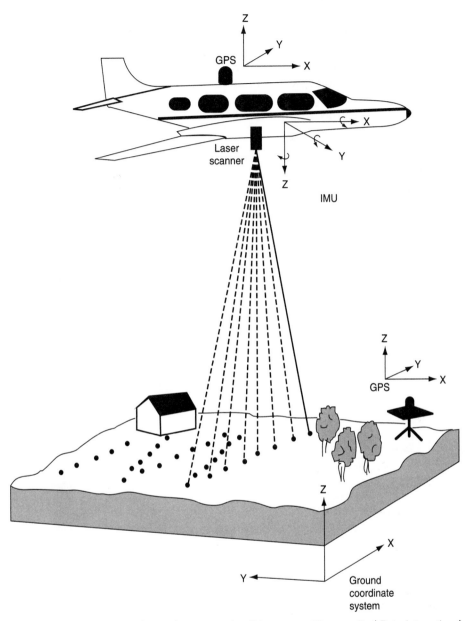

Figure 8.61   Components of an airborne scanning lidar system. (Courtesy EarthData International and Spencer B. Gross, Inc.)

tial postprocessing corrections. In addition, a calibrated alignment process for the GPS position of the sensor and the orientation parameters is required to assure and verify the accuracy of the lidar data sets.

In addition to rapid pulsing, modern systems are able to record up to five returns per pulse, which demonstrates the ability of lidar to discriminate not only such features as a forest canopy and bare ground but also surfaces in between (such as the intermediate forest structure and understory). In urban

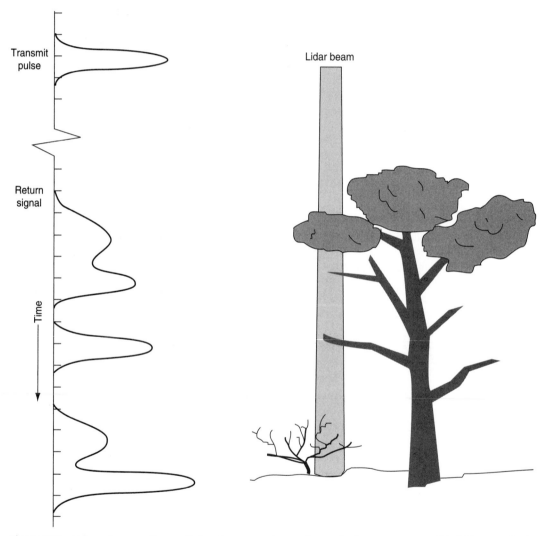

Figure 8.62 Lidar pulse recording multiple returns as various surfaces of a forest canopy are "hit." (Courtesy Earth-Data International.)

areas, the first return of lidar data typically measures the elevations of tree canopies, building roofs, and other unobstructed surfaces. These data sets are easily acquired, processed, and made quickly available to support orthophoto development, high resolution contour production, and bare-earth surface evaluation determination for DEM construction. A distinct advantage to lidar is that all the data are georeferenced from inception, making them inherently compatible with GIS applications. Figure 8.62 illustrates a theoretical pulse emitted from a lidar system and traveling along a line of sight through a forest canopy and recording multiple returns as various surfaces are "hit."

Following the initial postprocessing and positional assurance, the lidar data are filtered for noise removal and are prepared as a file of x, y, z points. Depending on the surface complexity (e.g., variable vegetation heights, terrain changes), the data sets can be remarkably large. Typical suburban-area lidar acquisition produces in excess of 250,000 points per square kilometer (640,000 points per square mile); a forest of relatively open-spaced tall trees with an understory may result in 600,000 points per square kilometer (1.5 million points per square mile) from discrete multiple-return lidar systems.

Plate 40 shows a side view of the "point cloud" of lidar returns from a Douglas-fir forest in the northwestern United States. The data were acquired by an AeroScan lidar system, with recording of up to five returns per pulse. The horizontal spacing of postings on the ground was 1.5 m. In this graphic rendition, the height of each point is indicated both by its three-dimensional perspective position and by its color, ranging from violet at the lowest elevations to red at the highest points. Many of the initial returns in this data set represent points within the forest canopy; close examination shows that the shapes of individual tree crowns can be discerned, primarily in green, yellow, and red at the center of this image. Beneath the tree crowns, the underlying topography at this site consists of a swale or small valley and is indicated by the final lidar return from each pulse.

Software and expertise for defining the first surface (or canopy layer) and the ground are delivering proven results. These lidar applications are particularly well suited for generation of digital DEMs, topographic contouring, and automatic feature extraction. Figure 8.63 illustrates a canopy layer separated from the bare ground to reveal hydrological data hidden beneath the forest canopy.

Applications for forestry assessment of canopy attributes are being established, and research continues for evaluation of crown diameter, canopy closure, and other biophysical properties. Additional applications range from wireless communication design to coastal engineering and survey assessments and volumetric calculations. At the same time, there is increasing use of ground-based lidar scanning systems as well. One such system is the Cyrax 2500 laser scanner manufactured by Cyra Technologies, Inc. This system has a 40° vertical and horizontal field of view, a measurement spot size of 6 mm at a range of 50 m, and a 3D positional accuracy of approximately 6 mm. Such

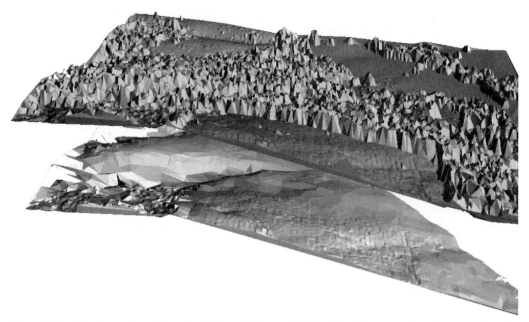

Figure 8.63 Lidar data collected over the McDonald Forest, OR. Note the ability to separate the returns from the forest canopy from those from the ground, revealing terrain information that would normally be obscured by the canopy with other sensing systems. (Courtesy Spencer B. Gross, Inc.)

systems are ideal for characterizing the three-dimensional micro-topography of complex objects and sites (e.g., bridges, buildings, streambeds).

Modern lidar systems have the capability to capture reflectance data from the returning pulses, in addition to the three-dimensional coordinates of the returns. Like the strength of radar returns, the intensity of lidar "echoes" varies with the wavelength of the source energy and the composition of the material returning the incoming signal. For example, commercial lidar systems frequently utilize a 1.064 μm near-infrared wavelength pulse. At this wavelength, the reflectance of snow is 70 to 90 percent, mixed forest cover reflectance is 50 to 60 percent, and black asphalt has a reflectance near 5 percent. The lidar measurements of these reflectance percentage values are referred to as *lidar intensity* and may be processed to produce a georeferenced raster image. The spatial resolution of an intensity image is determined by the lidar point spacing, while the radiometric characteristics of the image are influenced by the shape of the laser footprint and the scan angle of the pulse (e.g., increasing scan angles increase the amount of canopy viewed in a forested area). These intensity images are useful for identification of broad land cover types and can serve as ancillary data for postprocessing.

Figure 8.64 is an example of a lidar intensity image. The image covers a portion of the Lewiston/Nez Perce County Airport, Idaho. It was flown with a

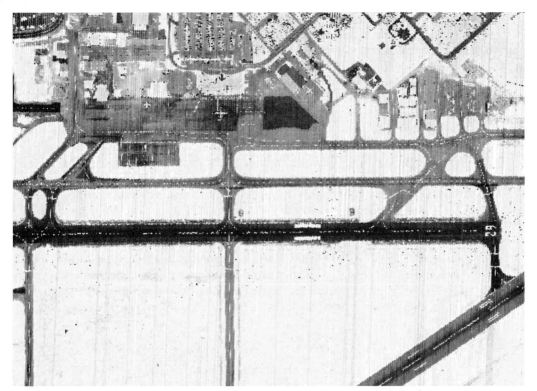

Figure 8.64 Lidar intensity image of a portion of the Lewiston/Nez Perce County Airport, Idaho. The data were acquired by a Leica Geosystems ALS 40 Lidar System at 24,000 pulses per second and 2-m post spacing. (Courtesy Spencer B. Gross, Inc.)

Leica Geosystems ALS 40 Lidar System at 24,000 pulses per second and 2-m post spacing.

The first free-flying satellite lidar system was launched successfully on January 12, 2003. This system, the *Ice, Cloud, and Land Elevation Satellite (ICESat)*, is the latest component of NASA's Earth Observing System (Section 6.18). ICESat carries a single instrument, the *Geoscience Laser Altimeter System (GLAS)*, which includes lidars operating in the near infrared and visible, at 1.064 and 0.532 μm, respectively. GLAS transmits 40 times per second, at an eye-safe level of intensity and measures the timing of the received return pulses to within 1 nsec. The laser footprints at ground level are approximately 70 m in diameter and are spaced 170 m apart along-track.

The primary purpose of ICESat is to collect precise measurements of the mass balance of polar ice sheets and to study how the earth's climate affects ice sheets and sea level. The ICESat lidar can measure very small changes in ice sheets when data are averaged over large areas. This should permit scientists to determine the contribution of the ice sheets of Greenland and Antarctica to global sea level change to within 0.1 cm per decade. In addition to measurements of surface elevation, GLAS is designed to collect vertical profiles of clouds and aerosols within the atmosphere, using sensitive detectors to record backscatter from the 0.532 μm wavelength laser.

Planning has been underway for several years for another satellite lidar system, the *Vegetation Canopy Lidar (VCL)*. One of a series of spacecraft in NASA's Earth System Science Pathfinder (ESSP) program, VCL was scheduled for launch in August 2000 but has been delayed and is currently (2002) on hold. If launched as planned, the system would have five separate lasers operating with a wavelength of 1.064 μm; each laser has a footprint on the ground that is approximately 25 m in diameter. In the along-track direction, the data points would be nearly contiguous, but in the across-track direction there would be a 2-km spacing between footprints. Over a 2-year period, VCL would produce measurements of the ground surface elevation and the top of the vegetation canopy, both with an accuracy of ±1 m, over much of the globe between latitudes 65° N and 65° S. In addition to ground and canopy-top elevations, VCL would measure the vertical distribution of canopy elements (or other surfaces located above the ground surface). Plans call for the data to be archived and distributed at the USGS EROS Data Center in Sioux Falls, South Dakota.

If its mission is successful, VCL will produce a wealth of data on the worldwide distribution of land cover and vegetation structure, a critical component for many earth system models and other global-scale investigations. The more than 1 billion individual data points produced by VCL would also be particularly useful as topographic control points for photogrammetry, radar interferometry, and other topographic mapping applications. It is thus fitting to note that the field of remote sensing is benefitting from the synergistic interactions of multiple sensors, based on diverse principles and technologies, helping us improve our understanding of the earth as a system and our own "position" in the environment.

SELECTED BIBLIOGRAPHY

Abdelsalam, M.G., et al., "Applications of Orbital Imaging Radar for Geologic Studies in Arid Regions: The Saharan Testimony," *Photogrammetric Engineering and Remote Sensing*, vol. 66, no. 6, 2000, pp. 717–726.

Ackermann, F., "Airborne Laser Scanning—Present Status and Future Expectations," *ISPRS Journal of Photogrammetry and Remote Sensing*, vol. 54, nos. 2/3, 1999, pp. 64–67.

Ahmed, S., et al., "The Radarsat System," *IEEE Transactions on Geoscience and Remote Sensing*, vol. 28, no. 4, 1990, pp. 598–602.

Amelung, F., et al., "Sensing the Ups and Downs of Las Vegas: InSAR Reveals Structural Control of Land Subsidence and Aquifer-System Deformation," *Geology*, vol. 27, 1999, pp. 483–486.

American Society for Photogrammetry and Remote Sensing (ASPRS), *Principles and Applications of Imaging Radar*, Manual of Remote Sensing, 3rd ed., Vol. 2, Wiley, New York, 1998.

American Society for Photogrammetry and Remote Sensing (ASPRS), *Digital Elevation Model Technologies and Applications: The DEM Users Manual*, ASPRS, Bethesda, MD, 2001.

Angelis, C.F., et al., "Multitemporal Analysis of Land Use/Land Cover JERS-1 Backscatter in the Brazilian Tropical Rainforest," *International Journal of Remote Sensing*, vol. 23, no. 7, 2002, pp. 1231–1240.

Arzandeh, S., and J. Wang, "Texture Evaluation of RADARSAT Imagery for Wetland Mapping," *Canadian Journal of Remote Sensing*, vol. 28, no. 5, 2002, pp. 653–666.

Baltsavias, E.P., "A Comparison between Photogrammetry and Laser Scanning," *ISPRS Journal of Photogrammetry and Remote Sensing*, vol. 54, nos. 2/3, 1999, pp. 83–94.

Banner, A.V., and F.J. Ahern, "Incidence Angle Effects on the Interpretability of Forest Clearcuts Using Airborne C-HH SAR Imagery," *Canadian Journal of Remote Sensing*, vol. 21, no. 1, 1995, pp. 64–66.

Barber, D.G., J.J. Yackel, and J.M. Hanesiak, "Sea Ice, RADARSAT-1 and Arctic Climate Processes: A Review and Update," *Canadian Journal of Remote Sensing*, vol. 27, no. 1, 2001, pp. 51–61.

Bergen, K.M., et al., "Characterizing Carbon in a Northern Forest by Using SIR-C/X-SAR Imagery," *Remote Sensing of Environment*, vol. 63, no. 1, 1998, pp. 24–39.

Boerner, W.-M., "Polarimetry in Radar Remote Sensing: Basic and Applied Concepts," *Principles and Applications of Imaging Radar*, Manual of Remote Sensing, 3rd ed., Vol. 2, Wiley, New York, 1998, pp. 271–357.

Boisvert, J.B., et al., "Potential of Synthetic Aperture Radar for Large-Scale Soil Moisture Monitoring: A Review," *Canadian Journal of Remote Sensing*, vol. 22, no. 1, 1996, pp. 2–13.

Brown, R.J., "RADARSAT Applications: Review of GlobeSAR Program," *Canadian Journal of Remote Sensing*, vol. 22, no. 4, 1996, pp. 404–419.

Burke, E.J., and L.P. Simmonds, "Passive Microwave Emission from Smooth Bare Soils: Developing a Simple Model to Predict Near Surface Water Content," *International Journal of Remote Sensing*, vol. 22, no. 18, 2001, pp. 3747–3761.

Campbell, B.A., *Radar Remote Sensing of Planetary Surfaces*, Cambridge University Press, New York, 2002.

Canadian Journal of Remote Sensing, Special Issue on the Applications of RADARSAT-1 Data in Data in Geology, vol. 25, no. 3, 1999.

Capstick, D., and R. Harris, "The Effects of Speckle Reduction on Classification of ERS SAR Data," *International Journal of Remote Sensing*, vol. 22, no. 18, 2001, pp. 3627–3641.

Castel, T., et al., "Retrieval Biomass of a Large Venezuelan Pine Plantation Using JERS-1 SAR Data. Analysis of Forest Structure Impact on Radar Signature," *Remote Sensing of Environment*, vol. 79, no. 1, 2002, pp. 30–41.

Chauvaud, S., C. Bouchon, and R. Manière, "Remote Sensing Techniques Adapted to High Resolution Mapping of Tropical Coastal Marine Ecosystems (Coral Reefs, Seagrass Beds, and Mangrove)," *International Journal of Remote Sensing*, vol. 19, no. 18, 1998, pp. 3625–3640.

Cherny, I.V., and V.Y. Raizer, *Passive Microwave Remote Sensing of Oceans*, Wiley, New York, 1998.

Chipman, J.W., et al., "Spaceborne Imaging Radar in Support of Forest Resource Management," *Photogrammetric Engineering and Remote Sensing*, vol. 66, no. 11, 2000, pp. 1357–1366.

Chorowicz, J., et al., "Observation of Recent and Active Landslides from SAR ERS-1 and JERS-1 Imagery Using a Stereo-Simulation Approach: Example of the Chicamocha Valley in Colombia," *International Journal of Remote Sensing*, vol. 19, no. 16, 1998, pp. 3187–3196.

Cihlar, J., et al., Review Article, "Procedures for the Description of Agricultural Crops and Soils in Optical and Microwave Remote Sensing Studies," *International Journal of Remote Sensing*, vol. 8, no. 3, 1987, pp. 441–448.

Cimino, J.B., and C. Elachi, *Shuttle Imaging Radar-A (SIR-A) Experiment*, JPL Publ. 82–77, NASA Jet Propulsion Laboratory, Pasadena, CA, 1982.

Cloude, S.R., "Polarimetric SAR Interferometry," *IEEE Transactions on Geoscience and Remote Sensing*, vol. 36, no. 5, 1998, pp. 1551–1565.

Curlander, J.C., and R.N. McDonough, *Synthetic Aperture Radar: Systems and Signal Processing*, Wiley, New York, 1991.

Dammert, P.B.G., J. Lepparanta, and J. Askne, "SAR Interferometry Over Baltic Sea Ice," *International Journal of Remote Sensing*, vol. 19, no. 16, 1998, pp. 3019–3037.

Del Frate, F., et al., "Neural Networks for Oil Spill Detection Using ERS-SAR Data," *IEEE Transactions on Geoscience and Remote Sensing*, vol. 38, no. 5, 2000, pp. 2282–2287.

deSouze, D.R., et al., "Restoration of Corrupted Optical Fuyo-1 (JERS-1) Data Using Frequency Domain

Techniques," *Photogrammetric Engineering and Remote Sensing*, vol. 62, no. 9, 1996, pp. 1037–1047.

Drobat, S.D., and D.G. Barber, "Towards Development of a Snow Water Equivalence (SWE) Algorithm Using Microwave Radiometry over Snow Covered First-Year Sea Ice," *Photogrammetric Engineering and Remote Sensing*, vol. 64, no. 5, 1998, pp. 415–423.

Elachi, C., *Introduction to the Physics and Techniques of Remote Sensing*, Wiley, New York, 1987a.

Elachi, C., *Spaceborne Radar Remote Sensing: Applications and Techniques*, IEEE Press, New York, 1987b.

Engeset, R.V., et al., "Change Detection and Monitoring of Glacier Mass Balance and Facies Using ERS SAR Winter Images Over Svalbard," *International Journal of Remote Sensing*, vol. 23, no. 10, 2002, pp. 2023–2050.

Estes, J.E., M.R. Mel, and J.O. Hooper, "Measuring Soil Moisture with an Airborne Imaging Passive Microwave Radiometer," *Photogrammetric Engineering and Remote Sensing*, vol. 43, no. 10, 1977, pp. 1273–1281.

Fingas, M.F., and C.E. Brown, "Review of Ship Detection from Airborne Platforms," *Canadian Journal of Remote Sensing*, vol. 27, no. 4, 2001, pp. 379–385.

Flood, M., "Laser Altimetry: From Science to Commercial Lidar Mapping," *Photogrammetric Engineering and Remote Sensing*, vol. 67, no. 11, 2001, p. 1209.

Forget, P., and P. Broche, "Slicks, Waves, and Fronts Observed in a Sea Coastal Area by an X-Band Airborne Synthetic Aperture Radar," *Remote Sensing of Environment*, vol. 57, no. 1, 1996, pp. 1–12.

Franceschetti, G., and R. Lanari, *Synthetic Aperture Radar Processing*, CRC Press, Boca Raton, FL, 1999.

Freeman, A., B. Chapman, and P. Siqueira, "The JERS-1 Amazon Multi-Season Mapping Study (JAMMS): Science Objectives and Implications for Future Missions," *International Journal of Remote Sensing*, vol. 23, no. 7, 2002, pp. 1447–1460.

Friedman, K.S., et al., "Routine Monitoring of Changes in the Columbia Glacier, Alaska, with Synthetic Aperture Radar," *Remote Sensing of Environment*, vol. 70, no. 3, 1999, pp. 257–264.

Fu, L., and B. Holt, *Seasat Views Oceans and Sea Ice with Synthetic-Aperture Radar*, JPL Publ. 81–120, NASA Jet Propulsion Laboratory, Pasadena, CA, 1982.

Gamba, P., and B. Houshmand, "Digital Surface Models and Building Extraction: A Comparison of IFSAR and LIDAR Data," *IEEE Transactions on Geoscience and Remote Sensing*, vol. 38, no. 4, 2000, pp.1959–1968.

Gens, R., and J.L. VanGenderen, "SAR Interferometry—Issues, Techniques, Applications," *International Journal of Remote Sensing*, vol. 17, no. 10, 1996, pp. 1803–1835.

Geudtner, D., et al., "Interferometric Analysis of RADARSAT Strip-Map Mode Data," *Canadian Journal of Remote Sensing*, vol. 27, no. 2, 2001, pp. 95–108.

Guenther, G.C., M.W. Brooks, and P.E. LaRocque, "New Capabilities of the "SHOALS" Airborne Lidar Bathymeter," *Remote Sensing of Environment*, vol. 73, no. 2, 2000, pp. 247–255.

Gupta, R.P., *Remote Sensing Geology*, Springer, Berlin, 1991.

Haack, B.N., N.D. Herold, and M.A. Bechdol, "Radar and Optical Data Integration for Land-Use/Land-Cover Mapping," *Photogrammetric Engineering and Remote Sensing*, vol. 66, no. 6, 2000, pp. 709–716.

Hall, D.K., and J. Martinee, *Remote Sensing of Ice and Snow*, Chapman and Hall, New York, 1985.

Hellwich, O., I. Laptev, and H. Mayer, "Extraction of Linear Objects from Interferometric SAR Data." *International Journal of Remote Sensing*, vol. 23, no. 3, 2002, pp. 461–475.

Helm, A., et al., "Calibration of the Shuttle Radar Topography Mission X-SAR Instrument Using a Synthetic Altimetry Data Model," *Canadian Journal of Remote Sensing*, vol. 28, no. 4, 2002, pp. 573–580.

Hepner, G.F., et al., "Investigation of the Integration of AVIRIS and IFSAR for Urban Analysis," *Photogrammetric Engineering and Remote Sensing*, vol. 64, no. 8, 1998, pp. 813–820.

Hill, M.J., G.E. Donald, and P.J. Vickery, "Relating Radar Backscatter to Biophysical Properties of Temperate Perennial Grassland," *Remote Sensing of Environment*, vol. 67, no. 1, 1999, pp. 15–31.

Hill, R.A., et al., "Landscape Modelling Using Integrated Airborne Multi-Spectral and Laser Scanning Data," *International Journal of Remote Sensing*, vol. 23, no. 11, 2002, pp. 2327–2334.

Hoffer, R.M., P.W. Mueller, and D.F. Lozano-Garcia, "Multiple Incidence Angle Shuttle Imaging Radar Data for Discriminating Forest Cover Types," *Technical Papers of the American Society for Photogrammetry and Remote Sensing*, ACSM-ASPRS Fall Technical Meeting, September 1985, pp. 476–485.

Horgam, G., "Wavelets for SAR Image Smoothing," *Photogrammetric Engineering and Remote Sensing*, vol. 64, no. 12, 1998, pp. 1171–1177.

Huadong, G., *Radar Remote Sensing Applications in China*, Taylor & Francis, New York, 2001.

Imhoff, M.L., et al., "Remotely Sensed Indicators of Habitat Heterogeneity: Use of Synthetic Aperture Radar in Mapping Vegetation Structure and Bird Habitat," *Remote Sensing of Environment*, vol. 60, no. 3, 1997, pp. 217–227.

International Journal of Remote Sensing, Special Issue on Advances in Shuttle Imaging Radar-B Research, vol. 9, no. 5, 1988.

Jones, B., "A Comparison of Visual Observations of Surface Oil with Synthetic Aperture Radar Imagery of the Sea Empress Oil Spill," *International Journal of Remote Sensing*, vol. 22, no. 9, 2001, pp. 1619–1638.

Jordan, A.K., ed., "Electromagnetic Properties of Sea Ice (EMPOSI)," Special Section, *IEEE Transactions on Geoscience and Remote Sensing*, vol. 36, no. 5, 1998.

Kasischke, E.S., and L.L. Bourgeau-Chavez, "Monitoring South Florida Wetlands Using ERS-1 SAR Imagery," *Photogrammetric Engineering and Remote Sensing*, vol. 63, no. 3, 1997, pp. 281–291.

Kendra, J.R., K. Sarabandi, and F.T. Ulaby, "Radar Measurements of Snow: Experiment and Analysis," *IEEE Transactions on Geoscience and Remote Sensing*, vol. 36, no. 3, 1998, pp. 864–879.

Kimura, H., and Y. Yamaguchi, "Detection of Landslide Areas Using Satellite Radar Interferometry," *Photogrammetric Engineering and Remote Sensing*, vol. 66, no. 3, 2000, pp. 337–344.

Koskinen, J., et al., "Snow Monitoring Using Radar and Optical Satellite Data," *Remote Sensing of Environment*, vol. 69, no. 1, 1999, pp. 16–29.

Kunch, T., et al., "Satellite Radar Assessment of Winter Cover Types," *Canadian Journal of Remote Sensing*, vol. 27, no. 6, 2001, pp. 603–615.

Kurvonen, L., J. Pulliainen, and M. Hallikainen, "Retrieval of Biomass in Boreal Forests from Multitemporal ERS-1 and JERS-1 SAR Images," *IEEE Transactions on Geoscience and Remote Sensing*, vol. 37, no. 1, 1999, pp. 198–205.

Kux, H.J.H., F.J. Ahern, and R.W. Pietsch, "Evaluation of Radar Remote Sensing for Natural Resource Management in the Tropical Rainforests of Acre State, Brazil," *Canadian Journal of Remote Sensing*, vol. 21, no. 4, 1995, pp. 430–440.

Leberl, F.W., *Radargrammetric Image Processing*, Artech House, Norwood, MA, 1990.

Leberl, F.W., "Radargrammetry," *Principles and Applications of Imaging Radar*, Manual of Remote Sensing, 3rd ed., Vol. 2, Wiley, New York, 1998, pp. 183–269.

Lee, H., and J.G. Liu, "Analysis of Topographic Decorrelation in SAR Interferometry Using Ratio Coherence Imagery," *IEEE Transactions on Geoscience and Remote Sensing*, vol. 39, no. 2, 2001, pp. 223–232.

Lefsky, M.A., et al., "Lidar Remote Sensing of the Canopy Structure and Biophysical Properties of Douglas-Fir Western Hemlock Forests," *Remote Sensing of Environment*, vol. 70, no. 3, 1999, pp. 339–361.

Lewis, A.J., ed., "Geoscience Applications of Imaging Radar Systems," *Remote Sensing of the Electromagnetic Spectrum*, vol. 3, no. 3, 1976.

Lewis, A.J., F.M. Henderson, and D.W. Holcomb, "Radar Fundamentals: The Geoscience Perspective," *Principles and Applications of Imaging Radar*, Manual of Remote Sensing, 3rd ed., Vol. 2, Wiley, New York, 1998, pp. 131–181.

Lu, Z., and D.J. Meyer, "Study of High SAR Backscattering Caused by an Increase of Soil Moisture over a Sparsely Vegetated Area: Implications for Characteristics of Backscattering," *International Journal of Remote Sensing*, vol. 23, no. 6, 2002, pp. 1063–1074.

Luckman, A., et al., "A Study of the Relationship Between Radar Backscatter and Regenerating Tropical Forest Biomass for Spaceborne SAR Instruments," *Remote Sensing of Environment*, vol. 60, no. 1, 1997, pp. 1–13.

Luckman, A., et al., "Tropical Forest Biomass Density Estimation Using JERS-1 SAR: Seasonal Variation, Confidence Limits, and Application to Image Mosaics," *Remote Sensing of Environment*, vol. 63, no. 2, 1998, pp. 126–139.

Magagi, R.D., Y.H. Kerr, and J.C. Meunier, "Results of Combining L- and C-Band Passive Microwave Airborne Data over the Sahelian Area," *IEEE Transactions on Geoscience and Remote Sensing*, vol. 38, no. 4, 2000, pp. 1997–2008.

Mahafza, B.R., *Introduction to Radar Analysis*, CRC Press, Boca Raton, FL, 1998.

Massom, R.A., et al., "Regional Classes of Sea Ice Cover in the East Antarctic Pack Observed from Satellite and In Situ Data during a Winter Time Period," *Remote Sensing of Environment*, vol. 68, no. 1, 1999, pp. 61–76.

McCauley, J.F., et al., "Subsurface Valleys and Geoarcheology of the Eastern Sahara Revealed by Shuttle Radar," *Science*, vol. 218, no. 4576, December 3, 1982, pp. 1004–1020.

McIntosh, K., and A. Krupnik, "Integration of Laser-Derived DSMs and Matched Image Edges for Generating an Accurate Surface Model," *ISPRS Journal of Photogrammetry and Remote Sensing*, vol. 56, no. 3, 2002, pp. 167–176.

McNairn, H., et al., "Defining the Sensitivity of Multi-Frequency and Multi-Polarized Radar Backscatter to Post-Harvest Crop Residue," *Canadian Journal of Remote Sensing*, vol. 27, no. 3, 2001, pp. 247–262.

Means, J.E., et al., "Use of Large-Footprint Scanning Airborne Lidar to Estimate Forest Stand Characteristics in the Western Cascades of Oregon," *Remote Sensing of Environment*, vol. 67, no. 3, 1999, pp. 298–308.

Measures, R.M., *Laser Remote Sensing*, Wiley, New York, 1984.

Menges, C.H., et al., "Incidence Angle Correction of AirSAR Data to Facilitate Land-Cover Classification," *Photogrammetric Engineering and Remote Sensing*, vol. 67, no. 4, 2001, pp. 479–489.

Metternicht, G.I., and J.A. Zinck, "Evaluating the Information Content of JERS-1 SAR and Landsat TM Data for Discrimination of Soil Erosion Features," *ISPRS Journal of Photogrammetry and Remote Sensing*, vol. 53, no. 3, 1998, pp. 143–153.

Michelson, D.B., B.M. Liljeberg, and P. Pilesjo, "Comparison of Algorithms for Classifying Swedish Landcover Using Landsat TM and ERS-1 SAR Data," *Remote Sensing of Environment*, vol. 71, no. 1, 2000, pp. 1–15.

Milne, A.K., G. Horn, and M. Finlayson, "Monitoring Wetlands Inundation Patterns Using RADARSAT Multitemporal Data," *Canadian Journal of Remote Sensing*, vol. 26, No. 2, 2000, pp. 133–141.

Milne, A.K., and Y. Dong, "Vegetation Mapping Using JERS-1 SAR Mosaic for Northern Australia," *International Journal of Remote Sensing*, vol. 23, no. 7, 2002, pp. 1475–1486.

Moran, M.S., et al., "Combining Multifrequency Microwave and Optical Data for Crop Management," *Remote Sensing of Environment*, vol. 61, no. 1, 1997, pp. 96–109.

Moran, M.S., et al., "Ku- and C-Band SAR for Discriminating Agricultural Crop and Soil Conditions," *IEEE Transactions on Geoscience and Remote Sensing*, vol. 36, no. 1, 1998, pp. 265–272.

Murphy, M.A., I.P. Martini, and R. Protz, "Seasonal Changes in Subarctic Wetlands and River Ice Break-up Detectable on RADARSAT Images, Southern Hudson Bay Lowland, Ontario, Canada," *Canadian Journal of Remote Sensing*, vol. 27, no. 2, 2001, pp. 143–158.

Musick, H., G.S. Schaber, and C.S. Breed, "AIRSAR Studies of Woody Shrub Density in Semiarid Rangeland: Jornada del Muerto, New Mexico," *Remote Sensing of Environment*, vol. 66, 1998, pp. 29–40.

Naesset, E., "Estimating Timber Volume of Forest Stands Using Airborne Laser Scanner Data," *Remote Sensing of Environment*, vol. 61, no. 2, 1997, pp. 246–253.

Nagler, T., H. Rott, and G. Glendinning, "Snowmelt Runoff Modelling by Means of RADARSAT and ERS SAR," *Canadian Journal of Remote Sensing*, vol. 26, no. 6, 2000, pp. 512–520.

Narayanan, R.M., and P.P. Hirsave, "Soil Moisture Estimation Models Using SIR-C SAR Data: A Case Study in New Hampshire, USA," *Remote Sensing of Environment*, vol. 75, no. 3, 2001, pp. 385–396.

NASA-JPL, *SIRCED03: Educational CD-ROM*, with Companion World Wide Web Site http://southport.jpl.nasa.gov/companion/, NASA Jet Propulsion Laboratory, Pasadena, CA, 1997.

Nelson, R., W. Krabill, and J. Tonelli, "Estimating Forest Biomass and Volume Using Airborne Laser Data," *Remote Sensing of Environment*, vol. 24, no. 2, 1988, pp. 247–267.

Nilsson, M., "Estimation of Tree Heights and Stand Volume Using an Airborne Lidar System," *Remote Sensing of Environment*, vol. 56, no. 1, 1996, pp. 1–7.

Paganelli, F., and B. Rivard, "Contribution of the Synergy of RADARSAT-1 and Seismic Imagery Interpretation in the Structural Geology of the Central Alberta Foothills, Canada, as an Aid for Oil and Gas Exploration," *Canadian Journal of Remote Sensing*, vol. 28, no. 5, 2002, pp. 686–700.

Paradella, W.R., et al., "Geological Investigation Using RADARSAT-1 Images in the Tropical Rain Forest Environment of Brazil," *Canadian Journal of Remote Sensing*, vol. 26, no. 2, 2000, pp. 82–90.

Parmuchi, M.G., H. Karszenbaum, and P. Kandus, "Mapping Wetlands Using Multi-Temporal RADARSAT-1 Data and a Decision-Based Classifier," *Canadian Journal of Remote Sensing*, vol. 28, no. 2, 2002, pp. 175–186.

Pierce, L.E., et al., "Multitemporal Land-Cover Classification using SIR-C/X-SAR Imagery," *Remote Sensing of Environment*, vol. 64, no. 1, 1998, pp. 20–33.

Plaster, R.L., "Lidar: Transitioning Technologies from Lab to Applications around the Globe," *Photogrammetric Engineering and Remote Sensing*, vol. 67, no. 11, 2001, pp. 1219.

Power, D., et al., "Iceberg Detection Capabilities of RADARSAT Synthetic Aperture Radar," *Canadian Journal of Remote Sensing*, vol. 27, no. 5, 2001, pp. 476–486.

Pulliainen, J., and M. Hallikainen, "Retrieval of Regional Snow Water Equivalent from Space-Borne Passive Microwave Observations," *Remote Sensing of Environment*, vol. 75, no. 1, 2001, pp. 76–85.

Ramadan, T.M., M.G. Abdelsalam, and R.J. Stern, "Mapping Gold-Bearing Massive Sulfide Deposits in the Neoproterozoic Allaqi Suture, Southeast Egypt with Landsat TM and SIR-C/X SAR Images," *Photogrammetric Engineering and Remote Sensing*, vol. 67, no. 4, 2001, pp. 491–497.

Ramsey, B., et al., "Use of Radarsat Data in the Canadian Ice Service, *Canadian Journal of Remote Sensing*, vol. 24, no. 1, 1998, pp. 36–42.

Raney, R.K., "Radar Fundamentals: Technical Perspective," *Principles and Applications of Imaging Radar*, Manual of Remote Sensing, 3rd ed., Vol. 2, Wiley, New York, 1998, pp. 9–130.

Ranson, K.J., and G. Sun, "Effects of Environmental Conditions on Boreal Forest Classification and Biomass Estimates with SAR," *IEEE Transactions on Geoscience and Remote Sensing*, vol. 38, no. 3, 2000, pp. 1242–1252.

Remote Sensing of Environment, Special Issue on Spaceborne Imaging Radar Mission, vol. 59, no. 2, 1997.

Ribbes, F., and T. Le Toan, "Rice Field Mapping and Monitoring with RADARSAT Data," *International Journal of Remote Sensing*, vol. 20, no. 4, 1999, pp. 745–765.

Rio, J.N.R., and D.F. Lozano-Garcia, "Spatial Filtering of Radar Data (RADARSAT) for Wetlands (Brackish Marshes) Classification," *Remote Sensing of Environment*, vol.73, no. 2, 2000, pp. 143–151.

Sabins, F.F., Jr., *Remote Sensing—Principles and Interpretation*, 3rd ed., W.H. Freeman, New York, 1997.

Saich, P., W.G. Rees, and M. Borgeaud, "Detecting Pollution Damage to Forests in the Kola Peninsula Using the ERS SAR," *Remote Sensing of Environment*, vol. 75, no. 1, 2001, pp. 22–28.

Salgado, H., et al., "Surface Soil Moisture Estimation in Argentina Using RADARSAT-1 Imagery," *Canadian Journal of Remote Sensing*, vol. 27, no. 6, 2001, pp. 685–690.

Saraf, A.K., et al., "Passive Microwave Data for Snow-Depth and Snow-Extent Estimations in the Himalayan Mountains," *International Journal of Remote Sensing*, vol. 20, no. 1, 1999, pp. 83–96.

Sardar, A.M., "The Evolution of Space-Borne Imaging Radar Systems: A Chronological History," *Canadian Journal of Remote Sensing*, vol. 23, no. 3, 1997, pp. 276–280.

Schaber, G.G., "SAR Studies in the Yuma Desert, Arizona: Sand Penetration, Geology, and the Detection of Military Ordnance Debris," *Remote Sensing of Environment*, vol. 67, no. 3, 1999, pp. 320–347.

Schaber, G.G., and C.S. Breed, "The Importance of SAR Wavelength in Penetrating Blow Sand in Northern Arizona," *Remote Sensing of Environment*, vol. 69, no. 2, 1999, pp. 87–104.

Schmullius, C.C., and D.L. Evans, "Synthetic Aperture Radar (SAR) Frequency and Polarization Requirements for Applications in Ecology, Geology, Hydrology, and Oceanography: A Tabular Status Quo After SIR-C/X-SAR," *International Journal of Remote Sensing*, vol. 18, no. 13, 1997, pp. 2713–2722.

Schoups, G., P.A. Troch, and N. Verhoest, "Soil Moisture Influences on the Radar Backscattering of Sugar Beet Fields," *Remote Sensing of Environment*, vol. 65, no. 2, 1998, pp. 184–194.

Schuler, D.L., et al., "Topographic Mapping Using Polarimetric SAR Data," *International Journal of Remote Sensing*, vol. 19, no. 1, 1998, pp. 141–160.

Shokr, M.E., B. Ramsay, and J.C. Falkingham, "Operational Use of ERS-1 SAR Images in the Canadian Ice Monitoring Programme," *International Journal of Remote Sensing*, vol. 17, no. 4, 1996, pp. 667–682.

Short, N.H., E.L. Simms, and J.D. Jacobs, "RADARSAT SAR Applied to Ice Margin Mapping: The Barnes Ice Cap, Nunavut, Canada," *Canadian Journal of Remote Sensing*, vol. 26, no. 2, 2000, pp. 91–102.

Siegal, B.S., and A.R. Gillespie, eds., *Remote Sensing in Geology*, Wiley, New York, 1980.

Siqueira, P., et al., "A Continental-Scale Mosaic of the Amazon Basin Using JERS-1 SAR," *IEEE Transactions on Geoscience and Remote Sensing*, vol. 38, no. 6, 2000, pp. 2638–2644.

Strozzi, T., et al., "Land Subsidence Monitoring with Differential SAR Interferometry," *Photogrammetric Engineering and Remote Sensing*, vol. 67, no, 11, 2001, pp. 1261–1270.

Tait, A.B., "Estimation of Snow Water Equivalent Using Passive Microwave Radiation Data," *Remote Sensing of Environment*, vol. 64, no. 3, 1998, pp. 286–291.

Tait, A.B., et al., "High Frequency Passive Microwave Radiometry over a Snow-Covered Surface in Alaska," *Photogrammetric Engineering and Remote Sensing*, vol. 65, no. 6, 1999a, pp. 689–695.

Tait, A., et al., "Detection of Snow Cover Using Millimeter-Wave Imaging Radiometer (MIR) Data," *Remote Sensing of Environment*, vol. 68, no. 1, 1999b, pp. 53–60.

Tang, C.L., "Ice Patches on the Northeast Newfoundland Shelf Observed by RADARSAT," *Canadian Journal of Remote Sensing*, vol. 27, no. 1, 2001, pp. 44–50.

Taylor, G.R., et al., "Characterization of Saline Soils Using Airborne Radar Imagery," *Remote Sensing of Environment*, vol. 57, no. 3, 1996, pp. 127–142.

Toutin, T., and S. Amaral, "Stereo RADARSAT Data for Canopy Height in Brazilian Forests," *Canadian Journal of Remote Sensing*, vol. 26, no. 3, 2000, pp. 189–199.

Toutin, T., and L. Gray, "State-of-the-Art of Elevation Extraction from Satellite SAR Data," *ISPRS Journal of Photogrammetry and Remote Sensing*, vol. 55, no. 1, 2000, pp. 13–33.

Townsend, P.A., "Mapping Seasonal Flooding in Forested Wetlands Using Multi-Temporal Radarsat SAR," *Photogrammetric Engineering and Remote Sensing*, vol. 67, no. 7, 2001, pp. 857–864.

Treitz, P.M., et al., "Agricultural Crop Classification Using SAR Tone and Texture Statistics," *Canadian Journal of Remote Sensing*, vol. 26, no. 1, 2000, pp. 18–29.

Tso, B., and P.M. Mather, "Crop Discrimination Using Multi-Temporal SAR Imagery," *International Journal of Remote Sensing*, vol. 20, no. 12, 1999, pp. 2443–2460.

Vachon, P.W., et al., "Differential SAR Interferometry Measurements of Athabasca and Saskatchewan Glacier Flow Rate," *Canadian Journal of Remote Sensing*, vol. 22, no. 3, 1996, pp. 287–296.

Vainio, J., M. Similä, and H. Grönvall, "Operational Use of RADARSAT SAR Data as Aid to Winter Navigation in the Baltic Sea," *Canadian Journal of Remote Sensing*, vol. 26, no. 4, 2000, pp. 314–317.

van der Sanden, J.J., and D.H. Hockman, "Potential of Airborne Radar to Support the Assessment of Land Cover in a Tropical Rain Forest Environment," *Remote Sensing of Environment*, vol. 68, no. 1, 1999, pp. 26–40.

Wacherman, C.C., et al., "Automatic Detection of Ships in RADARSAT-1 SAR Imagery," *Canadian Journal of Remote Sensing*, vol. 27, no. 5, 2001, pp. 568–577.

Wang, Y., J.L. Day, and F.W. Davis, "Sensitivity of Modeled C- and L-Band Radar to Ground Surface Parameters in Loblolly Pine Forest," *Remote Sensing of Environment*, vol. 66, no. 3, 1998, pp. 331–342.

Wehr, A., and U. Lohr, "Airborne Laser Scanning—An Introduction and Overview," *ISPRS Journal of Photogrammetry and Remote Sensing*, vol. 54, nos. 2/3, 1999, pp. 68–82.

Wismann, V., et al., "Radar Signatures of Marine Mineral Oil Spills Measured by an Airborne Multi-Frequency Radar," *International Journal of Remote Sensing*, vol. 19, no. 18, 1998, pp. 3607–3624.

Yatabe, S.M., and D.G. Leckie, "Clearcut and Forest-Type Discrimination in Satellite SAR Imagery," *Canadian Journal of Remote Sensing*, vol. 21, no. 4, 1995, pp. 455–467.

Yocky, D.A., and B.F. Johnson, "Repeat-Pass Dual-Antenna Synthetic Aperture Radar Interferometric Change-Detection Post-Processing," *Photogrammetric Engineering and Remote Sensing*, vol. 64, no. 5, 1998, pp. 425–429.

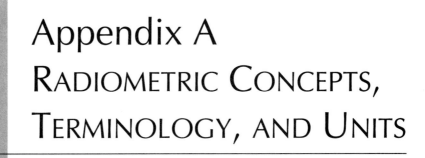

Appendix A
RADIOMETRIC CONCEPTS,
TERMINOLOGY, AND UNITS

GEOMETRIC CHARACTERISTICS OF RADIATION MEASUREMENT

In most applications of remote sensing, radiation is measured in *radiometric* units. Which particular terms and units are used in any given situation is often determined by the angular nature of the measurements involved. Figure A.1 illustrates the two broad categories of angular measurement commonly used in remote sensing. These are *hemispherical* measurement (*a*) and *directional* measurement (*b*).

Hemispherical measurements account for all energy contained within a hemisphere above a surface or object of interest. The term *irradiance* refers to the hemispherical radiation that is incident on an object or area. The hemispherical radiation that leaves an object or area is described by the term *radiant exitance*.

Directional radiation measurements account for the radiation measured in a particular direction of illuminating or viewing. An important aspect of directional measurement is the *cosine effect*. This concept is illustrated in Figure A.2 for an object that does not fill the field of view of an optical system. When such an object is viewed in a perpendicular, or normal, direction, it will

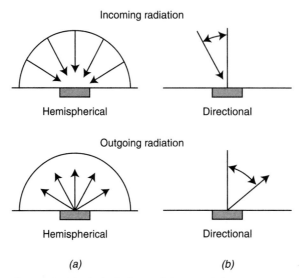

Figure A.1 Hemispherical (*a*) and directional (*b*) measurement of incoming and outgoing radiation.

appear to the optical system to have a certain area (*A*). If the optical system is moved away from the direction of the surface normal in an angular fashion, the apparent area of the surface decreases according to the cosine of the viewing angle (θ). The apparent area projected in the direction of observation equals $A \cos \theta$.

Also central to understanding directional radiation measurement are the concepts of *plane-angle measurement* and *solid-angle measurement*. These two

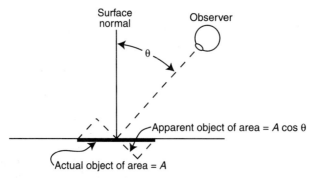

Figure A.2 Projected area cosine effect in a viewing direction other than normal. (Adapted from Swain and Davis, 1978; see Chapter 1 Selected Bibliography herein.)

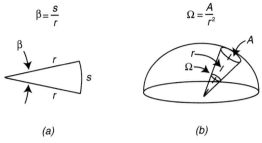

$$\beta = \frac{s}{r} \qquad\qquad \Omega = \frac{A}{r^2}$$

(a) *(b)*

Figure A.3 Plane-angle (a) and solid-angle (b) definitions used in radiation measurement.

forms of angular measurement are illustrated in Figure A.3. Shown in (a) is a plane angle that is measured in two dimensions and usually expressed in *radians*. The size of a plane angle (β) expressed in radians is equal to the length of the arc of a circle subtended by the angle divided by the radius of the circle (when the arc length exactly equals the radius, the size of the angle is exactly 1 radian).

Shown in Figure A.3*b* is a solid angle that is measured in three dimensions and expressed in units of *steradians*. The size of a solid angle (Ω) expressed in steradians is equal to the area subtended by the surface of a sphere divided by the square of the radius of the sphere. (When the subtended spherical area exactly equals the square of the radius, the size of the solid angle is 1 steradian).

RADIOMETRIC TERMINOLOGY AND UNITS

Figure A.4 and Table A.1 illustrate the basic relationship among the various terms that are used in both hemispherical and directional radiation measurement. We define here only a small subset of the complete terminology.

Radiant energy, Q, is the energy carried by an electromagnetic wave and a measure of the capacity of the wave to do work, measured in joules.

Radiant flux, Φ, is the amount of radiant energy emitted, transmitted, or received per unit time, measured in watts (joules per second).

Radiant flux density is the radiant flux at a surface divided by the area of the surface. Again, the density for flux incident upon a surface is *irradiance, E.* The density for flux leaving a surface is *radiant exitance, M.* Both are measured in watts per square meter.

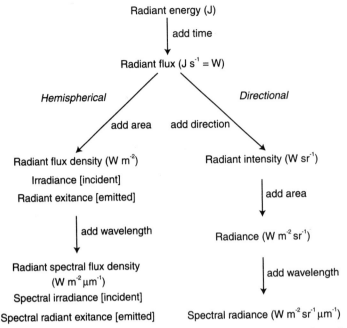

Figure A.4 Relationship among the various terms used in hemispherical and directional radiation measurement. (Adapted from Campbell and Norman, 1997; see Chapter 1 Selected Bibliography herein.)

Radiant spectral flux density is the radiant flux density per unit wavelength interval. Both *spectral irradiance, E_λ,* and *spectral radiant exitance, M_λ* are measured in units of watts per square meter per micrometer.

Radiant intensity, I, is the flux emanating from a point source per unit solid angle in the direction considered, measured in watts per steradian.

TABLE A.1 Radiation Terms and Units

Term	Symbol	Units
Radiant energy	Q	J
Radiant flux	Φ	W
Radiant flux density	E (irradiance)	Wm^{-2}
	M (radiant exitance)	Wm^{-2}
Radiant spectral flux density	E_λ (spectral irradiance)	$Wm^{-2}\,\mu m^{-1}$
	M_λ (spectral radiant exitance)	$Wm^{-2}\,\mu m^{-1}$
Radiant intensity	I	Wsr^{-1}
Radiance	L	$Wm^{-2}\,sr^{-1}$
Spectral radiance	L_λ	$Wm^{-2}\,sr^{-1}\,\mu m^{-1}$

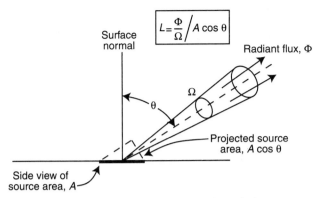

Figure A.5 Concept of radiance. (Adapted from Elachi, 1987; see Chapter 1 Selected Bibliography herein.)

Radiance, L, is the radiant flux per unit solid angle emanating from a surface in a given direction per unit of projected surface in the direction considered. Figure A.5 illustrates the concept of radiance. Radiance is measured in units of watts per square meter per steradian.

Spectral radiance, L_λ, is the radiance per wavelength interval, measured in watts per square meter per steradian per micrometer.

Appendix B
REMOTE SENSING DATA
AND INFORMATION
RESOURCES

SOURCES OF REMOTE SENSING DATA AND INFORMATION

Listed below are major sources of remote sensing data and information. Some data types are distributed by several sources.

The URLs listed below were accurate as of early 2003 and are subject to change.

ASCS (FSA) and USFS photographs
http://www.apfo.usda.gov

ASTER data
http://edc.usgs.gov

AVHRR data
http://edc.usgs.gov

Canada Centre for Remote Sensing
http://www.ccrs.nrcan.gc.ca

Corona satellite photography
http://edc.usgs.gov/products/satellite/declass1.html

Declassified satellite imagery

Declass-1 data: CORONA , ARGON, and LANYARD photographic imagery from 1959 to 1972.
http://edc.usgs.gov/products/satellite/declass1.html

Declass-2 data: KH-7 and KH-9 photographic imagery from 1963 to 1980.
http://edc.usgs.gov/products/satellite/declass2.html

Defense Meteorological Satellite Program data
http://dmsp.ngdc.noaa.gov/dmsp.html

EarthExplorer (USGS)
http://earthexplorer.usgs.gov

EROS Data Center
http://edc.usgs.gov

ERS radar data
http://www.esa.int
See also Radarsat International, Inc. (below)

Global Land Information System (USGS)
See *EarthExplorer (USGS)*

Global Land Cover Characterization study data and information
http://edcdaac.usgs.gov/glcc/glcc.html

IKONOS data
http://www.spaceimaging.com

IRS data
http://www.spaceimaging.com

JERS data
http://www.eorc.nasda.go.jp/JERS-1

Kodak aerial film data
http://www.kodak.com

Landsat data
Space Imaging
http://www.spaceimaging.com

Landsat data are also available at:
http://edc.usgs.gov

Microsoft Terraserver
http://terraserver.microsoft.com

MODIS data
http://edc.usgs.gov

Multi-Resolution Land Characteristics Consortium
http://www.epa.gov/mrlc

NAPP photographs
http://edc.usgs.gov

NASA: photographs
http://edc.usgs.gov

NASA: Earth Observatory
http://earthobservatory.nasa.gov

NASA: Natural Hazards
http://earthobservatory.nasa.gov/NaturalHazards

NASA: Visible Earth
http://visibleearth.nasa.gov

National Resources Inventory data and information (USDA)
http://www.nrcs.usda.gov

National Soil Information System
http://nasis.nrcs.usda.gov

NHAP photographs
http://edc.usgs.gov

NOAA: natural hazards
http://www.ngdc.noaa.gov/seg/hazard/hazards.shtml

OrbView data
http://www.orbimage.com

QuickBird satellite data
http://www.digitalglobe.com

Radarsat data
RADARSAT International, Inc.
http://www.rsi.ca

Canada Centre for Remote Sensing
http://www.ccrs.nrcan.gc.ca

SeaWiFS data
http://seawifs.gsfc.nasa.gov

Shuttle Imaging Radar data
http://edc.usgs.gov
http://southport.jpl.nasa.gov

Spectral reflectance curve data sets
ASTER spectral library
http://speclib.jpl.nasa.gov

U.S. Army Topographic Engineering Center Spectral Library
http://www.tec.army.mil/Hypercube

USGS Digital Spectral Library
http://speclab.cr.usgs.gov/spectral-lib.html

SPIN-2 data
http://terraserver.com/home.asp

SPOT data
SPOT Image (France)
http://www.spotimage.fr/home/

SPOT Image Corporation (USA)
http://www.spot.com

The National Map (USGS)
http://nationalmap.usgs.gov

USGS: Digital Orthophoto Program
http://mapping.usgs.gov/www/ndop

USGS Earth Science Information Center
http://ask.usgs.gov

USGS: National Gap Analysis Program (GAP)
http://biology.usgs.gov/state.partners/gap.html

USGS: Hazards
http://www.usgs.gov/themes/hazard.html

USGS: Photographs
http://edc.usgs.gov

USGS-EPA North American Landscape Characterization Project (NALC) data and information
http://eosims.cr.usgs.gov:5725/
CAMPAIGN_DOCS/nalc_proj_camp.html

Information on cartographic data throughout the United States, including maps, charts, airphotos, space images, geodetic control, digital elevation and planimetric data, and related information from federal, state, and some private sources can be obtained from

Earth Science Information Center (ESIC)
http://ask.usgs.gov

Information on older data is also available from

http://www.nara.gov

Additional information about federal, state, regional, and commercial sources of aerial photography in the United States is contained in

Collins, M. R., *The Aerial Photo Sourcebook*, Scarecrow Press, Lanham, MD, 1998.

REMOTE SENSING PERIODICALS

Listed below are major remote sensing journals:

Canadian Journal of Remote Sensing
Canadian Aeronautics and Space Institute
130 Slater Street, National Building, Suite 818
Ottawa, Ontario K1P 6E2
Canada

Geocarto International
Geocarto International Centre
G.P.O. Box 4122
Hong Kong

IEEE Transactions on Geoscience and Remote Sensing
IEEE Geoscience and Remote Sensing Society
Institute of Electrical and Electronics Engineers
445 Hoes Lane
Piscataway, NJ 08854

International Journal of Remote Sensing
Remote Sensing Society
c/o Taylor and Francis, Ltd.
Rankine Road
Basingstoke, Hants RG24 OPR
United Kingdom

ISPRS Journal of Photogrammetry and Remote Sensing
Elsevier Science Publishers
P.O. Box 1930, 1000 BX Amsterdam
The Netherlands

ITC Journal
International Institute for Aerial Survey and Earth Sciences
Enschede
The Netherlands

Photogrammetric Engineering and Remote Sensing
American Society for Photogrammetry and Remote Sensing
5410 Grosvenor Lane, Suite 210
Bethesda, MD 20814-2160
http://www.asprs.org

Remote Sensing of Environment
Elsevier Publishing Co.
52 Vanderbilt Avenue
New York, NY 10017

Remote Sensing Reviews
Taylor & Francis
http://www.tandf.co.uk/journals/default.html

REMOTE SENSING GLOSSARIES

Canada Centre for Remote Sensing Glossary
http://www.ccrs.nrcan.gc.ca/ccrs/learn/terms/glossary/glossary_e.html

EOSDIS Glossary
http://www-v0ims.gsfc.nasa.gov/v0ims/glossary.of.terms.html

NASA Earth Observatory Glossary
http://earthobservatory.nasa.gov/Library/glossary.php3

ONLINE REMOTE SENSING COURSES AND TUTORIALS

AGFAnet Photo Guide: Beginners Photo Course,
Classical Photo Course, and Digital Photo Course
  http://www.agfanet.com/en/cafe/photocourse/
  cont_index.php3

Canada Centre for Remote Sensing, Remote Sensing
Tutorials
  http://www.ccrs.nrean.gc.cn/ccrs/learn/
  learn_e.html

Digital Photography, The Textbook, A Free On-line
Course by Dennis P. Curtin
  http://www.photocourse.com

Geospatial Workforce Development Mc
  http://geoworkforce.olemiss.edu

Kodak Digital Learning Center
  http://www.kodak.com/US/en/digital/dlc

The Remote Sensing Core Curriculum (ASPRS and
others)
  http://www.research.umbc.edu/~tbenja1

The Remote Sensing Tutorial (NASA)
  http://rst.gsfc.nasa.gov

Appendix C
SAMPLE COORDINATE TRANSFORMATION AND RESAMPLING PROCEDURES

TWO-DIMENSIONAL AFFINE COORDINATE TRANSFORMATION

Frequently, two-dimensional ground coordinates (X, Y) are related to image coordinates (x, y) using an affine coordinate transformation. This particular transformation accounts for six potential factors when relating the ground and image coordinate systems: scale in the x direction, scale in the y direction, skew (or nonperpendicularity of the axis system), rotation between the systems, translation of the origins in the x direction, and translation of the origins in the y direction. When these six factors are combined, they result in coordinate transformation equations of the form

$$x = a_0 + a_1 X + a_2 Y \qquad (C.1)$$
$$x = b_0 + b_1 X + b_2 Y$$

where

$$(x, y) = \text{image coordinates (column, row)}$$
$$(X, Y) = \text{ground (or map) coordinates}$$
$$a_0, a_1, a_2, b_0, b_1, b_2 = \text{transformation parameters}$$

748

Initially, the six transformation parameters are unknown. However, two equations of the above form (C.1) can be written for each ground control point [i.e., the xy and XY coordinates for ground control points (GCPs) are known]. Hence, if the locations of three GCPs are known in both coordinate systems, a total of six equations of the above form (C.1) can be generated and solved simultaneously to obtain the values for the six transformation parameters. When more than three GCPs are used, redundancy exists in that there are more equations than unknowns (e.g., five GCPs would yield 10 equations containing the six unknowns). In this case, the transformation parameters can be computed using a least squares solution.

The efficacy of an affine transformation between ground and image coordinates is, among other things, a function of the number of GCPs used, the particular form of distortion present in the image data, and the degree of topographic relief displacement present in the imagery. For moderate distortions over relatively small ground areas an affine transformation often suffices. More complex transformations must be applied when large distortions and geographic areas are involved.

RESAMPLING PROCEDURES

Equation C.1 above can be used to find the image column and row (x, y) coordinates that correspond to a set of ground or map coordinates (X, Y), based on a mathematical relationship between the two coordinate systems. However, when these equations are applied to a given set of map coordinates, the resulting image coordinates typically do not correspond to the exact "center" of a pixel in the original image (Figure C.1). In this case, there is no single solution to the question of what digital number (DN) best represents the image data at the specified location. As we discuss in Section 7.2, several techniques are available to "resample" the original image; that is, to assign a new, estimated DN to the computed location. The three most widely used resampling methods are *nearest neighbor*, *bilinear interpolation*, and *cubic convolution* (also known as *bicubic interpolation*). These resampling methods may be applied to individual points (e.g., to determine the image brightness value at a given GPS-identified ground location) or used to transform an entire image into a new coordinate system.

No one resampling method is optimal for all situations, so it is necessary to understand the advantages and disadvantages of each method. These advantages and disadvantages are discussed in Section 7.2, and examples of the results are shown in Figure 7.2. The following discussion is intended to provide a more detailed conceptual explanation of the three most common resampling methods, along with the equations used to perform them.

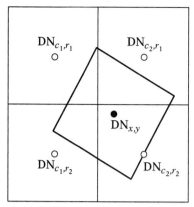

Figure C.1   Four image pixels surrounding the location of an output pixel, resulting from the transformation of ground (or map) coordinates (*X, Y*) back to image column and row coordinates (*x, y*).

The simplest resampling method, nearest neighbor resampling, involves assigning the DN from the nearest input pixel center to the output coordinates. In Figure C.1, the output pixel would be assigned the value of $DN_{2,2}$ because the center of the input pixel at that location is closest to the desired output pixel coordinates. Implementing nearest neighbor resampling can be as simple as using Equation C.1 to compute the image row and column coordinates for the desired output pixel location and then rounding to the nearest whole row and column number. For example, in Figure C.1, the computed coordinates for the output pixel location are (1.67, 1.58), which round to (2, 2).

Bilinear interpolation provides an estimated value that represents a weighted average of the four input pixels surrounding the output pixel location (Figure C.2*a*). First, two "vertical" interpolations are performed, along the columns to the left and right of the output pixel location:

$$I_1 = DN_{c_1,r_1} + (y - r_1)(DN_{c_1,r_2} - DN_{c_1,r_1})$$
$$I_2 = DN_{c_2,r_1} + (y - r_1)(DN_{c_2,r_2} - DN_{c_2,r_1})$$

(C.2)

where

c_1, c_2 = column numbers in the original image
r_1, r_2 = row numbers in the original image
y = row number for the output location (1.58 in Figure C.1)
$DN_{c,r}$ = digital number for pixel (*c,r*) in the original image
I_1, I_2 = interpolated digital numbers along columns c_1 and c_2

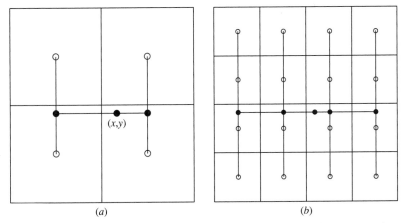

Figure C.2 Interpolation methods for resampling. (*a*) Bilinear interpolation among four pixels, beginning with interpolation down the two columns, followed by a third interpolation across the row. (*b*) Bicubic interpolation among 16 pixels.

Finally, a third ("horizontal") interpolation is performed, between the results of the first two interpolations:

$$\text{DN}_{x,y} = I_1 + (x - c_1)(I_2 - I_1) \tag{C.3}$$

where

    c_1 = column number in original image
     x = column number for output location (1.67 in Figure C.1)
  I_1, I_2 = interpolated DNs along columns c_1 and c_2
  $\text{DN}_{x,y}$ = interpolated digital number for location (x,y)

The third resampling method, known as bicubic interpolation or cubic convolution, is based on the assumption that the underlying signal at a given point along any single row or column of the image can be represented by the *sinc function*:

$$\text{sinc}(x) = \frac{\sin(\pi x)}{\pi x} \tag{C.4}$$

where x is the distance (in radians) from the point at which the underlying signal is being estimated.

Rather than using the sinc function directly, however, bicubic interpolation employs a cubic spline as an approximation to this function. The form of this cubic spline is given in Equation C.5 (adapted from Wolf and Dewitt, 2000; see Chapter 3 Selected Bibliography herein):

$$f(x) = (a + 2)|x|^3 - (a + 3)|x|^2 + 1 \quad \text{for} \quad 0 \le |x| \le 1$$
$$f(x) = a|x|^3 - 5a|x|^2 + 8a|x| - 4a \quad \text{for} \quad 1 < |x| \le 2 \quad \quad \text{(C.5)}$$
$$f(x) = 0 \quad \text{for} \quad |x| > 2$$

where

x = distance (in rows or columns and fractional rows or columns) from the point being interpolated

a = parameter corresponding to slope of weighting function at $x = 1$

$f(x)$ = weighting factor for pixel at distance x

Typically, the value for a in Equation C.5 is set to -0.5 (Schowengert, 1997; see Chapter 7 Selected Bibliography herein).

As with the bilinear interpolation described above, this bicubic interpolation is implemented by first interpolating along columns, and then interpolating across the rows. Unlike bilinear interpolation, however, four rows and four columns are used; that is, the estimated DN for the output location is the result of interpolation among the 16 closest neighbors (Figure C.2b). The four column interpolations are computed as follows:

$$I_1 = f(r_1)DN_{c_1,r_1} + f(r_2)DN_{c_1,r_2} + f(r_3)DN_{c_1,r_3} + f(r_4)DN_{c_1,r_4}$$
$$I_2 = f(r_1)DN_{c_2,r_1} + f(r_2)DN_{c_2,r_2} + f(r_3)DN_{c_2,r_3} + f(r_4)DN_{c_2,r_4} \quad \text{(C.6)}$$
$$I_3 = f(r_1)DN_{c_3,r_1} + f(r_2)DN_{c_3,r_2} + f(r_3)DN_{c_3,r_3} + f(r_4)DN_{c_3,r_4}$$
$$I_4 = f(r_1)DN_{c_4,r_1} + f(r_2)DN_{c_4,r_2} + f(r_3)DN_{c_4,r_3} + f(r_4)DN_{c_4,r_4}$$

where

$f(r)$ = weighting factor for pixel in row r, derived from Equation C.5

$DN_{c,r}$ = digital number for pixel (c,r) in original image

I_1, I_2, I_3, I_4 = interpolated digital numbers along columns c_1, c_2, c_3, c_4

Finally, the estimated output DN is determined by interpolating among the results of the four column interpolations:

$$DN_{x,y} = f(c_1)I_1 + f(c_2)I_2 + f(c_3)I_3 + f(c_4)I_4 \quad \text{(C.7)}$$

where

$f(c)$ = weighting factor for pixel in column c, derived from Equation C.5

I_1, I_2, I_3, I_4 = interpolated digital numbers along columns c_1, c_2, c_3, c_4

$DN_{x,y}$ = interpolated digital number for pixel (x,y)

INDEX